Introduction to Probability, Statistics, and Random Processes

Hossein Pishro-Nik

University of Massachusetts Amherst

Contents

Preface

Introduction and Goals

For years, I have been joking with my students that I would teach probability with the same level of excitement even if I were woken up in the middle of the night and asked to teach it. Years later, as a new father, I started writing this book when it became clear to me that I would not be sleeping at night for the foreseeable future.

This book is intended for undergraduate and first-year graduate-level courses in probability, statistics, and random processes. My goal has been to provide a clear and intuitive approach to these topics while maintaining an acceptable level of mathematical accuracy.

I have been teaching two courses on this subject for several years at the University of Massachusetts Amherst. While one of these courses is an undergraduate course taken by juniors, the other is a graduate-level course taken by our first-year Masters and PhD students.

My goal throughout this process has been to write a textbook that has the flexibility to be used in *both* courses while sacrificing neither the quality nor the presentational needs of either course. To achieve such a goal, I have tried to minimize the dependency between different sections of the book. In particular, when a small part from a different section of the book is useful elsewhere within the text, I have repeated said part rather than simply referring to it. My reasoning for doing so is twofold. Firstly, this format should make it easier for students to read the book and, secondly, this format should allow instructors the flexibility to select individual sections from the book more easily.

Additionally, I wanted the book to be easy to read and accessible as a self-study reference. It was also imperative that the book be available to anyone in the world, and as such the book in its entirety can be found online at www.probabilitycourse.com.

The book contains a large number of solved exercises. In addition to the examples found within the text, there is a set of solved problems at the end of each section. Detailed and step-by-step solutions to these problems are provided to help students learn problem-solving techniques. The solutions to the end-of-chapter problems, however, are available only to instructors.

Lastly, throughout the book, some examples of applications—such as engineering, finance, everyday life, etc.—are provided to aid in motivating the subject. These examples have been worded to be understandable to all students. As such, some technical issues have been left out.

Coverage

After a brief review of set theory and other required mathematical concepts, the text covers topics as follows:

- Chapters 1 and 2: basic concepts such as random experiments, probability axioms, conditional probability, law of total probability, Bayes' rule, and counting methods;

- Chapters 3 through 6: single and multiple random variables (discrete, continuous, and mixed), as well as moment-generating functions, characteristics functions, random vectors, and inequalities;

- Chapter 7: limit theorems and convergence;

- Chapters 8 and 9: Bayesian and classical statistics;

- Chapters 10: Introduction to random processes, processing of random signals;

- Chapter 11: Poisson processes, discrete-time Markov chains, continuous-time Markov chains, and Brownian motion;

- Chapter 12: basic methods of generating random variables and simulating probabilistic systems (using MATLAB);

- Chapter 13: basic methods of generating random variables and simulating probabilistic systems (using R);

- Chapter 14: recursive methods;

All chapters are available at www.probabilitycourse.com. Chapters 12 through 14 are available as PDFs and are downloadable from the textbook website. Chapters 12 and 13 cover the same material. The difference is that the codes in chapter 12 are provided in MATLAB while the codes in Chapter 13 are provided in R. The reason for this again is to give flexibility to instructors and students to choose whichever they prefer. Nevertheless, students who are unfamiliar with MATLAB and R should still be able to understand the algorithms.

Required Background

The majority of the text does not require any previous knowledge apart from a one-semester course in calculus. The exceptions to this statement are as follows:

- Sections 5.2 (Two Continuous Random Variables) and 6.1 (Methods for More Than Two Random Variables) both require a light introduction to double integrals and partial derivatives;

- Section 6.1.5 (Random Vectors) uses a few concepts from linear algebra;

- Section 10.2 (Processing of Random Signals) requires familiarity with the Fourier transform.

Acknowledgements

This project grew out of my educational activities regarding my National Science Foundation CAREER award. I am very thankful to the people in charge of the Open Education Initiative at the University of Massachusetts Amherst. In particular, I am indebted to Charlotte Roh and Marilyn Billings at the UMass Amherst library for all of their help and support.

I am grateful to my colleagues Dennis Goeckel and Patrick Kelly, who generously provided their lecture notes to me when I first joined UMass. These notes proved to be very useful in developing my course materials and, eventually, in writing this book. I am also thankful to Mario Parente—who used an early version of this book in his course—for very useful discussions.

Many people provided comments and suggestions. I would like to especially thank Hamid Saeedi for reading the manuscript in its entirety and providing very valuable comments. I am indebted to Evan Ray and Michael Miller for their helpful comments and suggestions, as well as to Eliza Mitchell and Linnea Duley for their detailed review and comments. I am thankful to Alexandra Saracino for her help regarding the figures and illustrations in this book. I would also like to thank Ali Rakhshan, who coauthored the chapter on simulation and who, along with Ali Eslami, helped me with my LaTeX problems. I am grateful to Sofya Vorotnikova, Stephen Donahue, Andrey Smirnov, and Elnaz Jedari Fathi for their help with the website. I would also like to thank Elnaz Jedari Fathi for designing the book cover.

I am indebted to all of my students in my classes, who not only encouraged me with their positive feedback to continue this project, but who also found many typographical errors in the early versions of this book. I am thankful to all of my teaching assistants who helped in various aspects of both the course and the book.

Last—but certainly not least—I would like to thank my family for their patience and support.

Chapter 1

Basic Concepts

In this chapter, we will provide some basic concepts and definitions. We will first begin with a brief discussion of what probability is, then we will review some mathematical foundations that are needed for developing probability theory. Next, we will discuss the concept of random experiments and the axioms of probability. We will then introduce discrete and continuous probability models. Finally, we will discuss conditional probability.

1.1   Introduction: What is Probability?

Randomness and uncertainty exist in our daily lives as well as in every discipline in science, engineering, and technology. Probability theory, the subject of the first part of this book, is a mathematical framework that allows us to describe and analyze random phenomena in the world around us. By random phenomena, we mean events or experiments whose outcomes we can't predict with certainty.

Let's consider a couple of specific applications of probability in order to get some intuition. First, let's think more carefully about what we mean by the terms "randomness" and "probability" in the context of one of the simplest possible random experiments: flipping a fair coin.

One way of thinking about "randomness" is that it's a way of expressing what we don't know. Perhaps, if we knew more about the force I flipped the coin with, the initial orientation of the coin, the impact point between my finger and the coin, the turbulence in the air, the surface smoothness of the table the coin lands on, the material characteristics of the coin and the table, and so on, we would be able to definitively say whether the coin would come up heads or tails. However, in the absence of all that information, we cannot predict the outcome of the coin flip. When we say that something is random, we are saying that our knowledge about the outcome is limited, so we can't be certain what will happen.

Since the coin is fair, if we don't know anything about how it was flipped, the probability that it will come up heads is 50%, or $\frac{1}{2}$. What exactly do we mean by this? There are two common interpretations of the word "probability." One is in terms of **relative frequency**. In other words, if we flip the coin a very large number of times, it will come up heads about $\frac{1}{2}$ of the time. As the number of coin flips increases, the proportion that comes up heads will tend to get closer and closer to $\frac{1}{2}$. In fact, this intuitive understanding of probability is a special case of the **law of large numbers**, which we will state and prove formally in later chapters of the book.

A second interpretation of probability is that it is a quantification of our degree of **subjective personal belief** that something will happen. To get a sense of what we mean by this, it may be helpful to consider a second example: predicting the weather. When we think about the chances that it will rain today, we consider things like whether or not there are clouds in the sky, and the humidity. However, the beliefs that we form based on these factors may vary from person to person – different people may make different estimates of the probability that it will rain. Often these two interpretations of probability coincide – for instance, we may base our personal beliefs about the chance that it will rain on an assessment of the relative frequency of rain on days with conditions like today.

The beauty of probability theory is that it is applicable regardless of the interpretation of probability that we use (i.e., in terms of long-run frequency or degree of belief). Probability theory provides a solid framework to study random phenomena. It starts by assuming **axioms of probability**, and then builds the entire theory using mathematical arguments.

Before delving into studying probability theory, let us briefly look at an example showing how probability theory has been applied in a real-life system.

1.1.1   An Example: Communication Systems

Communication systems play a central role in our lives. Everyday, we use our cell phones, access the internet, use our TV remote controls, and so on. Each of these systems relies on transferring information from one place to another. For example, when you talk on the phone, what you say is converted to a sequence of 0's or 1's called *information bits*. These information bits are then transmitted by your cell phone antenna to a nearby cell tower as shown in Figure 1.1.

The problem that communication engineers must consider is that the transmission is always affected by **noise**. That is, some of the bits received at the cell tower are incorrect. For example, your cell phone may transmit the sequence "010**0**10 $\cdots$," while the sequence "010**1**10 $\cdots$" might be received at the cell tower. In this case, the fourth bit is incorrect. Errors like this could affect the quality of the audio in your phone conversation.

The noise in the transmission is a random phenomenon. Before sending the transmission we do not know which bits will be affected. It is as if someone tosses a (biased) coin for each bit and decides whether or not that bit will be received in error. Probability theory is used extensively in the design of modern communication systems in order to understand the behavior of noise in these systems and take measures to correct the errors.

This example shows just one application of probability. You can pick almost any discipline and find many applications in which probability is used as a major tool. Randomness is prevalent everywhere, and probability theory has proven to be a powerful way to understand and manage its effects.

1.2   Review of Set Theory

Probability theory uses the language of sets. As we will see later, probability is defined and calculated for sets. Thus, here, we will briefly review some basic concepts from set theory that are used in this book. We will discuss set notations, definitions, and operations (such

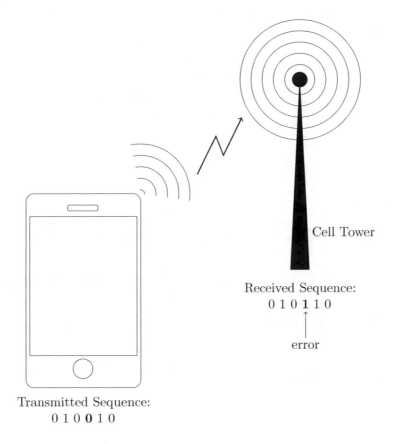

Cell Tower

Received Sequence:
0 1 0 **1** 1 0

↑

error

Transmitted Sequence:
0 1 0 **0** 1 0

Figure 1.1: Transmission of data from a cell phone to a cell tower.

as intersections and unions). We will then introduce countable and uncountable sets. Finally, we will briefly discuss functions. This section may seem somewhat theoretical and thus less interesting than the rest of the book, but it lays the foundation for what is to come.

A **set** is a collection of some items (elements). We often use capital letters to denote a set. To define a set we can simply list all the elements in curly brackets, for example to define a set A that consists of the two elements ♣ and ◇, we write $A = \{♣, ◇\}$. To say that ◇ belongs to A, we write $◇ \in A$, where "$\in$" is pronounced "belongs to." To say that an element does not belong to a set, we use $\notin$. For example, we may write $♡ \notin A$.

A **set** is a collection of things (elements).

Note that ordering does not matter, so the two sets $\{♣, ◇\}$ and $\{◇, ♣\}$ are equal. We often work with sets of numbers. Some important sets are given in the following example.

Example 1.1. The following sets are used in this book:

- The set of natural numbers, $\mathbb{N} = \{1, 2, 3, \cdots\}$.

- The set of integers, $\mathbb{Z} = \{\cdots, -3, -2, -1, 0, 1, 2, 3, \cdots\}$.

- The set of rational numbers $\mathbb{Q}$.

- The set of real numbers $\mathbb{R}$.

- Closed intervals on the real line. For example, $[2, 3]$ is the set of all real numbers x such that $2 \leq x \leq 3$.

- Open intervals on the real line. For example $(-1, 3)$ is the set of all real numbers x such that $-1 < x < 3$.

- Similarly, $[1, 2)$ is the set of all real numbers x such that $1 \leq x < 2$.

- The set of complex numbers $\mathbb{C}$ is the set of numbers in the form of $a + bi$, where $a, b \in \mathbb{R}$, and $i = \sqrt{-1}$.

We can also define a set by mathematically stating the properties satisfied by the elements in the set. In particular, we may write

$$A = \{x | x \text{ satisfies some property}\} \text{ or } A = \{x : x \text{ satisfies some property}\}.$$

The symbols "|" and " : " are pronounced "such that."

Example 1.2. Here are some examples of sets defined by stating the properties satisfied by the elements:

- If the set C is defined as $C = \{x | x \in \mathbb{Z}, -2 \leq x < 10\}$, then $C = \{-2, -1, 0, \cdots, 9\}$.

- If the set D is defined as $D = \{x^2 | x \in \mathbb{N}\}$, then $D = \{1, 4, 9, 16, \cdots\}$.

- The set of rational numbers can be defined as $\mathbb{Q} = \{\frac{a}{b} | a, b \in \mathbb{Z}, b \neq 0\}$.

- For real numbers a and b, where $a < b$, we can write $(a, b] = \{x \in \mathbb{R} \mid a < x \leq b\}$.

- $\mathbb{C} = \{a + bi \mid a, b \in \mathbb{R}, i = \sqrt{-1}\}$.

Set A is a **subset** of set B if every element of A is also an element of B. We write $A \subset B$, where "$\subset$" indicates "subset." Equivalently, we say B is a **superset** of A, or $B \supset A$.

Example 1.3. Here are some examples of sets and their subsets:

- If $E = \{1, 4\}$ and $C = \{1, 4, 9\}$, then $E \subset C$.

- $\mathbb{N} \subset \mathbb{Z}$.

- $\mathbb{Q} \subset \mathbb{R}$.

Two sets are equal if they have the exact same elements. Thus, $A = B$ if and only if $A \subset B$ and $B \subset A$. For example, $\{1, 2, 3\} = \{3, 2, 1\}$, and $\{a, a, b\} = \{a, b\}$. The set with no elements, i.e., $\emptyset = \{\}$ is the **null set** or the **empty set**. For any set A, $\emptyset \subset A$.

The **universal set** is the set of all things that we could possibly consider in the context we are studying. Thus every set A is a subset of the universal set. In this book, we often denote the universal set by S (As we will see, in the language of probability theory, the universal set is called the *sample space*.) For example, if we are discussing rolling of a die, our universal set may be defined as $S = \{1, 2, 3, 4, 5, 6\}$, or if we are discussing tossing of a coin once, our universal set might be $S = \{H, T\}$ (H for heads and T for tails).

1.2.1 Venn Diagrams

Venn diagrams are very useful in visualizing relations between sets. In a **Venn diagram** any set is depicted by a closed region. Figure 1.2 shows an example of a Venn diagram. In this figure, the big rectangle shows the universal set S. The shaded area shows another set A.

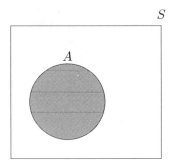

Figure 1.2: Venn Diagram

Figure 1.3 shows two sets A and B, where $B \subset A$.

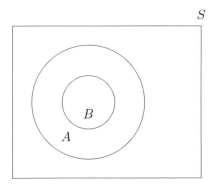

Figure 1.3: Venn Diagram for two sets A and B, where $B \subset A$.

1.2.2   Set Operations

The **union** of two sets is a set containing all elements that are in A <u>or</u> in B (possibly both). For example, $\{1,2\} \cup \{2,3\} = \{1,2,3\}$. Thus, we can write $x \in (A \cup B)$ if and only if $(x \in A)$ or $(x \in B)$. Note that $A \cup B = B \cup A$. In Figure 1.4, the union of sets A and B is shown by the shaded area in the Venn diagram.

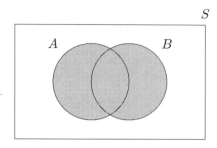

Figure 1.4: The shaded area shows the set $B \cup A$.

Similarly we can define the union of three or more sets. In particular, if $A_1, A_2, A_3, \cdots, A_n$ are n sets, their union $A_1 \cup A_2 \cup A_3 \cdots \cup A_n$ is a set containing all elements that are in at least one of the sets. We can write this union more compactly by

$$\bigcup_{i=1}^{n} A_i.$$

For example, if $A_1 = \{a,b,c\}, A_2 = \{c,h\}, A_3 = \{a,d\}$, then $\bigcup_i A_i = A_1 \cup A_2 \cup A_3 = \{a,b,c,h,d\}$. We can similarly define the union of infinitely many sets $A_1 \cup A_2 \cup A_3 \cup \cdots$.

The **intersection** of two sets A and B, denoted by $A \cap B$, consists of all elements that are both in A <u>and</u> B. For example, $\{1,2\} \cap \{2,3\} = \{2\}$. In Figure 1.5, the intersection of sets A and B is shown by the shaded area using a Venn diagram. More generally, for sets $A_1, A_2, A_3, \cdots$, their intersection $\bigcap_i A_i$ is defined as the set consisting of the elements that are in all A_i's. Figure 1.6 shows the intersection of three sets.

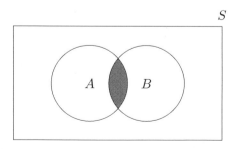

Figure 1.5: The shaded area shows the set $B \cap A$.

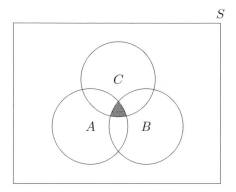

Figure 1.6: The shaded area shows the set $A \cap B \cap C$.

The **complement** of a set A, denoted by A^c or $\bar{A}$, is the set of all elements that are in the universal set S but are not in A. In Figure 1.7, $\bar{A}$ is shown by the shaded area of the Venn diagram.

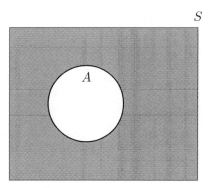

Figure 1.7: The shaded area shows the set $\bar{A} = A^c$.

The **difference (subtraction)** is defined as follows. The set $A - B$ consists of elements that are in A but not in B. For example if $A = \{1, 2, 3\}$ and $B = \{3, 5\}$, then $A - B = \{1, 2\}$. In Figure 1.8, $A - B$ is shown by the shaded area of the Venn diagram. Note that $A - B = A \cap B^c$.

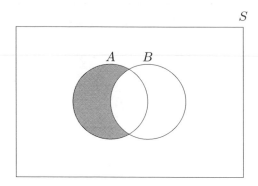

Figure 1.8: The shaded area shows the set $A - B$.

Two sets A and B are **mutually exclusive** or **disjoint** if they do not have any shared elements, i.e., their intersection is the empty set, $A \cap B = \emptyset$. More generally, several sets are called disjoint if they are pairwise disjoint, i.e., no two of them share a common element. Figure 1.9 shows three disjoint sets.

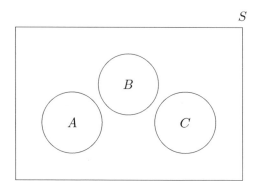

Figure 1.9: Sets A, B, and C are disjoint.

If the earth's surface is our sample space, we might want to partition it to the different continents. Similarly, a country can be partitioned to different provinces. In general, a collection of nonempty sets $A_1, A_2, \cdots$ is a **partition** of a set A if they are disjoint and their union is A. In Figure 1.10, the sets A_1, A_2, A_3 and A_4 form a partition of the universal set S.

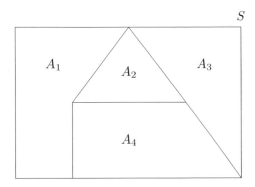

Figure 1.10: The collection of sets A_1, A_2, A_3 and A_4 is a partition of S.

Here are some rules that are often useful when working with sets. We will see examples of their usage shortly.

Theorem 1.1. (De Morgan's law) For any sets A_1, A_2, $\cdots$, A_n, we have

(a) $(A_1 \cup A_2 \cup A_3 \cup \cdots A_n)^c = A_1^c \cap A_2^c \cap A_3^c \cdots \cap A_n^c$;

(b) $(A_1 \cap A_2 \cap A_3 \cap \cdots A_n)^c = A_1^c \cup A_2^c \cup A_3^c \cdots \cup A_n^c$.

Theorem 1.2. (Distributive law) For any sets A, B, and C we have

(a) $A \cap (B \cup C) = (A \cap B) \cup (A \cap C)$;

(b) $A \cup (B \cap C) = (A \cup B) \cap (A \cup C)$.

Example 1.4. If the universal set is given by $S = \{1, 2, 3, 4, 5, 6\}$, and $A = \{1, 2\}$, $B = \{2, 4, 5\}, C = \{1, 5, 6\}$ are three sets, find the following sets:

(a) $A \cup B$

(b) $A \cap B$

(c) $\overline{A}$

(d) $\overline{B}$

(e) Check De Morgan's law by finding $(A \cup B)^c$ and $A^c \cap B^c$.

(f) Check the distributive law by finding $A \cap (B \cup C)$ and $(A \cap B) \cup (A \cap C)$.

Solution:

(a) $A \cup B = \{1, 2, 4, 5\}$.

(b) $A \cap B = \{2\}$.

(c) $\overline{A} = \{3, 4, 5, 6\}$ ($\overline{A}$ consists of elements that are in S but not in A).

(d) $\overline{B} = \{1, 3, 6\}$.

(e) We have

$$(A \cup B)^c = \{1, 2, 4, 5\}^c = \{3, 6\},$$

which is the same as

$$A^c \cap B^c = \{3, 4, 5, 6\} \cap \{1, 3, 6\} = \{3, 6\}.$$

(f) We have

$$A \cap (B \cup C) = \{1, 2\} \cap \{1, 2, 4, 5, 6\} = \{1, 2\},$$

which is the same as

$$(A \cap B) \cup (A \cap C) = \{2\} \cup \{1\} = \{1, 2\}.$$

A **Cartesian product** of two sets A and B, written as $A \times B$, is the set containing **ordered** pairs from A and B. That is, if $C = A \times B$, then each element of C is of the form (x, y), where $x \in A$ and $y \in B$:

$$A \times B = \{(x, y) | x \in A \text{ and } y \in B\}.$$

For example if $A = \{1, 2, 3\}$ and $B = \{H, T\}$ then

$$A \times B = \{(1, H), (1, T), (2, H), (2, T), (3, H), (3, T)\}.$$

Note that here the pairs are ordered, so for example, $(1, H) \neq (H, 1)$. Thus $A \times B$ is **not** the same as $B \times A$.

If you have two finite sets A and B, where A has M elements and B has N elements, then $A \times B$ has $M \times N$ elements. This rule is called the **multiplication principle** and is very useful in counting the numbers of elements in sets. The number of elements in a set is denoted by $|A|$, so here we write $|A| = M, |B| = N$, and $|A \times B| = MN$. In the above example, $|A| = 3, |B| = 2$, thus $|A \times B| = 3 \times 2 = 6$. We can similarly define the Cartesian product of n sets $A_1, A_2, \cdots, A_n$ as

$$A_1 \times A_2 \times A_3 \times \cdots \times A_n = \{(x_1, x_2, \cdots, x_n) | x_1 \in A_1 \text{ and } x_2 \in A_2 \text{ and } \cdots x_n \in A_n\}.$$

The multiplication principle states that for finite sets $A_1, A_2, \cdots, A_n$, if $|A_1| = M_1, |A_2| = M_2, \cdots, |A_n| = M_n$, then $| A_1 \times A_2 \times A_3 \times \cdots \times A_n | = M_1 \times M_2 \times M_3 \times \cdots \times M_n$.

An important example of sets obtained using a Cartesian product is $\mathbb{R}^n$, where n is a natural number. For $n = 2$, we have

$$\mathbb{R}^2 = \mathbb{R} \times \mathbb{R}$$
$$= \{(x, y) | x \in \mathbb{R}, y \in \mathbb{R}\}.$$

Thus, $\mathbb{R}^2$ is the set consisting of all points in the two-dimensional plane. Similarly, $\mathbb{R}^3 = \mathbb{R} \times \mathbb{R} \times \mathbb{R}$ and so on.

1.2.3 Cardinality: Countable and Uncountable Sets

Here, we need to talk about **cardinality** of a set, which is basically the size of the set. The cardinality of a set is denoted by $|A|$. We will first discuss cardinality for finite sets and then talk about infinite sets.

Finite Sets:

Consider a set A. If A has only a finite number of elements, its cardinality is simply the number of elements in A. For example, if $A = \{2, 4, 6, 8, 10\}$, then $|A| = 5$. Before discussing infinite sets, which is the main discussion of this section, we would like to talk about a very useful rule: the **inclusion-exclusion principle**. For two finite sets A and B, we have

$$|A \cup B| = |A| + |B| - |A \cap B|.$$

To see this, note that when we add $|A|$ and $|B|$, we are counting the elements in $|A \cap B|$ twice, thus by subtracting it from $|A| + |B|$, we obtain the number of elements in $|A \cup B|$ (you can refer to Figure 1.16 to see this pictorially). We can extend the same idea to three or more sets.

Inclusion-exclusion principle:

1. $|A \cup B| = |A| + |B| - |A \cap B|$,

2. $|A \cup B \cup C| = |A| + |B| + |C| - |A \cap B| - |A \cap C| - |B \cap C| + |A \cap B \cap C|$.

Generally, for n finite sets $A_1, A_2, A_3, \cdots, A_n$, we can write

---

$$\left| \bigcup_{i=1}^{n} A_i \right| = \sum_{i=1}^{n} |A_i| - \sum_{i<j} |A_i \cap A_j|$$
$$+ \sum_{i<j<k} |A_i \cap A_j \cap A_k| - \cdots + (-1)^{n+1} |A_1 \cap \cdots \cap A_n|.$$

Example 1.5. In a party,

- there are 10 people with white shirts and 8 people with red shirts;

- 4 people have black shoes and white shirts;

- 3 people have black shoes and red shirts;

- the total number of people with white or red shirts or black shoes is 21.

How many people have black shoes?

Solution: Let W, R, and B, be the number of people with white shirts, red shirts, and black shoes respectively. Then, here is the summary of the available information:

$$|W| = 10, |R| = 8, |W \cap B| = 4, |R \cap B| = 3, |W \cup B \cup R| = 21.$$

Also, it is reasonable to assume that W and R are disjoint, $|W \cap R| = 0$. Thus by applying the inclusion-exclusion principle we obtain

$$\begin{aligned}
|W \cup R \cup B| = 21 &= |W| + |R| + |B| \\
&\quad - |W \cap R| - |W \cap B| - |R \cap B| + |W \cap R \cap B| \\
&= 10 + 8 + |B| - 0 - 4 - 3 + 0.
\end{aligned}$$

Thus

$$|B| = 10.$$

Note that another way to solve this problem is using a Venn diagram as shown in Figure 1.11.

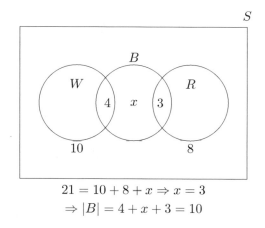

$$21 = 10 + 8 + x \Rightarrow x = 3$$
$$\Rightarrow |B| = 4 + x + 3 = 10$$

Figure 1.11: Using Venn diagrams

Infinite Sets:

What if A is an infinite set? It turns out we need to distinguish between two types of infinite sets, where one type is significantly "larger" than the other. In particular, one type is called **countable**, while the other is called **uncountable**. Sets such as $\mathbb{N}$ and $\mathbb{Z}$ are called countable, but "bigger" sets such as $\mathbb{R}$ are called uncountable. The difference between the two types is that you can list the elements of a countable set A, i.e., you can write $A = \{a_1, a_2, \cdots\}$, but you cannot list the elements in an uncountable set. For example, you can write

- $\mathbb{N} = \{1, 2, 3, \cdots\}$,

- $\mathbb{Z} = \{0, 1, -1, 2, -2, 3, -3, \cdots\}$.

The fact that you can list the elements of a countably infinite set means that the set can be put in one-to-one correspondence with natural numbers $\mathbb{N}$. On the other hand, you cannot list the elements in $\mathbb{R}$, so it is an uncountable set. To be precise, here is the definition.

Definition 1.1. Set A is called countable if one of the following is true:

(a) if it is a finite set, $\mid A \mid < \infty$; or

(b) it can be put in one-to-one correspondence with natural numbers $\mathbb{N}$, in which case the set is said to be countably infinite.

A set is called uncountable if it is not countable.

Here is a simple guideline for deciding whether a set is countable or not. As far as applied probability is concerned, this guideline should be sufficient for most cases.

(i) $\mathbb{N}, \mathbb{Z}, \mathbb{Q}$, and any of their subsets are countable.

(ii) Any set containing an interval on the real line such as $[a, b], (a, b], [a, b)$, or (a, b), where $a < b$ is uncountable.

The above rule is usually sufficient for the purpose of this book. For the interested reader, we now provide some results that help us prove whether a set is countable or not. If you are less interested in proofs, you may decide to skip them.

Theorem 1.3. Any subset of a countable set is countable. Any superset of an uncountable set is uncountable.

Proof. The intuition behind this theorem is the following: If a set is countable, then any "smaller" set should also be countable, so a subset of a countable set should be countable as well. To provide a proof, we can argue in the following way.

Let A be a countable set and $B \subset A$. If A is a finite set, then $|B| \leq |A| < \infty$, thus B is countable. If A is countably infinite, then we can list the elements in A, then by removing the elements in the list that are not in B, we can obtain a list for B, thus B is countable.

The second part of the theorem can be proved using the first part. Assume B is uncountable. If $B \subset A$ and A is countable, by the first part of the theorem B is also a countable set which is a contradiction. ∎

Theorem 1.4. If $A_1, A_2, \cdots$ is a list of countable sets, then the set $\bigcup_i A_i = A_1 \cup A_2 \cup A_3 \cdots$ is also countable.

Proof. It suffices to create a list of elements in $\bigcup_i A_i$. Since each A_i is countable we can list its elements: $A_i = \{a_{i1}, a_{i2}, \cdots\}$. Thus, we have

- $A_1 = \{a_{11}, a_{12}, \cdots\}$,

- $A_2 = \{a_{21}, a_{22}, \cdots\}$,

- $A_3 = \{a_{31}, a_{32}, \cdots\}$,

- ...

Now we need to make a list that contains all the above lists. This can be done in different ways. One way to do this is to use the ordering shown in Figure 1.12 to make a list. Here, we can write

$$\bigcup_i A_i = \{a_{11}, a_{12}, a_{21}, a_{31}, a_{22}, a_{13}, a_{14}, \cdots\} \tag{1.1}$$

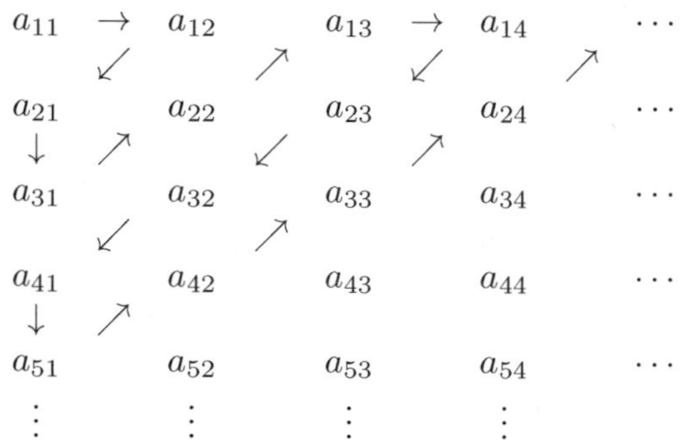

$$
\begin{array}{cccccc}
a_{11} & \rightarrow & a_{12} & a_{13} & \rightarrow & a_{14} & \cdots \\
 & \swarrow & & \nearrow & \swarrow & & \nearrow \\
a_{21} & & a_{22} & a_{23} & & a_{24} & \cdots \\
\downarrow & \nearrow & & \swarrow & & \nearrow \\
a_{31} & & a_{32} & a_{33} & & a_{34} & \cdots \\
 & \swarrow & & \nearrow \\
a_{41} & & a_{42} & a_{43} & & a_{44} & \cdots \\
\downarrow & \nearrow \\
a_{51} & & a_{52} & a_{53} & & a_{54} & \cdots \\
\vdots & & \vdots & \vdots & & \vdots
\end{array}
$$

Figure 1.12: Ordering to make a list.

We have been able to create a list that contains all the elements in $\bigcup_i A_i$, so this set is countable.

$\blacksquare$

Theorem 1.5. If A and B are countable, then $A \times B$ is also countable.

Proof. The proof of this theorem is very similar to the previous theorem. Since A and B are countable, we can write

$$
\begin{aligned}
A &= \{a_1, a_2, a_3, \cdots\}, \\
B &= \{b_1, b_2, b_3, \cdots\}.
\end{aligned}
$$

Now, we create a list containing all elements in $A \times B = \{(a_i, b_j) | i, j = 1, 2, 3, \cdots\}$. The idea is exactly the same as before. Figure 1.13 shows one possible ordering.

$$(a_1, b_1) \quad \rightarrow \quad (a_1, b_2) \qquad (a_1, b_3) \quad \rightarrow \quad (a_1, b_4) \qquad \cdots$$
$$\swarrow \qquad \nearrow \qquad \swarrow$$
$$(a_2, b_1) \qquad (a_2, b_2) \qquad (a_2, b_3) \qquad (a_2, b_4) \qquad \cdots$$
$$\downarrow \qquad \nearrow \qquad \swarrow$$
$$(a_3, b_1) \qquad (a_3, b_2) \qquad (a_3, b_3) \qquad (a_3, b_4) \qquad \cdots$$
$$\swarrow$$
$$(a_4, b_1) \qquad (a_4, b_2) \qquad (a_4, b_3) \qquad (a_4, b_4) \qquad \cdots$$
$$\downarrow$$
$$\vdots \qquad\qquad \vdots \qquad\qquad \vdots \qquad\qquad \vdots$$

Figure 1.13: Ordering to make a list.

■

The above arguments can be repeated for any set C in the form of

$$C = \bigcup_i \bigcup_j \{a_{ij}\},$$

where indices i and j belong to some countable sets. Thus, any set in this form is countable. For example, a consequence of this is that the set of rational numbers $\mathbb{Q}$ is countable. This is because we can write

$$\mathbb{Q} = \bigcup_{i \in \mathbb{Z}} \bigcup_{j \in \mathbb{N}} \{\frac{i}{j}\}.$$

The above theorems confirm that sets such as $\mathbb{N}, \mathbb{Z}, \mathbb{Q}$ and their subsets are countable. However, as we mentioned, intervals in $\mathbb{R}$ are uncountable. Thus, you can never provide a list in the form of $\{a_1, a_2, a_3, \cdots\}$ that contains all the elements in, say $[0, 1]$. This fact can be proved using a so-called diagonal argument, and we omit the proof here as it is not instrumental for the rest of the book.

1.2.4 Functions

We often need the concept of functions in probability. A function f is a rule that takes an input from a specific set, called the **domain**, and produces an output from another set, called the **co-domain**. Thus, a function *maps* elements from the domain set to elements in the co-domain with the property that each input is mapped to exactly one output. For a function f, if x is an element in the domain, then the function value (the output of the function) is shown by $f(x)$. If A is the domain and B is the co-domain for the function f, we use the following notation:

$$f : A \rightarrow B.$$

Example 1.6.

- Consider the function $f : \mathbb{R} \to \mathbb{R}$, defined as $f(x) = x^2$. This function takes any real number x and outputs x^2. For example, $f(2) = 4$.

- Consider the function $g : \{H, T\} \to \{0, 1\}$, defined as $g(H) = 0$ and $g(T) = 1$. This function can only take two possible inputs H or T, where H is mapped to 0 and T is mapped to 1.

The output of a function $f : A \to B$ always belongs to the co-domain B. However, not all values in the co-domain are always covered by the function. In the above example, $f : \mathbb{R} \to \mathbb{R}$, the function value is always a positive number $f(x) = x^2 \geq 0$. We define the **range** of a function as the set containing all the possible values of $f(x)$. Thus, the range of a function is always a subset of its co-domain. For the above function $f(x) = x^2$, the range of f is given by

$$\text{Range}(f) = \mathbb{R}^+ = \{x \in \mathbb{R} | x \geq 0\}.$$

Figure 1.14 pictorially shows a function, its domain, co-domain, and range. The figure shows that an element x in the domain is mapped to $f(x)$ in the range.

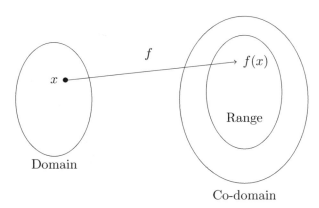

Figure 1.14: Function $f : A \to B$, the range is always a subset of the co-domain.

1.2.5 Solved Problems

1. Let A, B, C be three sets. For each of the following sets, draw a Venn diagram and shade the area representing the given set.

   (a) $A \cup B \cup C$

   (b) $A \cap B \cap C$

   (c) $A \cup (B \cap C)$

   (d) $A - (B \cap C)$

   (e) $A \cup (B \cap C)^c$

Solution: Figure 1.15 shows Venn diagrams for these sets.

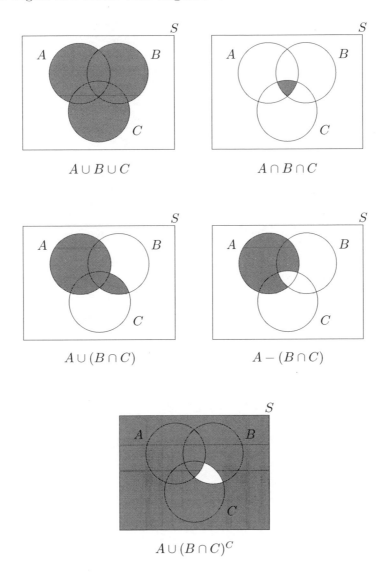

Figure 1.15: Venn diagram for different sets.

2. Using Venn diagrams, verify the following identities.

(a) $A = (A \cap B) \cup (A - B)$

(b) If A and B are finite sets, we have

$$|A \cup B| = |A| + |B| - |A \cap B| \qquad (1.2)$$

Solution: Figure 1.16 pictorially verifies the given identities. Note that in the second identity, we show the number of elements in each set by the corresponding shaded area.

3. Let $S = \{1, 2, 3\}$. Write all the possible partitions of S.

Solution: Remember that a partition of S is a collection of nonempty sets that are disjoint and their union is S. There are 5 possible partitions for $S = \{1, 2, 3\}$:

  - $\{1\},\{2\},\{3\}$;
  - $\{1,2\},\{3\}$;
  - $\{1,3\},\{2\}$;
  - $\{2,3\},\{1\}$;
  - $\{1,2,3\}$.

4. Determine whether each of the following sets is countable or uncountable.

(a) $A = \{x \in \mathbb{Q} | -100 \le x \le 100\}$

(b) $B = \{(x, y) | x \in \mathbb{N}, y \in \mathbb{Z}\}$

(c) $C = (0, 0.1]$

(d) $D = \{\frac{1}{n} | n \in \mathbb{N}\}$

Solution:

(a) $A = \{x \in \mathbb{Q} | -100 \le x \le 100\}$ is **countable** since it is a subset of a countable set, $A \subset \mathbb{Q}$.

(b) $B = \{(x, y) | x \in \mathbb{N}, y \in \mathbb{Z}\}$ is **countable** because it is the Cartesian product of two countable sets, i.e., $B = \mathbb{N} \times \mathbb{Z}$.

(c) $C = (0, .1]$ is **uncountable** since it is an interval of the form $(a, b]$, where $a < b$.

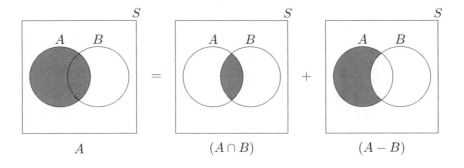

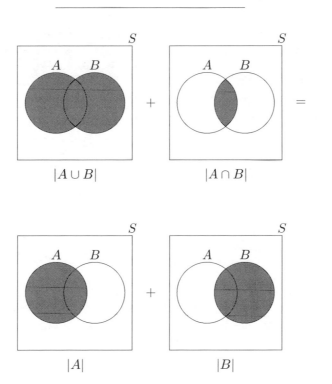

Figure 1.16: Venn diagram for some identities.

(d) $D = \{\frac{1}{n} | n \in \mathbb{N}\}$ is **countable** since it is in one-to-one correspondence with the set of natural numbers. In particular, you can list all the elements in the set D, $D = \{1, \frac{1}{2}, \frac{1}{3}, \cdots\}$.

5. Find the range of the function $f : \mathbb{R} \to \mathbb{R}$ defined as $f(x) = \sin(x)$.

   Solution: For any real value x, $-1 \leq \sin(x) \leq 1$. Also, all values in $[-1, 1]$ are covered by $\sin(x)$. Thus, Range(f) = $[-1, 1]$.

1.3   Random Experiments and Probabilities

1.3.1   Random Experiments

Before rolling a die you do not know the result. This is an example of a **random experiment**. In particular, a random experiment is a process by which we observe something uncertain. After the experiment, the result of the random experiment is known. An **outcome** is a result of a random experiment. The set of all possible outcomes is called the **sample space**. Thus in the context of a random experiment, the sample space is our *universal set*. Here are some examples of random experiments and their sample spaces:

1. Random experiment: toss a coin; sample space: $S = \{heads, tails\}$ or, as we usually write it, $\{H, T\}$.

2. Random experiment: roll a die; sample space: $S = \{1, 2, 3, 4, 5, 6\}$.

3. Random experiment: observe the number of iPhones sold by an Apple store in Boston in 2015; sample space: $S = \{0, 1, 2, 3, \cdots\}$.

4. Random experiment: observe the number of goals in a soccer match; sample space: $S = \{0, 1, 2, 3, \cdots\}$.

When we repeat a random experiment several times, we call each one of them a **trial**. Thus, a trial is a particular performance of a random experiment. In the example of tossing a coin, each trial will result in either heads or tails. Note that the sample space is defined based on how you define your random experiment. For example,

Example 1.7. We toss a coin three times and observe the sequence of heads/tails. The sample space here may be defined as

$$S = \{(H, H, H), (H, H, T), (H, T, H), (T, H, H),$$
$$(H, T, T), (T, H, T), (T, T, H), (T, T, T)\}.$$

Our goal is to assign probability to certain **events**. For example, suppose that we would like to know the probability that the outcome of rolling a fair die is an even number. In this case, our event is the set $E = \{2, 4, 6\}$. If the result of our random experiment belongs to the set E, we say that the event E has occurred. Thus an event is a collection of possible outcomes. In other words, an event is a subset of the sample space to which we assign a probability. Although we have not yet discussed how to find the probability of an event, you might be able to guess that the probability of $\{2,4,6\}$ is 50 percent which is the same as $\frac{1}{2}$ in the probability theory convention.

Outcome: A result of a random experiment.
Sample Space: The set of all possible outcomes.
Event: A subset of the sample space.

Union and Intersection: If A and B are events, then $A \cup B$ and $A \cap B$ are also events. By remembering the definition of union and intersection, we observe that $A \cup B$ occurs if A <u>or</u> B occur. Similarly, $A \cap B$ occurs if both A <u>and</u> B occur. Similarly, if $A_1, A_2, \cdots, A_n$ are events, then the event $A_1 \cup A_2 \cup A_3 \cdots \cup A_n$ occurs if <u>at least</u> one of $A_1, A_2, \cdots, A_n$ occurs. The event $A_1 \cap A_2 \cap A_3 \cdots \cap A_n$ occurs if <u>all of</u> $A_1, A_2, \cdots, A_n$ occur. It can be helpful to remember that the key words "or" and "at least" correspond to unions and the key words "and" and "all of" correspond to intersections.

1.3.2 Probability

We assign a **probability** measure $P(A)$ to an event A. This is a value between 0 and 1 that shows how likely the event is. If $P(A)$ is close to 0, it is very unlikely that the event A occurs. On the other hand, if $P(A)$ is close to 1, A is very likely to occur. The main subject of probability theory is to develop tools and techniques to calculate probabilities of different events. Probability theory is based on some axioms that act as the foundation for the theory, so let us state and explain these axioms.

Axioms of Probability:

1. Axiom 1: For any event A, $P(A) \geq 0$.

2. Axiom 2: Probability of the sample space S is $P(S) = 1$.

3. Axiom 3: If $A_1, A_2, A_3, \cdots$ are disjoint events, then $P(A_1 \cup A_2 \cup A_3 \cdots) = P(A_1) + P(A_2) + P(A_3) + \cdots$

Let us take a few moments and make sure we understand each axiom thoroughly. The first axiom states that probability cannot be negative. The smallest value for $P(A)$ is zero and if $P(A) = 0$, then the event A will never happen. The second axiom states that the probability of the whole sample space is equal to one; i.e., 100 percent. The reason for this is that the sample space S contains all possible outcomes of our random experiment. Thus, the outcome of each trial always belongs to S, i.e., the event S always occurs and $P(S) = 1$. In the example of rolling a die, $S = \{1, 2, 3, 4, 5, 6\}$, and since the outcome is always among the numbers 1 through 6, $P(S) = 1$.

The third axiom is probably the most interesting one. The basic idea is that, if some events are disjoint (i.e., there is no overlap between them), then the probability of their union must be the summations of their probabilities. Another way to think about this is to imagine the probability of a set as the area of that set in the Venn diagram. If several sets are disjoint such as the ones shown Figure 1.9, then the total area of their union is the sum of individual areas. The following example illustrates the idea behind the third axiom.

Example 1.8. In a presidential election, there are four candidates. Call them A, B, C, and D. Based on our polling analysis, we estimate that A has a 20 percent chance of winning the election, while B has a 40 percent chance of winning. What is the probability that A **or** B win the election?

Solution: Notice that the events that {A wins}, {B wins}, {C wins}, and {D wins} are disjoint since more than one of them cannot occur at the same time. For example, if A wins, then B cannot win. From the third axiom of probability, the probability of the union of two disjoint events is the summation of individual probabilities. Therefore,

$$P(\text{A wins } \underline{\text{or }} \text{B wins}) = P(\{\text{A wins}\} \cup \{\text{B wins}\})$$
$$= P(\{\text{A wins}\}) + P(\{\text{B wins}\})$$
$$= 0.2 + 0.4$$
$$= 0.6$$

In summary, if A_1 and A_2 are disjoint events, then $P(A_1 \cup A_2) = P(A_1) + P(A_2)$. The same argument is true when you have n disjoint events $A_1, A_2, \cdots, A_n$:

$$P(A_1 \cup A_2 \cup A_3 \cdots \cup A_n) = P(A_1) + P(A_2) + \cdots + P(A_n), \text{ if } A_1, A_2, \cdots, A_n \text{ are disjoint.}$$

In fact, the third axiom goes beyond that and states that the same is true even for a countably infinite number of disjoint events. We will see more examples of how we use the third axiom shortly.

As we have seen, when working with events, *intersection* means *"and"*, and *union* means *"or"*. The probability of intersection of A and B, $P(A \cap B)$, is sometimes shown by $P(A, B)$ or $P(AB)$.

Notation:

1. $P(A \cap B) = P(A \text{ and } B) = P(A, B)$,

2. $P(A \cup B) = P(A \text{ or } B)$.

1.3.3 Finding Probabilities

Suppose that we are given a random experiment with a sample space S. To find the probability of an event, there are usually two steps: first, we use the specific information that we have about the random experiment. Second, we use the probability axioms. Let's look at an example. Although this is a simple example and you might be tempted to write the answer without following the steps, we encourage you to follow them.

Example 1.9. You roll a fair die. What is the probability of $E = \{1, 5\}$?

Solution: Let's first use the specific information that we have about the random experiment. The problem states that the die is fair, which means that all six possible outcomes are equally likely, i.e.,

$$P(\{1\}) = P(\{2\}) = \cdots = P(\{6\}).$$

Now we can use the axioms of probability. In particular, since the events $\{1\}, \{2\}, \cdots, \{6\}$ are disjoint we can write

$$1 = P(S) = P\Big(\{1\} \cup \{2\} \cup \cdots \cup \{6\}\Big)$$
$$= P(\{1\}) + P(\{2\}) + \cdots + P(\{6\})$$
$$= 6P(\{1\}).$$

Thus,

$$P(\{1\}) = P(\{2\}) = \cdots = P(\{6\}) = \frac{1}{6}.$$

Again, since $\{1\}$ and $\{5\}$ are disjoint, we have

$$P(E) = P(\{1, 5\}) = P(\{1\}) + P(\{5\}) = \frac{2}{6} = \frac{1}{3}.$$

It is worth noting that we often write $P(1)$ instead of $P(\{1\})$ to simplify the notation, but we should emphasize that probability is defined for sets (events) not for individual outcomes. Thus, when we write $P(2) = \frac{1}{6}$, what we really mean is that $P(\{2\}) = \frac{1}{6}$.

We will see that the two steps explained above can be used to find probabilities for much more complicated events and random experiments. Let us now practice using the axioms by proving some useful facts.

Example 1.10. Using the axioms of probability, prove the following:

(a) For any event A, $P(A^c) = 1 - P(A)$.

(b) The probability of the empty set is zero, i.e., $P(\emptyset) = 0$.

(c) For any event A, $P(A) \leq 1$.

(d) $P(A - B) = P(A) - P(A \cap B)$.

(e) $P(A \cup B) = P(A) + P(B) - P(A \cap B)$, (inclusion-exclusion principle for $n = 2$).

(f) If $A \subset B$ then $P(A) \leq P(B)$.

Solution:

(a) This states that the probability that A does not occur is $1 - P(A)$. To prove it using the axioms, we can write

$$1 = P(S) \qquad\qquad\qquad\qquad \text{(by axiom 2)}$$
$$= P(A \cup A^c) \qquad\qquad \text{(definition of complement)}$$
$$= P(A) + P(A^c) \qquad\qquad \text{(since } A \text{ and } A^c \text{ are disjoint)}.$$

(b) Since $\emptyset = S^c$, we can use part (a) to see that $P(\emptyset) = 1 - P(S) = 0$. Note that this makes sense as by definition: an event happens if the outcome of the random experiment belongs to that event. Since the empty set does not have any element, the outcome of the experiment never belongs to the empty set.

(c) From part (a), $P(A) = 1 - P(A^c)$ and since $P(A^c) \geq 0$ (the first axiom), we have $P(A) \leq 1$.

(d) We show that $P(A) = P(A \cap B) + P(A - B)$. Note that the two sets $A \cap B$ and $A - B$ are disjoint and their union is A (Figure 1.17). Thus, by the third axiom of probability

$$P(A) = P\big((A \cap B) \cup (A - B)\big) \qquad\qquad (\text{ since } A = (A \cap B) \cup (A - B))$$
$$= P(A \cap B) + P(A - B) \qquad\qquad (\text{since } A \cap B \text{ and } A - B \text{ are disjoint}).$$

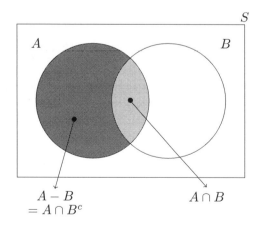

Figure 1.17: $P(A) = P(A \cap B) + P(A - B)$

Note that since $A - B = A \cap B^c$, we have shown

$$P(A) = P(A \cap B) + P(A \cap B^c).$$

Note also that the two sets B and B^c form a partition of the sample space (since they are disjoint and their union is the whole sample space). This is a simple form of law of total probability that we will discuss shortly and is a very useful rule in finding probability of some events.

(e) Note that A and $B - A$ are disjoint sets and their union is $A \cup B$. Thus,

$$P(A \cup B) = P(A \cup (B - A)) \qquad \qquad (A \cup B = A \cup (B - A))$$
$$= P(A) + P(B - A) \qquad \qquad \text{(since } A \text{ and } B - A \text{ are disjoint)}$$
$$= P(A) + P(B) - P(A \cap B) \qquad \qquad \text{(by part (d))}.$$

(f) Note that $A \subset B$ means that whenever A occurs B occurs, too. Thus intuitively we expect that $P(A) \leq P(B)$. Again the proof is similar as before. If $A \subset B$, then $A \cap B = A$. Thus,

$$P(B) = P(A \cap B) + P(B - A) \qquad \qquad \text{(by part (d))}$$
$$= P(A) + P(B - A) \qquad \qquad \text{(since } A = A \cap B)$$
$$\geq P(A) \qquad \qquad \text{(by axiom 1)}.$$

Example 1.11. Suppose we have the following information:

1. There is a 60 percent chance that it will rain today.

2. There is a 50 percent chance that it will rain tomorrow.

3. There is a 30 percent chance that it does not rain on either day.

Find the following probabilities:

(a) The probability that it will rain today or tomorrow.

(b) The probability that it will rain today and tomorrow.

(c) The probability that it will rain today but not tomorrow.

(d) The probability that it either will rain today or tomorrow, but not both.

Solution: An important step in solving problems like this is to correctly convert them to probability language. This is especially useful when the problems become complex. For this problem, let's define A as the event that it will rain today, and B as the event that it will rain tomorrow. Then, let's summarize the available information:

1. $P(A) = 0.6$,

2. $P(B) = 0.5$,

3. $P(A^c \cap B^c) = 0.3$.

Now that we have summarized the information, we should be able to use them alongside probability rules to find the requested probabilities:

(a) The probability that it will rain today or tomorrow: this is $P(A \cup B)$. To find this we notice that

$$P(A \cup B) = 1 - P\Big((A \cup B)^c\Big) \qquad \qquad \text{by Example 1.10}$$
$$= 1 - P(A^c \cap B^c) \qquad \text{by De Morgan's Law (Theorem 1.1)}$$
$$= 1 - 0.3$$
$$= 0.7$$

(b) The probability that it will rain today and tomorrow: this is $P(A \cap B)$. To find this we note that

$$P(A \cap B) = P(A) + P(B) - P(A \cup B) \qquad \text{by Example 1.10}$$
$$= 0.6 + 0.5 - 0.7$$
$$= 0.4$$

(c) The probability that it will rain today but not tomorrow: this is $P(A \cap B^c)$.

$$P(A \cap B^c) = P(A - B)$$
$$= P(A) - P(A \cap B) \qquad \text{by Example 1.10}$$
$$= 0.6 - 0.4$$
$$= 0.2$$

(d) The probability that it either will rain today or tomorrow but not both: this is $P(A - B) + P(B - A)$. We have already found $P(A - B) = 0.2$. Similarly, we can find $P(B - A)$:

$$P(B - A) = P(B) - P(B \cap A) \qquad \text{by Example 1.10}$$
$$= 0.5 - 0.4$$
$$= 0.1$$

Thus,

$$P(A - B) + P(B - A) = 0.2 + 0.1 = 0.3$$

In this problem, it is stated that there is a 50 percent chance that it will rain tomorrow. You might have heard this information from news on the TV. A more interesting question is how the number 50 is obtained. This is an example of a real-life problem in which tools from probability and statistics are used. As you read more chapters from the book, you will learn many of these tools that are frequently used in practice.

Inclusion-Exclusion Principle:

The formula $P(A \cup B) = P(A) + P(B) - P(A \cap B)$ that we proved in Example 1.10 is a simple form of the inclusion-exclusion principle. We can extend it to the union of three or more sets.

Inclusion-exclusion principle:

$$P(A \cup B) = P(A) + P(B) - P(A \cap B),$$

$$P(A \cup B \cup C) = P(A) + P(B) + P(C) - $$
$$P(A \cap B) - P(A \cap C) - P(B \cap C) + P(A \cap B \cap C).$$

Generally for n events $A_1, A_2, \cdots, A_n$, we have

$$P\left(\bigcup_{i=1}^{n} A_i\right) = \sum_{i=1}^{n} P(A_i) - \sum_{i<j} P(A_i \cap A_j)$$
$$+ \sum_{i<j<k} P(A_i \cap A_j \cap A_k) - \cdots + (-1)^{n-1} P\left(\bigcap_{i=1}^{n} A_i\right).$$

1.3.4 Discrete Probability Models

Here, we distinguish between two different types of sample spaces, discrete and continuous. We will discuss the difference more in detail later on, when we discuss random variables. The basic idea is that in discrete probability models we can compute the probability of events by adding all the corresponding outcomes, while in continuous probability models we need to use integration instead of summation.

Consider a sample space S. If S is a *countable* set, this refers to a **discrete** probability model. In this case, since S is countable, we can list all the elements in S:

$$S = \{s_1, s_2, s_3, \cdots\}.$$

If $A \subset S$ is an event, then A is also countable, and by the third axiom of probability we can write

$$P(A) = P(\bigcup_{s_j \in A} \{s_j\}) = \sum_{s_j \in A} P(s_j).$$

Thus, in a countable sample space, to find probability of an event, all we need to do is sum the probability of individual elements in that set.

Example 1.12. I play a gambling game in which I will win $k - 2$ dollars with probability $\frac{1}{2^k}$ for any $k \in \mathbb{N}$. That is,

- with probability $\frac{1}{2}$, I lose 1 dollar;

- with probability $\frac{1}{4}$, I win 0 dollar;

- with probability $\frac{1}{8}$, I win 1 dollar;

- with probability $\frac{1}{16}$, I win 2 dollars;

- with probability $\frac{1}{32}$, I win 3 dollars;

- $\cdots$

What is the probability that I win more than or equal to 1 dollar and less than 4 dollars? What is the probability that I win more than 2 dollars?

Solution: In this problem, the random experiment is the gambling game and the outcomes are the amount in dollars that I win (lose). Thus we may write

$$S = \{-1, 0, 1, 2, 3, 4, 5, \cdots\}.$$

As we see this is an infinite but countable set. The problem also states that

$$P(k) = P(\{k\}) = \frac{1}{2^{k+2}}, \text{for } k \in S.$$

First, let's check that this is a valid probability measure. To do so, we should check if all probabilities add up to one, i.e., $P(S) = 1$. We have

$$
\begin{aligned}
P(S) &= \sum_{k=-1}^{\infty} P(k) \\
&= \sum_{k=-1}^{\infty} \frac{1}{2^{k+2}} \\
&= \frac{1}{2} + \frac{1}{4} + \frac{1}{8} + \cdots \qquad \text{(geometric sum)} \\
&= 1.
\end{aligned}
$$

Now let's solve the problem. Let's define A as the event that I win more than or equal to 1 dollar and less than 4 dollars, and B as the event that I win more than 2 dollars. Thus,

$$A = \{1, 2, 3\}, B = \{3, 4, 5, \cdots\}.$$

Then

$$
\begin{aligned}
P(A) &= P(1) + P(2) + P(3) \\
&= \frac{1}{8} + \frac{1}{16} + \frac{1}{32} \\
&= \frac{7}{32} \\
&\approx 0.219
\end{aligned}
$$

Similarly,

$$
\begin{aligned}
P(B) &= P(3) + P(4) + P(5) + P(6) + \cdots \\
&= \frac{1}{32} + \frac{1}{64} + \frac{1}{128} + \frac{1}{256} + \cdots \qquad \text{(geometric sum)} \\
&= \frac{1}{16} \\
&= 0.0625
\end{aligned}
$$

Note that another way to find $P(B)$ is to write

$$
\begin{aligned}
P(B) &= 1 - P(B^c) \\
&= 1 - P(\{-1, 0, 1, 2\}) \\
&= 1 - \left(P(-1) + P(0) + P(1) + P(2) \right) \\
&= 1 - \left(\frac{1}{2} + \frac{1}{4} + \frac{1}{8} + \frac{1}{16} \right) \\
&= 1 - \frac{15}{16} \\
&= \frac{1}{16} \\
&= 0.0625
\end{aligned}
$$

Note: Here we have used the geometric series sum formula. In particular, for any $a, x \in \mathbb{R}$, we have

$$
a + ax + ax^2 + ax^3 + \cdots + ax^{n-1} = \sum_{k=0}^{n-1} ax^k = a \frac{1 - x^n}{1 - x} \tag{1.3}
$$

Moreover, if $|x| < 1$, then we have

$$
a + ax + ax^2 + ax^3 + \cdots = \sum_{k=0}^{\infty} ax^k = a \frac{1}{1 - x} \tag{1.4}
$$

Finite Sample Spaces with Equally Likely Outcomes:

An important special case of discrete probability models is when we have a finite sample space S, where each outcome is equally likely, i.e.,

$$
S = \{s_1, s_2, \cdots, s_N\}, \text{ where } P(s_i) = P(s_j) \text{ for all } i, j \in \{1, 2, \cdots, N\}.
$$

Rolling a fair die is an instance of such a probability model. Since all outcomes are equally likely, we must have

$$
P(s_i) = \frac{1}{N}, \text{ for all } i \in \{1, 2, \cdots, N\}.
$$

In such a model, if A is any event with cardinality $|A| = M$, we can write

$$
P(A) = \sum_{s_j \in A} P(s_j) = \sum_{s_j \in A} \frac{1}{N} = \frac{M}{N} = \frac{|A|}{|S|}.
$$

Thus, finding probability of A reduces to a *counting* problem in which we need to count how many elements are in A and S.

Example 1.13. I roll a fair die twice and obtain two numbers: $X_1 = $ result of the first roll, and $X_2 = $ result of the second roll. Write down the sample space S, and assuming that all outcomes are equally likely (because the die is fair) find the probability of the event A, defined as the event that $X_1 + X_2 = 8$.

Solution: The sample space S can be written as

$$S = \{(1,1), (1,2), (1,3), (1,4), (1,5), (1,6),$$
$$(2,1), (2,2), (2,3), (2,4), (2,5), (2,6),$$
$$(3,1), (3,2), (3,3), (3,4), (3,5), (3,6),$$
$$(4,1), (4,2), (4,3), (4,4), (4,5), (4,6),$$
$$(5,1), (5,2), (5,3), (5,4), (5,5), (5,6),$$
$$(6,1), (6,2), (6,3), (6,4), (6,5), (6,6)\}.$$

As we see, there are $|S| = 36$ elements in S. To find probability of A, all we need to do is find $M = |A|$. In particular, A is defined as

$$A = \{(X_1, X_2) | X_1 + X_2 = 8, X_1, X_2 \in \{1, 2, \cdots, 6\}\}$$
$$= \{(2,6), (3,5), (4,4), (5,3), (6,2)\}.$$

Thus, $|A| = 5$, which means that

$$P(A) = \frac{|A|}{|S|} = \frac{5}{36}.$$

A very common mistake is not distinguishing between, say $(2,6)$ and $(6,2)$. It is important to note that these are two different outcomes: $(2,6)$ means that the first roll is a 2 and the second roll is a 6, while $(6,2)$ means that the first roll is a 6 and the second roll is a 2. Note that it is very common to write $P(X_1 + X_2 = 8)$ when referring to $P(A)$ as defined above. In fact, X_1 and X_2 are examples of *random variables* that will be discussed in detail later on.

In a finite sample space S, where all outcomes are equally likely, the probability of any event A can be found by

$$P(A) = \frac{|A|}{|S|}.$$

The formula $P(A) = \frac{|A|}{|S|}$ suggests that it is important to be able to count elements in sets. If sets are small, this is an easy task; however, if the sets are large and defined implicitly, this could be a difficult job. That is why we discuss counting methods later on.

1.3.5   Continuous Probability Models

Consider a scenario where your sample space S is, for example, $[0, 1]$. This is an uncountable set; we cannot list the elements in the set. At this time, we have not yet developed the tools needed to deal with continuous probability models, but we can provide some intuition by looking at a simple example.

Example 1.14. Your friend tells you that she will stop by your house sometime after or equal to 1 p.m. and before 2 p.m., but she cannot give you any more information as her schedule is quite hectic. Your friend is very dependable, so you are sure that she will stop by your house, but other than that we have no information about the arrival time. Thus, we assume that the arrival time is completely random in the 1 p.m. and 2 p.m. interval. (As we will see, in the language of probability theory, we say that the arrival time is "uniformly" distributed on the $[1,2)$ interval). Let T be the arrival time.

(a) What is the sample space S?

(b) What is the probability of $P(1.5)$? Why?

(c) What is the probability of $T \in [1, 1.5)$?

(d) For $1 \le a \le b \le 2$, what is $P(a \le T \le b) = P([a,b])$?

   Solution:

(a) Since any real number in $[1,2)$ is a possible outcome, the sample space is indeed $S = [1,2)$.

(b) Now, let's look at $P(1.5)$. A reasonable guess would be $P(1.5) = 0$. But can we provide a reason for that? Let us divide the $[1,2)$ interval to $2N + 1$ equal-length and disjoint intervals, $[1, 1 + \frac{1}{2N+1}), [1 + \frac{1}{2N+1}, 1 + \frac{2}{2N+1}), \cdots, [1 + \frac{N}{2N+1}, 1 + \frac{N+1}{2N+1}), \cdots, [1 + \frac{2N}{2N+1}, 2)$. See Figure 1.18. Here, N could be any positive integer.

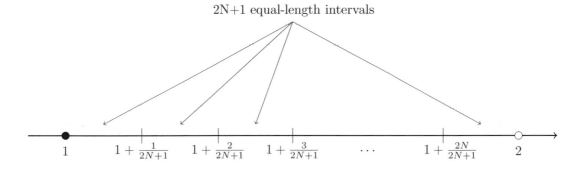

Figure 1.18: Dividing the interval $[1,2)$ to $2N + 1$ equal-length intervals

The only information that we have is that the arrival time is "uniform" on the $[1,2)$ interval. Therefore, all of the above intervals should have the same probability, and since their union is S we conclude that

$$P\left(\left[1, 1 + \frac{1}{2N+1}\right)\right) = P\left(\left[1 + \frac{1}{2N+1}, 1 + \frac{2}{2N+1}\right)\right) = \cdots$$

$$\cdots = P\left(\left[1 + \frac{N}{2N+1}, 1 + \frac{N+1}{2N+1}\right)\right) = \cdots$$

$$\cdots = P\left(\left[1 + \frac{2N}{2N+1}, 2\right)\right) = \frac{1}{2N+1}.$$

In particular, by defining $A_N = \left[1 + \frac{N}{2N+1}, 1 + \frac{N+1}{2N+1}\right)$, we conclude that

$$P(A_N) = P\left(\left[1 + \frac{N}{2N+1}, 1 + \frac{N+1}{2N+1}\right)\right) = \frac{1}{2N+1}.$$

Now note that for any positive integer N, $1.5 \in A_N$. Thus, $\{1.5\} \subset A_N$, so

$$P(1.5) \leq P(A_N) = \frac{1}{2N+1}, \qquad \text{for all } N \in \mathbb{N}.$$

Note that as N becomes large, $P(A_N)$ approaches 0. Since $P(1.5)$ cannot be negative, we conclude that $P(1.5) = 0$. Similarly, we can argue that $P(x) = 0$ for all $x \in [1, 2)$.

(c) Next, we find $P([1, 1.5))$. This is the first half of the entire sample space $S = [1, 2)$ and because of uniformity, its probability must be 0.5. In other words,

$$P([1, 1.5)) = P([1.5, 2)) \qquad\qquad \text{(by uniformity)}$$
$$P([1, 1.5)) + P([1.5, 2)) = P(S) = 1.$$

Thus

$$P([1, 1.5)) = P([1.5, 2)) = \frac{1}{2}.$$

(d) The same uniformity argument suggests that all intervals in $[1, 2)$ with the same length must have the same probability. In particular, the probability of an interval is proportional to its length. For example, since

$$[1, 1.5) = [1, 1.25) \cup [1.25, 1.5).$$

Thus, we conclude

$$P\big([1, 1.5)\big) = P\big([1, 1.25)\big) + P\big([1.25, 1.5)\big)$$
$$= 2P\big([1, 1.25)\big).$$

And finally, since $P\big([1, 2)\big) = 1$, we conclude

$$P([a, b]) = b - a, \qquad \text{for } 1 \leq a \leq b < 2.$$

The above example was a somewhat simple situation in which we have a continuous sample space. In reality, the probability might not be uniform, so we need to develop tools that help us deal with general distributions of probabilities. These tools will be introduced in the coming chapters.

Discussion: You might ask why $P(x) = 0$ for all $x \in [1, 2)$, but at the same time, the outcome of the experiment is always a number in $[1, 2)$. We can answer this question from different points of view. From a mathematical point of view, we can explain this issue by using the following analogy: consider a line segment of length one. This line segment consists of points of length zero. Nevertheless, these zero-length points as a whole constitute a line segment of length one. From a practical point of view, we can provide the following explanation: our observed outcome is not all real values in $[1, 2)$. That is, if we are observing time, our measurement might be accurate up to minutes, or seconds, or milliseconds, etc. Our continuous probability model is a limit of a discrete probability model, when the precision becomes infinitely accurate. Thus, in reality, we are always interested in the probability of some intervals rather than a specific point x. For example, when we say, "What is the probability that your friend shows up at $1 : 32$ p.m.?", what we may mean is, "What is the probability that your friend shows up between $1 : 32 : 00$ p.m. and $1 : 32 : 59$ p.m.?" This probability is nonzero as it refers to an interval with a one-minute length. Thus, in some sense, a continuous probability model can be looked at as the "limit" of a discrete space. Remembering from calculus, we note that integrals are defined as the limits of sums. That is why we use integrals to find probabilities for continuous probability models, as we will see later.

1.3.6 Solved Problems

1. Consider a sample space S and three events A, B, and C. For each of the following events draw a Venn diagram representation as well as a set expression.

   (a) Among A, B, and C, only A occurs.

   (b) At least one of the events A, B, or C occurs.

   (c) A or C occurs, but not B.

   (d) At most two of the events A, B, or C occur.

   Solution:

   (a) Among A, B, and C, only A occurs: $A - B - C = A - (B \cup C)$.

   (b) At least one of the events A, B, or C occurs: $A \cup B \cup C$.

   (c) A or C occurs, but not B: $(A \cup C) - B$.

   (d) At most two of the events A, B, or C occur: $(A \cap B \cap C)^c = A^c \cup B^c \cup C^c$.

   The Venn diagrams are shown in Figure 1.19.

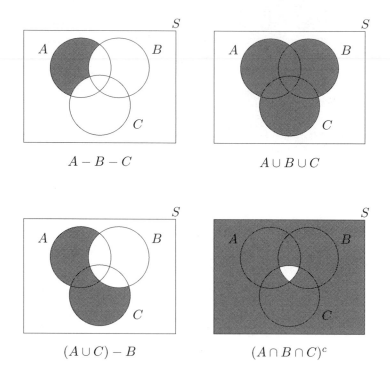

Figure 1.19: Venn diagrams for Solved Problem 1

2. Write the sample space S for the following random experiments.

   (a) We toss a coin until we see two consecutive tails. We record the total number of coin tosses.

   (b) A bag contains 4 balls: one is red, one is blue, one is white, and one is green. We choose two distinct balls and record their color in order.

   (c) A customer arrives at a bank and waits in the line. We observe T, which is the total time (in hours) that the customer waits in the line. The bank has a strict policy that no customer waits more than 20 minutes under any circumstances.

   Solution: Remember that the sample space is the set of all possible outcomes. Usually, when you have a random experiment, there are different ways to define the sample space S depending on what you observe as the outcome. In this problem, for each experiment it is stated what outcomes we observe in order to help you write down the sample space S.

(a) We toss a coin until we see two consecutive tails. We record the total number of coin tosses: Here, the total number of coin tosses is a natural number larger than or equal to 2. The sample space is

$$S = \{2, 3, 4, \cdots\}.$$

(b) A bag contains 4 balls: one is red, one is blue, one is white, and one is green. We choose two distinct balls and record their color in order: The sample space can be written as

$$S = \{(R, B), (B, R), (R, W), (W, R), (R, G), (G, R),$$
$$(B, W), (W, B), (B, G), (G, B), (W, G), (G, W)\}.$$

(c) A customer arrives at a bank and waits in the line. We observe T...: In theory, T can be any real number between 0 and $\frac{1}{3} = 20$ minutes. Thus,

$$S = \left[0, \frac{1}{3}\right] = \left\{x \in \mathbb{R} | 0 \leq x \leq \frac{1}{3}\right\}.$$

3. Let A, B, and C be three events in the sample space S. Suppose we know

- $A \cup B \cup C = S$,
- $P(A) = \frac{1}{2}$,
- $P(B) = \frac{2}{3}$,
- $P(A \cup B) = \frac{5}{6}$.

Answer the following questions:

(a) Find $P(A \cap B)$.

(b) Do A, B, and C form a partition of S?

(c) Find $P(C - (A \cup B))$.

(d) If $P(C \cap (A \cup B)) = \frac{5}{12}$, find $P(C)$.

Solution: As before, it is always useful to draw a Venn diagram; however, here we provide the solution without using a Venn diagram.

(a) Using the inclusion-exclusion principle, we have

$$P(A \cup B) = P(A) + P(B) - P(A \cap B).$$

Thus,

$$P(A \cap B) = P(A) + P(B) - P(A \cup B)$$
$$= \frac{1}{2} + \frac{2}{3} - \frac{5}{6}$$
$$= \frac{1}{3}.$$

(b) No since $A \cap B \neq \emptyset$.

(c) We can write

$$C - (A \cup B) = \left(C \cup (A \cup B)\right) - (A \cup B)$$
$$= S - (A \cup B) \qquad\qquad (\text{since } A \cup B \cup C = S)$$
$$= (A \cup B)^c.$$

Thus

$$P\big(C - (A \cup B)\big) = P\big((A \cup B)^c\big)$$
$$= 1 - P(A \cup B)$$
$$= \frac{1}{6}.$$

(d) We have

$$P(C) = P(C \cap (A \cup B)) + P(C - (A \cup B)) = \frac{5}{12} + \frac{1}{6} = \frac{7}{12}.$$

4. I roll a fair die twice and obtain two numbers: $X_1 =$ result of the first roll, and $X_2 =$ result of the second roll. Find the probability of the following events:

(a) A defined as "$X_1 < X_2$";

(b) B defined as "You observe a 6 at least once."

Solution: As we saw before, the sample space S has 36 elements.

(a) We have

$$A = \{(1,2),(1,3),(1,4),(1,5),(1,6),(2,3),(2,4),(2,5),$$
$$(2,6),(3,4),(3,5),(3,6),(4,5),(4,6),(5,6)\}.$$

Then, we obtain

$$P(A) = \frac{|A|}{|S|} = \frac{15}{36} = \frac{5}{12}.$$

(b) We have

$$B = \{(6,1),(6,2),(6,3),(6,4),(6,5),(6,6),(1,6),(2,6),(3,6),(4,6),(5,6)\}.$$

We obtain

$$P(B) = \frac{|B|}{|S|} = \frac{11}{36}.$$

5. You purchase a certain product. The manual states that the lifetime T of the product, defined as the amount of time (in years) the product works properly until it breaks down, satisfies

$$P(T \geq t) = e^{-\frac{t}{5}}, \text{ for all } t \geq 0.$$

For example, the probability that the product lasts for more than (or equal to) 2 years is $P(T \geq 2) = e^{-\frac{2}{5}} = 0.6703$.

(a) This is an example of a continuous probability model. Write down the sample space S.

(b) Check that the statement in the manual makes sense by finding $P(T \geq 0)$ and $\lim_{t \to \infty} P(T \geq t)$.

(c) Also check that if $t_1 < t_2$, then $P(T \geq t_1) \geq P(T \geq t_2)$. Why does this need to be true?

(d) Find the probability that the product breaks down within three years of the purchase time.

(e) Find the probability that the product breaks down in the second year, i.e., find $P(1 \leq T < 2)$.

Solution:

(a) The sample space S is the set of all possible outcomes. Here, the possible outcomes are the possible values for T which can be any real number larger than or equal to zero. Thus

$$S = [0, \infty).$$

(b) We have

$$P(T \geq 0) = e^{-\frac{0}{5}} = 1,$$

$$\lim_{t \to \infty} P(T \geq t) = e^{-\infty} = 0,$$

which is what we expect. In particular, T is always larger than or equal to zero, thus we expect $P(T \geq 0) = 1$. Also, since the product will eventually fail at some point, we expect that $P(T \geq t)$ approaches zero as t goes to infinity.

(c) First note that if $t_1 < t_2$, then $P(T \geq t_1) = e^{-\frac{t_1}{5}} > e^{-\frac{t_2}{5}} = P(T \geq t_2)$ (since $f(x) = e^x$ is an increasing function). Here we have two events, A is the event that $T \geq t_1$ and B is the event that $T \geq t_2$. That is,

$$A = [t_1, \infty), B = [t_2, \infty).$$

Since B is a subset of A, $B \subset A$, we must have $P(B) \leq P(A)$, thus

$$P(A) = P(T \geq t_1) \geq P(T \geq t_2) = P(B).$$

(d) The probability that the product breaks down within three years of the purchase time is

$$P(T < 3) = 1 - P(T \geq 3) = 1 - e^{-\frac{3}{5}} \approx 0.4512$$

(e) Note that if $A \subset B$, then

$$P(B - A) = P(B) - P(B \cap A)$$
$$= P(B) - P(A) \qquad\qquad \text{(since } A \subset B\text{).}$$

Choosing $A = [2, \infty)$ and $B = [1, \infty)$, we can write

$$P(1 \leq T < 2) = P(T \geq 1) - P(T \geq 2)$$
$$= e^{-\frac{1}{5}} - e^{-\frac{2}{5}} = 0.1484$$

6. I first saw this question in a math contest many years ago: You get a stick and break it randomly into three pieces. What is the probability that you can make a triangle using the three pieces? You can assume the break points are chosen completely at random, i.e. if the length of the original stick is 1 unit, and x, y, z are the lengths of the three pieces, then (x, y, z) are uniformly chosen from the set

$$\{(x, y, z) \in \mathbb{R}^3 | x + y + z = 1, x, y, z \geq 0\}.$$

Solution: This is again a problem on a continuous probability space. The basic idea is pretty simple. First, we need to identify the sample space S. In this case the sample space is going to be a two-dimensional set. Second, we need to identify the set A that contains the favorable outcomes (the set of (x, y, z) in S that form a triangle). And finally, since the space is uniform, we will divide area of set A by the area of S to obtain $P(A)$.

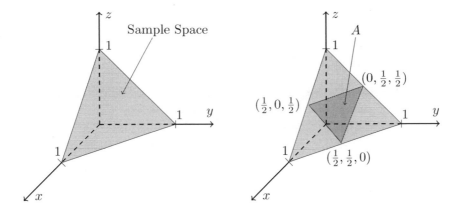

Figure 1.20: The sample space and set A for Problem 6

First, we need to find the sets S and A. This is basically a geometry problem. The two sets, S and A, are shown in Figure 1.20.

Note that in $\mathbb{R}^3$, $x + y + z = 1$ represents a plane that goes through the points $(1, 0, 0)$, $(0, 1, 0)$, $(0, 0, 1)$. To find the sample space S, note that $S = \{(x, y, z) \in \mathbb{R}^3 | x + y + z = 1, x, y, z \geq 0\}$, thus S is the part of the plane that is shown in Figure 1.20. To find the set A, note that we need (x, y, z) to satisfy the triangle inequality

$$x + y > z,$$
$$y + z > x,$$
$$x + z > y.$$

Note that since $x + y + z = 1$, we can equivalently write the three equations as

$$x < \frac{1}{2},$$
$$y < \frac{1}{2},$$
$$z < \frac{1}{2}.$$

Thus, we conclude that the set A is the area shown in Figure 1.20. In particular, we note that the set S consists of four triangles with equal areas. Therefore, its area is four times the area of A, and we have

$$P(A) = \frac{\text{Area of } A}{\text{Area of } S} = \frac{1}{4}.$$

1.4 Conditional Probability

In this section, we will discuss one of the most fundamental concepts in probability theory. Here is the question: As you obtain additional information, how should you update probabilities of events? For example, suppose that in a certain city, 23 percent of the days are rainy. Thus, if you pick a random day, the probability that it rains that day is 23 percent:

$$P(R) = 0.23, \text{where } R \text{ is the event that it rains on the randomly chosen day.}$$

Now suppose that I pick a random day, but I also tell you that it is cloudy on the chosen day. Now that you have this extra piece of information, how do you update the chance that it rains on that day? In other words, what is the probability that it rains **given that** it is cloudy? If C is the event that it is cloudy, then we write this as $P(R|C)$, the *conditional probability of R given that C has occurred*. It is reasonable to assume that in this example, $P(R|C)$ should be larger than the original $P(R)$, which is called the **prior probability** of R. But what exactly should $P(R|C)$ be? Before providing a general formula, let's look at a simple example.

Example 1.15. I roll a fair die. Let A be the event that the outcome is an odd number, i.e., $A = \{1,3,5\}$. Also let B be the event that the outcome is less than or equal to 3, i.e., $B = \{1,2,3\}$. What is the probability of A, $P(A)$? What is the probability of A given B, $P(A|B)$?

   Solution: This is a finite sample space, so

$$P(A) = \frac{|A|}{|S|} = \frac{|\{1,3,5\}|}{6} = \frac{1}{2}.$$

Now, let's find the conditional probability of A given that B occurred. If we know B has occurred, the outcome must be among $\{1,2,3\}$. For A to also happen the outcome must be in $A \cap B = \{1,3\}$. Since all die rolls are equally likely, we argue that $P(A|B)$ must be equal to

$$P(A|B) = \frac{|A \cap B|}{|B|} = \frac{2}{3}.$$

   Now let's see how we can generalize the above example. We can rewrite the calculation by dividing the numerator and denominator by $|S|$ in the following way:

$$P(A|B) = \frac{|A \cap B|}{|B|} = \frac{\frac{|A \cap B|}{|S|}}{\frac{|B|}{|S|}} = \frac{P(A \cap B)}{P(B)}.$$

Although the above calculation has been done for a finite sample space with equally likely outcomes, it turns out the resulting formula is quite general and can be applied in any setting. Below, we formally provide the formula and then explain the intuition behind it.

If A and B are two events in a sample space S, then the **conditional probability of A given B** is defined as

$$P(A|B) = \frac{P(A \cap B)}{P(B)}, \text{ when } P(B) > 0.$$

Here is the intuition behind the formula. When we know that B has occurred, every outcome that is outside B should be discarded. Thus, *our sample space is reduced to the set B*, Figure 1.21. Now the only way that A can happen is when the outcome belongs to the set $A \cap B$. We divide $P(A \cap B)$ by $P(B)$, so that the conditional probability of the new sample space becomes 1, i.e., $P(B|B) = \frac{P(B \cap B)}{P(B)} = 1$.

Note that conditional probability of $P(A|B)$ is undefined when $P(B) = 0$. That is okay because if $P(B) = 0$, it means that the event B never occurs so it does not make sense to talk about the probability of A given B.

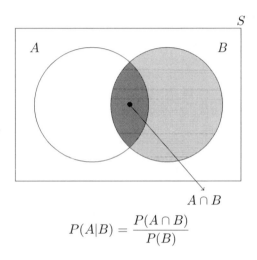

$$P(A|B) = \frac{P(A \cap B)}{P(B)}$$

Figure 1.21: Venn diagram for conditional probability, $P(A|B)$.

It is important to note that conditional probability itself is a probability measure, so it satisfies probability axioms. In particular,

1. Axiom 1: For any event A, $P(A|B) \geq 0$.

2. Axiom 2: Conditional probability of B given B is 1, i.e., $P(B|B) = 1$.

3. Axiom 3: If $A_1, A_2, A_3, \cdots$ are disjoint events, then $P(A_1 \cup A_2 \cup A_3 \cdots |B) = P(A_1|B) + P(A_2|B) + P(A_3|B) + \cdots$.

In fact, all rules that we have learned so far can be extended to conditional probability. For example, the formulas given in Example 1.10 can be rewritten:

Example 1.16. For three events, A, B, and C, with $P(C) > 0$, we have

- $P(A^c|C) = 1 - P(A|C)$;

- $P(\emptyset|C) = 0$;

- $P(A|C) \leq 1$;

- $P(A - B|C) = P(A|C) - P(A \cap B|C)$;

- $P(A \cup B|C) = P(A|C) + P(B|C) - P(A \cap B|C)$;
- if $A \subset B$, then $P(A|C) \leq P(B|C)$.

Let's look at some special cases of conditional probability:

- When A and B are disjoint: In this case $A \cap B = \emptyset$, so

$$P(A|B) = \frac{P(A \cap B)}{P(B)}$$
$$= \frac{P(\emptyset)}{P(B)}$$
$$= 0.$$

  This makes sense. In particular, since A and B are disjoint they cannot both occur at the same time. Thus, given that B has occurred, the probability of A must be zero.

- When B is a subset of A: If $B \subset A$, then whenever B happens, A also happens. Thus, given that B occurred, we expect that probability of A be one. In this case $A \cap B = B$, so

$$P(A|B) = \frac{P(A \cap B)}{P(B)}$$
$$= \frac{P(B)}{P(B)} =$$
$$= 1.$$

- When A is a subset of B: In this case $A \cap B = A$, so

$$P(A|B) = \frac{P(A \cap B)}{P(B)}$$
$$= \frac{P(A)}{P(B)}.$$

Example 1.17. I roll a fair die twice and obtain two numbers $X_1 = $ result of the first roll, and $X_2 = $ result of the second roll. Given that I know $X_1 + X_2 = 7$, what is the probability that $X_1 = 4$ or $X_2 = 4$?

Solution: Let A be the event that $X_1 = 4$ or $X_2 = 4$, and B be the event that $X_1 + X_2 = 7$. we are interested in $P(A|B)$, so we can use

$$P(A|B) = \frac{P(A \cap B)}{P(B)}.$$

We note that

$$A = \{(4,1), (4,2), (4,3), (4,4), (4,5), (4,6), (1,4), (2,4), (3,4), (5,4), (6,4)\};$$
$$B = \{(6,1), (5,2), (4,3), (3,4), (2,5), (1,6)\};$$
$$A \cap B = \{(4,3), (3,4)\}.$$

We conclude

$$P(A|B) = \frac{P(A \cap B)}{P(B)}$$

$$= \frac{\frac{2}{36}}{\frac{6}{36}}$$

$$= \frac{1}{3}.$$

Let's look at a famous probability problem, called the two-child problem. Many versions of this problem have been discussed [1] in the literature and we will review a few of them in this chapter. We suggest that you try to guess the answers before solving the problem using probability formulas.

Example 1.18. Consider a family that has two children. We are interested in the children's genders. Our sample space is $S = \{(G,G), (G,B), (B,G), (B,B)\}$. Also assume that all four possible outcomes are equally likely.

(a) What is the probability that both children are girls, given that the first child is a girl?

(b) We ask the father: "Do you have at least one daughter?" He responds "Yes!" Given this extra information, what is the probability that both children are girls? In other words, what is the probability that both children are girls given that we know at least one of them is a girl?

Solution: Let A be the event that both children are girls, i.e., $A = \{(G,G)\}$. Let B be the event that the first child is a girl, i.e., $B = \{(G,G), (G,B)\}$. Finally, let C be the event that at least one of the children is a girl, i.e., $C = \{(G,G), (G,B), (B,G)\}$. Since the outcomes are equally likely, we can write

$$P(A) = \frac{1}{4},$$

$$P(B) = \frac{2}{4} = \frac{1}{2},$$

$$P(C) = \frac{3}{4}.$$

(a) What is the probability that both children are girls given that the first child is a girl? This is $P(A|B)$, thus we can write

$$P(A|B) = \frac{P(A \cap B)}{P(B)}$$

$$= \frac{P(A)}{P(B)} \qquad \text{(since } A \subset B\text{)}$$

$$= \frac{\frac{1}{4}}{\frac{1}{2}} = \frac{1}{2}.$$

(b) What is the probability that both children are girls given that we know at least one of them is a girl? This is $P(A|C)$, thus we can write

$$P(A|C) = \frac{P(A \cap C)}{P(C)}$$

$$= \frac{P(A)}{P(C)} \qquad \text{(since } A \subset C)$$

$$= \frac{\frac{1}{4}}{\frac{3}{4}} = \frac{1}{3}.$$

Discussion: Asked to guess the answers in the above example, many people would guess that both $P(A|B)$ and $P(A|C)$ should be 50 percent. However, as we see $P(A|B)$ is 50 percent, while $P(A|C)$ is only 33 percent. This is an example where the answers might seem counterintuitive. To understand the results of this problem, it is helpful to note that the event B is a subset of the event C. In fact, it is strictly smaller: it does not include the element (B, G), while C has that element. Thus the set C has more outcomes that are not in A than B, which means that $P(A|C)$ should be smaller than $P(A|B)$.

It is often useful to think of probability as percentages. For example, to better understand the results of this problem, let us imagine that there are 4000 families that have two children. Since the outcomes $(G, G), (G, B), (B, G)$, and (B, B) are equally likely, we will have roughly 1000 families associated with each outcome as shown in Figure 1.22. To find probability $P(A|C)$, we are performing the following experiment: We choose a random family from the families with at least one daughter. These are the families shown in the box. From these families, there are 1000 families with two girls and there are 2000 families with exactly one girl. Thus, the probability of choosing a family with two girls is $\frac{1}{3}$.

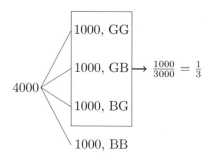

Figure 1.22: An example to help the understanding of $P(A|C)$ in Example 1.18.

Chain rule for conditional probability:

Let us write the formula for conditional probability in the following format

$$P(A \cap B) = P(A)P(B|A) = P(B)P(A|B) \qquad (1.5)$$

This format is particularly useful in situations when we know the conditional probability, but we are interested in the probability of the intersection. We can interpret this formula using a tree diagram such as the one shown in Figure 1.23. In this figure, we obtain the probability at each point by multiplying probabilities on the branches leading to that point. This type of diagram can be very useful for some problems.

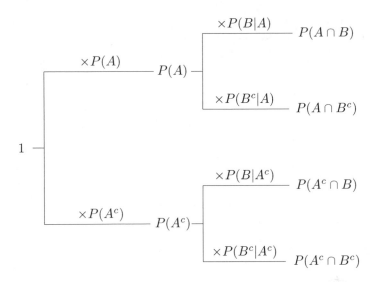

Figure 1.23: A tree diagram.

Now we can extend this formula to three or more events:

$$P(A \cap B \cap C) = P\big(A \cap (B \cap C)\big) = P(A)P(B \cap C|A) \qquad (1.6)$$

From Equation 1.5,

$$P(B \cap C) = P(B)P(C|B).$$

Conditioning both sides on A, we obtain

$$P(B \cap C|A) = P(B|A)P(C|A, B) \qquad (1.7)$$

Combining Equations 1.6 and 1.7 we obtain the following chain rule:

$$P(A \cap B \cap C) = P(A)P(B|A)P(C|A, B).$$

The point here is understanding how you can derive these formulas and trying to have intuition about them rather than memorizing them. You can extend the tree in Figure 1.23 to this case. Here the tree will have eight leaves. A general statement of the chain rule for n events is as follows:

Chain rule for conditional probability:

$$P(A_1 \cap A_2 \cap \cdots \cap A_n) = P(A_1)P(A_2|A_1)P(A_3|A_2, A_1) \cdots P(A_n|A_{n-1}A_{n-2} \cdots A_1)$$

Example 1.19. In a factory there are 100 units of a certain product, 5 of which are defective. We pick 3 units from the 100 units at random. What is the probability that none of them are defective?

Solution: Let us define A_i as the event that the ith chosen unit is not defective, for $i = 1, 2, 3$. We are interested in $P(A_1 \cap A_2 \cap A_3)$. Note that

$$P(A_1) = \frac{95}{100}.$$

Given that the first chosen item was good, the second item will be chosen from 94 good units and 5 defective units, thus

$$P(A_2|A_1) = \frac{94}{99}.$$

Given that the first and second chosen items were okay, the third item will be chosen from 93 good units and 5 defective units, thus

$$P(A_3|A_2, A_1) = \frac{93}{98}.$$

Thus, we have

$$P(A_1 \cap A_2 \cap A_3) = P(A_1)P(A_2|A_1)P(A_3|A_2, A_1)$$
$$= \frac{95}{100}\frac{94}{99}\frac{93}{98}$$
$$= 0.8560$$

As we will see later on, another way to solve this problem is to use counting arguments.

1.4.1   Independence

Let A be the event that it rains tomorrow, and suppose that $P(A) = \frac{1}{3}$. Also suppose that I toss a fair coin; let B be the event that it lands heads up. We have $P(B) = \frac{1}{2}$.

Now I ask you, what is $P(A|B)$? What is your guess? You probably guessed that $P(A|B) = P(A) = \frac{1}{3}$. You are right! The result of my coin toss does not have anything to do with tomorrow's weather. Thus, no matter if B happens or not, the probability of A should not change. This is an example of two **independent** events. Two events are independent if one does not convey any information about the other. Let us now provide a formal definition of independence.

Two events A and B are independent if and only if $P(A \cap B) = P(A)P(B)$.

Now, let's first reconcile this definition with what we mentioned earlier, $P(A|B) = P(A)$. If two events are independent, then $P(A \cap B) = P(A)P(B)$, so

$$P(A|B) = \frac{P(A \cap B)}{P(B)}$$
$$= \frac{P(A)P(B)}{P(B)}$$
$$= P(A).$$

Thus, if two events A and B are independent and $P(B) \neq 0$, then $P(A|B) = P(A)$. To summarize, we can say "independence means we can multiply the probabilities of events to obtain the probability of their intersection," or equivalently, "independence means that conditional probability of one event given another is the same as the original (prior) probability."

Sometimes the independence of two events is quite clear because the two events seem not to have any physical interaction with each other (such as the two events discussed above). At other times, it is not as clear and we need to check if they satisfy the independence condition. Let's look at an example.

Example 1.20. I pick a random number from $\{1, 2, 3, \cdots, 10\}$, and call it N. Suppose that all outcomes are equally likely. Let A be the event that N is less than 7, and let B be the event that N is an even number. Are A and B independent?

Solution: We have $A = \{1, 2, 3, 4, 5, 6\}$, $B = \{2, 4, 6, 8, 10\}$, and $A \cap B = \{2, 4, 6\}$. Then,

$$P(A) = 0.6,$$
$$P(B) = 0.5,$$
$$P(A \cap B) = 0.3$$

Therefore, $P(A \cap B) = P(A)P(B)$, so A and B are independent. This means that knowing that B has occurred does not change our belief about the probability of A. In this problem, the two events are about the same random number, but they are still independent because they satisfy the definition.

The definition of independence can be extended to the case of three or more events.

Three events A, B, and C are independent if **all** of the following conditions hold:

$$P(A \cap B) = P(A)P(B),$$
$$P(A \cap C) = P(A)P(C),$$
$$P(B \cap C) = P(B)P(C),$$
$$P(A \cap B \cap C) = P(A)P(B)P(C).$$

Note that all four of the stated conditions must hold for three events to be independent. In particular, you can find situations in which three of them hold, but the fourth one does not. In general, for n events $A_1, A_2, \cdots, A_n$ to be independent we must have

$$P(A_i \cap A_j) = P(A_i)P(A_j), \text{ for all distinct } i, j \in \{1, 2, \cdots, n\};$$
$$P(A_i \cap A_j \cap A_k) = P(A_i)P(A_j)P(A_k), \text{ for all distinct } i, j, k \in \{1, 2, \cdots, n\};$$

$$P(A_1 \cap A_2 \cap A_3 \cdots \cap A_n) = P(A_1)P(A_2)P(A_3) \cdots P(A_n).$$

This might look like a difficult definition, but we can usually argue that the events are independent in a much easier way. For example, we might be able to justify independence by looking at the way the random experiment is performed. A simple example of an independent event is when you toss a coin repeatedly. In such an experiment, the results of any subset of the coin tosses do not have any impact on the other ones.

Example 1.21. I toss a coin repeatedly until I observe the first tails, at which point I stop. Let X be the total number of coin tosses. Find $P(X = 5)$.

Solution: Here, the outcome of the random experiment is a number X. The goal is to find $P(A) = P(5)$. But what does $X = 5$ mean? It means that the first 4 coin tosses result in heads and the fifth one results in tails. Thus the problem is to find the probability of the sequence $HHHHT$ when tossing a coin five times. Note that $HHHHT$ is a shorthand for the event "(The first coin toss results in heads) and (The second coin toss results in heads) and (The third coin toss results in heads) and (The fourth coin toss results in heads) and (The fifth coin toss results in tails)." Since all the coin tosses are independent, we can write

$$P(HHHHT) = P(H)P(H)P(H)P(H)P(T)$$
$$= \frac{1}{2} \cdot \frac{1}{2} \cdot \frac{1}{2} \cdot \frac{1}{2} \cdot \frac{1}{2}$$
$$= \frac{1}{32}.$$

Discussion: Some people find it more understandable if you look at the problem in the following way: I never stop tossing the coin. So the outcome of this experiment is always an infinite sequence of heads or tails. The value X (which we are interested in) is just a function of the beginning part of the sequence until you observe a tails. If you think about the problem this way, you should not worry about the stopping time. For this problem it might not make a big difference conceptually, but for some similar problems this way of thinking might be beneficial.

We have seen that two events A and B are independent if $P(A \cap B) = P(A)P(B)$. In the next two results, we examine what independence can tell us about other set operations such as complements and unions.

Lemma 1.1. If A and B are independent then

- A and B^c are independent,

- A^c and B are independent,

- A^c and B^c are independent.

Proof. We prove the first one as the others can be concluded from the first one immediately. We have

$$
\begin{aligned}
P(A \cap B^c) &= P(A - B) \\
&= P(A) - P(A \cap B) \\
&= P(A) - P(A)P(B) \qquad \text{(since } A \text{ and } B \text{ are independent)} \\
&= P(A)(1 - P(B)) \\
&= P(A)P(B^c).
\end{aligned}
$$

Thus, A and B^c are independent. ∎

Sometimes we are interested in the probability of the union of several independent events $A_1, A_2, \cdots, A_n$. For independent events, we know how to find the probability of intersection easily, but not the union. It is helpful in these cases to use De Morgan's Law:

$$
A_1 \cup A_2 \cup \cdots \cup A_n = (A_1^c \cap A_2^c \cap \cdots \cap A_n^c)^c.
$$

Thus we can write

$$
\begin{aligned}
P(A_1 \cup A_2 \cup \cdots \cup A_n) &= 1 - P(A_1^c \cap A_2^c \cap \cdots \cap A_n^c) \\
&= 1 - P(A_1^c)P(A_2^c) \cdots P(A_n^c) \qquad \text{(since } A_i\text{'s are independent)} \\
&= 1 - (1 - P(A_1))(1 - P(A_2)) \cdots (1 - P(A_n)).
\end{aligned}
$$

If $A_1, A_2, \cdots, A_n$ are independent then

$$
P(A_1 \cup A_2 \cup \cdots \cup A_n) = 1 - (1 - P(A_1))(1 - P(A_2)) \cdots (1 - P(A_n)).
$$

Example 1.22. Suppose that the probability of being killed in a single flight is $p_c = \frac{1}{4 \times 10^6}$ based on available statistics. Assume that different flights are independent. If a businessman takes 20 flights per year, what is the probability that he is killed in a plane crash within the next 20 years? (Let's assume that he will not die because of another reason within the next 20 years.)

Solution: The total number of flights that he will take during the next 20 years is $N = 20 \times 20 = 400$. Let p_s be the probability that he survives a given single flight. Then we have

$$
p_s = 1 - p_c.
$$

Since these flights are independent, the probability that he will survive all $N = 400$ flights is

$$
P(\text{Survive } N \text{ flights}) = p_s \times p_s \times \cdots \times p_s = p_s^N = (1 - p_c)^N.
$$

Let A be the event that the businessman is killed in a plane crash within the next 20 years. Then

$$P(A) = 1 - (1 - p_c)^N = 9.9995 \times 10^{-5} \approx \frac{1}{10000}.$$

Warning! One common mistake is to confuse <u>independence</u> and <u>being disjoint</u>. These are completely different concepts. When two events A and B are disjoint it means that if one of them occurs, the other one cannot occur, i.e., $A \cap B = \emptyset$. Thus, event A usually gives a lot of information about event B which means that they cannot be independent. Let's make it precise.

Lemma 1.2. Consider two events A and B, with $P(A) \neq 0$ and $P(B) \neq 0$. If A and B are disjoint, then they are **not** independent.

Proof. Since A and B are disjoint, we have

$$P(A \cap B) = 0 \neq P(A)P(B).$$

Thus, A and B are not independent.   ∎

Table 1.1 summarizes the two concepts of disjointness and independence.

Table 1.1: Differences between disjointness and independence

Concept	Meaning	Formulas
Disjoint	A and B cannot occur at the same time	$A \cap B = \emptyset$, $P(A \cup B) = P(A) + P(B)$
Independent	A does not give any information about B	$P(A\|B) = P(A), P(B\|A) = P(B)$ $P(A \cap B) = P(A)P(B)$

Example 1.23. [1] Two basketball players play a game in which they alternately shoot a basketball at a hoop. The first one to make a basket wins the game. On each shot, Player 1 (the one who shoots first) has probability p_1 of success, while Player 2 has probability p_2 of success (assume $0 < p_1, p_2 < 1$). The shots are assumed to be independent.

(a) Find $P(W_1)$, the probability that Player 1 wins the game.

(b) For what values of p_1 and p_2 is this a fair game, i.e., each player has a 50 percent chance of winning the game?

Solution: In this game, the event W_1 can happen in many different ways. We calculate the probability of each of these ways and then add them up to find the total probability of winning. In particular, Player 1 may win on her first shot, or her second shot, and so on. Define A_i as the event that Player 1 wins on her ith shot. What is the probability of A_i? A_i happens if

---

[1]A similar problem is given in [6].

Player 1 is unsuccessful at her first $i - 1$ shots and successful at her ith shot, while Player 2 is unsuccessful at her first $i - 1$ shots. Since different shots are independent, we obtain

$$P(A_1) = p_1,$$
$$P(A_2) = (1 - p_1)(1 - p_2)p_1,$$
$$P(A_3) = (1 - p_1)(1 - p_2)(1 - p_1)(1 - p_2)p_1,$$
$$\cdots$$
$$P(A_k) = \left[(1 - p_1)(1 - p_2)\right]^{k-1}p_1,$$
$$\cdots$$

Note that $A_1, A_2, A_3, \cdots$ are disjoint events, because if one of them occurs the other ones cannot occur. The event that Player 1 wins is the union of the A_i's, and since the A_i's are disjoint, we have

$$
\begin{aligned}
P(W_1) &= P(A_1 \cup A_2 \cup A_3 \cup \cdots) \\
&= P(A_1) + P(A_2) + P(A_3) + \cdots \\
&= p_1 + (1 - p_1)(1 - p_2)p_1 + \left[(1 - p_1)(1 - p_2)\right]^2 p_1 + \cdots \\
&= p_1 \left[1 + (1 - p_1)(1 - p_2) + \left[(1 - p_1)(1 - p_2)\right]^2 + \cdots\right].
\end{aligned}
$$

Note that since $0 < p_1, p_2 < 1$, for $x = (1 - p_1)(1 - p_2)$ we have $0 < x < 1$. Thus, using the geometric sum formula ($\sum_{k=0}^{\infty} ax^k = a\frac{1}{1-x}$ for $|x| < 1$), we obtain

$$P(W_1) = \frac{p_1}{1 - (1 - p_1)(1 - p_2)} = \frac{p_1}{p_1 + p_2 - p_1 p_2}.$$

It is always a good idea to look at limit cases to check our answer. For example, if we plug in $p_1 = 0, p_2 \neq 0$, we obtain $P(W_1) = 0$, which is what we expect. Similarly, if we let $p_2 = 0, p_1 \neq 0$, we obtain $P(W_1) = 1$, which again makes sense.

Now, to make this a fair game (in the sense that $P(W_1) = .5$), we have

$$P(W_1) = \frac{p_1}{p_1 + p_2 - p_1 p_2} = 0.5,$$

and we obtain

$$p_1 = \frac{p_2}{1 + p_2}.$$

Note that this means that $p_1 < p_2$, which makes sense intuitively. Since Player 1 has the advantage of starting the game, she should have a smaller success rate so that the whole game is fair.

1.4.2 Law of Total Probability

Let us start this section by asking a very simple question: In a certain country there are three provinces, call them B_1, B_2, and B_3 (i.e., the country is partitioned into three disjoint sets B_1, B_2, and B_3). We are interested in the total forest area in the country. Suppose that we know that the forest area in B_1, B_2, and B_3 are $100km^2$, $50km^2$, and $150km^2$, respectively. What is the total forest area in the country? If your answer is

$$100km^2 + 50km^2 + 150km^2 = 300km^2,$$

you are right. That is, you can simply add forest areas in each province (partition) to obtain the forest area in the whole country. This is the idea behind the law of total probability, in which the *area of forest* is replaced by *probability of an event A*. In particular, if you want to find $P(A)$, you can look at a partition of S, and add the amount of probability of A that falls in each partition. We have already seen the special case where the partition is B and B^c: we saw that for any two events A and B,

$$P(A) = P(A \cap B) + P(A \cap B^c)$$

and using the definition of conditional probability, $P(A \cap B) = P(A|B)P(B)$, we can write

$$P(A) = P(A|B)P(B) + P(A|B^c)P(B^c).$$

We can state a more general version of this formula which applies to a general partition of the sample space S.

Law of Total Probability:

If $B_1, B_2, B_3, \cdots$ is a partition of the sample space S, then for any event A we have
$$P(A) = \sum_i P(A \cap B_i) = \sum_i P(A|B_i)P(B_i).$$

Using a Venn diagram, we can pictorially see the idea behind the law of total probability. In Figure 1.24, we have

$$A_1 = A \cap B_1,$$
$$A_2 = A \cap B_2,$$
$$A_3 = A \cap B_3.$$

As it can be seen from the figure, A_1, A_2, and A_3 form a partition of the set A, and thus by the third axiom of probability

$$P(A) = P(A_1) + P(A_2) + P(A_3).$$

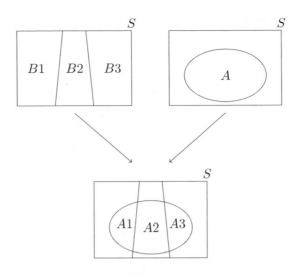

Figure 1.24: Law of total probability

Here is a proof of the law of total probability using probability axioms:

Proof. Since $B_1, B_2, B_3, \cdots$ is a partition of the sample space S, we can write

$$S = \bigcup_i B_i$$

$$A = A \cap S$$
$$= A \cap \left(\bigcup_i B_i\right)$$
$$= \bigcup_i (A \cap B_i) \qquad \text{by the distributive law (Theorem 1.2).}$$

Now note that the sets $A \cap B_i$ are disjoint (since the B_i's are disjoint). Thus, by the third probability axiom,

$$P(A) = P\left(\bigcup_i (A \cap B_i)\right) = \sum_i P(A \cap B_i) = \sum_i P(A|B_i)P(B_i).$$

■

Here is a typical scenario in which we use the law of total probability. We are interested in finding the probability of an event A, but we don't know how to find $P(A)$ directly. Instead, we know the conditional probability of A given some events B_i, where the B_i's form a partition of the sample space. Thus, we will be able to find $P(A)$ using the law of total probability, $P(A) = \sum_i P(A|B_i)P(B_i)$.

Example 1.24. I have three bags that each contain 100 marbles:

- Bag 1 has 75 red and 25 blue marbles;

- Bag 2 has 60 red and 40 blue marbles;

- Bag 3 has 45 red and 55 blue marbles.

I choose one of the bags at random and then pick a marble from the chosen bag, also at random. What is the probability that the chosen marble is red?

Solution: Let R be the event that the chosen marble is red. Let B_i be the event that I choose Bag i. We already know that

$$P(R|B_1) = 0.75,$$
$$P(R|B_2) = 0.60,$$
$$P(R|B_3) = 0.45$$

We choose our partition as B_1, B_2, B_3. Note that this is a valid partition because, firstly, the B_i's are disjoint (only one of them can happen), and secondly, because their union is the entire sample space, as one of the bags will be chosen for sure (i.e., $P(B_1 \cup B_2 \cup B_3) = 1$). Using the law of total probability, we can write

$$P(R) = P(R|B_1)P(B_1) + P(R|B_2)P(B_2) + P(R|B_3)P(B_3)$$
$$= (0.75)\frac{1}{3} + (0.60)\frac{1}{3} + (0.45)\frac{1}{3},$$
$$= 0.60$$

1.4.3   Bayes' Rule

Now we are ready to state one of the most useful results in conditional probability: Bayes' rule. Suppose that we know $P(A|B)$, but we are interested in the probability $P(B|A)$. Using the definition of conditional probability, we have

$$P(A|B)P(B) = P(A \cap B) = P(B|A)P(A).$$

Dividing by $P(A)$, we obtain

$$P(B|A) = \frac{P(A|B)P(B)}{P(A)},$$

which is the famous Bayes' rule. Often, in order to find $P(A)$ in Bayes' formula we need to use the law of total probability, so sometimes Bayes' rule is stated as

$$P(B_j|A) = \frac{P(A|B_j)P(B_j)}{\sum_i P(A|B_i)P(B_i)},$$

where $B_1, B_2, \cdots, B_n$ forms a partition of the sample space.

Bayes' Rule

- For any two events A and B, where $P(A) \neq 0$, we have

$$P(B|A) = \frac{P(A|B)P(B)}{P(A)}.$$

- If $B_1, B_2, B_3, \cdots$ form a partition of the sample space S, and A is any event with $P(A) \neq 0$, we have

$$P(B_j|A) = \frac{P(A|B_j)P(B_j)}{\sum_i P(A|B_i)P(B_i)}.$$

Example 1.25. In Example 1.24, suppose we observe that the chosen marble is red. What is the probability that Bag 1 was chosen?

Solution: Here we know $P(R|B_i)$ but we are interested in $P(B_1|R)$, so this is a scenario in which we can use Bayes' rule. We have

$$P(B_1|R) = \frac{P(R|B_1)P(B_1)}{P(R)}$$
$$= \frac{0.75 \times \frac{1}{3}}{0.6}$$
$$= \frac{5}{12}.$$

$P(R)$ was obtained using the law of total probability in Example 1.24, thus we did not have to recompute it here. Also, note that $P(B_1|R) = \frac{5}{12} > \frac{1}{3}$. This makes sense intuitively because Bag 1 is the bag with the highest number of red marbles. Thus if the chosen marble is red, it is more likely that Bag 1 was chosen.

Example 1.26. (False positive paradox [5]) A certain disease affects about 1 out of 10,000 people. There is a test to check whether the person has the disease. The test is quite accurate. In particular, we know that

- the probability that the test result is positive (suggesting the person has the disease), given that the person does not have the disease, is only 2 percent;

- the probability that the test result is negative (suggesting the person does not have the disease), given that the person has the disease, is only 1 percent.

A random person gets tested for the disease and the result comes back positive. What is the probability that the person has the disease?

Solution: Let D be the event that the person has the disease, and let T be the event that the test result is positive. We know

$$P(D) = \frac{1}{10,000},$$
$$P(T|D^c) = 0.02,$$
$$P(T^c|D) = 0.01$$

What we want to compute is $P(D|T)$. Again, we use Bayes' rule:

$$
\begin{aligned}
P(D|T) &= \frac{P(T|D)P(D)}{P(T|D)P(D) + P(T|D^c)P(D^c)} \\
&= \frac{(1 - 0.01) \times 0.0001}{(1 - 0.01) \times 0.0001 + 0.02 \times (1 - 0.0001)} \\
&= 0.0049
\end{aligned}
$$

This means that there is less than half a percent chance that the person has the disease.

Discussion: This might seem somewhat counterintuitive as we know the test is quite accurate. The point is that the disease is also very rare. Thus, there are two competing forces here, and since the rareness of the disease (1 out of 10,000) is stronger than the accuracy of the test (98 or 99 percent), there is still good chance that the person does not have the disease.

Another way to think about this problem is illustrated in the tree diagram in Figure 1.25. Suppose 1 million people get tested for the disease. Out of the one million people, about 100 of them have the disease, while the other $999,900$ do not have the disease. Out of the 100 people who have the disease $100 \times .99 = 99$ people will have positive test results. However, out of the people who do not have the disease $999,900 \times .02 = 19998$ people will have positive test results. Thus in total there are $19998 + 99$ people with positive test results, and only 99 of them actually have the disease. Therefore, the probability that a person from the "positive test result" group actually has the disease is

$$P(D|T) = \frac{99}{19998 + 99} = .0049$$

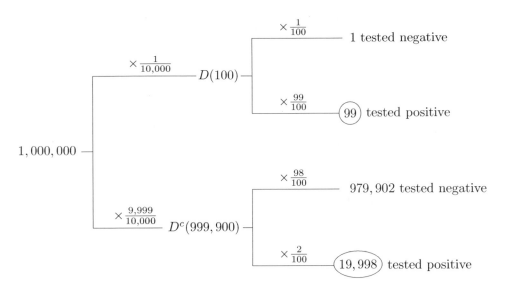

Figure 1.25: Tree diagram for Example 1.26

1.4.4 Conditional Independence

As we mentioned earlier, almost any concept that is defined for probability can also be extended to conditional probability. Remember that two events A and B are independent if

$$P(A \cap B) = P(A)P(B), \text{ or equivalently, } P(A|B) = P(A).$$

We can extend this concept to conditionally independent events. In particular,

Definition 1.2. Two events A and B are **conditionally independent** given an event C with $P(C) > 0$ if

$$P(A \cap B|C) = P(A|C)P(B|C). \tag{1.8}$$

Recall that from the definition of conditional probability,

$$P(A|B) = \frac{P(A \cap B)}{P(B)},$$

if $P(B) > 0$. By conditioning on C, we obtain

$$P(A|B, C) = \frac{P(A \cap B|C)}{P(B|C)}$$

if $P(B|C), P(C) \neq 0$. If A and B are conditionally independent given C, we obtain

$$P(A|B, C) = \frac{P(A \cap B|C)}{P(B|C)}$$
$$= \frac{P(A|C)P(B|C)}{P(B|C)}$$
$$= P(A|C).$$

Thus, if A and B are conditionally independent given C, then

$$P(A|B,C) = P(A|C) \tag{1.9}$$

Thus, Equations 1.8 and 1.9 are equivalent statements of the definition of conditional independence. Now let's look at an example.

Example 1.27. A box contains two coins: a regular coin and one fake two-headed coin ($P(H) = 1$). I choose a coin at random and toss it twice. Define the following events:

- A= First coin toss results in an H.

- B= Second coin toss results in an H.

- C= Coin 1 (regular) has been selected.

Find $P(A|C), P(B|C), P(A \cap B|C), P(A), P(B)$, and $P(A \cap B)$. Note that A and B are NOT independent, but they are *conditionally* independent given C.

Solution: We have $P(A|C) = P(B|C) = \frac{1}{2}$. Also, given that Coin 1 is selected, we have $P(A \cap B|C) = \frac{1}{2} \cdot \frac{1}{2} = \frac{1}{4}$. To find $P(A), P(B)$, and $P(A \cap B)$, we use the law of total probability:

$$
\begin{aligned}
P(A) &= P(A|C)P(C) + P(A|C^c)P(C^c) \\
&= \frac{1}{2} \cdot \frac{1}{2} + 1 \cdot \frac{1}{2} \\
&= \frac{3}{4}.
\end{aligned}
$$

Similarly, $P(B) = \frac{3}{4}$. For $P(A \cap B)$, we have

$$
\begin{aligned}
P(A \cap B) =& P(A \cap B|C)P(C) + P(A \cap B|C^c)P(C^c) \\
=& P(A|C)P(B|C)P(C) + \\
& P(A|C^c)P(B|C^c)P(C^c) \qquad \text{(by conditional independence of A and B)} \\
=& \frac{1}{2} \cdot \frac{1}{2} \cdot \frac{1}{2} + 1 \cdot 1 \cdot \frac{1}{2} \\
=& \frac{5}{8}.
\end{aligned}
$$

As we see, $P(A \cap B) = \frac{5}{8} \neq P(A)P(B) = \frac{9}{16}$, which means that A and B are not independent. We can also justify this intuitively. For example, if we know A has occurred (i.e., the first coin toss has resulted in heads), we would guess that it is more likely that we have chosen Coin 2 than Coin 1. This in turn increases the conditional probability that B occurs. This suggests that A and B are not independent. On the other hand, given C (Coin 1 is selected), A and B are independent.

One important lesson here is that, generally speaking, conditional independence neither implies (nor is it implied by) independence. Thus, we can have two events that are conditionally independent but not unconditionally independent (such as A and B above). Also, we can have two events that are independent but are not conditionally independent, given an event C. Here is a simple example regarding this case. Consider rolling a die and let

$$A = \{1, 2\},$$
$$B = \{2, 4, 6\},$$
$$C = \{1, 4\}.$$

Then, we have

$$P(A) = \frac{1}{3}, P(B) = \frac{1}{2};$$
$$P(A \cap B) = \frac{1}{6} = P(A)P(B).$$

Thus, A and B are independent. But we have

$$P(A|C) = \frac{1}{2}, P(B|C) = \frac{1}{2};$$
$$P(A \cap B|C) = P(\{2\}|C) = 0.$$

Thus

$$P(A \cap B|C) \neq P(A|C)P(B|C),$$

which means A and B are not conditionally independent given C.

1.4.5 Solved Problems

In die and coin problems, unless stated otherwise, it is assumed coins and dice are fair and repeated trials are independent.

1. You purchase a certain product. The manual states that the lifetime T of the product, defined as the amount of time (in years) the product works properly until it breaks down, satisfies

$$P(T \geq t) = e^{-\frac{t}{5}}, \text{ for all } t \geq 0.$$

For example, the probability that the product lasts more than (or equal to) 2 years is $P(T \geq 2) = e^{-\frac{2}{5}} = 0.6703$. I purchase the product and use it for two years without any problems. What is the probability that it breaks down in the third year?

Solution: Let A be the event that a purchased product breaks down in the third year. Also, let B be the event that a purchased product does not break down in the first two years. We are interested in $P(A|B)$. We have

$$P(B) = P(T \geq 2)$$
$$= e^{-\frac{2}{5}}.$$

We also have

$$P(A) = P(2 \leq T \leq 3)$$
$$= P(T \geq 2) - P(T \geq 3)$$
$$= e^{-\frac{2}{5}} - e^{-\frac{3}{5}}.$$

Finally, since $A \subset B$, we have $A \cap B = A$. Therefore,

$$P(A|B) = \frac{P(A \cap B)}{P(B)}$$
$$= \frac{P(A)}{P(B)}$$
$$= \frac{e^{-\frac{2}{5}} - e^{-\frac{3}{5}}}{e^{-\frac{2}{5}}}$$
$$= 0.1813$$

2. You toss a fair coin three times:

   (a) What is the probability of three heads, HHH?

   (b) What is the probability that you observe exactly one heads?

   (c) Given that you have observed *at least* one heads, what is the probability that you observe at least two heads?

   Solution: We assume that the coin tosses are independent.

   (a) $P(HHH) = P(H) \cdot P(H) \cdot P(H) = 0.5^3 = \frac{1}{8}$.

   (b) To find the probability of exactly one heads, we can write

   $$P(\text{One heads}) = P(HTT \cup THT \cup TTH)$$
   $$= P(HTT) + P(THT) + P(TTH)$$
   $$= \frac{1}{8} + \frac{1}{8} + \frac{1}{8}$$
   $$= \frac{3}{8}.$$

   (c) Given that you have observed *at least* one heads, what is the probability that you observe at least two heads? Let A_1 be the event that you observe at least one heads, and A_2 be the event that you observe at least two heads. Then

   $$A_1 = S - \{TTT\}, \text{ and } P(A_1) = \frac{7}{8};$$
   $$A_2 = \{HHT, HTH, THH, HHH\}, \text{ and } P(A_2) = \frac{4}{8}.$$

Thus, we can write

$$P(A_2|A_1) = \frac{P(A_2 \cap A_1)}{P(A_1)}$$

$$= \frac{P(A_2)}{P(A_1)}$$

$$= \frac{4}{8} \cdot \frac{8}{7} = \frac{4}{7}.$$

3. For three events A, B, and C, we know that

  - A and C are independent,
  - B and C are independent,
  - A and B are disjoint,
  - $P(A \cup C) = \frac{2}{3}, P(B \cup C) = \frac{3}{4}, P(A \cup B \cup C) = \frac{11}{12}$.

Find $P(A), P(B)$, and $P(C)$.

Solution: We can use the Venn diagram in Figure 1.26 to better visualize the events in this problem. We assume $P(A) = a, P(B) = b$, and $P(C) = c$. Note that the assumptions about independence and disjointness of sets are already included in the figure.

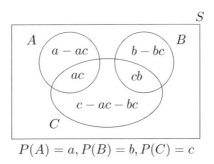

$$P(A) = a, P(B) = b, P(C) = c$$

Figure 1.26: Venn diagram for Problem 3

Now we can write

$$\begin{cases} P(A \cup C) = a + c - ac = \frac{2}{3} \\ P(B \cup C) = b + c - bc = \frac{3}{4} \\ P(A \cup B \cup C) = a + b + c - ac - bc = \frac{11}{12} \end{cases}$$

By subtracting the third equation from the sum of the first and second equations, we immediately obtain $c = \frac{1}{2}$, which then gives $a = \frac{1}{3}$ and $b = \frac{1}{2}$.

4. Let $C_1, C_2, \cdots, C_M$ be a partition of the sample space S, and A and B be two events. Suppose we know that

  - A and B are conditionally independent given C_i, for all $i \in \{1, 2, \cdots, M\}$;
  - B is independent of all C_i's.

Prove that A and B are independent.

Solution: Since the C_i's form a partition of the sample space, we can apply the law of total probability for $A \cap B$:

$$P(A \cap B) = \sum_{i=1}^{M} P(A \cap B | C_i) P(C_i)$$

$$= \sum_{i=1}^{M} P(A|C_i) P(B|C_i) P(C_i) \quad (A \text{ and } B \text{ are conditionally independent})$$

$$= \sum_{i=1}^{M} P(A|C_i) P(B) P(C_i) \quad\quad (B \text{ is independent of all the } C_i\text{'s})$$

$$= P(B) \sum_{i=1}^{M} P(A|C_i) P(C_i)$$

$$= P(B) P(A) \quad\quad\quad\quad\quad\quad\quad (\text{law of total probability}).$$

5. In my town, it's rainy one third of the days. Given that it is rainy, there will be heavy traffic with probability $\frac{1}{2}$, and given that it is not rainy, there will be heavy traffic with probability $\frac{1}{4}$. If it's rainy and there is heavy traffic, I arrive late for work with probability $\frac{1}{2}$. On the other hand, the probability of being late is reduced to $\frac{1}{8}$ if it is not rainy and there is no heavy traffic. In other situations (rainy and no traffic, not rainy and traffic) the probability of being late is 0.25. You pick a random day.

  (a) What is the probability that it's not raining and there is heavy traffic, and I am not late?

  (b) What is the probability that I am late?

  (c) Given that I arrived late at work, what is the probability that it rained that day?

Solution: Let R be the event that it's rainy, T be the event that there is heavy traffic, and L be the event that I am late for work. As it is seen from the problem statement, we are given conditional probabilities in a chain format. Thus, it is useful to draw a tree diagram. Figure 1.27 shows a tree diagram for this problem. In this figure, each leaf in the tree corresponds to a single outcome in the sample space. We can calculate the probabilities of each outcome in the sample space by multiplying the probabilities on the edges of the tree that lead to the corresponding outcome.

(a) The probability that it's not raining and there is heavy traffic and I am not late can be found using the tree diagram which is in fact applying the chain rule:

$$P(R^c \cap T \cap L^c) = P(R^c)P(T|R^c)P(L^c|R^c \cap T)$$
$$= \frac{2}{3} \cdot \frac{1}{4} \cdot \frac{3}{4}$$
$$= \frac{1}{8}.$$

(b) The probability that I am late can be found from the tree. All we need to do is sum the probabilities of the outcomes that correspond to me being late. In fact, we are using the law of total probability here.

$$P(L) = P(R,T,L) + P(R,T^c,L) + P(R^c,T,L) + P(R^c,T^c,L)$$
$$= \frac{1}{12} + \frac{1}{24} + \frac{1}{24} + \frac{1}{16}$$
$$= \frac{11}{48}.$$

(c) We can find $P(R|L)$ using $P(R|L) = \frac{P(R \cap L)}{P(L)}$. We have already found $P(L) = \frac{11}{48}$, and we can find $P(R \cap L)$ similarly by adding the probabilities of the outcomes that belong to $R \cap L$. In particular,

$$P(R \cap L) = P(R,T,L) + P(R,T^c,L)$$
$$= \frac{1}{12} + \frac{1}{24}$$
$$= \frac{1}{8}.$$

Thus, we obtain

$$P(R|L) = \frac{P(R \cap L)}{P(L)}$$
$$= \frac{1}{8} \cdot \frac{48}{11}$$
$$= \frac{6}{11}.$$

6. A box contains three coins: two regular coins and one fake two-headed coin ($P(H) = 1$),

   (a) You pick a coin at random and toss it. What is the probability that it lands heads up?

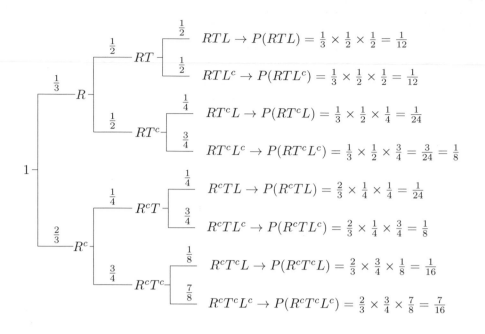

Figure 1.27: Tree diagram for Problem 5

(b) You pick a coin at random and toss it, and get heads. What is the probability that it is the two-headed coin?

Solution: This is another typical problem for which the law of total probability is useful. Let C_1 be the event that you choose a regular coin, and let C_2 be the event that you choose the two-headed coin. Note that C_1 and C_2 form a partition of the sample space. We already know that

$$P(H|C_1) = 0.5,$$
$$P(H|C_2) = 1.$$

(a) Thus, we can use the law of total probability to write

$$P(H) = P(H|C_1)P(C_1) + P(H|C_2)P(C_2)$$
$$= \frac{1}{2} \cdot \frac{2}{3} + 1 \cdot \frac{1}{3}$$
$$= \frac{2}{3}.$$

(b) Now, for the second part of the problem, we are interested in $P(C_2|H)$. We use

Bayes' rule

$$P(C_2|H) = \frac{P(H|C_2)P(C_2)}{P(H)}$$

$$= \frac{1.\frac{1}{3}}{\frac{2}{3}}$$

$$= \frac{1}{2}.$$

7. Here is another variation of the family-with-two-children problem [1,7]: A family has two children. We ask from the father, "Do you have at least one daughter named Lilia?" He replies, "Yes!" What is the probability that both children are girls? In other words, we want to find the probability that both children are girls, given that the family has at least one daughter named Lilia. Here you can assume that if a child is a girl, her name will be Lilia with probability $\alpha \ll 1$ independently from other children's names. If the child is a boy, his name will not be Lilia. Compare your result with the second part of Example 1.18.

Solution: Here we have four possibilities, $GG =$ (girl, girl)$, GB, BG, BB$, and $P(GG) = P(GB) = P(BG) = P(BB) = \frac{1}{4}$. Let also L be the event that the family has at least one child named Lilia. We have

$$P(L|BB) = 0,$$
$$P(L|BG) = P(L|GB) = \alpha,$$
$$P(L|GG) = \alpha(1 - \alpha) + (1 - \alpha)\alpha + \alpha^2 = 2\alpha - \alpha^2.$$

We can use Bayes' rule to find $P(GG|L)$:

$$P(GG|L) = \frac{P(L|GG)P(GG)}{P(L)}$$

$$= \frac{P(L|GG)P(GG)}{P(L|GG)P(GG) + P(L|GB)P(GB) + P(L|BG)P(BG) + P(L|BB)P(BB)}$$

$$= \frac{(2\alpha - \alpha^2)\frac{1}{4}}{(2\alpha - \alpha^2)\frac{1}{4} + \alpha\frac{1}{4} + \alpha\frac{1}{4} + 0.\frac{1}{4}}$$

$$= \frac{2 - \alpha}{4 - \alpha} \approx \frac{1}{2}.$$

Let's compare the result with part (b) of Example 1.18. Amazingly, we notice that the extra information about the name of the child increases the conditional probability of GG from $\frac{1}{3}$ to about $\frac{1}{2}$. How can we explain this intuitively? Here is one way to look at the problem. In part (b) of Example 1.18, we know that the family has at least one girl.

Thus, the sample space reduces to three equally likely outcomes: GG, GB, BG. Therefore, the conditional probability of GG is one third in this case. On the other hand, in this problem, the available information is that the event L has occurred. The conditional sample space here still is GG, GB, BG, but these events are not equally likely anymore. A family with two girls is more likely to name at least one of them Lilia than a family who has only one girl ($P(L|BG) = P(L|GB) = \alpha$, $P(L|GG) = 2\alpha - \alpha^2$). Thus, in this case, the conditional probability of GG is higher. We would like to mention here that these problems are confusing and counterintuitive to most people. So, do not be disappointed if they seem confusing to you. We seek several goals by including such problems.

First, we would like to emphasize that we should not rely too much on our intuition when solving probability problems. Intuition is useful, but at the end, we must use laws of probability to solve problems. Second, after obtaining counterintuitive results, you are encouraged to think deeply about them to explain your confusion. This thinking process can be very helpful to improve our understanding of probability. Finally, I personally think these paradoxical-looking problems make probability more interesting.

8. If you are not yet confused, let's look at another family-with-two-children problem! I know that a family has two children. I see one of the children in the mall and notice that she is a girl. What is the probability that both children are girls? Again, compare your result with the second part of Example 1.18. Note: Let's agree on what precisely the problem statement means. Here is a more precise statement of the problem: "A family has two children. We choose one of them at random and find out that she is a girl. What is the probability that both children are girls?"

Solution: Here again, we have four possibilities, $GG = $ (girl, girl)$, GB, BG, BB$, and $P(GG) = P(GB) = P(BG) = P(BB) = \frac{1}{4}$. Now, let G_r be the event that a randomly chosen child is a girl. Then we have

$$P(G_r|GG) = 1$$
$$P(G_r|GB) = P(G_r|BG) = \frac{1}{2}$$
$$P(G_r|BB) = 0.$$

We can use Bayes' rule to find $P(GG|G_r)$:

$$P(GG|G_r) = \frac{P(G_r|GG)P(GG)}{P(G_r)}$$

$$= \frac{P(G_r|GG)P(GG)}{P(G_r|GG)P(GG) + P(G_r|GB)P(GB) + P(G_r|BG)P(BG) + P(G_r|BB)P(BB)}$$

$$= \frac{1.\frac{1}{4}}{1.\frac{1}{4} + \frac{1}{2}\frac{1}{4} + \frac{1}{2}\frac{1}{4} + 0.\frac{1}{4}}$$

$$= \frac{1}{2}.$$

So the answer again is different from the second part of Example 1.18. This is surprising to most people. The two problem statements look very similar but the answers are completely different. This is again similar to the previous problem (please read the explanation there). The conditional sample space here still is GG, GB, BG, but the point here is that these are not equally likely as in Example 1.18. The probability that a randomly chosen child from a family with two girls is a girl is one, while this probability for a family who has only one girl is $\frac{1}{2}$. Thus, intuitively, the conditional probability of the outcome GG in this case is higher than GB and BG, and thus this conditional probability must be larger than one third.

9. Okay, another family-with-two-children problem. Just kidding! This problem has nothing to do with the two previous problems. I toss a coin repeatedly. The coin is unfair and $P(H) = p$. The game ends the first time that two consecutive heads (HH) or two consecutive tails (TT) are observed. I win if HH is observed and lose if TT is observed. For example if the outcome is $HTH\underline{TT}$, I lose. On the other hand, if the outcome is $THTHT\underline{HH}$, I win. Find the probability that I win.

Solution: Let W be the event that I win. We can write down the set W by listing all the different sequences that result in my winning. It is cleaner if we divide W into two parts depending on the result of the first coin toss:

$$W = \{HH, HTHH, HTHTHH, \cdots\} \cup \{THH, THTHH, THTHTHH, \cdots\}.$$

Let $q = 1 - p$. Then

$$\begin{aligned}
P(W) &= P(\{HH, HTHH, HTHTHH, \cdots\}) \\
&\quad + P(\{THH, THTHH, THTHTHH, \cdots\}) \\
&= p^2 + p^3 q + p^4 q^2 + \cdots \\
&\quad + p^2 q + p^3 q^2 + p^4 q^3 + \cdots \\
&= p^2(1 + pq + (pq)^2 + (pq)^3 + \cdots) \\
&\quad + p^2 q(1 + pq + (pq)^2 + (pq)^3 + \cdots) \\
&= p^2(1 + q)(1 + pq + (pq)^2 + (pq)^3 + \cdots) \\
&= \frac{p^2(1 + q)}{1 - pq}, \text{ Using the geometric series formula (Equation 1.4)} \\
&= \frac{p^2(2 - p)}{1 - p + p^2}.
\end{aligned}$$

1.5   End of Chapter Problems

1. Suppose that the universal set S is defined as $S = \{1, 2, \cdots, 10\}$ and $A = \{1, 2, 3\}$, $B = \{x \in S : 2 \leq x \leq 7\}$, and $C = \{7, 8, 9, 10\}$.
   (a) Find $A \cup B$.
   (b) Find $(A \cup C) - B$.
   (c) Find $\bar{A} \cup (B - C)$.
   (d) Do A, B, and C form a partition of S?

2. When working with real numbers, our universal set is $\mathbb{R}$. Find each of the following sets.
   (a) $[6, 8] \cup [2, 7)$
   (b) $[6, 8] \cap [2, 7)$
   (c) $[0, 1]^c$
   (d) $[6, 8] - (2, 7)$

3. For each of the following Venn diagrams, write the set denoted by the shaded area.
   (a)

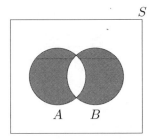

   (b)

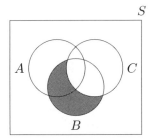

   (c)

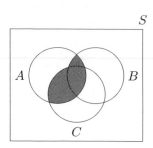

(d)

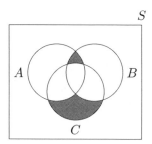

4. A coin is tossed twice. Let S be the set of all possible pairs that can be observed, i.e., $S = \{H, T\} \times \{H, T\} = \{(H, H), (H, T), (T, H), (T, T)\}$. Write the following sets by listing their elements.
   (a) A: The first coin toss results in head.
   (b) B: At least one tail is observed.
   (c) C: The two coin tosses result in different outcomes.

5. Let $A = \{1, 2, \cdots, 100\}$. For any $i \in \mathbb{N}$, define A_i as the set of numbers in A that are divisible by i. For example,
   $A_2 = \{2, 4, 6, \cdots, 100\}$
   $A_3 = \{3, 6, 9, \cdots, 99\}$.
   (a) Find $|A_2|, |A_3|, |A_4|, |A_5|$.
   (b) Find $|A_2 \cup A_3 \cup A_5|$.

6. Suppose that A_1, A_2, A_3 form a partition of the universal set S. Let B be an arbitrary set. Assume that we know:

$$|B \cap A_1| = 10$$
$$|B \cap A_2| = 20$$
$$|B \cap A_3| = 15.$$

   Find $|B|$.

7. Determine whether each of the following sets is countable or uncountable.
   (a) $A = \{1, 2, \cdots, 10^{10}\}$.

(b) $B = \{a + b\sqrt{2} \mid a, b \in \mathbb{Q}\}$.

(c) $C = \{(X, Y) \in \mathbb{R}^2 \mid x^2 + y^2 \leq 1\}$.

8. * Let $A_n = \left[0, \frac{n-1}{n}\right) = \{x \in \mathbb{R} \mid 0 \leq x < \frac{n-1}{n}\}$, for $n = 2, 3, \cdots$. Define

$$A = \bigcup_{n=1}^{\infty} A_n = A_1 \cup A_2 \cup A_3 \cdots$$

Find A.

9. * Let $A_n = \left[0, \frac{1}{n}\right) = \{x \in \mathbb{R} \mid 0 \leq x < \frac{1}{n}\}$ for $n = 1, 2, \cdots$. Define

$$A = \bigcap_{n=1}^{\infty} A_n = A_1 \cap A_2 \cap \cdots$$

Find A.

10. * In this problem our goal is to show that sets that are not in the form of intervals may also be uncountable. In particular, consider the set A defined as the set of all subsets of $\mathbb{N}$:

$$A = \{B : B \subset \mathbb{N}\}.$$

We usually denote this set by $A = 2^{\mathbb{N}}$.

(a) Show that $2^{\mathbb{N}}$ is in one-to-one correspondence with the set of all (infinite) binary sequences:

$$C = \{b_1, b_2, b_3, \cdots \mid b_i \in \{0, 1\}\}.$$

(b) Show that C is in one-to-one correspondence with $[0, 1]$. From (a) and (b) we conclude that the set $2^{\mathbb{N}}$ is uncountable.

11. * Show the set $[0, 1)$ is uncountable. That is, you can never provide a list in the form of $\{a_1, a_2, a_3, \cdots\}$ that contains all the elements in $[0, 1)$.

12. Recall that

$$\{H, T\}^3 = \{H, T\} \times \{H, T\} \times \{H, T\}$$
$$= \{(H, H, H), (H, H, T), \cdots, (T, T, T)\}.$$

Consider the following function:

$$f : \{H, T\}^3 \longrightarrow \mathbb{N} \cup \{0\},$$

defined as

$$f(x) = \text{the number of H's in } x.$$

For example,

$$f(HTH) = 2.$$

(a) Determine the domain and co-domain for f.

(b) Find range of f:Range(f).

(c) If we know $f(x) = 2$, what can we say about x?

13. Two teams A and B play a soccer match, and we are interested in the winner. The sample space can be defined as

$$S = \{a, b, d\},$$

where a shows the outcome that A wins, b shows the outcome that B wins, and d shows the outcome that they draw. Suppose we know that:
(1) the probability that A wins is $P(a) = P(\{a\}) = 0.5$;
(2) the probability of a draw is $P(d) = P(\{d\}) = 0.25$.

  (a) Find the probability that B wins.

  (b) Find the probability that B wins or a draw occurs.

14. Let A and B be two events such that

$$P(A) = 0.4, P(B) = 0.7, P(A \cup B) = 0.9$$

(a) Find $P(A \cap B)$.
(b) Find $P(A^c \cap B)$.
(c) Find $P(A - B)$.
(d) Find $P(A^c - B)$.
(e) Find $P(A^c \cup B)$.
(f) Find $P(A \cap (B \cup A^c))$.

15. I roll a fair die twice and obtain two numbers: X_1 = result of the first roll, X_2 = result of the second roll.
(a) Find the probability that $X_2 = 4$.
(b) Find the probability that $X_1 + X_2 = 7$.
(c) Find the probability that $X_1 \neq 2$ and $X_2 \geq 4$.

16. Consider a random experiment with a sample space.

$$S = \{1, 2, 3, \cdots\}.$$

Suppose that we know

$$P(k) = P(\{k\}) = \frac{c}{3^k} \quad \text{for} \quad k = 1, 2, \cdots,$$

where c is a constant number.
(a) Find c.
(b) Find $P(\{2, 4, 6\})$.
(c) Find $P(\{3, 4, 5, \cdots\})$.

17. Four teams A, B, C, and D compete in a tournament and exactly one of them will win the tournament. Teams A and B have the same chance of winning the tournament. Team C is twice as likely to win the tournament as team D. The probability that either team A or team C wins the tournament is 0.6. Find the probabilities of each team winning the tournament.

18. Let T be the time needed to complete a job at a certain factory. By using the historical data, we know that

$$P(T \leq t) = \begin{cases} \frac{1}{16}t^2 & \text{for} \quad 0 \leq t \leq 4 \\ 1 & \text{for} \quad t \geq 4 \end{cases}$$

   (a) Find the probability that the job is completed in less than one hour, i.e., find $P(T \leq 1)$.
   (b) Find the probability that the job needs more than 2 hours.
   (c) Find the probability that $1 \leq T \leq 3$.

19. * You choose a point (A, B) uniformly at random in the unit square $\{(x, y) : x, y \in [0, 1]\}$.

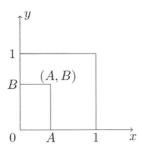

   What is the probability that the equation

$$AX^2 + X + B = 0$$

   has real solutions?

20. * (continuity of probability)
   (a) Let $A_1, A_2, A_3, \cdots$ be a sequence of increasing events, that is,

$$A_1 \subset A_2 \subset A_3 \subset \cdots$$

   Show that

$$P\left(\bigcup_{i=1}^{\infty} A_i\right) = \lim_{n \to \infty} P(A_n).$$

   (b) Using part(a), show that if $A_1, A_2, \cdots$ is a decreasing sequence of events, i.e.,

$$A_1 \supset A_2 \supset A_3 \supset \cdots .$$

   Then

$$P\left(\bigcap_{i=1}^{\infty} A_i\right) = \lim_{n \to \infty} P(A_n).$$

21. * (continuity of probability) For any sequence of events $A_1, A_2, A_3, \cdots$, prove

$$P\left(\bigcup_{i=1}^{\infty} A_i\right) = \lim_{n\to\infty} P\left(\bigcup_{i=1}^{n} A_i\right)$$

$$P\left(\bigcap_{i=1}^{\infty} A_i\right) = \lim_{n\to\infty} P\left(\bigcap_{i=1}^{n} A_i\right).$$

22. Suppose that, of all the customers at a coffee shop,

   -70% purchase a cup of coffee;
   -40% purchase a piece of cake;
   -20% purchase both a cup of coffee and a piece of cake.
   Given that a randomly chosen customer has purchased a piece of cake, what is the probability that he/she has also purchased a cup of coffee?

23. Let A, B, and C be three events with probabilities given below:

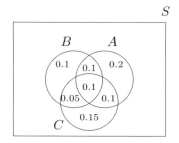

   a) Find $P(A|B)$.
   b) Find $P(C|B)$.
   c) Find $P(B|A \cup C)$.
   d) Find $P(B|A, C) = P(B|A \cap C)$.

24. A **real** number X is selected uniformly at random in the continuous interval $[0, 10]$. (For example, X could be 3.87.)
   (a) Find $P(2 \leq X \leq 5)$.
   (b) Find $P(X \leq 2|X \leq 5)$.
   (c) Find $P(3 \leq X \leq 8|X \geq 4)$.

25. A professor thinks students who live on campus are more likely to get As in the probability course. To check this theory, the professor combines the data from the past few years:

   (a) 600 students have taken the course;

   (b) 120 students have gotten As;

   (c) 200 students lived on campus;

(d) 80 students lived off campus and got *A*s.

Does this data suggest that "getting an *A*" and "living on campus" are dependent or independent?

26. I roll a die n times, $n \in \mathbb{N}$. Find the probability that numbers 1 and 6 are both observed at least once.

27. Consider a communication system. At any given time, the communication channel is in good condition with probability 0.8, and is in bad condition with probability 0.2. An error occurs in a transmission with probability 0.1 if the channel is in good condition, and with probability 0.3 if the channel is in bad condition. Let G be the event that the channel is in good condition and E be the event that there is an error in transmission.
(a) Complete the following tree diagram:

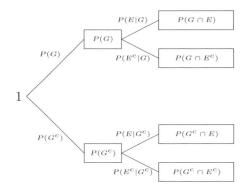

(b) Using the tree find $P(E)$.
(c) Using the tree find $P(G|E^c)$.

28. In a factory there are 100 units of a certain product, 5 of which are defective. We pick three units from the <u>100</u> units at random. What is the probability that exactly one of them is defective?

29. Reliability:
Real-life systems often are composed of several components. For example, a system may consist of two components that are connected in parallel as shown in Figure 1.28. When the system's components are connected in parallel, the system works if <u>at least one</u> of the components is functional. The components might also be connected in series as shown in Figure 1.28. When the system's components are connected in series, the system works if <u>all</u> of the components are functional.

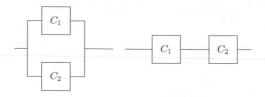

Figure 1.28: In left figure, Components C_1 and C_2 are connected in parallel. The system is functional if at least one of the C_1 and C_2 is functional. In right figure, Components C_1 and C_2 are connected in series. The system is functional only if both C_1 and C_2 are functional.

For each of the following systems, find the probability that the system is functional. Assume that component k is functional with probability P_k independent of other components.

a)

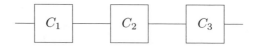

b)

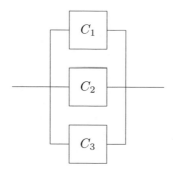

c)

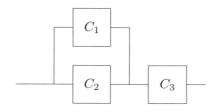

d)

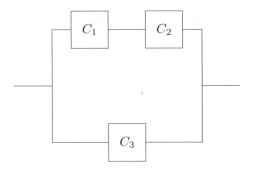

e)

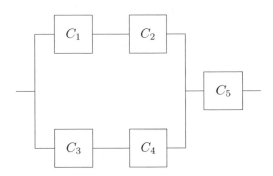

30. You choose a point (X, Y) uniformly at random in the unit square

$$S = \{(x, y) \in \mathbb{R}^2 : 0 \leq x \leq 1, 0 \leq y \leq 1\}.$$

Let A be the event $\{(x, y) \in S : |x - y| \leq \frac{1}{2}\}$ and B be the event $\{(x, y) \in S : y \geq x\}$.
(a) Show sets A and B in the x-y plane.
(b) Find $P(A)$ and $P(B)$.
(c) Are A and B independent?

31. One way to design a spam filter is to look at the words in an email. In particular, some words are more frequent in spam emails. Suppose that we have the following information:

(a) 50% of emails are spam;

(b) 1% of spam emails contain the word "refinance";

(c) 0.001% of non-spam emails contain the word "refinance."

Suppose that an email is checked and found to contain the word "refinance." What is the probability that the email is spam?

32. You would like to go from point A to point B in Figure 1.29. There are 5 bridges on different branches of the river as shown in Figure 1.29.

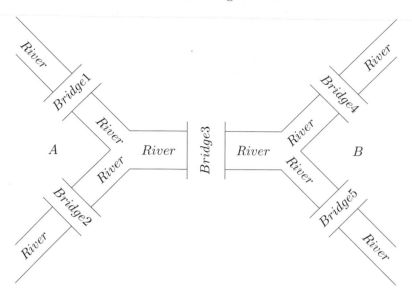

Figure 1.29: Problem 32

Bridge i is open with probability P_i, $i = 1, 2, 3, 4, 5$. Let A be the event that there is a path from A to B and let B_k be the event that k^{th} bridge is open.

(a) Find $P(A)$.

(b) Find $P(B_3|A)$.

33. * (The Monty Hall Problem [2]) You are in a game show, and the host gives you the choice of three doors. Behind one door is a car and behind the others are goats. You pick a door, say Door 1. The host who knows what is behind the doors opens a different door and reveals a goat (the host can always open such a door because there is only one door behind which is a car). The host then asks you: "Do you want to switch?" The question is, is it to your advantage to switch your choice?

| 1 | 2 | Goat |

34. I toss a fair die twice, and obtain two numbers X and Y. Let A be the event that $X = 2$, B be the event that $X + Y = 7$, and C be the event that $Y = 3$.

(a) Are A and B independent?

---

[2]http://en.wikipedia.org/wiki/Monty_Hall_problem

(b) Are A and C independent?

(c) Are B and C independent?

(d) Are A, B, and C are independent?

35. You and I play the following game: I toss a coin repeatedly. The coin is unfair and $P(H) = p$. The game ends the first time that two consecutive heads (HH) or two consecutive tails (TT) are observed. I win if (HH) is observed and you win if (TT) is observed. Given that I won the game, find the probability that the first coin toss resulted in heads?

36. * A box contains two coins: a regular coin and one fake two-headed coin (P(H)=1). I choose a coin at random and toss it n times. If the first n coin tosses result in heads, what is the probability that the $(n + 1)^{th}$ coin toss will also result in heads?

37. * A family has n children, $n \geq 2$. We ask the father: "Do you have at least one daughter?" He responds "Yes!" Given this extra information, what is the probability that all n children are girls? In other words, what is the probability that all of their children are girls, given that at least one of them is a girl?

38. * A family has n children, $n \geq 2$. We ask from the father, "Do you have at least one daughter named Lilia?" He replies, "Yes!" What is the probability that all of their children are girls? In other words, we want to find the probability that all n children are girls, given that the family has at least one daughter named Lilia. Here you can assume that if a child is a girl, her name will be Lilia with probability $\alpha \ll 1$ independently from other children's names. If the child is a boy, his name will not be Lilia.

39. * A family has n children. We pick one of them at random and find out that she is a girl. What is the probability that all their children are girls?

Chapter 2

Combinatorics: Counting Methods

2.1 Finding Probabilities Using Counting Methods

In this chapter, we will discuss counting methods that can be used in sample spaces with equally likely outcomes. Remember that for a finite sample space S with equally likely outcomes, the probability of an event A is given by

$$P(A) = \frac{|A|}{|S|} = \frac{M}{N}.$$

Thus, finding probability of A reduces to a **counting** problem in which we need to count how many elements are in A and S. In this section, we will discuss ways to count the number of elements in a set in an efficient manner. Counting is an area of its own and there are books on this subject alone. Here, we will provide a basic introduction to the material that is usually needed in probability. Almost everything that we need about counting is the result of the **multiplication principle**. We previously saw the multiplication principle when we were talking about Cartesian products. Here, we will look at it from a different perspective. Let us look at a simple example.

Example 2.1. Suppose that I want to purchase a tablet computer. I can choose either a large or a small screen; a 64GB, 128GB, or 256GB storage capacity; and a black or white cover. How many different options do I have?

Solution: Here are the options:

1. L-64-B,

2. L-64-W,

3. L-128-B,

4. L-128-W,

5. L-256-B,

6. L-256-W,

7. S-64-B,

8. S-64-W,

9. S-128-B,

10. S-128-W,

11. S-256-B,

12. S-256-W.

Thus, there are 12 possible options. The multiplication principle states that we can simply multiply the number of options in each category (screen size, memory, color) to get the total number of possibilities, i.e., the answer is $2 \times 3 \times 2 = 12$. Here is a formal statement of the multiplication principle.

Multiplication Principle

Suppose that we perform r experiments such that the kth experiment has n_k possible outcomes, for $k = 1, 2, \cdots, r$. Then there are a total of $n_1 \times n_2 \times n_3 \times \cdots \times n_r$ possible outcomes for the sequence of r experiments.

Example 2.2. I need to choose a password for a computer account. The rule is that the password must consist of two lowercase letters (a to z) followed by one capital letter (A to Z) followed by four digits $(0, 1, \cdots, 9)$. For example, the following is a valid password

$$ejT3018.$$

- Find the total number of possible passwords, N.

- A hacker has been able to write a program that randomly and independently generates 10^8 passwords according to the above rule. Note that the same password could be generated more than once. If one of the randomly chosen passwords matches my password, then he can access my account information. What is the probability that he is successful in accessing my account information?

Solution: To choose a password, I need to first choose a lowercase letter, then another lowercase letter, then one capital letter, and then 4 digits. There are 26 lowercase letters, 26 capital letters, and 10 digits. Thus, by the multiplication principle, the total number of possible valid passwords is

$$N = 26 \times 26 \times 26 \times 10 \times 10 \times 10 \times 10 = 26^3 \times 10^4.$$

Let G_i denote the event that the hacker's ith guess matches mine, for $i = 1, 2, \cdots, 10^8$. The probability that the ith randomly chosen password matches mine is

$$P(G_i) = \frac{1}{N}.$$

Now let p_{hack} be the probability that the hacker is successful, that is at least one of the randomly chosen passwords matches mine. Recall that "at least" means union:

$$p_{hack} = P\left(\bigcup_i G_i\right).$$

Note that the events G_i are independent since the guesses are independently generated, but they are not disjoint since multiple guesses could be correct if the hacker's program generates the same password. Therefore in this case it is easier to work with intersections than unions, so we will find the probability of the complement event first:

$$P\left(\bigcup_i G_i\right)^c = P\left(\bigcap_i G_i^c\right)$$

$$= \prod_{i=1}^{N} P(G_i^c) \qquad \text{(by independence)}$$

$$= \left(1 - \frac{1}{N}\right)^{10^8}.$$

Therefore,

$$p_{hack} = 1 - \left(1 - \frac{1}{N}\right)^{10^8}$$

$$= 1 - \left(1 - \frac{1}{26^3 \times 10^4}\right)^{10^8}$$

$$= 0.4339$$

Example 2.3. Let A be a set with $|A| = n < \infty$. How many distinct subsets does A have?

Solution: Let's assume $A = \{a_1, a_2, a_3, \cdots, a_n\}$. We can look at this problem in the following way: To choose a subset B, we perform the following experiment. First we decide whether or not $a_1 \in B$ (two choices), then we decide whether or not $a_2 \in B$ (two choices), then we decide whether or not $a_3 \in B$ (two choices), ..., and finally we decide whether or not $a_n \in B$ (two choices). By the multiplication principle, the total number of subsets is then given by $2 \times 2 \times 2 \times \cdots \times 2 = 2^n$. To check our answer, let's assume $A = \{1, 2\}$. Then our formula states that there are 4 possible subsets. Indeed, the subsets are

- $\{\} = \emptyset$,
- $\{1\}$,
- $\{2\}$,
- $\{1, 2\}$.

Here, we would like to provide some general terminology for the counting problems that show up in probability to make sure that the language that we use is precise and clear.

- **Sampling**: Sampling from a set means choosing an element from that set. We often **draw** a sample at random from a given set in which each element of the set has equal chance of being chosen.

- **With or without replacement:** Usually we draw multiple samples from a set. If we put each object back after each draw, we call this **sampling with replacement**. In this case, a single object can be possibly chosen multiple times. For example, if $A = \{a_1, a_2, a_3, a_4\}$ and we pick 3 elements with replacement, a possible choice might be (a_3, a_1, a_3). Thus "with replacement" means "repetition is allowed." On the other hand, if repetition is not allowed, we call it **sampling without replacement**.

- **Ordered or unordered:** If ordering matters (i.e., $a_1, a_2, a_3 \neq a_2, a_3, a_1$), this is called **ordered sampling**. Otherwise, it is called **unordered**.

Thus, when we talk about sampling from sets, we can talk about four possibilities:

- ordered sampling with replacement;

- ordered sampling without replacement;

- unordered sampling without replacement;

- unordered sampling with replacement.

We will discuss each of these in detail and indeed will provide a formula for each. The formulas will be summarized at the end in Table 2.1. Nevertheless, the best approach here is to understand how to derive these formulas. You do not actually need to memorize them if you understand the way they are obtained.

2.1.1   Ordered Sampling with Replacement

Here, we have a set with n elements (e.g.: $A = \{1, 2, 3, \cdots .n\}$), and we want to draw k samples from the set such that ordering matters and repetition is allowed. For example, if $A = \{1, 2, 3\}$ and $k = 2$, there are 9 different possibilities:

1. (1,1);

2. (1,2);

3. (1,3);

4. (2,1);

5. (2,2);

6. (2,3);

7. (3,1);

8. (3,2);

9. (3,3).

In general, we can argue that there are k positions in the chosen list: (Position 1, Position 2, ..., Position k). There are n options for each position. Thus, when ordering matters and repetition is allowed, the total number of ways to choose k objects from a set with n elements is

$$n \times n \times ... \times n = n^k.$$

Note that this is a special case of the multiplication principle where there are k "experiments" and each experiment has n possible outcomes.

2.1.2 Ordered Sampling without Replacement: Permutations

Consider the same setting as above, but now repetition is not allowed. For example, if $A = \{1, 2, 3\}$ and $k = 2$, there are 6 different possibilities:

1. (1,2);

2. (1,3);

3. (2,1);

4. (2,3);

5. (3,1);

6. (3,2).

In general, we can argue that there are k positions in the chosen list: (Position 1, Position 2, ..., Position k). There are n options for the first position, $(n-1)$ options for the second position (since one element has already been allocated to the first position and cannot be chosen here), $(n-2)$ options for the third position, ... $(n-k+1)$ options for the kth position. Thus, when ordering matters and repetition is not allowed, the total number of ways to choose k objects from a set with n elements is

$$n \times (n-1) \times ... \times (n-k+1).$$

Any of the chosen lists in the above setting (choose k elements, ordered and no repetition) is called a k-permutation of the elements in set A. We use the following notation to show the number of k-permutations of an n-element set:

$$P_k^n = n \times (n-1) \times ... \times (n-k+1).$$

Note that if k is larger than n, then $P_k^n = 0$. This makes sense, since if $k > n$ there is no way to choose k distinct elements from an n-element set. Let's look at a very famous problem, called the birthday problem, or the birthday paradox.

Example 2.4. If k people are at a party, what is the probability that at least two of them have the same birthday? Suppose that there are $n = 365$ days in a year and all days are equally likely to be the birthday of a specific person.

Solution: Let A be the event that at least two people have the same birthday. First note that if $k > n$, then $P(A) = 1$; so, let's focus on the more interesting case where $k \leq n$. Again, the phrase "at least" suggests that it might be easier to find the probability of the complement event, $P(A^c)$. This is the event that no two people have the same birthday, and we have

$$P(A) = 1 - \frac{|A^c|}{|S|}.$$

Thus, to solve the problem it suffices to find $|A^c|$ and $|S|$. Let's first find $|S|$. What is the total number of possible sequences of birthdays of k people? Well, there are $n = 365$ choices for the first person, $n = 365$ choices for the second person,... $n = 365$ choices for the kth person. Thus there are

$$n^k$$

possibilities. This is, in fact, an ordered sampling with replacement problem, and as we have discussed, the answer should be n^k (here we draw k samples, birthdays, from the set $\{1, 2, ..., n = 365\}$). Now let's find $|A^c|$. If no birthdays are the same, this is similar to finding $|S|$ with the difference that repetition is not allowed, so we have

$$|A^c| = P_k^n = n \times (n - 1) \times ... \times (n - k + 1).$$

You can see this directly by noting that there are $n = 365$ choices for the first person, $n-1 = 364$ choices for the second person,..., $n - k + 1$ choices for the kth person. Thus the probability of A can be found as

$$P(A) = 1 - \frac{|A^c|}{|S|}$$
$$= 1 - \frac{P_k^n}{n^k}.$$

Discussion: The reason this is called a paradox is that $P(A)$ is numerically different from what most people expect. For example, if there are k=23 people in the party, what do you guess is the probability that at least two of them have the same birthday, $P(A)$? The answer is .5073, which is much higher than what most people guess. The probability crosses 99 percent when the number of peoples reaches 57. But why is the probability higher than what we expect?

It is important to note that in the birthday problem, neither of the two people are chosen beforehand. To better answer this question let us look at a different problem: I am at a party with $k - 1$ people. What is the probability that at least one person at the party has the same birthday as mine? Well, we need to choose the birthdays of $k - 1$ people, the total number of ways to do this is n^{k-1}. The total number of ways to choose the birthdays so that no one has my birthday is $(n-1)^{k-1}$. Thus, the probability that at least one person has the same birthday as mine is

$$P(B) = 1 - \left(\frac{n-1}{n}\right)^{k-1}.$$

Now, if $k = 23$, this probability is only $P(B) = 0.0586$, which is much smaller than the corresponding $P(A) = 0.5073$. The reason is that event B is looking only at the case where one person in the party has the same birthday as I. This is a much smaller event than event A which looks at all possible pairs of people. Thus, $P(A)$ is much larger than $P(B)$. We might guess that the value of $P(A)$ is much lower than it actually is, because we might confuse it with $P(B)$.

Permutations of n elements: An n-permutation of n elements is just called a permutation of those elements. In this case, $k = n$ and we have

$$P_n^n = n \times (n - 1) \times ... \times (n - n + 1)$$
$$= n \times (n - 1) \times ... \times 1,$$

which is denoted by $n!$, pronounced "n factorial". Thus $n!$ is simply the total number of permutations of n elements, i.e., the total number of ways you can order n different objects. To make our formulas consistent, we define $0! = 1$.

Example 2.5. Shuffle a deck of 52 cards. How many outcomes are possible? (In other words, how many different ways can you order 52 distinct cards? How many different permutations of 52 distinct cards exist?) The answer is 52!.

Now, using the definition of $n!$, we can rewrite the formula for P_k^n as:

$$P_k^n = \frac{n!}{(n - k)!}.$$

The number of k-permutations of n distinguishable objects is given by

$$P_k^n = \frac{n!}{(n - k)!}, \quad \text{for } 0 \le k \le n.$$

Note: There are several different common notations that are used to show the number of k-permutations of an n-element set including $P_{n,k}, P(n, k), nPk$, etc. In this book, we always use P_k^n.

2.1.3 Unordered Sampling without Replacement: Combinations

Here, we have a set with n elements (e.g., $A = \{1, 2, 3, ...n\}$), and we want to draw k samples from the set such that ordering does not matter and repetition is not allowed. Thus, we basically want to choose a k-element subset of A, which we also call a k-**combination** of the set A. For example if $A = \{1, 2, 3\}$ and $k = 2$, there are 3 different possibilities:

1. $\{1,2\}$;

2. $\{1,3\}$;

3. $\{2,3\}$.

We show the number of k-element subsets of A by

$$\binom{n}{k}.$$

This is read "n choose k." A typical scenario here is that we have a group of n people, and we would like to choose k of them to serve on a committee. A simple way to find $\binom{n}{k}$ is to compare it with P_k^n. Note that the difference between the two is ordering. In fact, for any k-element subset of $A = \{1, 2, 3, ...n\}$, we can order the elements in $k!$ ways, thus we can write

$$P_k^n = \binom{n}{k} \times k!$$

Therefore,

$$\binom{n}{k} = \frac{n!}{k!(n-k)!}.$$

Note that if k is an integer larger than n, then $\binom{n}{k} = 0$. This makes sense, since if $k > n$ there is no way to choose k distinct elements from an n-element set.

The number of k-combinations of an n-element set is given by

$$\binom{n}{k} = \frac{n!}{k!(n-k)!}, \text{ for } 0 \leq k \leq n.$$

$\binom{n}{k}$ is also called the **binomial coefficient**. This is because the coefficients in the binomial theorem are given by $\binom{n}{k}$. In particular, the binomial theorem states that for an integer $n \geq 0$, we have

$$(a+b)^n = \sum_{k=0}^{n} \binom{n}{k} a^k b^{n-k}.$$

Note: There are several different common notations that are used to show the number of k-combinations of an n-element set including $C_{n,k}, C(n,k), C_k^n, nCk$, etc. In this book, we always use $\binom{n}{k}$.

Example 2.6. I choose 3 cards from the standard deck of cards. What is the probability that these cards contain at least one ace?

Solution: Again the phrase "at least" suggests that it might be easier to first find $P(A^c)$, the probability that there is no ace. Here, the sample space contains all possible ways to choose 3 cards from 52 cards, thus

$$|S| = \binom{52}{3}.$$

There are $52 - 4 = 48$ non-ace cards, so we have

$$|A^c| = \binom{48}{3}.$$

Thus

$$P(A) = 1 - \frac{\binom{48}{3}}{\binom{52}{3}}.$$

Example 2.7. How many distinct sequences can we make using 3 letter "A"s and 5 letter "B"s? (AAABBBBB, AABABBBB, etc.)

Solution: You can think of this problem in the following way. You have $3+5 = 8$ positions to fill with letters A or B. From these 8 positions, you need to choose 3 of them for As. Whatever is left will be filled with Bs. Thus the total number of ways is

$$\binom{8}{3}.$$

Now, you could have equivalently chosen the locations for Bs, so the answer would have been

$$\binom{8}{5}.$$

Thus, we conclude that

$$\binom{8}{3} = \binom{8}{5}.$$

The same argument can be repeated for general n and k to conclude

$$\binom{n}{k} = \binom{n}{n-k}.$$

You can check this identity directly algebraically, but the way we showed it here is interesting in the sense that you do not need any algebra. This is sometimes a very effective way of proving some identities of binomial coefficients. This is proof by *combinatorial interpretation*. The basic idea is that you count the same thing twice, each time using a different method and then conclude that the resulting formulas must be equal. Let us look at some other examples.

Example 2.8. Show the following identities for non-negative integers k, m, and n, using combinatorial interpretation arguments.

1. We have $\sum_{k=0}^{n} \binom{n}{k} = 2^n$.

2. For $0 \le k < n$, we have $\binom{n+1}{k+1} = \binom{n}{k+1} + \binom{n}{k}$.

3. We have $\binom{m+n}{k} = \sum_{i=0}^{k} \binom{m}{i}\binom{n}{k-i}$ (Vandermonde's identity).

Solution:

1. To show this identity, we count the total number of subsets of an n-element set A. We have already seen that this is equal to 2^n in Example 2.3. Another way to count the number of subsets is to first count the subsets with 0 elements, and then add the number of subsets with 1 element, and then add the number of subsets with 2 elements, etc. But we know that the number of k-element subsets of A is $\binom{n}{k}$, thus we have

$$2^n = \text{Number of subsets of } A$$

$$= \sum_{k=0}^{n} \text{Number of } k\text{-element subsets of } A \qquad (2.1)$$

$$= \sum_{k=0}^{n} \binom{n}{k}.$$

We can also prove this identity algebraically, using the binomial theorem, $(a + b)^n = \sum_{k=0}^{n} \binom{n}{k} a^k b^{n-k}$. If we let $a = b = 1$, we obtain $2^n = \sum_{k=0}^{n} \binom{n}{k}$.

2. To show this identity, let's assume that we have an arbitrary set A with $n + 1$ distinct elements:

$$A = \{a_1, a_2, a_3, ..., a_n, a_{n+1}\}.$$

We would like to choose a $k + 1$-element subset B. We know that we can do this in $\binom{n+1}{k+1}$ ways (the left hand side of the identity). Another way to count the number of $k + 1$-element subsets B is to divide them into two non-overlapping categories based on whether or not they contain a_{n+1}. In particular, if $a_{n+1} \notin B$, then we need to choose $k+1$ elements from $\{a_1, a_2, a_3, ..., a_n\}$, which we can do in $\binom{n}{k+1}$ different ways. If, on the other hand, $a_{n+1} \in B$, then we need to choose another k elements from $\{a_1, a_2, a_3, ..., a_n\}$ to complete B, and we can do this in $\binom{n}{k}$ different ways. Thus, we have shown that the total number of $k + 1$-element subsets of an $n + 1$-element set is equal to $\binom{n}{k+1} + \binom{n}{k}$.

3. Here, we assume that we have a set A that has $m + n$ elements:

$$A = \{a_1, a_2, a_3, ..., a_m, b_1, b_2, ..., b_n\}.$$

We would like to count the number of k-element subsets of A. This is $\binom{m+n}{k}$. Another way to do this is to first choose i elements from $\{a_1, a_2, a_3, ..., a_m\}$ and then $k - i$ elements from $\{b_1, b_2, ..., b_n\}$. This can be done in $\binom{m}{i}\binom{n}{k-i}$ number of ways. But i can be any number from 0 to k, so we conclude $\binom{m+n}{k} = \sum_{i=0}^{k} \binom{m}{i}\binom{n}{k-i}$.

Let us now provide another interpretation of $\binom{n}{k}$. Suppose that we have a group of n people and we would like to divide them two groups A and B such that group A consists of k people and group B consists of $n - k$ people. To do this, we just simply need to choose k people and put them in group A, and whoever is left will be in group B. Thus, the total number of ways to do this is $\binom{n}{k}$.

> The total number of ways to divide n distinct objects into two groups A and B such that group A consists of k objects and group B consists of $n - k$ objects is $\binom{n}{k}$.

Note: For the special case when $n = 2k$ and we do not particularly care about group names A and B, the number of ways to do this division is $\frac{1}{2}\binom{n}{k}$ to avoid double counting. For example, if 22 players want to play a soccer game and we need to divide them into two groups of 11 players, there will be $\frac{1}{2}\binom{22}{11}$ ways to do this. The reason for this is that, if we label the players 1 to 22, then the two choices

$$A = \{1, 2, 3, ..., 11\} \text{ and } B = \{12, 13, 14, ..., 22\};$$
$$A = \{12, 13, 14, ..., 22\} \text{ and } B = \{1, 2, 3, ..., 11\}.$$

are essentially the same.

For example, we can solve Example 2.7 in the following way: We have 8 blank positions to be filled with letters "A" or "B." We need to divide them into two groups A and B such that group A consists of three blank positions and group B consists of 5 blank spaces. The elements in group A show the positions of "A"s and the elements in group B show the positions of "B"s. Therefore the total number of possibilities is $\binom{8}{3}$.

Bernoulli Trials and Binomial Distribution:

Now, we are ready to discuss an important class of random experiments that appear frequently in practice. First, we define Bernoulli trials and then discuss the binomial distribution. A **Bernoulli Trial** is a random experiment that has two possible outcomes which we can label as "success" and "failure," such as the following:

- You toss a coin. The possible outcomes are "heads" and "tails." You can define "heads" as success and "tails" as failure here.

- You take a pass-fail test. The possible outcomes are "pass" and "fail."

We usually denote the probability of success by p and probability of failure by $q = 1 - p$. If we have an experiment in which we perform n <u>independent</u> Bernoulli trials and count the total number of successes, we call it a **binomial** experiment. For example, you may toss a coin n times repeatedly and be interested in the total number of heads.

Example 2.9. Suppose that I have a coin for which $P(H) = p$ and $P(T) = 1 - p$. I toss the coin 5 times.

(a) What is the probability that the outcome is $THHHH$?

(b) What is the probability that the outcome is $HTHHH$?

(c) What is the probability that the outcome is $HHTHH$?

(d) What is the probability that I will observe exactly four heads and one tails?

(e) What is the probability that I will observe exactly three heads and two tails?

(f) If I toss the coin n times, what is the probability that I observe exactly k heads and $n - k$ tails?

Solution:

(a) To find the probability of the event $A = \{THHHH\}$, we note that A is the intersection of 5 independent events: $A \equiv$ first coin toss is tails, and the next four coin tosses result in heads. Since the individual coin tosses are independent, we obtain

$$P(THHHH) = p(T) \times p(H) \times p(H) \times p(H) \times p(H)$$
$$= (1 - p)p^4.$$

(b) Similarly,

$$P(HTHHH) = p(H) \times p(T) \times p(H) \times p(H) \times p(H)$$
$$= (1 - p)p^4.$$

(c) Similarly,

$$P(HHTHH) = p(H) \times p(H) \times p(T) \times p(H) \times p(H)$$
$$= (1 - p)p^4.$$

(d) Let B be the event that I observe exactly one tails and four heads. Then

$$B = \{THHHH, HTHHH, HHTHH, HHHTH, HHHHT\}.$$

Thus

$$P(B) = P(THHHH) + P(HTHHH) + P(HHTHH) + P(HHHTH) + P(HHHHT)$$
$$= (1 - p)p^4 + (1 - p)p^4 + (1 - p)p^4 + (1 - p)p^4 + (1 - p)p^4$$
$$= 5p^4(1 - p).$$

(e) Let C be the event that I observe exactly three heads and two tails. Then

$$C = \{TTHHH, THTHH, THHHTH, ..., HHHTT\}.$$

Thus

$$P(C) = P(TTHHH) + P(THTHH) + P(THHTH) + ... + P(HHHTT)$$
$$= (1 - p)^2 p^3 + (1 - p)^2 p^3 + (1 - p)^2 p^3 + ... + (1 - p)^2 p^3$$
$$= |C| p^3 (1 - p)^2.$$

But what is $|C|$? Luckily, we know how to find $|C|$. This is the total number of distinct sequences that you can create using two tails and three heads. This is exactly the same as Example 2.7. The idea is that we have 5 positions to fill with letters H or T. From these 5 positions, you need to choose 3 of them for Hs. Whatever is left is going to be filled with Ts. Thus the total number of elements in C is $\binom{5}{3}$, and

$$P(C) = \binom{5}{3} p^3 (1 - p)^2.$$

(f) Finally, we can repeat the same argument when we toss the coin n times and obtain

$$P(k \text{ heads and } n - k \text{ tails}) = \binom{n}{k} p^k (1-p)^{n-k}.$$

Note that here, instead of writing $P(k \text{ heads and } n - k \text{ tails})$, we can just write $P(k \text{ heads})$.

Binomial Formula:

For n independent Bernoulli trials where each trial has success probability p, the probability of k successes is given by

$$P(k) = \binom{n}{k} p^k (1-p)^{n-k}.$$

Multinomial Coefficients:

The interpretation of the binomial coefficient $\binom{n}{k}$ as the number of ways to divide n objects into two groups of size k and $n - k$ has the advantage of being generalizable to dividing objects into more than two groups.

Example 2.10. Ten people have a potluck. Five people will be selected to bring a main dish, three people will bring drinks, and two people will bring dessert. How many ways can they be divided into these three groups?

Solution: We can solve this problem in the following way. First, we can choose 5 people for the main dish. This can be done in $\binom{10}{5}$ ways. From the remaining 5 people, we then choose 3 people for drinks, and finally the remaining 2 people will bring desert. Thus, by the multiplication principle, the total number of ways is given by

$$\binom{10}{5} \binom{5}{3} \binom{2}{2} = \frac{10!}{5!5!} \cdot \frac{5!}{3!2!} \cdot \frac{2!}{2!0!} = \frac{10!}{5!3!2!}.$$

This argument can be generalized for the case when we have n people and would like to divide them to r groups. The number of ways in this case is given by the **multinomial** coefficients. In particular, if $n = n_1 + n_2 + ... + n_r$, where all $n_i \geq 0$ are integers, then the number of ways to divide n distinct objects to r distinct groups of sizes $n_1, n_2, ..., n_r$ is given by

$$\binom{n}{n_1, n_2, ..., n_r} = \frac{n!}{n_1! n_2! ... n_r!}.$$

We can also state the general format of the binomial theorem, which is called the multinomial theorem:

$$(x_1 + x_2 + \cdots + x_r)^n = \sum_{n_1 + n_2 + \cdots + n_r = n} \binom{n}{n_1, n_2, \ldots, n_r} x_1^{n_1} x_2^{n_2} ... x_r^{n_r} \qquad (2.2)$$

Finally, the binomial formula for Bernoulli trials can also be extended to the case where each trial has more than two possible outcomes.

Example 2.11. I roll a die 18 times. What is the probability that each number appears exactly 3 times?

Solution: First of all, each sequence of outcomes in which each number appears 3 times has probability

$$\left(\frac{1}{6}\right)^3 \times \left(\frac{1}{6}\right)^3 \times \left(\frac{1}{6}\right)^3 \times \left(\frac{1}{6}\right)^3 \times \left(\frac{1}{6}\right)^3 \times \left(\frac{1}{6}\right)^3$$
$$= \left(\frac{1}{6}\right)^{18}.$$

How many distinct sequences are there with three 1's, three 2's, ..., and three 6's? Each sequence has 18 positions which we need to fill with the digits. To obtain a sequence, we need to choose three positions for 1's, three positions for 2's, ..., and three positions for 6's. The number of ways to do this is given by the multinomial coefficient

$$\binom{18}{3, 3, 3, 3, 3, 3} = \frac{18!}{3!3!3!3!3!3!}.$$

Thus the total probability is

$$\frac{18!}{(3!)^6} \left(\frac{1}{6}\right)^{18}.$$

We now state the general form of the multinomial formula. Suppose that an experiment has r possible outcomes, so the sample space is given by

$$S = \{s_1, s_2, ..., s_r\}.$$

Also suppose that $P(s_i) = p_i$ for $i = 1, 2, ..., r$. Then for $n = n_1 + n_2 + ... + n_r$ independent trials of this experiment, the probability that each s_i appears n_i times is given by

$$\binom{n}{n_1, n_2, ..., n_r} p_1^{n_1} p_2^{n_2} ... p_r^{n_r} = \frac{n!}{n_1! n_2! ... n_r!} p_1^{n_1} p_2^{n_2} ... p_r^{n_r}.$$

2.1.4 Unordered Sampling with Replacement

Among the four possibilities we listed for ordered/unordered sampling with/without replacement, unordered sampling with replacement is the most challenging one. Suppose that we want to sample from the set $A = \{a_1, a_2, ..., a_n\}$ k times such that repetition is allowed and ordering does not matter. For example, if $A = \{1, 2, 3\}$ and $k = 2$, then there are 6 different ways of doing this:

- 1,1;

- 1,2;

- 1,3;

- 2,2;

- 2,3;

- 3,3.

How can we get the number 6 without actually listing all the possibilities? One way to think about this is to note that any of the pairs in the above list can be represented by the number of 1's, 2's and 3's it contains. That is, if x_1 is the number of ones, x_2 is the number of twos, and x_3 is the number of threes, we can equivalently represent each pair by a vector (x_1, x_2, x_3), i.e.,

- 1,1 $\rightarrow$ $(x_1, x_2, x_3) = (2, 0, 0)$;

- 1,2 $\rightarrow$ $(x_1, x_2, x_3) = (1, 1, 0)$;

- 1,3 $\rightarrow$ $(x_1, x_2, x_3) = (1, 0, 1)$;

- 2,2 $\rightarrow$ $(x_1, x_2, x_3) = (0, 2, 0)$;

- 2,3 $\rightarrow$ $(x_1, x_2, x_3) = (0, 1, 1)$;

- 3,3 $\rightarrow$ $(x_1, x_2, x_3) = (0, 0, 2)$.

Note that here $x_i \geq 0$ are integers and $x_1 + x_2 + x_3 = 2$. Thus, we can claim that the number of ways we can sample two elements from the set $A = \{1, 2, 3\}$ such that ordering does not matter and repetition is allowed is the same as solutions to the following equation

$$x_1 + x_2 + x_3 = 2, \text{ where } x_i \in \{0, 1, 2\}.$$

This is an interesting observation and in fact using the same argument we can make the following statement for general k and n.

Lemma 2.1. The total number of distinct k samples from an n-element set such that repetition is allowed and ordering does not matter is the same as the number of distinct solutions to the equation

$$x_1 + x_2 + ... + x_n = k, \text{ where } x_i \in \{0, 1, 2, 3, ...\}.$$

So far we have seen the number of unordered k-samples from an n element set is the same as the number of solutions to the above equation. But how do we find the number of solutions to that equation?

Theorem 2.1. The number of distinct solutions to the equation

$$x_1 + x_2 + ... + x_n = k, \text{ where } x_i \in \{0, 1, 2, 3, ...\} \tag{2.3}$$

is equal to

$$\binom{n + k - 1}{k} = \binom{n + k - 1}{n - 1}.$$

Proof. Let us first define the following simple mapping in which we replace an integer $x_i \geq 0$ with x_i vertical lines, i.e.,

$$1 \rightarrow |$$
$$2 \rightarrow ||$$
$$3 \rightarrow |||$$

$$...$$

Now suppose we have a solution to the Equation 2.3. We can replace the x_i's by their equivalent vertical lines. Thus, for example if we have $x_1 + x_2 + x_3 + x_4 = 3 + 0 + 2 + 1$, we can equivalently write $||| + +|| + |$. Thus, we claim that for each solution to the Equation 2.3, we have unique representation using vertical lines ('|') and plus signs ('+'). Indeed, each solution can be represented by k vertical lines (since the x_i sum to k) and $n-1$ plus signs. Now, this is exactly the same as Example 2.7: how many distinct sequences you can make using k vertical lines (|) and $n-1$ plus signs (+)? The answer as we have seen is

$$\binom{n+k-1}{k} = \binom{n+k-1}{n-1}.$$

■

Example 2.12. Ten passengers get on an airport shuttle at the airport. The shuttle has a route that includes 5 hotels, and each passenger gets off the shuttle at his/her hotel. The driver records how many passengers leave the shuttle at each hotel. How many different possibilities exist?

Solution: Let x_i be the number of passengers that get off the shuttle at Hotel i. Then we have

$$x_1 + x_2 + x_3 + x_4 + x_5 = 10, \text{where } x_i \in \{0, 1, 2, 3, ...\} \tag{2.4}$$

Thus, the number of solutions is

$$\binom{5+10-1}{10} = \binom{5+10-1}{5-1} = \binom{14}{4}.$$

Let's summarize the formulas for the four categories of sampling. Assuming that we have a set with n elements, and we want to draw k samples from the set, then the total number of ways we can do this is given in Table 2.1.

2.1.5   Solved Problems

1. Let A and B be two finite sets, with $|A| = m$ and $|B| = n$. How many distinct functions (mappings) can you define from set A to set B, $f : A \rightarrow B$?

Solution: We can solve this problem using the multiplication principle. Let

$$A = \{a_1, a_2, a_3, ..., a_m\},$$
$$B = \{b_1, b_2, b_3, ..., b_n\}.$$

Table 2.1: Counting results for different sampling methods.

ordered sampling with replacement	n^k
ordered sampling without replacement	$P_k^n = \frac{n!}{(n-k)!}$
unordered sampling without replacement	$\binom{n}{k} = \frac{n!}{k!(n-k)!}$
unordered sampling with replacement	$\binom{n+k-1}{k}$

Note that to define a mapping from A to B, we have n options for $f(a_1)$, i.e., $f(a_1) \in B = \{b_1, b_2, b_3, ..., b_n\}$. Similarly we have n options for $f(a_2)$, and so on. Thus by the multiplication principle, the total number of distinct functions $f : A \to B$ is

$$n \cdot n \cdot n \cdots n = n^m.$$

2. A function is said to be **one-to-one** if, for all $x_1 \neq x_2$, we have $f(x_1) \neq f(x_2)$. Equivalently, we can say a function is one-to-one if, whenever $f(x_1) = f(x_2)$, then $x_1 = x_2$. Let A and B be two finite sets, with $|A| = m$ and $|B| = n$. How many distinct one-to-one functions (mappings) can you define from set A to set B, $f : A \to B$?

Solution: Again let

$$A = \{a_1, a_2, a_3, ..., a_m\}$$
$$B = \{b_1, b_2, b_3, ..., b_n\}.$$

To define a one-to-one mapping from A to B, we have n options for $f(a_1)$, i.e., $f(a_1) \in B = \{b_1, b_2, b_3, ..., b_n\}$. Given $f(a_1)$, we have $n - 1$ options for $f(a_2)$, and so on. Thus by the multiplication principle, the total number of distinct functions $f : A \to B$, is

$$n \cdot (n - 1) \cdot (n - 2) \cdots (n - m + 1) = P_m^n.$$

Thus, in other words, choosing a one-to-one function from A to B is equivalent to choosing an m-permutation from the n-element set B (ordered sampling without replacement) and as we have seen there are P_m^n ways to do that.

3. An urn contains 30 red balls and 70 green balls. What is the probability of getting exactly k red balls in a sample of size 20 if the sampling is done **with** replacement (repetition allowed)? Assume $0 \leq k \leq 20$.

Solution: Here, any time we take a sample from the urn we put it back before the next sample (sampling with replacement). Thus in this experiment each time we sample, the probability of choosing a red ball is $\frac{30}{100}$, and we repeat this in 20 independent trials. This is exactly the binomial experiment. Thus, using the binomial formula we obtain

$$P(k \text{ red balls}) = \binom{20}{k}(0.3)^k(0.7)^{20-k}.$$

4. An urn consists of 30 red balls and 70 green balls. What is the probability of getting exactly k red balls in a sample of size 20 if the sampling is done **without** replacement (repetition not allowed)?

Solution: Let A be the event (set) of getting exactly k red balls. To find $P(A) = \frac{|A|}{|S|}$, we need to find $|A|$ and $|S|$. First, note that $|S| = \binom{100}{20}$. Next, to find $|A|$, we need to find out in how many ways we can choose k red balls and $20 - k$ green balls. Using the multiplication principle, we have

$$|A| = \binom{30}{k}\binom{70}{20-k}.$$

Thus, we have

$$P(A) = \frac{\binom{30}{k}\binom{70}{20-k}}{\binom{100}{20}}.$$

5. Assume that there are k people in a room and we know that

   - $k = 5$ with probability $\frac{1}{4}$;
   - $k = 10$ with probability $\frac{1}{4}$;
   - $k = 15$ with probability $\frac{1}{2}$.

   (a) What is the probability that at least two of them have been born in the same month? Assume that all months are equally likely.

   (b) Given that we already know there are at least two people that celebrate their birthday in the same month, what is the probability that $k = 10$?

Solution:

   (a) The first part of the problem is very similar to the birthday problem. One difference here is that $n = 12$ instead of 365. Let A_k be the event that at least two people out of k people have birthdays in the same month. We have

$$P(A_k) = 1 - \frac{P_k^{12}}{12^k}, \text{ for } k \in \{2, 3, 4, ..., 12\}.$$

Note that $P(A_k) = 1$ for $k > 12$. Let A be the event that at least two people in the room were born in the same month. Using the law of total probability, we have

$$P(A) = \frac{1}{4}P(A_5) + \frac{1}{4}P(A_{10}) + \frac{1}{2}P(A_{15})$$
$$= \frac{1}{4}\left(1 - \frac{P_5^{12}}{12^5}\right) + \frac{1}{4}\left(1 - \frac{P_{10}^{12}}{12^{10}}\right) + \frac{1}{2}.$$

(b) The second part of the problem asks for $P(k = 10|A)$. We can use Bayes' rule to write

$$P(k = 10|A) = \frac{P(A|k = 10)P(k = 10)}{P(A)}$$
$$= \frac{P(A_{10})}{4P(A)}$$
$$= \frac{1 - \frac{P_{10}^{12}}{12^{10}}}{(1 - \frac{P_5^{12}}{12^5}) + (1 - \frac{P_{10}^{12}}{12^{10}}) + 2}.$$

6. How many distinct solutions does the following equation have?

$$x_1 + x_2 + x_3 + x_4 = 100, \text{ such that}$$
$$x_1 \in \{1, 2, 3..\}, x_2 \in \{2, 3, 4, ..\}, x_3, x_4 \in \{0, 1, 2, 3, ...\}.$$

Solution: We already know that in general the number of solutions to the equation

$$x_1 + x_2 + ... + x_n = k, \text{ where } x_i \in \{0, 1, 2, 3, ...\}$$

is equal to

$$\binom{n + k - 1}{k} = \binom{n + k - 1}{n - 1}.$$

We need to convert the restrictions in this problem to match this general form. We are given that $x_1 \in \{1, 2, 3..\}$, so if we define

$$y_1 = x_1 - 1,$$

then $y_1 \in \{0, 1, 2, 3, ...\}$. Similarly define $y_2 = x_2 - 2$, so $y_2 \in \{0, 1, 2, 3, ...\}$. Now the question becomes equivalent to finding the number of solutions to the equation

$$y_1 + 1 + y_2 + 2 + x_3 + x_4 = 100, \text{ where } y_1, y_2, x_3, x_4 \in \{0, 1, 2, 3, ...\},$$

or equivalently, the number of solutions to the equation

$$y_1 + y_2 + x_3 + x_4 = 97, \text{ where } y_1, y_2, x_3, x_4 \in \{0, 1, 2, 3, ...\}$$

As we know, this is equal to

$$\binom{4+97-1}{3} = \binom{100}{3}.$$

7. (The matching problem) Here is a famous problem: N guests arrive at a party. Each person is wearing a hat. We collect all hats and then randomly redistribute the hats, giving each person one of the N hats randomly. What is the probability that at least one person receives his/her own hat? *Hint:* Use the inclusion-exclusion principle.

Solution: Let A_i be the event that ith person receives his/her own hat. Then we are interested in finding $P(E)$, where $E = A_1 \cup A_2 \cup A_3 \cup ... \cup A_N$. To find $P(E)$, we use the inclusion-exclusion principle. We have

$$P(E) = P\left(\bigcup_{i=1}^{N} A_i\right) = \sum_{i=1}^{N} P(A_i) - \sum_{i,j\,:\,i<j} P(A_i \cap A_j)$$

$$+ \sum_{i,j,k\,:\,i<j<k} P(A_i \cap A_j \cap A_k) - \cdots + (-1)^{N-1} P\left(\bigcap_{i=1}^{N} A_i\right).$$

Note that there is complete symmetry here, that is, we can write

$$P(A_1) = P(A_2) = P(A_3) = ... = P(A_N)$$
$$P(A_1 \cap A_2) = P(A_1 \cap A_3) = ... = P(A_2 \cap A_4) = ...$$
$$P(A_1 \cap A_2 \cap A_3) = P(A_1 \cap A_2 \cap A_4) = ... = P(A_2 \cap A_4 \cap A_5) = ...$$
...

Thus, we have

$$\sum_{i=1}^{N} P(A_i) = NP(A_1)$$

$$\sum_{i,j\,:\,i<j} P(A_i \cap A_j) = \binom{N}{2} P(A_1 \cap A_2)$$

$$\sum_{i,j,k\,:\,i<j<k} P(A_i \cap A_j \cap A_k) = \binom{N}{3} P(A_1 \cap A_2 \cap A_3)$$

...

Therefore, we have

$$P(E) = NP(A_1) - \binom{N}{2} P(A_1 \cap A_2)$$

$$+ \binom{N}{3} P(A_1 \cap A_2 \cap A_3) - ... + (-1)^{N-1} P(A_1 \cap A_2 \cap A_3 ... \cap A_N) \qquad (2.5)$$

Now, we only need to find $P(A_1)$, $P(A_1 \cap A_2)$, $P(A_1 \cap A_2 \cap A_3)$, etc. to finish solving the problem. To find $P(A_1)$, we have

$$P(A_1) = \frac{|A_1|}{|S|}.$$

Here, the sample space S consists of all possible permutations of N objects (hats). Thus, we have

$$|S| = N!$$

On the other hand, A_1 consists of all possible permutations of $N - 1$ objects (because the first object is fixed). Thus

$$|A_1| = (N - 1)!$$

Therefore, we have

$$P(A_1) = \frac{|A_1|}{|S|} = \frac{(N-1)!}{N!} = \frac{1}{N}.$$

Similarly, we have

$$|A_1 \cap A_2| = (N - 2)!$$

Thus,

$$P(A_1 \cap A_2) = \frac{|A_1 \cap A_2|}{|S|} = \frac{(N-2)!}{N!} = \frac{1}{P_{N-2}^N}.$$

Similarly,

$$P(A_1 \cap A_2 \cap A_3) = \frac{|A_1 \cap A_2 \cap A_3|}{|S|} = \frac{(N-3)!}{N!} = \frac{1}{P_{N-3}^N}$$

$$P(A_1 \cap A_2 \cap A_3 \cap A_4) = \frac{|A_1 \cap A_2 \cap A_3 \cap A_4|}{|S|} = \frac{(N-4)!}{N!} = \frac{1}{P_{N-4}^N}$$

$$\dots$$

Thus, using Equation 2.5 we have

$$P(E) = N.\frac{1}{N} - \binom{N}{2} \cdot \frac{1}{P_{N-2}^N}$$

$$+ \binom{N}{3} \cdot \frac{1}{P_{N-3}^N} - \dots + (-1)^{N-1}\frac{1}{N!} \tag{2.6}$$

By simplifying a little bit, we obtain

$$P(E) = 1 - \frac{1}{2!} + \frac{1}{3!} - \dots + (-1)^{N-1}\frac{1}{N!}.$$

We are done. It is interesting to note what happens when N becomes large. To see that, we should remember the Taylor series of e^x. In particular,

$$e^x = 1 + \frac{x}{1!} + \frac{x^2}{2!} + \frac{x^3}{3!} + \dots$$

Letting $x = -1$, we have

$$e^{-1} = 1 - \frac{1}{1!} + \frac{1}{2!} - \frac{1}{3!} + \dots$$

Thus, we conclude that as N becomes large, $P(E)$ approaches $1 - \frac{1}{e}$.

2.2 End of Chapter Problems

1. A coffee shop has 4 different types of coffee. You can order your coffee in a small, medium, or large cup. You can also choose whether you want to add cream, sugar, or milk (any combination is possible, for example, you can choose to add all three). In how many ways can you order your coffee?

2. Eight committee members are meeting in a room that has twelve chairs. In how many ways can they seat themselves in the chairs?

3. There are 20 black cell phones and 30 white cell phones in a store. An employee takes 10 phones at random. Find the probability that

   (a) there will be exactly 4 black cell phones among the chosen phones;

   (b) there will be less than 3 black cell phones among the chosen phones.

4. Five cards are dealt from a shuffled deck. What is the probability that the dealt hand contains

   (a) exactly one ace;

   (b) at least one ace?

5. Five cards are dealt from a shuffled deck. What is the probability that the dealt hand contains exactly two aces, given that we know it contains at least one ace?

6. The 52 cards in a shuffled deck are dealt equally among four players (call them A, B, C, and D). If A and B have exactly 7 spades, what is the probability that C has exactly 4 spades?

7. There are 50 students in a class and the professor chooses 15 students at random. What is the probability that you or your friend Joe are among the chosen students?

8. In how many ways can you arrange the letters in the word "Massachusetts"?

9. You have a biased coin for which $P(H) = p$. You toss the coin 20 times. What is the probability that

   (a) you observe 8 heads and 12 tails;

   (b) you observe more than 8 heads and more than 8 tails?

10. A wireless sensor grid consists of $21 \times 11 = 231$ sensor nodes that are located at points (i, j) in the plane such that $i \in \{0, 1, \cdots, 20\}$ and $j \in \{0, 1, 2, \cdots, 10\}$ as shown in Figure 2.1. The sensor node located at point $(0, 0)$ needs to send a message to a node located at $(20, 10)$. The messages are sent to the destination by going from each sensor to a neighboring sensor located above or to the right. That is, we assume that each node located at point (i, j) will only send messages to the nodes located at $(i+1, j)$ or $(i, j+1)$. How many different paths exist for sending the message from node $(0, 0)$ to node $(20, 10)$?

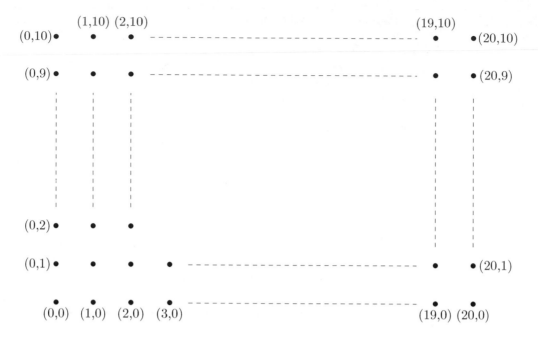

Figure 2.1: Figure for problem 10

11. In Problem 10 assume that all the appropriate paths are equally likely. What is the probability that the sensor located at point $(10, 5)$ receives the message? That is, what is the probability that a randomly chosen path from $(0, 0)$ to $(20, 10)$ goes through the point $(10, 5)$?

12. * In Problem 10 assume that if a sensor has a choice, it will send the message to the above sensor with probability p_a, and will send the message to the sensor to the right with probability $p_r = 1 - p_a$. What is the probability that the sensor located at point $(10, 5)$ receives the message?

13. There are two coins in a bag. For Coin 1, $P(H) = \frac{1}{2}$ and for Coin 2, $P(H) = \frac{1}{3}$. Your friend chooses one of the coins at random and tosses it 5 times.

    (a) What is the probability of observing at least 3 heads?

    (b) * You ask your friend, "Did you observe at least three heads?" Your friend replies, "Yes." What is the probability that Coin 2 had been chosen?

14. There are 15 people in a party, including Hannah and Sarah. We divide the 15 people into 3 groups, where each group has 5 people. What is the probability that Hannah and Sarah are in the same group?

15. You roll a die 5 times. What is the probability that at least one value is observed more than once?

16. I have 10 red and 10 blue cards. I shuffle the cards and then label the cards based on their orders: I write the number one on the first card, the number two on the second card, and so on. What is the probability that

    (a) all red cards are assigned numbers less than or equal to 15;

    (b) exactly 8 red cards are assigned numbers less than or equal to 15?

17. I have two bags. Bag 1 contains 10 blue marbles, while Bag 2 contains 15 blue marbles. I pick one of the bags at random, and throw 6 red marbles in it. Then I shake the bag and choose 5 marbles (without replacement) at random from the bag. If there are exactly 2 red marbles among the 5 chosen marbles, what is the probability that I have chosen Bag 1?

18. In a communication system, packets are transmitted from a sender to a receiver. Each packet is received with no error with probability p independently from other packets (with probability $1 - p$ the packet is lost). The receiver can decode the message as soon as it receives k packets with no error. Find the probability that the sender sends exactly n packets until the receiver can decode the message successfully.

19. How many distinct solutions does the following equation have such that all $x_i \in \mathbb{N}$?

$$x_1 + x_2 + x_3 + x_4 + x_5 = 100$$

20. How many distinct solutions does the following equation have?

$$x_1 + x_2 + x_3 + x_4 = 100, \text{ such that}$$
$$x_1 \in \{0, 1, 2, \cdots, 10\}, x_2, x_3, x_4 \in \{0, 1, 2, 3, ...\}.$$

21. For this problem suppose that x_i's must be non-negative integers, i.e., $x_i \in \{0, 1, 2, \cdots\}$ for $i = 1, 2, 3$. How many distinct solutions does the following equation have such that at least one of the x_i's is larger than 40?

$$x_1 + x_2 + x_3 = 100$$

Chapter 3

Discrete Random Variables

3.1 Basic Concepts

3.1.1 Random Variables

In general, to analyze random experiments, we usually focus on some numerical aspects of the experiment. For example, in a soccer game we may be interested in the number of goals, shots, shots on goal, corners kicks, fouls, etc. If we consider an entire soccer match as a random experiment, then each of these numerical results gives some information about the outcome of the random experiment. These are examples of *random variables*. In a nutshell, a random variable is a real-valued variable whose value is determined by an underlying random experiment. Let's look at an example.

Example 3.1. I toss a coin five times. This is a random experiment and the sample space can be written as

$$S = \{TTTTT, TTTTH, ..., HHHHH\}.$$

Note that here the sample space S has $2^5 = 32$ elements. Suppose that, in this experiment, we are interested in the number of heads. We can define a random variable X whose value is the number of observed heads. The value of X will be one of $0, 1, 2, 3, 4$ or 5 depending on the outcome of the random experiment.

In essence, a random variable is a real-valued function that assigns a numerical value to each possible outcome of the random experiment. For example, the random variable X defined above assigns the value 0 to the outcome $TTTTT$, the value 2 to the outcome $THTHT$, and so on. Hence, the random variable X is a function from the sample space $S=\{TTTTT, TTTTH, \cdots, HHHHH\}$ to the real numbers (for this particular random variable, the values are always integers between 0 and 5).

107

Random Variables: A random variable X is a function from the sample space to the real numbers.

$$X : S \to \mathbb{R}$$

We usually show random variables by capital letters such as X, Y, and Z. Since a random variable is a function, we can talk about its range. The range of a random variable X, shown by Range(X) or R_X, is the set of possible values for X. In the above example, Range$(X) = R_X = \{0, 1, 2, 3, 4, 5\}$.

The range of a random variable X, shown by Range(X) or R_X, is the set of possible values of X.

Example 3.2. Find the range for each of the following random variables.

1. I toss a coin 100 times. Let X be the number of heads I observe.

2. I toss a coin until the first heads appears. Let Y be the total number of coin tosses.

3. The random variable T is defined as the time (in hours) from now until the next earthquake occurs in a certain city.

Solution:

1. The random variable X can take any integer from 0 to 100, so $R_X = \{0, 1, 2, ..., 100\}$.

2. The random variable Y can take any positive integer, so $R_Y = \{1, 2, 3, ...\} = \mathbb{N}$.

3. The random variable T can in theory get any nonnegative real number, so $R_T = [0, \infty)$.

3.1.2   Discrete Random Variables

There are two important classes of random variables that we discuss in this book: *discrete random variables* and *continuous random variables*. We will discuss discrete random variables in this chapter and continuous random variables in Chapter 4. There will be a third class of random variables that are called *mixed random variables*. Mixed random variables, as the name suggests, can be thought of as mixture of discrete and continuous random variables. We will discuss mixed random variables in Chapter 4 as well.

Remember that a set A is countable if either

- A is a finite set such as $\{1, 2, 3, 4\}$, or

- it can be put in one-to-one correspondence with natural numbers (in this case, the set is said to be countably infinite).

In particular, as we discussed in Chapter 1, sets such as $\mathbb{N}, \mathbb{Z}, \mathbb{Q}$ and their subsets are countable, while sets such as nonempty intervals $[a, b]$ in $\mathbb{R}$ are uncountable. A random variable is discrete if its range is a countable set. In Example 3.2, the random variables X and Y are discrete, while the random variable T is not discrete.

X is a discrete random variable if its range is countable.

3.1.3 Probability Mass Function (PMF)

If X is a discrete random variable, then its range R_X is a countable set; so, we can list the elements in R_X. In other words, we can write

$$R_X = \{x_1, x_2, x_3, ...\}.$$

Note that here $x_1, x_2, x_3, ...$ are possible values of the random variable X. While random variables are usually denoted by capital letters, to represent the numbers in the range we usually use lowercase letters such as x, x_1, y, z, etc. For a discrete random variable X, we are interested in knowing the probabilities of $X = x_k$. Note that here, the event $A = \{X = x_k\}$ is defined as the set of outcomes s in the sample space S for which the corresponding value of X is equal to x_k. In particular,

$$A = \{s \in S | X(s) = x_k\}.$$

The probabilities of events $\{X = x_k\}$ are formally shown by the **probability mass function (PMF)** of X.

Definition 3.1. Let X be a discrete random variable with range $R_X = \{x_1, x_2, x_3, ...\}$ (finite or countably infinite). The function

$$P_X(x_k) = P(X = x_k), \text{ for } k = 1, 2, 3, ...,$$

is called the *probability mass function (PMF)* of X.

Thus, the PMF is a probability measure that gives us probabilities of the possible values for a random variable. While the above notation is the standard notation for the PMF of X, it might look confusing at first. The subscript X here indicates that this is the PMF of the random variable X. Thus, for example, $P_X(1)$ shows the probability that $X = 1$. To better understand all of the above concepts, let's look at some examples.

Example 3.3. I toss a fair coin twice, and let X be defined as the number of heads I observe. Find the range of X, R_X, as well as its probability mass function P_X.

Solution: Here, our sample space is given by

$$S = \{HH, HT, TH, TT\}.$$

The number of heads will be 0, 1 or 2. Thus

$$R_X = \{0, 1, 2\}.$$

Since this is a finite (and thus a countable) set, the random variable X is a discrete random variable. Next, we need to find the PMF of X. The PMF is defined as

$$P_X(k) = P(X = k) \text{ for } k = 0, 1, 2.$$

We have

$$P_X(0) = P(X = 0) = P(TT) = \frac{1}{4},$$

$$P_X(1) = P(X = 1) = P(\{HT, TH\}) = \frac{1}{4} + \frac{1}{4} = \frac{1}{2},$$

$$P_X(2) = P(X = 2) = P(HH) = \frac{1}{4}.$$

Although the PMF is usually defined for values in the range, it is sometimes convenient to extend the PMF of X to all real numbers. If $x \notin R_X$, we can simply write $P_X(x) = P(X = x) = 0$. Thus, in general we can write

$$P_X(x) = \begin{cases} P(X = x) & \text{if } x \text{ is in } R_X \\ 0 & \text{otherwise} \end{cases}$$

To better visualize the PMF, we can plot it. Figure 3.1 shows the PMF of the above random variable X. As we see, the random variable can take three possible values $0, 1$ and 2. The figure also clearly indicates that the event $X = 1$ is twice as likely as the other two possible values. The figure can be interpreted in the following way: If we repeat the random experiment (tossing a coin twice) a large number of times, then about half of the times we observe $X = 1$, about a quarter of times we observe $X = 0$, and about a quarter of times we observe $X = 2$. For discrete random variables, the PMF is also called the **probability distribution**. Thus, when asked to find the probability distribution of a discrete random variable X, we can do this by finding its PMF. [1]

Example 3.4. I have an unfair coin for which $P(H) = p$, where $0 < p < 1$. I toss the coin repeatedly until I observe a heads for the first time. Let Y be the total number of coin tosses. Find the distribution of Y.

Solution: First, we note that the random variable Y can potentially take any positive integer, so we have $R_Y = \mathbb{N} = \{1, 2, 3, ...\}$. To find the distribution of Y, we need to find

---

[1] The phrase *distribution function* is usually reserved exclusively for the cumulative distribution function CDF (as defined later in the book). The word *distribution*, on the other hand, is used in this book in a broader sense and could refer to PMF, probability density function (PDF), or CDF.

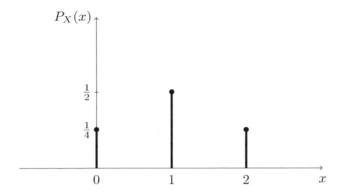

Figure 3.1: PMF for random Variable X in Example 3.3

$P_Y(k) = P(Y = k)$ for $k = 1, 2, 3, \cdots$. We have

$$P_Y(1) = P(Y = 1) = P(H) = p,$$
$$P_Y(2) = P(Y = 2) = P(TH) = (1 - p)p,$$
$$P_Y(3) = P(Y = 3) = P(TTH) = (1 - p)^2 p,$$

$$P_Y(k) = P(Y = k) = P(TT...TH) = (1 - p)^{k-1}p.$$

Thus, we can write the PMF of Y in the following way

$$P_Y(y) = \begin{cases} (1 - p)^{y-1}p & \text{for } y = 1, 2, 3, ... \\ 0 & \text{otherwise} \end{cases}$$

Consider a discrete random variable X with $\text{Range}(X) = R_X$. Note that by definition the PMF is a probability measure, so it satisfies all properties of a probability measure. In particular, we have

- $0 \le P_X(x) \le 1$ for all x, and

- $\sum_{x \in R_X} P_X(x) = 1$.

Also note that for any set $A \subset R_X$, we can find the probability that $X \in A$ using the PMF

$$P(X \in A) = \sum_{x \in A} P_X(x).$$

Properties of PMF:

- $0 \leq P_X(x) \leq 1$ for all x;

- $\sum_{x \in R_X} P_X(x) = 1$;

- for any set $A \subset R_X, P(X \in A) = \sum_{x \in A} P_X(x)$.

Example 3.5. For the random variable Y in Example 3.4,

- Check that $\sum_{y \in R_Y} P_Y(y) = 1$.

- If $p = \frac{1}{2}$, find $P(2 \leq Y < 5)$.

Solution: In Example 3.4, we obtained

$$P_Y(k) = P(Y = k) = (1 - p)^{k-1} p, \text{ for } k = 1, 2, 3, \ldots$$

Thus,

- To check that $\sum_{y \in R_Y} P_Y(y) = 1$, we have

$$\sum_{y \in R_Y} P_Y(y) = \sum_{k=1}^{\infty} (1 - p)^{k-1} p$$

$$= p \sum_{j=0}^{\infty} (1 - p)^j$$

$$= p \frac{1}{1 - (1 - p)} \qquad \text{(geomtric sum)}$$

$$= 1.$$

- If $p = \frac{1}{2}$, to find $P(2 \leq Y < 5)$, we can write

$$P(2 \leq Y < 5) = \sum_{k=2}^{4} P_Y(k)$$

$$= \sum_{k=2}^{4} (1 - p)^{k-1} p$$

$$= \frac{1}{2} \left(\frac{1}{2} + \frac{1}{4} + \frac{1}{8} \right)$$

$$= \frac{7}{16}.$$

3.1.4 Independent Random Variables

In real life, we usually need to deal with more than one random variable. For example, if you study physical characteristics of people in a certain area, you might pick a person at random and then look at his/her weight, height, etc. The weight of the randomly chosen person is one random variable, while his/her height is another one. Not only do we need to study each random variable separately, but also we need to consider if there is *dependence* (i.e., correlation) between them. Is it true that a taller person is more likely to be heavier or not? The issues of dependence between several random variables will be studied in detail later on, but here we would like to talk about a special scenario where two random variables are independent.

The concept of independent random variables is very similar to independent events. Remember, two events A and B are independent if we have $P(A, B) = P(A)P(B)$. (Also, remember that a comma means *and*, i.e., $P(A, B) = P(A \text{ and } B) = P(A \cap B)$.) Similarly, we have the following definition for independent discrete random variables.

Definition 3.2. Consider two discrete random variables X and Y. We say that X and Y are independent if

$$P\left(X = x, Y = y\right) = P(X = x)P(Y = y), \qquad \text{for all } x, y.$$

In general, if two random variables are independent, then you can write

$$P\left(X \in A, Y \in B\right) = P(X \in A)P(Y \in B), \qquad \text{for all sets } A \text{ and } B.$$

Intuitively, two random variables X and Y are independent if knowing the value of one of them does not change the probabilities for the other one. In other words, if X and Y are independent, we can write

$$P(Y = y | X = x) = P(Y = y), \text{ for all } x, y.$$

Similar to independent events, it is sometimes easy to argue that two random variables are independent simply because they do not have any physical interactions with each other. Here is a simple example: I toss a coin $2N$ times. Let X be the number of heads that I observe in the first N coin tosses and let Y be the number of heads that I observe in the second N coin tosses. Since X and Y are the result of independent coin tosses, the two random variables X and Y are independent. On the other hand, in other scenarios, it might be more complicated to show whether two random variables are independent.

Example 3.6. I toss a coin twice and define X to be the number of heads I observe. Then, I toss the coin two more times and define Y to be the number of heads that I observe this time. Find $P\left((X < 2) \text{ and } (Y > 1)\right)$.

Solution: Since X and Y are the result of different independent coin tosses, the two random variables X and Y are independent. Also, note that both random variables have the distribution we found in Example 3.3. We can write

$$P\Big((X < 2) \text{ and } (Y > 1)\Big) = P(X < 2)P(Y > 1) \qquad \text{(because } X \text{ and } Y \text{ are independent)}$$
$$= \big(P_X(0) + P_X(1)\big)P_Y(2)$$
$$= \left(\frac{1}{4} + \frac{1}{2}\right)\frac{1}{4}$$
$$= \frac{3}{16}.$$

We can extend the definition of independence to n random variables.

Definition 3.3. Consider n discrete random variables $X_1, X_2, X_3, ..., X_n$. We say that $X_1, X_2, X_3, ..., X_n$ are independent if

$$P\Big(X_1 = x_1, X_2 = x_2, ..., X_n = x_n\Big)$$
$$= P(X_1 = x_1)P(X_2 = x_2)...P(X_n = x_n), \qquad \text{for all } x_1, x_2, ..., x_n.$$

3.1.5   Special Distributions

As it turns out, there are some specific distributions that are used over and over in practice, thus they have been given special names. There is a random experiment behind each of these distributions. Since these random experiments model a lot of real-life phenomena, these special distributions are used frequently in different applications. That's why they have been given a name and we devote a section to study them. We will provide PMFs for all of these special random variables, but rather than trying to memorize the PMF, you should understand the random experiment behind each of them. If you understand the random experiments, you can simply derive the PMFs when you need them. Although it might seem that there are a lot of formulas in this section, there are in fact very few new concepts. Do not get intimidated by the large number of formulas, look at each distribution as a practice problem on discrete random variables.

Bernoulli Distribution:

What is the simplest discrete random variable (i.e., simplest PMF) that you can imagine? My answer to this question is a PMF that is nonzero at only one point. For example, if you define

$$P_X(x) = \begin{cases} 1 & \text{for } x = 1 \\ 0 & \text{otherwise} \end{cases}$$

then, X is a discrete random variable that can only take one value, i.e., $X = 1$ with a probability of one. But this is not a very interesting distribution because it is not actually random. Then, you might ask what is the next simplest discrete distribution. And my answer to that is the **Bernoulli** distribution. A Bernoulli random variable is a random variable that can only take two possible values, usually 0 and 1. This random variable models random experiments that have two possible outcomes, sometimes referred to as "success" and "failure." Here are some examples:

- You take a pass-fail exam. You either pass (resulting in $X = 1$) or fail (resulting in $X = 0$).

- You toss a coin. The outcome is either heads or tails.

- A child is born. The gender is either male or female.

Formally, the Bernoulli distribution is defined as follows:

Definition 3.4. A random variable X is said to be a *Bernoulli* random variable with *parameter* p, shown as $X \sim Bernoulli(p)$, if its PMF is given by

$$P_X(x) = \begin{cases} p & \text{for } x = 1 \\ 1 - p & \text{for } x = 0 \\ 0 & \text{otherwise} \end{cases}$$

where $0 < p < 1$.

Figure 3.2 shows the PMF of a *Bernoulli(p)* random variable.

A Bernoulli random variable is associated with a certain event A. If event A occurs (for example, if you pass the test), then $X = 1$; otherwise $X = 0$. For this reason, the Bernoulli random variable is also called the **indicator** random variable. In particular, the indicator random variable I_A for an event A is defined by

$$I_A = \begin{cases} 1 & \text{if the event } A \text{ occurs} \\ 0 & \text{otherwise} \end{cases}$$

The indicator random variable for an event A has Bernoulli distribution with parameter $p = P(A)$, so we can write

$$I_A \sim Bernoulli\big(P(A)\big).$$

Geometric Distribution:

The random experiment behind the geometric distribution is as follows. Suppose that I have a coin with $P(H) = p$. I toss the coin until I observe the first heads. We define X as the total number of coin tosses in this experiment. Then X is said to have geometric distribution

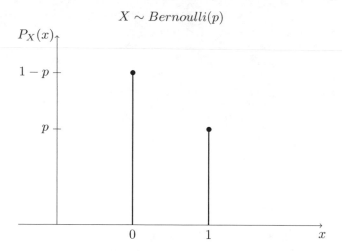

Figure 3.2: PMF of a *Bernoulli(p)* random variable

with parameter p. In other words, you can think of this experiment as repeating independent Bernoulli trials until observing the first success. This is exactly the same distribution that we saw in Example 3.4. The range of X here is $R_X = \{1, 2, 3, ...\}$. In Example 3.4, we obtained

$$P_X(k) = P(X = k) = (1 - p)^{k-1}p, \text{ for } k = 1, 2, 3, ...$$

We usually define $q = 1 - p$, so we can write $P_X(k) = pq^{k-1}$, for $k = 1, 2, 3, ...$ To say that a random variable has geometric distribution with parameter p, we write $X \sim Geometric(p)$. More formally, we have the following definition:

Definition 3.5. A random variable X is said to be a *geometric* random variable with *parameter* p, shown as $X \sim Geometric(p)$, if its PMF is given by

$$P_X(k) = \begin{cases} p(1-p)^{k-1} & \text{for } k = 1, 2, 3, ... \\ 0 & \text{otherwise} \end{cases}$$

where $0 < p < 1$.

Figure 3.3 shows the PMF of a *Geometric(0.3)* random variable.

We should note that some books define geometric random variables in slightly different ways. They define the geometric random variable X as the total number of failures before observing the first success. By this definition, the range of X is $R_X = \{0, 1, 2, ...\}$ and the PMF is given by

$$P_X(k) = \begin{cases} p(1-p)^{k} & \text{for } k = 0, 1, 2, 3, ... \\ 0 & \text{otherwise} \end{cases}$$

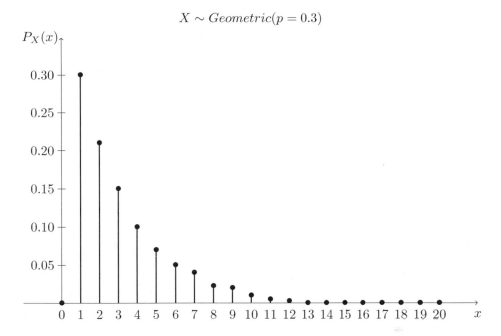

Figure 3.3: PMF of a *Geometric*(0.3) random variable

In this book, whenever we write $X \sim Geometric(p)$, we always mean X as the total number of trials as defined in Definition 3.5. Note that as long as you are consistent in your analysis, it does not matter which definition you use. That is why we emphasize that you should understand how to derive PMFs for these random variables rather than memorizing them.

Binomial Distribution:

The random experiment behind the binomial distribution is as follows. Suppose that I have a coin with $P(H) = p$. I toss the coin n times and define X to be the total number of heads that I observe. Then X is binomial with parameter n and p, and we write $X \sim Binomial(n, p)$. The range of X in this case is $R_X = \{0, 1, 2, ..., n\}$. As we have seen in Section 2.1.3, the PMF of X in this case is given by binomial formula

$$P_X(k) = \binom{n}{k} p^k (1-p)^{n-k}, \text{ for } k = 0, 1, 2, ..., n.$$

We have the following definition:

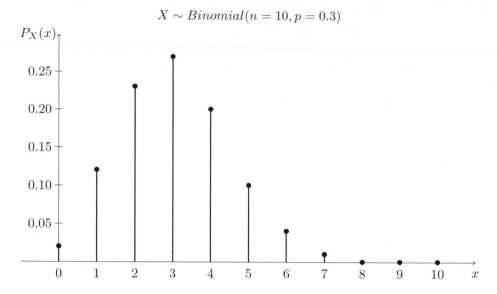

Figure 3.4: PMF of a $Binomial(10, 0.3)$ random variable

Definition 3.6. A random variable X is said to be a *binomial* random variable with parameters n and p, shown as $X \sim Binomial(n, p)$, if its PMF is given by

$$P_X(k) = \begin{cases} \binom{n}{k} p^k (1-p)^{n-k} & \text{for } k = 0, 1, 2, \cdots, n \\ 0 & \text{otherwise} \end{cases}$$

where $0 < p < 1$.

Figures 3.4 and 3.5 show the $Binomial(n, p)$ PMF for $n = 10, p = 0.3$ and $n = 20, p = 0.6$ respectively.

Binomial random variable as a sum of Bernoulli random variables: Here is a useful way of thinking about a binomial random variable. Note that a $Binomial(n, p)$ random variable can be obtained by n independent coin tosses. If we think of each coin toss as a $Bernoulli(p)$ random variable, the $Binomial(n, p)$ random variable is a sum of n independent $Bernoulli(p)$ random variables. This is stated more precisely in the following lemma.

Lemma 3.1. If $X_1, X_2, ..., X_n$ are independent $Bernoulli(p)$ random variables, then the random variable X defined by $X = X_1 + X_2 + ... + X_n$ has a $Binomial(n, p)$ distribution.

To generate a random variable $X \sim Binomial(n, p)$, we can toss a coin n times and count the number of heads. Counting the number of heads is exactly the same as finding $X_1 + X_2 + ... + X_n$, where each X_i is equal to one if the corresponding coin toss results in heads and zero otherwise. This interpretation of binomial random variables is sometimes very helpful. Let's look at an example.

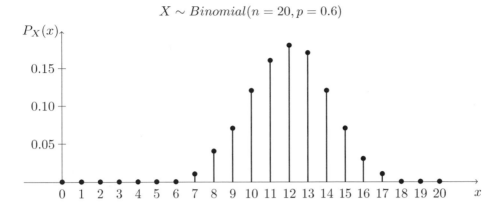

Figure 3.5: PMF of a $Binomial(20, 0.6)$ random variable

Example 3.7. Let $X \sim Binomial(n, p)$ and $Y \sim Binomial(m, p)$ be two independent random variables. Define a new random variable as $Z = X + Y$. Find the PMF of Z.

Solution: Since $X \sim Binomial(n, p)$, we can think of X as the number of heads in n independent coin tosses, i.e., we can write

$$X = X_1 + X_2 + ... + X_n,$$

where the X_i's are independent $Bernoulli(p)$ random variables. Similarly, since $Y \sim Binomial(m, p)$, we can think of Y as the number of heads in m independent coin tosses, i.e., we can write

$$Y = Y_1 + Y_2 + ... + Y_m,$$

where the Y_j's are independent $Bernoulli(p)$ random variables. Thus, the random variable $Z = X + Y$ will be the total number of heads in $n + m$ independent coin tosses:

$$Z = X + Y = X_1 + X_2 + ... + X_n + Y_1 + Y_2 + ... + Y_m,$$

where the X_i's and Y_j's are independent $Bernoulli(p)$ random variables. Thus, by Lemma 3.1, Z is a binomial random variable with parameters $m + n$ and p, i.e., $Binomial(m + n, p)$. Therefore, the PMF of Z is

$$P_Z(k) = \begin{cases} \binom{m+n}{k} p^k (1-p)^{m+n-k} & \text{for } k = 0, 1, 2, 3, ..., m + n \\ 0 & \text{otherwise} \end{cases}$$

The above solution is elegant and simple, but we may also want to directly obtain the PMF of Z using probability rules. Here is another method to solve Example 3.7. First, we note that $R_Z = \{0, 1, 2, ..., m + n\}$. For $k \in R_Z$, we can write

$$P_Z(k) = P(Z = k) = P(X + Y = k).$$

We will find $P(X+Y=k)$ by using conditioning and the law of total probability. In particular, we can write

$$P_Z(k) = P(X+Y=k)$$

$$= \sum_{i=0}^{n} P(X+Y=k|X=i)P(X=i) \qquad \text{(law of total probability)}$$

$$= \sum_{i=0}^{n} P(Y=k-i|X=i)P(X=i)$$

$$= \sum_{i=0}^{n} P(Y=k-i)P(X=i) \qquad \text{(since } X \text{ and } Y \text{ are independent)}$$

$$= \sum_{i=0}^{n} \binom{m}{k-i} p^{k-i}(1-p)^{m-k+i} \binom{n}{i} p^{i}(1-p)^{n-i} \qquad \text{(since } X \text{ and } Y \text{ are binomial)}$$

$$= \sum_{i=0}^{n} \binom{m}{k-i}\binom{n}{i} p^{k}(1-p)^{m+n-k}$$

$$= p^{k}(1-p)^{m+n-k} \sum_{i=0}^{n} \binom{m}{k-i}\binom{n}{i}$$

$$= \binom{m+n}{k} p^{k}(1-p)^{m+n-k} \qquad \text{by Example 2.8 (part 3).}$$

Thus, we have proved $Z \sim Binomial(m+n, p)$ by directly finding the PMF of Z.

Negative Binomial (Pascal) Distribution:

The negative binomial or Pascal distribution is a generalization of the geometric distribution. It relates to the random experiment of repeated independent trials until observing m successes. Again, different authors define the Pascal distribution slightly differently, and as we mentioned before if you understand one of them you can easily derive the other ones. Here is how we define the Pascal distribution in this book. Suppose that I have a coin with $P(H) = p$. I toss the coin until I observe m heads, where $m \in \mathbb{N}$. We define X as the total number of coin tosses in this experiment. Then X is said to have Pascal distribution with parameter m and p. We write $X \sim Pascal(m, p)$. Note that $Pascal(1, p)=Geometric(p)$. Note that by our definition the range of X is given by $R_X = \{m, m+1, m+2, m+3, \cdots\}$.

Let us derive the PMF of a $Pascal(m, p)$ random variable X. Suppose that I toss the coin until I observe m heads, and X is defined as the total number of coin tosses in this experiment. To find the probability of the event $A = \{X = k\}$, we argue as follows. By definition, event A can be written as $A = B \cap C$, where

- B is the event that we observe $m-1$ heads (successes) in the first $k-1$ trials, and

- C is the event that we observe a heads in the kth trial.

Note that B and C are independent events because they are related to different independent trials (coin tosses). Thus we can write

$$P(A) = P(B \cap C) = P(B)P(C).$$

Now, we have $P(C) = p$. Note also that $P(B)$ is the probability that I observe $m - 1$ heads in the $k - 1$ coin tosses. This probability is given by the binomial formula. In particular,

$$P(B) = \binom{k-1}{m-1} p^{m-1}(1-p)^{\left((k-1)-(m-1)\right)} = \binom{k-1}{m-1} p^{m-1}(1-p)^{k-m}.$$

Thus, we obtain

$$P(A) = P(B \cap C) = P(B)P(C) = \binom{k-1}{m-1} p^{m}(1-p)^{k-m}.$$

To summarize, we have the following definition for the Pascal random variable:

Definition 3.7. A random variable X is said to be a *Pascal* random variable with parameters m and p, shown as $X \sim Pascal(m, p)$, if its PMF is given by

$$P_X(k) = \begin{cases} \binom{k-1}{m-1} p^{m}(1-p)^{k-m} & \text{for } k = m, m+1, m+2, m+3, ... \\ 0 & \text{otherwise} \end{cases}$$

where $0 < p < 1$.

Figure 3.6 shows the PMF of a $Pascal(m, p)$ random variable with $m = 3$ and $p = 0.5$.

Hypergeometric Distribution:

Here is the random experiment behind the hypergeometric distribution. You have a bag that contains b blue marbles and r red marbles. You choose $k \leq b + r$ marbles at random (without replacement). Let X be the number of blue marbles in your sample. By this definition, we have $X \leq \min(k, b)$. Also, the number of red marbles in your sample must be less than or equal to r, so we conclude $X \geq \max(0, k - r)$. Therefore, the range of X is given by $R_X = \{\max(0, k - r), \max(0, k - r) + 1, \max(0, k - r) + 2, ..., \min(k, b)\}$.

To find $P_X(x)$, note that the total number of ways to choose k marbles from $b + r$ marbles is $\binom{b+r}{k}$. The total number of ways to choose x blue marbles and $k - x$ red marbles is $\binom{b}{x}\binom{r}{k-x}$. Thus, we have

$$P_X(x) = \frac{\binom{b}{x}\binom{r}{k-x}}{\binom{b+r}{k}}, \qquad \text{for } x \in R_X.$$

The following definition summarizes the discussion above.

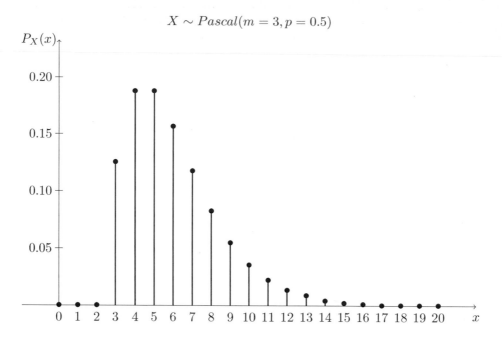

$$X \sim Pascal(m = 3, p = 0.5)$$

Figure 3.6: PMF of a $Pascal(3, 0.5)$ (negative binomial) random variable

---

Definition 3.8. A random variable X is said to be a *Hypergeometric* random variable with parameters b, r, and k, shown as $X \sim Hypergeometric(b, r, k)$, if its range is $R_X = \{\max(0, k - r), \max(0, k - r) + 1, \max(0, k - r) + 2, ..., \min(k, b)\}$, and its PMF is given by

$$P_X(x) = \begin{cases} \frac{\binom{b}{x}\binom{r}{k-x}}{\binom{b+r}{k}} & \text{for } x \in R_X \\ 0 & \text{otherwise} \end{cases}$$

Again, there is no point to memorizing the PMF. All you need to know is how to solve problems that can be formulated as a hypergeometric random variable.

Poisson Distribution:

The Poisson distribution is one of the most widely used probability distributions. It is usually used in scenarios where we are counting the occurrences of certain events in an interval of time or space. In practice, it is often an approximation of a real-life random variable. Here is an example of a scenario where a Poisson random variable might be used. Suppose that we are counting the number of customers who visit a certain store from 1 p.m. to 2 p.m. Based on data from previous days, we know that on average $\lambda = 15$ customers visit the store. Of course, there will be more customers some days and fewer on others. Here, we may model the random variable X showing the number customers as a Poisson random variable with parameter $\lambda = 15$. Let us introduce the Poisson PMF first, and then we will talk about more examples and interpretations of this distribution.

Definition 3.9. A random variable X is said to be a *Poisson* random variable with parameter λ, shown as $X \sim Poisson(\lambda)$, if its range is $R_X = \{0, 1, 2, 3, ...\}$, and its PMF is given by

$$P_X(k) = \begin{cases} \frac{e^{-\lambda}\lambda^k}{k!} & \text{for } k \in R_X \\ 0 & \text{otherwise} \end{cases}$$

Before going any further, let's check that this is a valid PMF. First, we note that $P_X(k) \geq 0$ for all k. Next, we need to check $\sum_{k \in R_X} P_X(k) = 1$. To do that, let us first remember the Taylor series for e^x: $e^x = \sum_{k=0}^{\infty} \frac{x^k}{k!}$. Now we can write

$$\sum_{k \in R_X} P_X(k) = \sum_{k=0}^{\infty} \frac{e^{-\lambda}\lambda^k}{k!}$$

$$= e^{-\lambda} \sum_{k=0}^{\infty} \frac{\lambda^k}{k!}$$

$$= e^{-\lambda}e^{\lambda} \qquad \text{(by Taylor series for } e^{\lambda}\text{)}$$

$$= 1.$$

Figures 3.7, 3.8, and 3.9 show the $Poisson(\lambda)$ PMF for $\lambda = 1$, $\lambda = 5$, and $\lambda = 10$ respectively.

Now let's look at an example.

Example 3.8. The number of emails that I get in a weekday can be modeled by a Poisson distribution with an average of 0.2 emails per minute.

1. What is the probability that I get no emails in an interval of length 5 minutes?

2. What is the probability that I get more than 3 emails in an interval of length 10 minutes?

Solution:

1. Let X be the number of emails that I get in the 5-minute interval. Then, by the assumption X is a Poisson random variable with parameter $\lambda = 5(0.2) = 1$,

$$P(X = 0) = P_X(0) = \frac{e^{-\lambda}\lambda^0}{0!} = \frac{e^{-1} \cdot 1}{1} = \frac{1}{e} \approx 0.3679$$

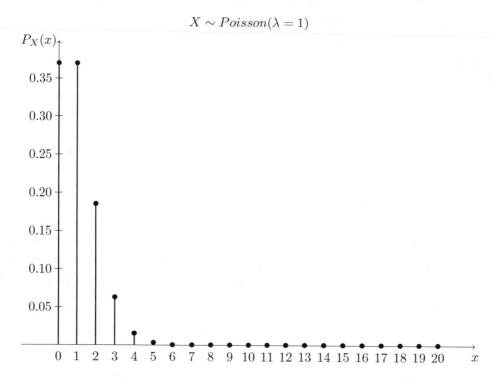

Figure 3.7: PMF of a *Poisson*(1) random variable

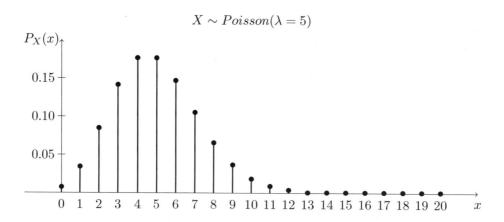

Figure 3.8: PMF of a *Poisson*(5) random variable

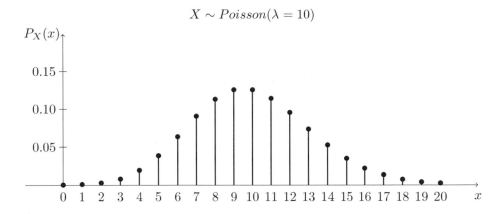

$$X \sim Poisson(\lambda = 10)$$

Figure 3.9: PMF of a $Poisson(10)$ random variable

2. Let Y be the number of emails that I get in the 10-minute interval. Then by the assumption Y is a Poisson random variable with parameter $\lambda = 10(0.2) = 2$,

$$
\begin{aligned}
P(Y > 3) &= 1 - P(Y \leq 3) \\
&= 1 - \big(P_Y(0) + P_Y(1) + P_Y(2) + P_Y(3)\big) \\
&= 1 - e^{-\lambda} - \frac{e^{-\lambda}\lambda}{1!} - \frac{e^{-\lambda}\lambda^2}{2!} - \frac{e^{-\lambda}\lambda^3}{3!} \\
&= 1 - e^{-2} - \frac{2e^{-2}}{1} - \frac{4e^{-2}}{2} - \frac{8e^{-2}}{6} \\
&= 1 - e^{-2}\left(1 + 2 + 2 + \frac{8}{6}\right) \\
&= 1 - \frac{19}{3e^2} \approx 0.1429
\end{aligned}
$$

Poisson as an approximation for binomial: The Poisson distribution can be viewed as the limit of binomial distribution. Suppose $X \sim Binomial(n, p)$ where n is very large and p is very small. In particular, assume that $\lambda = np$ is a positive constant. We show that the PMF of X can be approximated by the PMF of a $Poisson(\lambda)$ random variable. The importance of this is that Poisson PMF is much easier to compute than the binomial. Let us state this as a theorem.

Theorem 3.1. Let $X \sim Binomial(n, p = \frac{\lambda}{n})$, where $\lambda > 0$ is fixed. Then for any $k \in \{0, 1, 2, ...\}$, we have

$$\lim_{n \to \infty} P_X(k) = \frac{e^{-\lambda}\lambda^k}{k!}.$$

Proof. We have

$$\lim_{n\to\infty} P_X(k) = \lim_{n\to\infty} \binom{n}{k} \left(\frac{\lambda}{n}\right)^k \left(1 - \frac{\lambda}{n}\right)^{n-k}$$

$$= \lambda^k \lim_{n\to\infty} \frac{n!}{k!(n-k)!} \left(\frac{1}{n^k}\right) \left(1 - \frac{\lambda}{n}\right)^{n-k}$$

$$= \frac{\lambda^k}{k!} \cdot \lim_{n\to\infty} \left(\left[\frac{n(n-1)(n-2)...(n-k+1)}{n^k}\right] \left[\left(1 - \frac{\lambda}{n}\right)^n\right] \left[\left(1 - \frac{\lambda}{n}\right)^{-k}\right]\right).$$

Note that for a fixed k, we have

$$\lim_{n\to\infty} \frac{n(n-1)(n-2)...(n-k+1)}{n^k} = 1,$$

$$\lim_{n\to\infty} \left(1 - \frac{\lambda}{n}\right)^{-k} = 1,$$

$$\lim_{n\to\infty} \left(1 - \frac{\lambda}{n}\right)^n = e^{-\lambda}.$$

Thus, we conclude

$$\lim_{n\to\infty} P_X(k) = \frac{e^{-\lambda}\lambda^k}{k!}.$$

∎

3.1.6   Solved Problems

1. Let X be a discrete random variable with the following PMF

$$P_X(x) = \begin{cases} 0.1 & \text{for } x = 0.2 \\ 0.2 & \text{for } x = 0.4 \\ 0.2 & \text{for } x = 0.5 \\ 0.3 & \text{for } x = 0.8 \\ 0.2 & \text{for } x = 1 \\ 0 & \text{otherwise} \end{cases}$$

(a) Find R_X, the range of the random variable X.

(b) Find $P(X \le 0.5)$.

(c) Find $P(0.25 < X < 0.75)$.

(d) Find $P(X = 0.2 | X < 0.6)$.

Solution:

(a) The range of X can be found from the PMF. The range of X consists of possible values for X. Here we have

$$R_X = \{0.2, 0.4, 0.5, 0.8, 1\}.$$

(b) The event $X \leq 0.5$ can happen only if X is $0.2, 0.4$, or 0.5. Thus,

$$
\begin{aligned}
P(X \leq 0.5) &= P(X \in \{0.2, 0.4, 0.5\}) \\
&= P(X = 0.2) + P(X = 0.4) + P(X = 0.5) \\
&= P_X(0.2) + P_X(0.4) + P_X(0.5) \\
&= 0.1 + 0.2 + 0.2 = 0.5
\end{aligned}
$$

(c) Similarly, we have

$$
\begin{aligned}
P(0.25 < X < 0.75) &= P(X \in \{0.4, 0.5\}) \\
&= P(X = 0.4) + P(X = 0.5) \\
&= P_X(0.4) + P_X(0.5) \\
&= 0.2 + 0.2 = 0.4
\end{aligned}
$$

(d) This is a conditional probability problem, so we can use our famous formula $P(A|B) = \frac{P(A \cap B)}{P(B)}$. We have

$$
\begin{aligned}
P(X = 0.2 | X < 0.6) &= \frac{P\big((X = 0.2) \text{ and } (X < 0.6)\big)}{P(X < 0.6)} \\
&= \frac{P(X = 0.2)}{P(X < 0.6)} \\
&= \frac{P_X(0.2)}{P_X(0.2) + P_X(0.4) + P_X(0.5)} \\
&= \frac{0.1}{0.1 + 0.2 + 0.2} = 0.2
\end{aligned}
$$

2. I roll two dice and observe two numbers X and Y.

   (a) Find R_X, R_Y and the PMFs of X and Y.
   (b) Find $P(X = 2, Y = 6)$.
   (c) Find $P(X > 3 | Y = 2)$.
   (d) Let $Z = X + Y$. Find the range and PMF of Z.
   (e) Find $P(X = 4 | Z = 8)$.

   Solution:

   (a) We have $R_X = R_Y = \{1, 2, 3, 4, 5, 6\}$. Assuming the dice are fair, all values are equally likely so
   $$
   P_X(k) = \begin{cases} \frac{1}{6} & \text{for } k = 1, 2, 3, 4, 5, 6 \\ 0 & \text{otherwise} \end{cases}
   $$

Similarly for Y,

$$P_Y(k) = \begin{cases} \frac{1}{6} & \text{for } k = 1, 2, 3, 4, 5, 6 \\ 0 & \text{otherwise} \end{cases}$$

(b) Since X and Y are independent random variables, we can write

$$P(X = 2, Y = 6) = P(X = 2)P(Y = 6)$$
$$= \frac{1}{6} \cdot \frac{1}{6} = \frac{1}{36}.$$

(c) Since X and Y are independent, knowing the value of Y does not impact the probabilities for X,

$$P(X > 3 | Y = 2) = P(X > 3)$$
$$= P_X(4) + P_X(5) + P_X(6)$$
$$= \frac{1}{6} + \frac{1}{6} + \frac{1}{6} = \frac{1}{2}.$$

(d) First, we have $R_Z = \{2, 3, 4, ..., 12\}$. Thus, we need to find $P_Z(k)$ for $k = 2, 3, ..., 12$. We have

$$P_Z(2) = P(Z = 2) = P(X = 1, Y = 1)$$
$$= P(X = 1)P(Y = 1) \text{ (since } X \text{ and } Y \text{ are independent)}$$
$$= \frac{1}{6} \cdot \frac{1}{6} = \frac{1}{36};$$
$$P_Z(3) = P(Z = 3) = P(X = 1, Y = 2) + P(X = 2, Y = 1)$$
$$= P(X = 1)P(Y = 2) + P(X = 2)P(Y = 1)$$
$$= \frac{1}{6} \cdot \frac{1}{6} + \frac{1}{6} \cdot \frac{1}{6} = \frac{1}{18};$$
$$P_Z(4) = P(Z = 4) = P(X = 1, Y = 3) + P(X = 2, Y = 2) + P(X = 3, Y = 1)$$
$$= 3 \cdot \frac{1}{36} = \frac{1}{12}.$$

We can continue similarly:

$$P_Z(5) = \frac{4}{36} = \frac{1}{9};$$

$$P_Z(6) = \frac{5}{36};$$

$$P_Z(7) = \frac{6}{36} = \frac{1}{6};$$

$$P_Z(8) = \frac{5}{36};$$

$$P_Z(9) = \frac{4}{36} = \frac{1}{9};$$

$$P_Z(10) = \frac{3}{36} = \frac{1}{12};$$

$$P_Z(11) = \frac{2}{36} = \frac{1}{18};$$

$$P_Z(12) = \frac{1}{36}.$$

It is always a good idea to check our answers by verifying that $\sum_{z \in R_Z} P_Z(z) = 1$. Here, we have

$$\sum_{z \in R_Z} P_Z(z) = \frac{1}{36} + \frac{2}{36} + \frac{3}{36} + \frac{4}{36} + \frac{5}{36} + \frac{6}{36}$$
$$+ \frac{5}{36} + \frac{4}{36} + \frac{3}{36} + \frac{2}{36} + \frac{1}{36}$$
$$= 1.$$

(e) Note that here we cannot argue that X and Z are independent. Indeed, Z seems to completely depend on X, $Z = X + Y$. To find the conditional probability $P(X = 4|Z = 8)$, we use the formula for conditional probability

$$P(X = 4|Z = 8) = \frac{P(X = 4, Z = 8)}{P(Z = 8)}$$
$$= \frac{P(X = 4, Y = 4)}{P(Z = 8)}$$
$$= \frac{P(X = 4)P(Y = 4)}{P(Z = 8)} \quad \text{(since } X \text{ and } Y \text{ are independent)}$$
$$= \frac{\frac{1}{6} \cdot \frac{1}{6}}{\frac{5}{36}}$$
$$= \frac{1}{5}.$$

3. I roll a fair die repeatedly until a number larger than 4 is observed. If N is the total number of times that I roll the die, find $P(N = k)$, for $k = 1, 2, 3, ...$

Solution: In each trial, I may observe a number larger than 4 with probability $\frac{2}{6} = \frac{1}{3}$. Thus, you can think of this experiment as repeating a Bernoulli experiment with success probability $p = \frac{1}{3}$ until you observe the first success. Thus, N is a geometric random variable with parameter $p = \frac{1}{3}$, $N \sim Geometric(\frac{1}{3})$. Hence, we have

$$P_N(k) = \begin{cases} \frac{1}{3}(\frac{2}{3})^{k-1} & \text{for } k = 1, 2, 3, ... \\ 0 & \text{otherwise} \end{cases}$$

4. You take an exam that contains 20 multiple-choice questions. Each question has 4 possible options. You know the answer to 10 questions, but you have no idea about the other 10 questions so you choose answers randomly. Your score X on the exam is the total number of correct answers. Find the PMF of X. What is $P(X > 15)$?

Solution: Let's define the random variable Y as the number of your correct answers to the 10 questions you answer randomly. Then your total score will be $X = Y + 10$. First, let's find the PMF of Y. For each question your success probability is $\frac{1}{4}$. Hence, you perform 10 independent $Bernoulli(\frac{1}{4})$ trials and Y is the number of successes. Thus, we conclude $Y \sim Binomial(10, \frac{1}{4})$, so

$$P_Y(y) = \begin{cases} \binom{10}{y}(\frac{1}{4})^y(\frac{3}{4})^{10-y} & \text{for } y = 0, 1, 2, 3, ..., 10 \\ 0 & \text{otherwise} \end{cases}$$

Now we need to find the PMF of $X = Y + 10$. First note that $R_X = \{10, 11, 12, ..., 20\}$. We can write

$$P_X(10) = P(X = 10) = P(Y + 10 = 10)$$
$$= P(Y = 0) = \binom{10}{0}\left(\frac{1}{4}\right)^0\left(\frac{3}{4}\right)^{10-0} = \left(\frac{3}{4}\right)^{10};$$
$$P_X(11) = P(X = 11) = P(Y + 10 = 11)$$
$$= P(Y = 1) = \binom{10}{1}\left(\frac{1}{4}\right)^1\left(\frac{3}{4}\right)^{10-1} = 10\left(\frac{1}{4}\right)\left(\frac{3}{4}\right)^9.$$

So, you get the idea. In general for $k \in R_X = \{10, 11, 12, ..., 20\}$,

$$P_X(k) = P(X = k) = P(Y + 10 = k)$$

$$= P(Y = k - 10) = \binom{10}{k - 10}\left(\frac{1}{4}\right)^{k-10}\left(\frac{3}{4}\right)^{20-k}.$$

To summarize,

$$P_X(k) = \begin{cases} \binom{10}{k-10}\left(\frac{1}{4}\right)^{k-10}\left(\frac{3}{4}\right)^{20-k} & \text{for } k = 10, 11, 12, ..., 20 \\ 0 & \text{otherwise} \end{cases}$$

In order to calculate $P(X > 15)$, we know we should consider $y = 6, 7, 8, 9, 10$

$$P_Y(y) = \begin{cases} \binom{10}{y}\left(\frac{1}{4}\right)^y\left(\frac{3}{4}\right)^{10-y} & \text{for } y = 6, 7, 8, 9, 10 \\ 0 & \text{otherwise} \end{cases}$$

$$P_X(k) = \begin{cases} \binom{10}{k-10}\left(\frac{1}{4}\right)^{k-10}\left(\frac{3}{4}\right)^{20-k} & \text{for } k = 16, 17, ..., 20 \\ 0 & \text{otherwise} \end{cases}$$

$$P(X > 15) = P_X(16) + P_X(17) + P_X(18) + P_X(19) + P_X(20)$$

$$= \binom{10}{6}(\frac{1}{4})^6(\frac{3}{4})^4 + \binom{10}{7}(\frac{1}{4})^7(\frac{3}{4})^3 + \binom{10}{8}(\frac{1}{4})^8(\frac{3}{4})^2$$

$$+ \binom{10}{9}(\frac{1}{4})^9(\frac{3}{4})^1 + \binom{10}{10}(\frac{1}{4})^{10}(\frac{3}{4})^0.$$

5. Let $X \sim Pascal(m, p)$ and $Y \sim Pascal(l, p)$ be two independent random variables. Define a new random variable as $Z = X + Y$. Find the PMF of Z.

Solution: This problem is very similar to Example 3.7, and we can solve it using the same methods. We will show that $Z \sim Pascal(m + l, p)$. To see this, consider a sequence of Hs and Ts that is the result of independent coin tosses with $P(H) = p$, (Figure 3.10). If we define the random variable X as the number of coin tosses until the mth heads is observed, then $X \sim Pascal(m, p)$. Now, if we look at the rest of the sequence and count the number of heads until we observe l more heads, then the number of coin tosses in this part of the sequence is $Y \sim Pascal(l, p)$. Looking from the beginning, we have repeatedly tossed the coin until we have observed $m + l$ heads. Thus, we conclude the random variable Z defined as $Z = X + Y$ has a $Pascal(m + l, p)$ distribution.

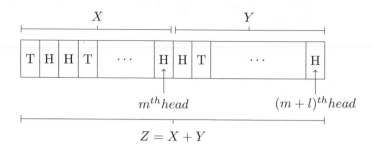

Figure 3.10: Sum of two Pascal random variables

In particular, remember that $Pascal(1, p) = Geometric(p)$. Thus, we have shown that if X and Y are two independent $Geometric(p)$ random variables, then $X+Y$ is a $Pascal(2, p)$ random variable. More generally, we can say that if $X_1, X_2, X_3, ..., X_m$ are m independent $Geometric(p)$ random variables, then the random variable X defined by $X = X_1 + X_2 + ... + X_m$ has a $Pascal(m, p)$ distribution.

6. The number of customers arriving at a grocery store is a Poisson random variable. On average 10 customers arrive per hour. Let X be the number of customers arriving from 10 a.m. to 11 : 30 a.m. What is $P(10 < X \leq 15)$?

Solution: We are looking at an interval of length 1.5 hours, so the number of customers in this interval is $X \sim Poisson(\lambda = 1.5 \times 10 = 15)$. Thus,

$$P(10 < X \leq 15) = \sum_{k=11}^{15} P_X(k)$$

$$= \sum_{k=11}^{15} \frac{e^{-15} 15^k}{k!}$$

$$= e^{-15} \left[\frac{15^{11}}{11!} + \frac{15^{12}}{12!} + \frac{15^{13}}{13!} + \frac{15^{14}}{14!} + \frac{15^{15}}{15!} \right]$$

$$= 0.4496$$

7. Let $X \sim Poisson(\alpha)$ and $Y \sim Poisson(\beta)$ be two independent random variables. Define a new random variable as $Z = X + Y$. Find the PMF of Z.

Solution: First note that since $R_X = \{0, 1, 2, ..\}$ and $R_Y = \{0, 1, 2, ..\}$, we can write $R_Z = \{0, 1, 2, ..\}$. We have

$$P_Z(k) = P(X + Y = k)$$

$$= \sum_{i=0}^{k} P(X + Y = k | X = i) P(X = i) \qquad \text{(law of total probability)}$$

$$= \sum_{i=0}^{k} P(Y = k - i | X = i) P(X = i)$$

$$= \sum_{i=0}^{k} P(Y = k - i) P(X = i) \qquad \text{(since } X \text{ and } Y \text{ are independent)}$$

$$= \sum_{i=0}^{k} \frac{e^{-\beta} \beta^{k-i}}{(k-i)!} \frac{e^{-\alpha} \alpha^{i}}{i!}$$

$$= e^{-(\alpha+\beta)} \sum_{i=0}^{k} \frac{\alpha^{i} \beta^{k-i}}{(k-i)! i!}$$

$$= \frac{e^{-(\alpha+\beta)}}{k!} \sum_{i=0}^{k} \frac{k!}{(k-i)! i!} \alpha^{i} \beta^{k-i}$$

$$= \frac{e^{-(\alpha+\beta)}}{k!} \sum_{i=0}^{k} \binom{k}{i} \alpha^{i} \beta^{k-i}$$

$$= \frac{e^{-(\alpha+\beta)}}{k!} (\alpha + \beta)^{k} \qquad \text{(by the binomial theorem).}$$

Thus, we conclude that $Z \sim Poisson(\alpha + \beta)$.

8. Let X be a discrete random variable with the following PMF

$$P_X(k) = \begin{cases} \frac{1}{4} & \text{for } k = -2 \\ \frac{1}{8} & \text{for } k = -1 \\ \frac{1}{8} & \text{for } k = 0 \\ \frac{1}{4} & \text{for } k = 1 \\ \frac{1}{4} & \text{for } k = 2 \\ 0 & \text{otherwise} \end{cases}$$

I define a new random variable Y as $Y = (X + 1)^2$.

(a) Find the range of Y.

(b) Find the PMF of Y.

Solution: Here, the random variable Y is a function of the random variable X. This means that we perform the random experiment and obtain $X = x$, and then the value of Y is determined as $Y = (x + 1)^2$. Since X is a random variable, Y is also a random variable.

(a) To find R_Y, we note that $R_X = \{-2, -1, 0, 1, 2\}$, and

$$R_Y = \{y = (x+1)^2 | x \in R_X\}$$
$$= \{0, 1, 4, 9\}.$$

(b) Now that we have found $R_Y = \{0, 1, 4, 9\}$, to find the PMF of Y we need to find $P_Y(0), P_Y(1), P_Y(4)$, and $P_Y(9)$:

$$P_Y(0) = P(Y = 0) = P((X + 1)^2 = 0)$$
$$= P(X = -1) = \frac{1}{8};$$
$$P_Y(1) = P(Y = 1) = P((X + 1)^2 = 1)$$
$$= P\big((X = -2) \text{ or } (X = 0)\big)$$
$$= P_X(-2) + P_X(0) = \frac{1}{4} + \frac{1}{8} = \frac{3}{8};$$
$$P_Y(4) = P(Y = 4) = P((X + 1)^2 = 4)$$
$$= P(X = 1) = \frac{1}{4};$$
$$P_Y(9) = P(Y = 9) = P((X + 1)^2 = 9)$$
$$= P(X = 2) = \frac{1}{4}.$$

Again, it is always a good idea to check that $\sum_{y \in R_Y} P_Y(y) = 1$. We have

$$\sum_{y \in R_Y} P_Y(y) = \frac{1}{8} + \frac{3}{8} + \frac{1}{4} + \frac{1}{4} = 1.$$

3.2   More about Discrete Random Variables

So far, we have discussed the PMF for discrete random variables, defined independence, and introduced some important distributions for discrete random variables. In this section, we will continue our discussion of discrete random variables. First, we will define the cumulative distribution function (CDF), which is another way to describe the distribution of a random variable. Then, we will discuss mean and variance. Next, we will discuss functions of random variables. Finally, we will talk about conditioning.

3.2.1 Cumulative Distribution Function (CDF)

The PMF is one way to describe the distribution of a discrete random variable. As we will see later on, PMF cannot be defined for continuous random variables. The cumulative distribution function (CDF) of a random variable is another method to describe the distribution of random variables. The advantage of the CDF is that it can be defined for any kind of random variable (discrete, continuous, and mixed).

Definition 3.10. The cumulative distribution function (CDF) of random variable X is defined as
$$F_X(x) = P(X \le x), \text{ for all } x \in \mathbb{R}.$$

Note that the subscript X indicates that this the CDF of the random variable X. Also, note that the CDF is defined for all $x \in \mathbb{R}$. Let us look at an example.

Example 3.9. I toss a coin twice. Let X be the number of observed heads. Find the CDF of X.

Solution: Note that here $X \sim Binomial(2, \frac{1}{2})$. The range of X is $R_X = \{0, 1, 2\}$ and its PMF is given by

$$P_X(0) = P(X = 0) = \frac{1}{4},$$
$$P_X(1) = P(X = 1) = \frac{1}{2},$$
$$P_X(2) = P(X = 2) = \frac{1}{4}.$$

To find the CDF, we argue as follows. First, note that if $x < 0$, then

$$F_X(x) = P(X \le x) = 0, \text{ for } x < 0.$$

Next, if $x \ge 2$,

$$F_X(x) = P(X \le x) = 1, \text{ for } x \ge 2.$$

Next, if $0 \le x < 1$,

$$F_X(x) = P(X \le x) = P(X = 0) = \frac{1}{4}, \text{ for } 0 \le x < 1.$$

Finally, if $1 \le x < 2$,

$$F_X(x) = P(X \le x) = P(X = 0) + P(X = 1) = \frac{1}{4} + \frac{1}{2} = \frac{3}{4}, \text{ for } 1 \le x < 2.$$

Thus, to summarize, we have

$$F_X(x) = \begin{cases} 0 & \text{for } x < 0 \\ \frac{1}{4} & \text{for } 0 \le x < 1 \\ \frac{3}{4} & \text{for } 1 \le x < 2 \\ 1 & \text{for } x \ge 2 \end{cases}$$

Note that when you are asked to find the CDF of a random variable, you need to find the function for the entire real line. Also, for discrete random variables, we must be mindful of when to use "<" or "≤". Figure 3.11 shows the graph of $F_X(x)$. Note that the CDF is flat between the points in R_X and jumps at each value in the range. The size of the jump at each point is equal to the probability at that point. For example, at point $x = 1$, the CDF jumps from $\frac{1}{4}$ to $\frac{3}{4}$. The size of the jump here is $\frac{3}{4} - \frac{1}{4} = \frac{1}{2}$ which is equal to $P_X(1)$. Also, note that the open and filled circles at point $x = 1$ indicate that $F_X(1) = \frac{3}{4}$ and not $\frac{1}{4}$.

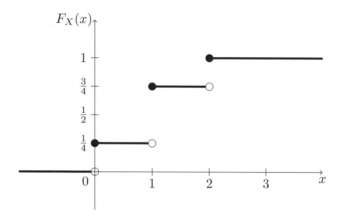

Figure 3.11: CDF for Example 3.9

In general, let X be a discrete random variable with range $R_X = \{x_1, x_2, x_3, ...\}$, such that $x_1 < x_2 < x_3 < ...$ [2] Figure 3.12 shows the general form of the CDF, $F_X(x)$, for such a random variable. We see that the CDF is in the form of a staircase. In particular, note that the CDF starts at 0, i.e., $F_X(-\infty) = 0$. Then, it jumps at each point in the range. In particular, the CDF stays flat between x_k and x_{k+1}, so we can write

$$F_X(x) = F_X(x_k), \text{ for } x_k \le x < x_{k+1}.$$

The CDF jumps at each x_k. In particular, we can write

$$F_X(x_k) - F_X(x_k - \epsilon) = P_X(x_k), \text{ for } \epsilon > 0 \text{ small enough.}$$

---

[2] Here, for simplicity, we assume that the range R_X is bounded from below, i.e., x_1 is the smallest value in R_X. If this is not the case then $F_X(x)$ approaches zero as $x \to -\infty$ rather than hitting zero.

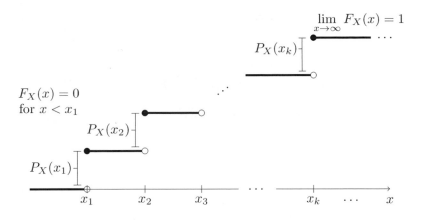

Figure 3.12: CDF of a discrete random variable

Thus, the CDF is always a non-decreasing function, i.e., if $y \geq x$ then $F_X(y) \geq F_X(x)$. Finally, the CDF approaches 1 as x becomes large. We can write

$$\lim_{x \to \infty} F_X(x) = 1.$$

Note that the CDF completely describes the distribution of a discrete random variable. In particular, we can find the PMF values by looking at the values of the jumps in the CDF function. Also, if we have the PMF, we can find the CDF from it. In particular, if $R_X = \{x_1, x_2, x_3, ...\}$, we can write

$$F_X(x) = \sum_{x_k \leq x} P_X(x_k).$$

Now, let us prove a useful formula.

For all $a \leq b$, we have
$$P(a < X \leq b) = F_X(b) - F_X(a) \tag{3.1}$$

To see this, note that for $a \leq b$ we have

$$P(X \leq b) = P(X \leq a) + P(a < X \leq b).$$

Thus,

$$F_X(b) = F_X(a) + P(a < X \leq b).$$

Again, pay attention to the use of "<" and "≤" as they could make a difference in the case of discrete random variables. We will see later that Equation 3.1 is true for all types of random variables (discrete, continuous, and mixed). Note that the CDF gives us $P(X \leq x)$. To find $P(X < x)$, for a discrete random variable, we can simply write

$$P(X < x) = P(X \leq x) - P(X = x) = F_X(x) - P_X(x).$$

Example 3.10. Let X be a discrete random variable with range $R_X = \{1, 2, 3, ...\}$. Suppose the PMF of X is given by

$$P_X(k) = \frac{1}{2^k} \text{ for } k = 1, 2, 3, ...$$

(a) Find and plot the CDF of X, $F_X(x)$.

(b) Find $P(2 < X \leq 5)$.

(c) Find $P(X > 4)$.

Solution: First, note that this is a valid PMF. In particular,

$$\sum_{k=1}^{\infty} P_X(k) = \sum_{k=1}^{\infty} \frac{1}{2^k} = 1 \text{ (geometric sum)}.$$

(a) To find the CDF, note that

$$\text{For } x < 1, \qquad F_X(x) = 0.$$
$$\text{For } 1 \leq x < 2, \ \ F_X(x) = P_X(1) = \frac{1}{2}.$$
$$\text{For } 2 \leq x < 3, \ \ F_X(x) = P_X(1) + P_X(2) = \frac{1}{2} + \frac{1}{4} = \frac{3}{4}.$$

In general we have

$$\text{For } 0 < k \leq x < k+1, \ \ F_X(x) = P_X(1) + P_X(2) + ... + P_X(k)$$
$$= \frac{1}{2} + \frac{1}{4} + ... + \frac{1}{2^k} = \frac{2^k - 1}{2^k}.$$

Figure 3.13 shows the CDF of X.

(b) To find $P(2 < X \leq 5)$, we can write

$$P(2 < X \leq 5) = F_X(5) - F_X(2) = \frac{31}{32} - \frac{3}{4} = \frac{7}{32}.$$

Or equivalently, we can write

$$P(2 < X \leq 5) = P_X(3) + P_X(4) + P_X(5) = \frac{1}{8} + \frac{1}{16} + \frac{1}{32} = \frac{7}{32},$$

which gives the same answer.

(c) To find $P(X > 4)$, we can write

$$P(X > 4) = 1 - P(X \leq 4) = 1 - F_X(4) = 1 - \frac{15}{16} = \frac{1}{16}.$$

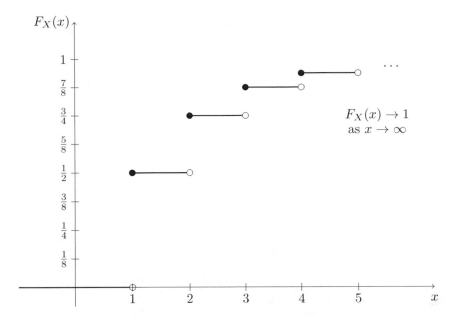

Figure 3.13: CDF of random variable given in Example 3.10

3.2.2 Expectation

If you have a collection of numbers $a_1, a_2, ..., a_N$, their average is a single number that describes the whole collection. Now, consider a random variable X. We would like to define its average, or as it is called in probability, its **expected value** or **mean**. The expected value is defined as the weighted average of the values in the range.

Expected value (=mean=average):

Definition 3.11. Let X be a discrete random variable with range $R_X = \{x_1, x_2, x_3, ...\}$ (finite or countably infinite). The *expected* value of X, denoted by EX, is defined as

$$EX = \sum_{x_k \in R_X} x_k P(X = x_k) = \sum_{x_k \in R_X} x_k P_X(x_k).$$

To understand the concept behind EX, consider a discrete random variable with range $R_X = \{x_1, x_2, x_3, ...\}$. This random variable is a result of random experiment. Suppose that we repeat this experiment a very large number of times, N, and that the trials are independent. Let N_1 be the number of times we observe x_1, N_2 be the number of times we observe x_2, ..., N_k be the number of times we observe x_k, and so on. Since $P(X = x_k) = P_X(x_k)$, we expect that

$$P_X(x_1) \approx \frac{N_1}{N},$$

$$P_X(x_2) \approx \frac{N_2}{N},$$

$$. \quad . \quad .$$

$$P_X(x_k) \approx \frac{N_k}{N},$$

$$. \quad . \quad .$$

In other words, we have $N_k \approx NP_X(x_k)$. Now, if we take the average of the observed values of X, we obtain

$$\begin{aligned} \text{Average} \ &= \ \frac{N_1x_1 + N_2x_2 + N_3x_3 + ...}{N} \\ &\approx \ \frac{x_1NP_X(x_1) + x_2NP_X(x_2) + x_3NP_X(x_3) + ...}{N} \\ &= \ x_1P_X(x_1) + x_2P_X(x_2) + x_3P_X(x_3) + ... = EX. \end{aligned}$$

Thus, the intuition behind EX is that if you repeat the random experiment independently N times and take the average of the observed data, the average gets closer and closer to EX as N gets larger and larger. We sometimes denote EX by μ_X.

Different notations for expected value of X: $EX = E[X] = E(X) = \mu_X$.

Let's compute the expected values of some well-known distributions.

Example 3.11. Let $X \sim Bernoulli(p)$. Find EX.

Solution: For the Bernoulli distribution, the range of X is $R_X = \{0, 1\}$, and $P_X(1) = p$ and $P_X(0) = 1 - p$. Thus,

$$\begin{aligned} EX &= 0 \cdot P_X(0) + 1 \cdot P_X(1) \\ &= 0 \cdot (1 - p) + 1 \cdot p = p. \end{aligned}$$

For a Bernoulli random variable, finding the expectation EX was easy. However, for some random variables, to find the expectation sum, you might need a little algebra. Let's look at another example.

Example 3.12. Let $X \sim Geometric(p)$. Find EX.

Solution: For the geometric distribution, the range is $R_X = \{1, 2, 3, ...\}$ and the PMF is given by

$$P_X(k) = q^{k-1}p, \text{ for } k = 1, 2, ...,$$

where $0 < p < 1$ and $q = 1 - p$. Thus, we can write

$$EX = \sum_{x_k \in R_X} x_k P_X(x_k)$$

$$= \sum_{k=1}^{\infty} kq^{k-1}p$$

$$= p\sum_{k=1}^{\infty} kq^{k-1}.$$

Now, we already know the geometric sum formula

$$\sum_{k=0}^{\infty} x^k = \frac{1}{1-x}, \qquad \text{for } |x| < 1.$$

But we need to find a sum $\sum_{k=1}^{\infty} kq^{k-1}$. Luckily, we can convert the geometric sum to the form we want by taking derivative with respect to x, i.e.,

$$\frac{d}{dx}\sum_{k=0}^{\infty} x^k = \frac{d}{dx}\frac{1}{1-x}, \qquad \text{for } |x| < 1.$$

Thus, we have

$$\sum_{k=0}^{\infty} kx^{k-1} = \frac{1}{(1-x)^2}, \qquad \text{for } |x| < 1.$$

To finish finding the expectation, we can write

$$EX = p\sum_{k=1}^{\infty} kq^{k-1}$$

$$= p\frac{1}{(1-q)^2}$$

$$= p\frac{1}{p^2} = \frac{1}{p}.$$

So, for $X \sim Geometric(p)$, $EX = \frac{1}{p}$. Note that this makes sense intuitively. The random experiment behind the geometric distribution was that we tossed a coin until we observed the first heads, where $P(H) = p$. Here, we found out that on average you need to toss the coin $\frac{1}{p}$ times in this experiment. In particular, if p is small (heads are unlikely), then $\frac{1}{p}$ is large, so you need to toss the coin a large number of times before you observe a heads. Conversely, for large p a few coin tosses usually suffices.

Example 3.13. Let $X \sim Poisson(\lambda)$. Find EX.

Before doing the math, we suggest that you try to guess what the expected value would be. It might be a good idea to think about the examples where the Poisson distribution is used. For the Poisson distribution, the range is $R_X = \{0, 1, 2, \cdots\}$ and the PMF is given by

$$P_X(k) = \frac{e^{-\lambda}\lambda^k}{k!}, \quad \text{for } k = 0, 1, 2, \ldots$$

Thus, we can write

$$EX = \sum_{x_k \in R_X} x_k P_X(x_k)$$

$$= \sum_{k=0}^{\infty} k \frac{e^{-\lambda} \lambda^k}{k!}$$

$$= e^{-\lambda} \sum_{k=1}^{\infty} \frac{\lambda^k}{(k-1)!}$$

$$= e^{-\lambda} \sum_{j=0}^{\infty} \frac{\lambda^{(j+1)}}{j!} \qquad \text{(by letting } j = k - 1)$$

$$= \lambda e^{-\lambda} \sum_{j=0}^{\infty} \frac{\lambda^j}{j!}$$

$$= \lambda e^{-\lambda} e^{\lambda} = \lambda \qquad \text{(Taylor series for } e^{\lambda}).$$

So the expected value is λ. Remember, when we first talked about the Poisson distribution, we introduced its parameter λ as the average number of events. So it is not surprising that the expected value is $EX = \lambda$.

Before looking at more examples, we would like to talk about an important property of expectation, which is *linearity*. Note that if X is a random variable, any function of X is also a random variable, so we can talk about its expected value. For example, if $Y = aX + b$, we can talk about $EY = E[aX + b]$. Or if you define $Y = X_1 + X_2 + \cdots + X_n$, where X_i's are random variables, we can talk about $EY = E[X_1 + X_2 + \cdots + X_n]$. The following theorem states that expectation is linear, which makes it easier to calculate the expected value of linear functions of random variables.

Expectation is linear:

Theorem 3.2. We have

- $E[aX + b] = aEX + b$, for all $a, b \in \mathbb{R}$;

- $E[X_1 + X_2 + \cdots + X_n] = EX_1 + EX_2 + \cdots + EX_n$, for any set of random variables $X_1, X_2, \cdots, X_n$.

We will prove this theorem later on in Chapter 5, but here we would like to emphasize its importance with an example.

Example 3.14. Let $X \sim Binomial(n, p)$. Find EX.

Solution: We provide two ways to solve this problem. One way is as before: we do the math and calculate $EX = \sum_{x_k \in R_X} x_k P_X(x_k)$ which will be a little tedious. A much faster way would be to use linearity of expectation. In particular, remember that if $X_1, X_2, ..., X_n$ are independent *Bernoulli(p)* random variables, then the random variable X defined by $X = X_1 + X_2 + ... + X_n$ has a *Binomial(n, p)* distribution. Thus, we can write

$$\begin{aligned} EX &= E[X_1 + X_2 + \cdots + X_n] \\ &= EX_1 + EX_2 + \cdots + EX_n \qquad \text{by linearity of expectation} \\ &= p + p + \cdots + p = np. \end{aligned}$$

We will provide the direct calculation of $EX = \sum_{x_k \in R_X} x_k P_X(x_k)$ in the Solved Problems section and as you will see it needs a lot more algebra than above. The bottom line is that linearity of expectation can sometimes make our calculations much easier. Let's look at another example.

Example 3.15. Let $X \sim Pascal(m, p)$. Find EX. (*Hint:* Try to write $X = X_1 + X_2 + \cdots + X_m$, such that you already know EX_i.)

Solution: We claim that if X_i's are independent and $X_i \sim Geometric(p)$, for $i = 1, 2, \cdots$, m, then the random variable X defined by $X = X_1 + X_2 + \cdots + X_m$ has $Pascal(m, p)$. To see this, you can look at Problem 5 in Section 3.1.6 and the discussion there. Now, since we already know $EX_i = \frac{1}{p}$, we conclude

$$\begin{aligned} EX &= E[X_1 + X_2 + \cdots + X_m] \\ &= EX_1 + EX_2 + \cdots + EX_m \qquad \text{by linearity of expectation} \\ &= \frac{1}{p} + \frac{1}{p} + \cdots + \frac{1}{p} = \frac{m}{p}. \end{aligned}$$

Again, you can try to find EX directly and as you will see, you need much more algebra compared to using the linearity of expectation.

3.2.3 Functions of Random Variables

If X is a random variable and $Y = g(X)$, then Y itself is a random variable. Thus, we can talk about its PMF, CDF, and expected value. First, note that the range of Y can be written as

$$R_Y = \{g(x) | x \in R_X\}.$$

If we already know the PMF of X, to find the PMF of $Y = g(X)$, we can write

$$\begin{aligned} P_Y(y) &= P(Y = y) \\ &= P(g(X) = y) \\ &= \sum_{x:g(x)=y} P_X(x). \end{aligned}$$

Let's look at an example.

Example 3.16. Let X be a discrete random variable with $P_X(k) = \frac{1}{5}$ for $k = -1, 0, 1, 2, 3$. Let $Y = 2|X|$. Find the range and PMF of Y.

Solution: First, note that the range of Y is

$$R_Y = \{2|x| \,|\, x \in R_X\}$$
$$= \{0, 2, 4, 6\}.$$

To find $P_Y(y)$, we need to find $P(Y = y)$ for $y = 0, 2, 4, 6$. We have

$$P_Y(0) = P(Y = 0) = P(2|X| = 0)$$
$$= P(X = 0) = \frac{1}{5};$$
$$P_Y(2) = P(Y = 2) = P(2|X| = 2)$$
$$= P\big((X = -1) \text{ or } (X = 1)\big)$$
$$= P_X(-1) + P_X(1) = \frac{1}{5} + \frac{1}{5} = \frac{2}{5};$$
$$P_Y(4) = P(Y = 4) = P(2|X| = 4)$$
$$= P(X = 2) + P(X = -2) = \frac{1}{5};$$
$$P_Y(6) = P(Y = 6) = P(2|X| = 6)$$
$$= P(X = 3) + P(X = -3) = \frac{1}{5}.$$

So, to summarize,

$$P_Y(k) = \begin{cases} \frac{1}{5} & \text{for } k = 0, 4, 6 \\ \frac{2}{5} & \text{for } k = 2 \\ 0 & \text{otherwise} \end{cases}$$

Expected Value of a Function of a Random Variable (LOTUS):

Let X be a discrete random variable with PMF $P_X(x)$, and let $Y = g(X)$. Suppose that we are interested in finding EY. One way to find EY is to first find the PMF of Y and then use the expectation formula $EY = E[g(X)] = \sum_{y \in R_Y} y P_Y(y)$. But there is another way which is usually easier. It is called the law of the unconscious statistician (LOTUS).

Law of the unconscious statistician (LOTUS) for discrete random variables:

$$E[g(X)] = \sum_{x_k \in R_X} g(x_k) P_X(x_k) \qquad (3.2)$$

You can prove this by writing $EY = E[g(X)] = \sum_{y \in R_Y} y P_Y(y)$ in terms of $P_X(x)$. In practice it is usually easier to use LOTUS than direct definition when we need $E[g(X)]$.

Example 3.17. Let X be a discrete random variable with range $R_X = \{0, \frac{\pi}{4}, \frac{\pi}{2}, \frac{3\pi}{4}, \pi\}$, such that $P_X(0) = P_X(\frac{\pi}{4}) = P_X(\frac{\pi}{2}) = P_X(\frac{3\pi}{4}) = P_X(\pi) = \frac{1}{5}$. Find $E[\sin(X)]$.

Solution: Using LOTUS, we have

$$E[g(X)] = \sum_{x_k \in R_X} g(x_k)P_X(x_k)$$

$$= \sin(0) \cdot \frac{1}{5} + \sin(\frac{\pi}{4}) \cdot \frac{1}{5} + \sin(\frac{\pi}{2}) \cdot \frac{1}{5} + \sin(\frac{3\pi}{4}) \cdot \frac{1}{5} + \sin(\pi) \cdot \frac{1}{5}$$

$$= 0 \cdot \frac{1}{5} + \frac{\sqrt{2}}{2} \cdot \frac{1}{5} + 1 \cdot \frac{1}{5} + \frac{\sqrt{2}}{2} \cdot \frac{1}{5} + 0 \cdot \frac{1}{5}$$

$$= \frac{\sqrt{2}+1}{5}.$$

Example 3.18. Prove $E[aX + b] = aEX + b$ (linearity of expectation).

Solution: Here $g(X) = aX + b$, so using LOTUS we have

$$E[aX + b] = \sum_{x_k \in R_X} (ax_k + b)P_X(x_k)$$

$$= \sum_{x_k \in R_X} ax_k P_X(x_k) + \sum_{x_k \in R_X} bP_X(x_k)$$

$$= a \sum_{x_k \in R_X} x_k P_X(x_k) + b \sum_{x_k \in R_X} P_X(x_k)$$

$$= aEX + b.$$

3.2.4 Variance

Consider two random variables X and Y with the following PMFs.

$$P_X(x) = \begin{cases} 0.5 & \text{for } x = -100 \\ 0.5 & \text{for } x = 100 \\ 0 & \text{otherwise} \end{cases} \tag{3.3}$$

$$P_Y(y) = \begin{cases} 1 & \text{for } y = 0 \\ 0 & \text{otherwise} \end{cases} \tag{3.4}$$

Note that $EX = EY = 0$. Although both random variables have the same mean value, their distribution is completely different. Y is always equal to its mean of 0, while X is either 100 or -100, quite far from its mean value. The **variance** is a measure of how spread out the distribution of a random variable is. Here, the variance of Y is quite small since its distribution is concentrated at a single value, while the variance of X will be larger since its distribution is more spread out.

The **variance** of a random variable X, with mean $EX = \mu_X$, is defined as

$$\text{Var}(X) = E\big[(X - \mu_X)^2\big].$$

By definition, the variance of X is the average value of $(X - \mu_X)^2$. Since $(X - \mu_X)^2 \geq 0$, the variance is always larger than or equal to zero. A large value of the variance means that $(X - \mu_X)^2$ is often large, so X often takes values far from its mean. This means that the distribution is very spread out. On the other hand, a low variance means that the distribution is concentrated around its average.

Note that if we did not square the difference between X and its mean, the result would be 0. That is

$$E[X - \mu_X] = EX - E[\mu_X] = \mu_X - \mu_X = 0.$$

X is sometimes below its average and sometimes above its average. Thus, $X - \mu_X$ is sometimes negative and sometimes positive, but on average it is zero.

To compute $Var(X) = E\big[(X - \mu_X)^2\big]$, note that we need to find the expected value of $g(X) = (X - \mu_X)^2$, so we can use LOTUS. In particular, we can write

$$\text{Var}(X) = E\big[(X - \mu_X)^2\big] = \sum_{x_k \in R_X} (x_k - \mu_X)^2 P_X(x_k).$$

For example, for X and Y defined in Equations 3.3 and 3.4, we have

$$\text{Var}(X) = (-100 - 0)^2(0.5) + (100 - 0)^2(0.5) = 10,000$$
$$\text{Var}(Y) = (0 - 0)^2(1) = 0.$$

As we expect, X has a very large variance while $\text{Var}(Y) = 0$.

Note that $\text{Var}(X)$ has a different unit than X. For example, if X is measured in *meters* then $\text{Var}(X)$ is in *meters*^{2}. To solve this issue, we define another measure, called the standard deviation, usually shown as σ_X, which is simply the square root of variance.

The **standard deviation** of a random variable X is defined as

$$\text{SD}(X) = \sigma_X = \sqrt{\text{Var}(X)}.$$

The standard deviation of X has the same unit as X. For X and Y defined in Equations 3.3 and 3.4, we have

$$\sigma_X = \sqrt{10,000} = 100$$
$$\sigma_X = \sqrt{0} = 0.$$

Here is a useful formula for computing the variance.

Computational formula for the variance:

$$\text{Var}(X) = E\left[X^2\right] - \left[EX\right]^2 \qquad (3.5)$$

To prove it note that

$$\text{Var}(X) = E\left[(X - \mu_X)^2\right] = E\left[X^2 - 2\mu_X X + \mu_X^2\right]$$
$$= E\left[X^2\right] - 2E\left[\mu_X X\right] + E\left[\mu_X^2\right] \quad \text{by linearity of expectation.}$$

Note that for a given random variable X, μ_X is just a constant real number. Thus, $E\left[\mu_X X\right] = \mu_X E[X] = \mu_X^2$, and $E[\mu_X^2] = \mu_X^2$, so we have

$$\text{Var}(X) = E\left[X^2\right] - 2\mu_X^2 + \mu_X^2$$
$$= E\left[X^2\right] - \mu_X^2.$$

Equation 3.5 is usually easier to work with compared to $\text{Var}(X) = E\left[(X - \mu_X)^2\right]$. To use this equation, we can find $E[X^2] = EX^2$ using LOTUS

$$EX^2 = \sum_{x_k \in R_X} x_k^2 P_X(x_k),$$

and then subtract μ_X^2 to obtain the variance.

Example 3.19. I roll a fair die and let X be the resulting number. Find EX, $\text{Var}(X)$, and σ_X.

Solution: We have $R_X = \{1, 2, 3, 4, 5, 6\}$ and $P_X(k) = \frac{1}{6}$ for $k = 1, 2, ..., 6$. Thus, we have

$$EX = 1 \cdot \frac{1}{6} + 2 \cdot \frac{1}{6} + 3 \cdot \frac{1}{6} + 4 \cdot \frac{1}{6} + 5 \cdot \frac{1}{6} + 6 \cdot \frac{1}{6} = \frac{7}{2};$$

$$EX^2 = 1 \cdot \frac{1}{6} + 4 \cdot \frac{1}{6} + 9 \cdot \frac{1}{6} + 16 \cdot \frac{1}{6} + 25 \cdot \frac{1}{6} + 36 \cdot \frac{1}{6} = \frac{91}{6}.$$

Thus

$$\text{Var}(X) = E\left[X^2\right] - \left(EX\right)^2 = \frac{91}{6} - \left(\frac{7}{2}\right)^2 = \frac{91}{6} - \frac{49}{4} \approx 2.92,$$

$$\sigma_X = \sqrt{\text{Var}(X)} \approx \sqrt{2.92} \approx 1.71$$

Note that variance is not a linear operator. In particular, we have the following theorem.

Theorem 3.3. For a random variable X and real numbers a and b,

$$\text{Var}(aX + b) = a^2\text{Var}(X) \tag{3.6}$$

Proof. If $Y = aX + b$, $EY = aEX + b$. Thus,

$$\begin{aligned}
\text{Var}(Y) &= E[(Y - EY)^2] \\
&= E[(aX + b - aEX - b)^2] \\
&= E[a^2(X - \mu_X)^2] \\
&= a^2 E[(X - \mu_X)^2] \\
&= a^2\text{Var}(X).
\end{aligned}$$

■

From Equation 3.6, we conclude that, for standard deviation, $\text{SD}(aX + b) = |a|\text{SD}(X)$. We mentioned that variance is NOT a linear operation. But there is a very important case, in which variance behaves like a linear operation and that is when we look at the sum of independent random variables.

Theorem 3.4. If $X_1, X_2, \cdots, X_n$ are independent random variables and $X = X_1 + X_2 + \cdots + X_n$, then

$$\text{Var}(X) = \text{Var}(X_1) + \text{Var}(X_2) + \cdots + \text{Var}(X_n) \tag{3.7}$$

We will prove this theorem in Chapter 6, but for now we can look at an example to see how we can use it.

Example 3.20. If $X \sim Binomial(n, p)$, find $\text{Var}(X)$.

Solution: We know that we can write a $Binomial(n, p)$ random variable as the sum of n **independent** $Bernoulli(p)$ random variables, i.e., $X = X_1 + X_2 + \cdots + X_n$. Thus, we conclude

$$\text{Var}(X) = \text{Var}(X_1) + \text{Var}(X_2) + \cdots + \text{Var}(X_n).$$

If $X_i \sim Bernoulli(p)$, then its variance is

$$\text{Var}(X_i) = E[X_i^2] - (EX_i)^2 = 1^2 \cdot p + 0^2 \cdot (1 - p) - p^2 = p(1 - p).$$

Thus,

$$\begin{aligned}
\text{Var}(X) &= p(1 - p) + p(1 - p) + \cdots + p(1 - p) \\
&= np(1 - p).
\end{aligned}$$

3.2.5   Solved Problems

1. Let X be a discrete random variable with the following PMF

$$P_X(x) = \begin{cases} 0.3 & \text{for } x = 3 \\ 0.2 & \text{for } x = 5 \\ 0.3 & \text{for } x = 8 \\ 0.2 & \text{for } x = 10 \\ 0 & \text{otherwise} \end{cases}$$

Find and plot the CDF of X.

Solution: The CDF is defined by $F_X(x) = P(X \leq x)$. We have

$$F_X(x) = \begin{cases} 0 & \text{for } x < 3 \\ P_X(3) = 0.3 & \text{for } 3 \leq x < 5 \\ P_X(3) + P_X(5) = 0.5 & \text{for } 5 \leq x < 8 \\ P_X(3) + P_X(5) + P_X(8) = 0.8 & \text{for } 8 \leq x < 10 \\ 1 & \text{for } x \geq 10 \end{cases}$$

2. Let X be a discrete random variable with the following PMF

$$P_X(k) = \begin{cases} 0.1 & \text{for } k = 0 \\ 0.4 & \text{for } k = 1 \\ 0.3 & \text{for } k = 2 \\ 0.2 & \text{for } k = 3 \\ 0 & \text{otherwise} \end{cases}$$

(a) Find EX.

(b) Find $\text{Var}(X)$.

(c) If $Y = (X - 2)^2$, find EY.

Solution:

(a)

$$\begin{aligned} EX &= \sum_{x_k \in R_X} x_k P_X(x_k) \\ &= 0(0.1) + 1(0.4) + 2(0.3) + 3(0.2) \\ &= 1.6 \end{aligned}$$

(b) We can use $\text{Var}(X) = EX^2 - (EX)^2 = EX^2 - (1.6)^2$. Thus we need to find EX^2. Using LOTUS (Equation 3.2), we have

$$EX^2 = 0^2(0.1) + 1^2(0.4) + 2^2(0.3) + 3^2(0.2) = 3.4$$

Thus, we have

$$\text{Var}(X) = (3.4) - (1.6)^2 = 0.84$$

(c) Again, using LOTUS, we have

$$E(X - 2)^2 = (0 - 2)^2(0.1) + (1 - 2)^2(0.4) + (2 - 2)^2(0.3) + (3 - 2)^2(0.2) = 1.$$

3. Let X be a discrete random variable with PMF

$$P_X(k) = \begin{cases} 0.2 & \text{for } k = 0 \\ 0.2 & \text{for } k = 1 \\ 0.3 & \text{for } k = 2 \\ 0.3 & \text{for } k = 3 \\ 0 & \text{otherwise} \end{cases}$$

Define $Y = X(X - 1)(X - 2)$. Find the PMF of Y.

Solution: First, note that $R_Y = \{x(x - 1)(x - 2) | x \in \{0, 1, 2, 3\}\} = \{0, 6\}$. Thus,

$$\begin{aligned} P_Y(0) &= P(Y = 0) = P\big((X = 0) \text{ or } (X = 1) \text{ or } (X = 2)\big) \\ &= P_X(0) + P_X(1) + P_X(2) \\ &= 0.7; \\ P_Y(6) &= P(X = 3) = 0.3 \end{aligned}$$

Thus,

$$P_Y(k) = \begin{cases} 0.7 & \text{for } k = 0 \\ 0.3 & \text{for } k = 6 \\ 0 & \text{otherwise} \end{cases}$$

4. Let $X \sim Geometric(p)$. Find $E\left[\frac{1}{2^X}\right]$.

Solution: The PMF of X is given by

$$P_X(k) = \begin{cases} pq^{k-1} & \text{for } k = 1, 2, 3, \dots \\ 0 & \text{otherwise} \end{cases}$$

where $q = 1 - p$. Thus,

$$
\begin{aligned}
E\left[\frac{1}{2^X}\right] &= \sum_{k=1}^{\infty} \frac{1}{2^k} P_X(k) \\
&= \sum_{k=1}^{\infty} \frac{1}{2^k} q^{k-1} p \\
&= \frac{p}{2} \sum_{k=1}^{\infty} \left(\frac{q}{2}\right)^{k-1} \\
&= \frac{p}{2} \frac{1}{1 - \frac{q}{2}} \\
&= \frac{p}{1+p}.
\end{aligned}
$$

5. If $X \sim Hypergeometric(b, r, k)$, find EX.

Solution: The PMF of X is given by

$$
P_X(x) = \begin{cases} \dfrac{\binom{b}{x}\binom{r}{k-x}}{\binom{b+r}{k}} & \text{for } x \in R_X \\ 0 & \text{otherwise} \end{cases}
$$

where $R_X = \{\max(0, k - r), \max(0, k - r) + 1, \max(0, k - r) + 2, ..., \min(k, b)\}$. Finding EX directly seems to be very complicated. So let's try to see if we can find an easier way to find EX. In particular, a powerful tool that we have is linearity of expectation. Can we write X as the sum of simpler random variables X_i? To do so, let's remember the random experiment behind the hypergeometric distribution. You have a bag that contains b blue marbles and r red marbles. You choose $k \leq b + r$ marbles at random (without replacement) and let X be the number of blue marbles in your sample. In particular, let's define the indicator random variables X_i as follows:

$$
X_i = \begin{cases} 1 & \text{if the } i\text{th chosen marble is blue} \\ 0 & \text{otherwise} \end{cases}
$$

Then, we can write

$$
X = X_1 + X_2 + \cdots + X_k.
$$

Thus,

$$
EX = EX_1 + EX_2 + \cdots + EX_k.
$$

To find $P(X_i = 1)$, we note that for any particular X_i all marbles are equally likely to be chosen. This is because of symmetry: no marble is more likely to be chosen as the ith marble than any other marble. Therefore,

$$P(X_i = 1) = \frac{b}{b+r}, \qquad \text{for all } i \in \{1, 2, \cdots, k\}.$$

We conclude

$$EX_i = 0 \cdot p(X_i = 0) + 1 \cdot P(X_i = 1)$$
$$= \frac{b}{b+r}.$$

Thus, we have

$$EX = \frac{kb}{b+r}.$$

6. In Example 3.14, we showed that if $X \sim Binomial(n, p)$, then $EX = np$. We found this by writing X as the sum of n $Bernoulli(p)$ random variables. Now, find EX directly using $EX = \sum_{x_k \in R_X} x_k P_X(x_k)$. *Hint:* Use $k\binom{n}{k} = n\binom{n-1}{k-1}$.

Solution: First note that we can prove $k\binom{n}{k} = n\binom{n-1}{k-1}$ by the following combinatorial interpretation: Suppose that from a group of n students we would like to choose a committee of k students, one of whom is chosen to be the committee chair. We can do this

- by choosing k people first (in $\binom{n}{k}$ ways), and then choosing one of them to be the chair (k ways), or

- by choosing the chair first (n possibilities and then choosing $k - 1$ students from the remaining $n - 1$ students (in $\binom{n-1}{k-1}$ ways)).

Thus, we conclude

$$k\binom{n}{k} = n\binom{n-1}{k-1}.$$

Now, let's find EX for $X \sim Binomial(n, p)$.

$$EX = \sum_{k=0}^{n} k \binom{n}{k} p^k q^{n-k}$$

$$= \sum_{k=1}^{n} k \binom{n}{k} p^k q^{n-k}$$

$$= \sum_{k=1}^{n} n \binom{n-1}{k-1} p^k q^{n-k}$$

$$= np \sum_{k=1}^{n} \binom{n-1}{k-1} p^{k-1} q^{n-k}$$

$$= np \sum_{l=0}^{n-1} \binom{n-1}{l} p^l q^{(n-1)-l}$$

$$= np.$$

Note that the last line is true because the $\sum_{l=0}^{n-1} \binom{n-1}{l} p^l q^{(n-1)-l}$ is equal to $\sum_{l=0}^{n-1} P_Y(l)$ for a random variable Y that has $Binomial(n-1, p)$ distribution, hence it is equal to 1.

7. Let X be a discrete random variable with $R_X \subset \{0, 1, 2, ...\}$. Prove

$$EX = \sum_{k=0}^{\infty} P(X > k).$$

Solution: Note that

$$P(X > 0) = P_X(1) + P_X(2) + P_X(3) + P_X(4) + \cdots ,$$
$$P(X > 1) = P_X(2) + P_X(3) + P_X(4) + \cdots ,$$
$$P(X > 2) = P_X(3) + P_X(4) + P_X(5) + \cdots .$$

Thus

$$\sum_{k=0}^{\infty} P(X > k) = P(X > 0) + P(X > 1) + P(X > 2) + \cdots$$

$$= P_X(1) + 2P_X(2) + 3P_X(3) + 4P_X(4) + \cdots$$
$$= EX.$$

8. If $X \sim Poisson(\lambda)$, find $\text{Var}(X)$.

Solution: We already know $EX = \lambda$, thus $\text{Var}(X) = EX^2 - \lambda^2$. You can find EX^2 directly using LOTUS; however, it is a little easier to find $E[X(X-1)]$ first. In particular, using LOTUS we have

$$E[X(X-1)] = \sum_{k=0}^{\infty} k(k-1)P_X(k)$$

$$= \sum_{k=0}^{\infty} k(k-1)e^{-\lambda}\frac{\lambda^k}{k!}$$

$$= e^{-\lambda}\sum_{k=2}^{\infty}\frac{\lambda^k}{(k-2)!}$$

$$= e^{-\lambda}\lambda^2\sum_{k=2}^{\infty}\frac{\lambda^{k-2}}{(k-2)!}$$

$$= e^{-\lambda}\lambda^2 e^{\lambda} = \lambda^2.$$

So, we have $\lambda^2 = E[X(X-1)] = EX^2 - EX = EX^2 - \lambda$. Thus, $EX^2 = \lambda^2 + \lambda$ and we conclude

$$\text{Var}(X) = EX^2 - (EX)^2$$
$$= \lambda^2 + \lambda - \lambda^2$$
$$= \lambda.$$

9. Let X and Y be two independent random variables. Suppose that we know $\text{Var}(2X - Y) = 6$ and $\text{Var}(X + 2Y) = 9$. Find $\text{Var}(X)$ and $\text{Var}(Y)$.

Solution: Let's first make sure we understand what $\text{Var}(2X - Y)$ and $\text{Var}(X + 2Y)$ mean. They are $\text{Var}(Z)$ and $\text{Var}(W)$, where the random variables Z and W are defined as $Z = 2X - Y$ and $W = X + 2Y$. Since X and Y are independent random variables, then $2X$ and $-Y$ are independent random variables. Also, X and $2Y$ are independent random variables. Thus, by using Equation 3.7, we can write

$$\text{Var}(2X - Y) = \text{Var}(2X) + \text{Var}(-Y) = 4\text{Var}(X) + \text{Var}(Y) = 6;$$
$$\text{Var}(X + 2Y) = \text{Var}(X) + \text{Var}(2Y) = \text{Var}(X) + 4\text{Var}(Y) = 9.$$

By solving for $\text{Var}(X)$ and $\text{Var}(Y)$, we obtain $\text{Var}(X) = 1$ and $\text{Var}(Y) = 2$.

3.3   End of Chapter Problems

1. Let X be a discrete random variable with the following PMF:

$$P_X(x) = \begin{cases} \frac{1}{2} & \text{for } x = 0 \\ \frac{1}{3} & \text{for } x = 1 \\ \frac{1}{6} & \text{for } x = 2 \\ 0 & \text{otherwise} \end{cases}$$

   (a) Find R_X, the range of the random variable X.
   (b) Find $P(X \geq 1.5)$.
   (c) Find $P(0 < X < 2)$.
   (d) Find $P(X = 0 | X < 2)$

2. Let X be the number of the cars being repaired at a repair shop. We have the following information:

   - At any time, there are at most 3 cars being repaired.
   - The probability of having 2 cars at the shop is the same as the probability of having one car.
   - The probability of having no car at the shop is the same as the probability of having 3 cars.
   - The probability of having 1 or 2 cars is half of the probability of having 0 or 3 cars.

   Find the PMF of X.

3. I roll two dice and observe two numbers X and Y. If $Z = X - Y$, find the range and PMF of Z.

4. Let X and Y be two independent discrete random variables with the following PMFs:

$$P_X(k) = \begin{cases} \frac{1}{4} & \text{for } k = 1 \\ \frac{1}{8} & \text{for } k = 2 \\ \frac{1}{8} & \text{for } k = 3 \\ \frac{1}{2} & \text{for } k = 4 \\ 0 & \text{otherwise} \end{cases}$$

   and

$$P_Y(k) = \begin{cases} \frac{1}{6} & \text{for } k = 1 \\ \frac{1}{6} & \text{for } k = 2 \\ \frac{1}{3} & \text{for } k = 3 \\ \frac{1}{3} & \text{for } k = 4 \\ 0 & \text{otherwise} \end{cases}$$

   (a) Find $P(X \leq 2 \text{ and } Y \leq 2)$.

   (b) Find $P(X > 2 \text{ or } Y > 2)$.

   (c) Find $P(X > 2|Y > 2)$.

   (d) Find $P(X < Y)$.

5. 50 students live in a dormitory. The parking lot has the capacity for 30 cars. If each student has a car with probability $\frac{1}{2}$ (independently from other students), what is the probability that there won't be enough parking spaces for all the cars?

6. (The Matching Problem) N guests arrive at a party. Each person is wearing a hat. We collect all the hats and then randomly redistribute the hats, giving each person one of the N hats randomly. Let X_N be the number of people who receive their own hats. Find the PMF of X_N.

   Hint: We previously found that (Problem 7 in Section 2.1.5)

$$P(X_N = 0) = \frac{1}{2!} - \frac{1}{3!} + \frac{1}{4!} - \cdots (-1)^N \frac{1}{N!}, \quad \text{for} \quad N = 1, 2, \cdots.$$

   Using this, find $P(X_N = k)$ for all $k \in \{0, 1, \cdots N\}$.

7. For each of the following random variables, find $P(X > 5)$, $P(2 < X \leq 6)$ and $P(X > 5|X < 8)$. You do not need to provide the numerical values for your answers. In other words, you can leave your answers in the form of sums.

   (a) $X \sim Geometric(\frac{1}{5})$

   (b) $X \sim Binomial(10, \frac{1}{3})$

   (c) $X \sim Pascal(3, \frac{1}{2})$

   (d) $X \sim Hypergeometric(10, 10, 12)$

   (e) $X \sim Poisson(5)$

8. Suppose you take a pass-fail test repeatedly. Let S_k be the event that you are successful in your k^{th} try, and F_k be the event that you fail the test in your k^{th} try. On your first try, you have a 50 percent chance of passing the test:

$$P(S_1) = 1 - P(F_1) = \frac{1}{2}.$$

Assume that as you take the test more often, your chance of failing the test goes down. In particular,

$$P(F_k) = \frac{1}{2} \cdot P(F_{k-1}), \quad \text{for } k = 2, 3, 4, \cdots$$

However, the result of different exams are independent. Suppose you take the test repeatedly until you pass the test for the first time. Let X be the total number of tests you take, so $Range(X) = \{1, 2, 3, \cdots\}$.

   (a) Find $P(X = 1), P(X = 2)$, and $P(X = 3)$.

   (b) Find a general formula for $P(X = k)$ for $k = 1, 2, \cdots$.

(c) Find the probability that you take the test more than 2 times.

(d) Given that you take the test more than once, find the probability that you take the test exactly twice.

9. In this problem, we would like to show that the geometric random variable is **memoryless**. Let $X \sim Geometric(p)$. Show that

$$P(X > m + l | X > m) = P(X > l), \qquad \text{for } m, l \in \{1, 2, 3, \cdots\}.$$

We can interpret this in the following way: Remember that a geometric random variable can be obtained by tossing a coin repeatedly until observing the first heads. If we toss the coin several times, and do not observe a heads, from now on it is like we start all over again. In other words, the failed coin tosses do not impact the distribution of waiting time from this point forward. The reason for this is that the coin tosses are independent.

10. An urn consists of 20 red balls and 30 green balls. We choose 10 balls at random from the urn. The sampling is done **without** replacement (repetition not allowed).

(a) What is the probability that there will be exactly 4 red balls among the chosen balls?

(b) Given that there are at least 3 red balls among the chosen balls, what is the probability that there are exactly 4 red balls?

11. The number of emails that I get in a weekday (Monday through Friday) can be modeled by a Poisson distribution with an average of $\frac{1}{6}$ emails per minute. The number of emails that I receive on weekends (Saturday and Sunday) can be modeled by a Poisson distribution with an average of $\frac{1}{30}$ emails per minute.

(a) What is the probability that I get no emails in an interval of length 4 hours on a Sunday?

(b) A random day is chosen (all days of the week are equally likely to be selected), and a random interval of length one hour is selected on the chosen day. It is observed that I did not receive any emails in that interval. What is the probability that the chosen day is a weekday?

12. Let X be a discrete random variable with the following PMF:

$$P_X(x) = \begin{cases} 0.2 & \text{for } x = -2 \\ 0.3 & \text{for } x = -1 \\ 0.2 & \text{for } x = 0 \\ 0.2 & \text{for } x = 1 \\ 0.1 & \text{for } x = 2 \\ 0 & \text{otherwise} \end{cases}$$

Find and plot the CDF of X.

13. Let X be a discrete random variable with the following CDF:

$$F_X(x) = \begin{cases} 0 & \text{for } x < 0 \\ \frac{1}{6} & \text{for } 0 \leq x < 1 \\ \frac{1}{2} & \text{for } 1 \leq x < 2 \\ \frac{3}{4} & \text{for } 2 \leq x < 3 \\ 1 & \text{for } x \geq 3 \end{cases}$$

Find the range and PMF of X.

14. Let X be a discrete random variable with the following PMF

$$P_X(x) = \begin{cases} 0.5 & \text{for } k = 1 \\ 0.3 & \text{for } k = 2 \\ 0.2 & \text{for } k = 3 \\ 0 & \text{otherwise} \end{cases}$$

(a) Find EX.

(b) Find $\text{Var}(X)$ and $SD(X)$.

(c) If $Y = \frac{2}{X}$, find EY.

15. Let $X \sim Geometric(\frac{1}{3})$, and let $Y = |X - 5|$. Find the range and PMF of Y.

16. Let X be a discrete random variable with the following PMF

$$P_X(k) = \begin{cases} \frac{1}{21} & \text{for } k \in \{-10, -9, \cdots, -1, 0, 1, \cdots, 9, 10\} \\ 0 & \text{otherwise} \end{cases}$$

The random variable $Y = g(X)$ is defined as

$$Y = g(X) = \begin{cases} 0 & \text{if } X \leq 0 \\ X & \text{if } 0 < X \leq 5 \\ 5 & \text{otherwise} \end{cases}$$

Find the PMF of Y.

17. Let $X \sim Geometric(p)$. Find $\text{Var}(X)$.

18. Let $X \sim Pascal(m, p)$. Find $\text{Var}(X)$.

19. Suppose that $Y = -2X + 3$. If we know $EY = 1$ and $EY^2 = 9$, find EX and $\text{Var}(X)$.

20. There are 1000 households in a town. Specifically, there are 100 households with one member, 200 households with 2 members, 300 households with 3 members, 200 households with 4 members, 100 households with 5 members, and 100 households with 6 members. Thus, the total number of people living in the town is

$$N = 100 \cdot 1 + 200 \cdot 2 + 300 \cdot 3 + 200 \cdot 4 + 100 \cdot 5 + 100 \cdot 6 = 3300 \qquad (3.8)$$

(a) We pick a household at random, and define the random variable X as the number of people in the chosen household. Find the PMF and the expected value of X.

(b) We pick a person in the town at random, and define the random variable Y as the number of people in the household where the chosen person lives. Find the PMF and the expected value of Y.

21. (Coupon collector's problem [8]) Suppose that there are N different types of coupons. Each time you get a coupon, it is equally likely to be any of the N possible types. Let X be the number of coupons you will need to get before having observed each coupon at least once.

(a) Show that you can write $X = X_0 + X_1 + \cdots + X_{N-1}$, where $X_i \sim Geometric(\frac{N-i}{N})$.

(b) Find EX.

22. (St. Petersburg Paradox) Here is a famous problem called the St. Petersburg Paradox. Wikipedia states the problem as follows [9]:"A casino offers a game of chance for a single player in which a fair coin is tossed at each stage. The pot starts at 1 dollar and is doubled every time a head appears. The first time a tail appears, the game ends and the player wins whatever is in the pot. Thus the player wins 1 dollar if a tail appears on the first toss, 2 dollars if a head appears on the first toss and a tail on the second, 4 dollars if a head appears on the first two tosses and a tail on the third, 8 dollars if a head appears on the first three tosses and a tail on the fourth, and so on. In short, the player wins 2^{k-1} dollars if the coin is tossed k times until the first tail appears. What would be a fair price to pay the casino for entering the game?"

(a) Let X be the amount of money (in dollars) that the player wins. Find EX.

(b) What is the probability that the player wins more than 65 dollars?

(c) Now suppose that the casino only has a finite amount of money. Specifically, suppose that the maximum amount of the money that the casino will pay you is 2^{30} dollars (around 1.07 billion dollars). That is, if you win more than 2^{30} dollars, the casino is going to pay you only 2^{30} dollars. Let Y be the money that the player wins in this case. Find EY.

23. Let X be a random variable with mean $EX = \mu$. Define the function $f(\alpha)$ as

$$f(\alpha) = E[(X - \alpha)^2].$$

Find the value of α that minimizes f.

24. You are offered to play the following game. You roll a fair die once and observe the result which is shown by the random variable X. At this point, you can stop the game and win X dollars. You can also choose to roll the die for the second time to observe the value Y. In this case, you will win Y dollars. Let W be the value that you win in this game. What strategy do you use to maximize EW? What is the maximum EW you can achieve using your strategy?

25. The **median** of a random variable X is defined as any number m that satisfies both of the following conditions:

$$P(X \geq m) \geq \frac{1}{2} \qquad \text{and} \qquad P(X \leq m) \geq \frac{1}{2}.$$

Note that the median of X is not necessarily unique. Find the median of X if

(a) The PMF of X is given by

$$P_X(k) = \begin{cases} 0.4 & \text{for } k = 1 \\ 0.3 & \text{for } k = 2 \\ 0.3 & \text{for } k = 3 \\ 0 & \text{otherwise} \end{cases}$$

(b) X is the result of a rolling of a fair die.

(c) $X \sim Geometric(p)$, where $0 < p < 1$.

Chapter 4

Continuous and Mixed Random Variables

Remember that discrete random variables can take only a countable number of possible values. On the other hand, a continuous random variable X has a range in the form of an interval or a union of non-overlapping intervals on the real line (possibly the whole real line). Also, for any $x \in \mathbb{R}$, $P(X = x) = 0$. Thus, we need to develop new tools to deal with continuous random variables. The good news is that the theory of continuous random variables is completely analogous to the theory of discrete random variables. Indeed, if we want to oversimplify things, we might say the following: take any formula about discrete random variables, and then replace *sums* with *integrals*, and replace *PMFs* with probability density functions (*PDFs*), and you will get the corresponding formula for continuous random variables. Of course, there is a little bit more to the story and that's why we need a chapter to discuss it. In this chapter, we will also introduce mixed random variables that are mixtures of discrete and continuous random variables.

4.1 Continuous Random Variables and their Distributions

We have in fact already seen examples of continuous random variables before, e.g., Example 1.14. Let us look at the same example with just a little bit different wording.

Example 4.1. I choose a real number uniformly at random in the interval $[a, b]$, and call it X. By uniformly at random, we mean all intervals in $[a, b]$ that have the same length must have the same probability. Find the CDF of X.

Solution: As we mentioned, this is almost exactly the same problem as Example 1.14, with the difference being, in that problem, we considered the interval from 1 to 2. In that example, we saw that all individual points have probability 0, i.e., $P(X = x) = 0$ for all x. Also, the uniformity implies that the probability of an interval of length l in $[a, b]$ must be proportional to its length:

$$P(X \in [x_1, x_2]) \propto (x_2 - x_1), \qquad \text{where } a \leq x_1 \leq x_2 \leq b.$$

Since $P(X \in [a, b]) = 1$, we conclude

$$P(X \in [x_1, x_2]) = \frac{x_2 - x_1}{b - a}, \qquad \text{where } a \leq x_1 \leq x_2 \leq b.$$

Now, let us find the CDF. By definition $F_X(x) = P(X \leq x)$, thus we immediately have

$$F_X(x) = 0, \qquad \text{for } x < a,$$
$$F_X(x) = 1, \qquad \text{for } x \geq b.$$

For $a \leq x \leq b$, we have

$$F_X(x) = P(X \leq x)$$
$$= P(X \in [a, x])$$
$$= \frac{x - a}{b - a}.$$

Thus, to summarize

$$F_X(x) = \begin{cases} 0 & \text{for } x < a \\ \frac{x-a}{b-a} & \text{for } a \leq x \leq b \\ 1 & \text{for } x > b \end{cases} \qquad (4.1)$$

Note that here it does not matter if we use "<" or "$\leq$", as each individual point has probability zero, so for example $P(X < 2) = P(X \leq 2)$. Figure 4.1 shows the CDF of X. As we expect the CDF starts at zero and ends at 1.

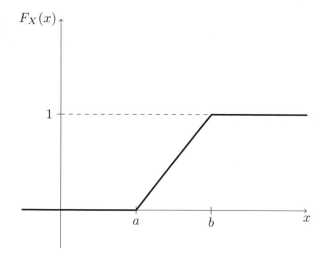

Figure 4.1: CDF for a continuous random variable uniformly distributed over $[a, b]$.

One big difference that we notice here as opposed to discrete random variables is that the CDF is a continuous function, i.e., it does not have any jumps. Remember that jumps in the

CDF correspond to points x for which $P(X = x) > 0$. Thus, the fact that the CDF does not have jumps is consistent with the fact that $P(X = x) = 0$ for all x. Indeed, we have the following definition for continuous random variables.

Definition 4.1. A random variable X with CDF $F_X(x)$ is said to be continuous if $F_X(x)$ is a continuous function for all $x \in \mathbb{R}$.

We will also assume that the CDF of a continuous random variable is differentiable almost everywhere in $\mathbb{R}$.

4.1.1 Probability Density Function (PDF)

To determine the distribution of a discrete random variable we can either provide its PMF or CDF. For continuous random variables, the CDF is well-defined so we can provide the CDF. However, the PMF does not work for continuous random variables because, for a continuous random variable, $P(X = x) = 0$ for all $x \in \mathbb{R}$. Instead, we can usually define the **probability density function (PDF)**. The PDF is the **density** of probability rather than the probability mass. The concept is very similar to mass density in physics: its unit is probability per unit length. To get a feeling for PDF, consider a continuous random variable X and define the function $f_X(x)$ as follows[1]:

$$f_X(x) = \lim_{\Delta \to 0^+} \frac{P(x < X \le x + \Delta)}{\Delta}.$$

The function $f_X(x)$ gives us the probability density at point x. It is the limit of the probability of the interval $(x, x + \Delta]$ divided by the length of the interval as the length of the interval goes to 0. Remember that

$$P(x < X \le x + \Delta) = F_X(x + \Delta) - F_X(x).$$

So, we conclude that

$$
\begin{aligned}
f_X(x) &= \lim_{\Delta \to 0} \frac{F_X(x + \Delta) - F_X(x)}{\Delta} \\
&= \frac{dF_X(x)}{dx} = F_X'(x), \qquad \text{if } F_X(x) \text{ is differentiable at } x.
\end{aligned}
$$

Thus, we have the following definition for the PDF of continuous random variables:

---

[1] wherever the limit exists

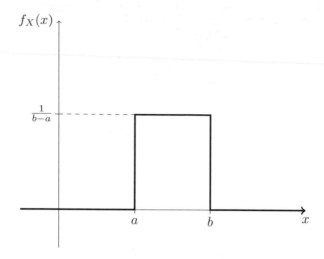

Figure 4.2: PDF for a continuous random variable uniformly distributed over $[a, b]$.

Definition 4.2. Consider a continuous random variable X with an absolutely continuous CDF $F_X(x)$. The function $f_X(x)$ defined by

$$f_X(x) = \frac{dF_X(x)}{dx} = F'_X(x), \qquad \text{if } F_X(x) \text{ is differentiable at } x$$

is called the probability density function (PDF) of X.

Let us find the PDF of the uniform random variable X discussed in Example 4.1. This random variable is said to have $Uniform(a, b)$ distribution. The CDF of X is given in Equation 4.1. By taking the derivative, we obtain

$$f_X(x) = \begin{cases} \frac{1}{b-a} & a < x < b \\ 0 & x < a \text{ or } x > b \end{cases}$$

Note that the CDF is not differentiable at points a and b. Nevertheless, as we will discuss later on, this is not important. Figure 4.2 shows the PDF of X. As we see, the value of the PDF is constant in the interval from a to b. That is why we say X is uniformly distributed over $[a, b]$.

The uniform distribution is the simplest continuous random variable you can imagine. For other types of continuous random variables the PDF is non-uniform. Note that for small values of δ we can write

$$P(x < X \leq x + \delta) \approx f_X(x)\delta.$$

Thus, if $f_X(x_1) > f_X(x_2)$, we can say $P(x_1 < X \leq x_1 + \delta) > P(x_2 < X \leq x_2 + \delta)$, i.e., the value of X is more likely to be around x_1 than x_2. This is another way of interpreting the PDF.

Since the PDF is the derivative of the CDF, the CDF can be obtained from PDF by integration (assuming absolute continuity):

$$F_X(x) = \int_{-\infty}^{x} f_X(u)du.$$

Also, we have

$$P(a < X \leq b) = F_X(b) - F_X(a) = \int_{a}^{b} f_X(u)du.$$

In particular, if we integrate over the entire real line, we must get 1, i.e.,

$$\int_{-\infty}^{\infty} f_X(u)du = 1.$$

That is, the area under the PDF curve must be equal to one. We can see that this holds for the uniform distribution since the area under the curve in Figure 4.2 is one. Note that $f_X(x)$ is density of probability, so it must be larger than or equal to zero, but it can be larger than 1. Let us summarize the properties of the PDF.

Consider a continuous random variable X with PDF $f_X(x)$. We have

1. $f_X(x) \geq 0$ for all $x \in \mathbb{R}$.

2. $\int_{-\infty}^{\infty} f_X(u)du = 1$.

3. $P(a < X \leq b) = F_X(b) - F_X(a) = \int_{a}^{b} f_X(u)du$.

4. More generally, for a set A, $P(X \in A) = \int_{A} f_X(u)du$.

In the last item above, the set A must satisfy some mild conditions which are almost always satisfied in practice. An example of set A could be a union of some disjoint intervals. For example, if you want to find $P(X \in [0,1] \cup [3,4])$, you can write

$$P(X \in [0,1] \cup [3,4]) = \int_{0}^{1} f_X(u)du + \int_{3}^{4} f_X(u)du.$$

Let us look at an example to practice the above concepts.

Example 4.2. Let X be a continuous random variable with the following PDF

$$f_X(x) = \begin{cases} ce^{-x} & x \geq 0 \\ 0 & \text{otherwise} \end{cases}$$

where c is a positive constant.

(a) Find c.

(b) Find the CDF of X, $F_X(x)$.

(c) Find $P(1 < X < 3)$.

  Solution:

(a) To find c, we can use Property 2 above, in particular

$$1 = \int_{-\infty}^{\infty} f_X(u)du = \int_{0}^{\infty} ce^{-u}du$$

$$= c\left[-e^{-x}\right]_{0}^{\infty} = c.$$

Thus, we must have $c = 1$.

(b) To find the CDF of X, we use $F_X(x) = \int_{-\infty}^{x} f_X(u)du$, so for $x < 0$, we obtain $F_X(x) = 0$. For $x \geq 0$, we have

$$F_X(x) = \int_{0}^{x} e^{-u}du = 1 - e^{-x}.$$

Thus,

$$F_X(x) = \begin{cases} 1 - e^{-x} & x \geq 0 \\ 0 & \text{otherwise} \end{cases}$$

(c) We can find $P(1 < X < 3)$ using either the CDF or the PDF. If we use the CDF, we have

$$P(1 < X < 3) = F_X(3) - F_X(1) = \left[1 - e^{-3}\right] - \left[1 - e^{-1}\right] = e^{-1} - e^{-3}.$$

  Equivalently, we can use the PDF. We have

$$P(1 < X < 3) = \int_{1}^{3} f_X(t)dt =$$

$$\int_{1}^{3} e^{-t}dt = e^{-1} - e^{-3}.$$

Range:

The range of a random variable X is the set of possible values of the random variable. If X is a continuous random variable, we can define the range of X as the set of real numbers x for which the PDF is larger than zero, i.e,[2]

$$R_X = \{x | f_X(x) > 0\}.$$

---

[2]The set R_X defined here might not exactly show all possible values of X, but the difference is practically unimportant.

4.1.2 Expected Value and Variance

As we mentioned earlier, the theory of continuous random variables is very similar to the theory of discrete random variables. In particular, usually summations are replaced by integrals and PMFs are replaced by PDFs. The proofs and ideas are very analogous to the discrete case, so sometimes we state the results without mathematical derivations for the purpose of brevity.

Remember that the expected value of a discrete random variable can be obtained as

$$EX = \sum_{x_k \in R_X} x_k P_X(x_k).$$

Now, by replacing the sum by an integral and PMF by PDF, we can write the definition of expected value of a continuous random variable as

$$EX = \int_{-\infty}^{\infty} x f_X(x) dx$$

Example 4.3. Let $X \sim Uniform(a, b)$. Find EX.

Solution: As we saw, the PDF of X is given by

$$f_X(x) = \begin{cases} \frac{1}{b-a} & a < x < b \\ 0 & x < a \text{ or } x > b \end{cases}$$

so to find its expected value, we can write

$$EX = \int_{-\infty}^{\infty} x f_X(x) dx$$
$$= \int_{a}^{b} x(\frac{1}{b-a}) dx$$
$$= \frac{1}{b-a} \left[\frac{1}{2} x^2 \right]_{a}^{b} dx = \frac{a+b}{2}.$$

This result is intuitively reasonable: since X is uniformly distributed over the interval $[a, b]$, we expect its mean to be the middle point, i.e., $EX = \frac{a+b}{2}$.

Example 4.4. Let X be a continuous random variable with PDF

$$f_X(x) = \begin{cases} 2x & 0 \le x \le 1 \\ 0 & \text{otherwise} \end{cases}$$

Find the expected value of X.

Solution: We have

$$EX = \int_{-\infty}^{\infty} x f_X(x) dx$$

$$= \int_0^1 x(2x) dx$$

$$= \int_0^1 2x^2 dx = \frac{2}{3}.$$

Expected Value of a Function of a Continuous Random Variable:

Remember the law of the unconscious statistician (LOTUS) for discrete random variables:

$$E[g(X)] = \sum_{x_k \in R_X} g(x_k) P_X(x_k) \tag{4.2}$$

Now, by changing the sum to integral and changing the PMF to PDF, we will obtain the similar formula for continuous random variables.

Law of the unconscious statistician (LOTUS) for continuous random variables:

$$E[g(X)] = \int_{-\infty}^{\infty} g(x) f_X(x) dx \tag{4.3}$$

As we have seen before, expectation is a linear operation, thus we always have

- $E[aX + b] = aEX + b$, for all $a, b \in \mathbb{R}$, and

- $E[X_1 + X_2 + ... + X_n] = EX_1 + EX_2 + ... + EX_n$, for any set of random variables $X_1, X_2, ..., X_n$.

Example 4.5. Let X be a continuous random variable with PDF

$$f_X(x) = \begin{cases} x + \frac{1}{2} & 0 \leq x \leq 1 \\ 0 & \text{otherwise} \end{cases}$$

Find $E(X^n)$, where $n \in \mathbb{N}$.
Solution: Using LOTUS we have

$$E[X^n] = \int_{-\infty}^{\infty} x^n f_X(x) dx$$

$$= \int_0^1 x^n (x + \frac{1}{2}) dx$$

$$= \left[\frac{1}{n+2} x^{n+2} + \frac{1}{2(n+1)} x^{n+1} \right]_0^1$$

$$= \frac{3n+4}{2(n+1)(n+2)}.$$

Variance:

Remember that the variance of any random variable is defined as

$$\text{Var}(X) = E\big[(X - \mu_X)^2\big] = EX^2 - (EX)^2.$$

So for a continuous random variable, we can write

$$\text{Var}(X) = E\big[(X - \mu_X)^2\big] = \int_{-\infty}^{\infty} (x - \mu_X)^2 f_X(x)dx$$

$$= EX^2 - (EX)^2 = \int_{-\infty}^{\infty} x^2 f_X(x)dx - \mu_X^2$$

Also remember that for $a, b \in \mathbb{R}$, we always have

$$\text{Var}(aX + b) = a^2 \text{Var}(X) \tag{4.4}$$

Example 4.6. Let X be a continuous random variable with PDF

$$f_X(x) = \begin{cases} \frac{3}{x^4} & x \geq 1 \\ 0 & \text{otherwise} \end{cases}$$

Find the mean and variance of X.
Solution:

$$E[X] = \int_{-\infty}^{\infty} x f_X(x)dx$$

$$= \int_{1}^{\infty} \frac{3}{x^3} dx$$

$$= \left[-\frac{3}{2}x^{-2} \right]_{1}^{\infty} = \frac{3}{2}.$$

Next, we find EX^2 using LOTUS,

$$E[X^2] = \int_{-\infty}^{\infty} x^2 f_X(x)dx$$

$$= \int_{1}^{\infty} \frac{3}{x^2} dx$$

$$= \left[-3x^{-1} \right]_{1}^{\infty} = 3.$$

Thus, we have

$$\text{Var}(X) = EX^2 - (EX)^2 = 3 - \frac{9}{4} = \frac{3}{4}.$$

4.1.3   Functions of Continuous Random Variables

If X is a continuous random variable and $Y = g(X)$ is a function of X, then Y itself is a random variable. Thus, we should be able to find the CDF and PDF of Y. It is usually more straightforward to start from the CDF and then to find the PDF by taking the derivative of the CDF. Note that before differentiating the CDF, we should check that the CDF is continuous. As we will see later, the function of a continuous random variable might be a non-continuous random variable. Let's look at an example.

Example 4.7. Let X be a $Uniform(0, 1)$ random variable, and let $Y = e^X$.

(a) Find the CDF of Y.

(b) Find the PDF of Y.

(c) Find EY.

Solution: First, note that we already know the CDF and PDF of X. In particular,

$$F_X(x) = \begin{cases} 0 & \text{for } x < 0 \\ x & \text{for } 0 \leq x \leq 1 \\ 1 & \text{for } x > 1 \end{cases}$$

It is a good idea to think about the range of Y before finding the distribution. Since e^x is an increasing function of x and $R_X = [0, 1]$, we conclude that $R_Y = [1, e]$. So we immediately know that

$$F_Y(y) = P(Y \leq y) = 0, \qquad \text{for } y < 1$$
$$F_Y(y) = P(Y \leq y) = 1, \qquad \text{for } y \geq e.$$

(a) To find $F_Y(y)$ for $y \in [1, e]$, we can write

$$\begin{aligned} F_Y(y) &= P(Y \leq y) \\ &= P(e^X \leq y) \\ &= P(X \leq \ln y) & \text{since } e^x \text{ is an increasing function} \\ &= F_X(\ln y) = \ln y & \text{since } 0 \leq \ln y \leq 1. \end{aligned}$$

To summarize

$$F_Y(y) = \begin{cases} 0 & \text{for } y < 1 \\ \ln y & \text{for } 1 \leq y < e \\ 1 & \text{for } y \geq e \end{cases}$$

(b) The above CDF is a continuous function, so we can obtain the PDF of Y by taking its derivative. We have

$$f_Y(y) = F'_Y(y) = \begin{cases} \frac{1}{y} & \text{for } 1 \leq y \leq e \\ 0 & \text{otherwise} \end{cases}$$

Note that the CDF is not technically differentiable at points 1 and e, but as we mentioned earlier we do not worry about this since this is a continuous random variable and changing the PDF at a finite number of points does not change probabilities.

(c) To find the EY, we can directly apply LOTUS,

$$E[Y] = E[e^X] = \int_{-\infty}^{\infty} e^x f_X(x)dx$$
$$= \int_0^1 e^x dx$$
$$= e - 1.$$

For this problem, we could also find EY using the PDF of Y,

$$E[Y] = \int_{-\infty}^{\infty} y f_Y(y)dy$$
$$= \int_1^e y\frac{1}{y}dy$$
$$= e - 1.$$

Note that since we have already found the PDF of Y it did not matter which method we used to find $E[Y]$. However, if the problem only asked for $E[Y]$ without asking for the PDF of Y, then using LOTUS would be much easier.

Example 4.8. Let $X \sim Uniform(-1, 1)$ and $Y = X^2$. Find the CDF and PDF of Y.

Solution: First, we note that $R_Y = [0, 1]$. As usual, we start with the CDF. For $y \in [0, 1]$, we have

$$F_Y(y) = P(Y \leq y)$$
$$= P(X^2 \leq y)$$
$$= P(-\sqrt{y} \leq X \leq \sqrt{y})$$
$$= \frac{\sqrt{y} - (-\sqrt{y})}{1 - (-1)} \qquad \text{since } X \sim Uniform(-1, 1)$$
$$= \sqrt{y}.$$

Thus, the CDF of Y is given by

$$F_Y(y) = \begin{cases} 0 & \text{for } y < 0 \\ \sqrt{y} & \text{for } 0 \leq y \leq 1 \\ 1 & \text{for } y > 1 \end{cases}$$

Note that the CDF is a continuous function of Y, so Y is a continuous random variable. Thus, we can find the PDF of Y by differentiating $F_Y(y)$,

$$f_Y(y) = F_Y'(y) = \begin{cases} \frac{1}{2\sqrt{y}} & \text{for } 0 \leq y \leq 1 \\ 0 & \text{otherwise} \end{cases}$$

The Method of Transformations:

So far, we have discussed how we can find the distribution of a function of a continuous random variable starting from finding the CDF. If we are interested in finding the PDF of $Y = g(X)$, and the function g satisfies some properties, it might be easier to use a method called the method of transformations. Let's start with the case where g is a function satisfying the following properties:

- $g(x)$ is differentiable;

- $g(x)$ is a strictly increasing function, that is, if $x_1 < x_2$, then $g(x_1) < g(x_2)$.

Now, let X be a continuous random variable and $Y = g(X)$. We will show that you can directly find the PDF of Y using the following formula.

$$f_Y(y) = \begin{cases} \frac{f_X(x_1)}{g'(x_1)} = f_X(x_1) . \frac{dx_1}{dy} & \text{where } g(x_1) = y \\ 0 & \text{if } g(x) = y \text{ does not have a solution.} \end{cases}$$

Note that since g is strictly increasing, its inverse function g^{-1} is well defined. That is, for each $y \in R_Y$, there exists a unique x_1 such that $g(x_1) = y$. We can write $x_1 = g^{-1}(y)$. To find the CDF of Y, we can write

$$\begin{aligned} F_Y(y) &= P(Y \leq y) \\ &= P(g(X) \leq y) \\ &= P(X < g^{-1}(y)) \qquad\qquad \text{since } g \text{ is strictly increasing} \\ &= F_X(g^{-1}(y)). \end{aligned}$$

To find the PDF of Y, we differentiate

$$\begin{aligned} f_Y(y) &= \frac{d}{dy} F_X(x_1) \qquad\qquad\qquad \text{where } g(x_1) = y \\ &= \frac{dx_1}{dy} \cdot F'_X(x_1) \\ &= f_X(x_1) \frac{dx_1}{dy} \\ &= \frac{f_X(x_1)}{g'(x_1)} \qquad\qquad\qquad \text{since } \frac{dx}{dy} = \frac{1}{\frac{dy}{dx}}. \end{aligned}$$

We can repeat the same argument for the case where g is strictly decreasing. In that case, $g'(x_1)$ will be negative, so we need to use $|g'(x_1)|$. Thus, we can state the following theorem for a strictly monotonic function. (A function $g : \mathbb{R} \to \mathbb{R}$ is called strictly monotonic if it is strictly increasing or strictly decreasing.)

Theorem 4.1. Suppose that X is a continuous random variable and $g : \mathbb{R} \to \mathbb{R}$ is a strictly monotonic differentiable function. Let $Y = g(X)$. Then the PDF of Y is given by

$$f_Y(y) = \begin{cases} \frac{f_X(x_1)}{|g'(x_1)|} = f_X(x_1).|\frac{dx_1}{dy}| & \text{where } g(x_1) = y \\ 0 & \text{if } g(x) = y \text{ does not have a solution} \end{cases} \tag{4.5}$$

To see how to use the formula, let's look at an example.

Example 4.9. Let X be a continuous random variable with PDF

$$f_X(x) = \begin{cases} 4x^3 & 0 < x \leq 1 \\ 0 & \text{otherwise} \end{cases}$$

and let $Y = \frac{1}{X}$. Find $f_Y(y)$.

Solution: First note that $R_Y = [1, \infty)$. Also, note that $g(x)$ is a strictly decreasing and differentiable function on $(0, 1]$, so we may use Equation 4.5. We have $g'(x) = -\frac{1}{x^2}$. For any $y \in [1, \infty)$, $x_1 = g^{-1}(y) = \frac{1}{y}$. So, for $y \in [1, \infty)$

$$f_Y(y) = \frac{f_X(x_1)}{|g'(x_1)|}$$

$$= \frac{4x_1^3}{|-\frac{1}{x_1^2}|}$$

$$= 4x_1^5 = \frac{4}{y^5}.$$

Thus, we conclude

$$f_Y(y) = \begin{cases} \frac{4}{y^5} & y \geq 1 \\ 0 & \text{otherwise} \end{cases}$$

Theorem 4.1 can be extended to a more general case. In particular, if g is not monotonic, we can usually divide it into a finite number of monotonic differentiable functions. Figure 4.3 shows a function g that has been divided into monotonic parts. We may state a more general form of Theorem 4.1.

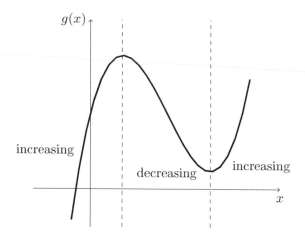

Figure 4.3: Partitioning a function to monotone parts

Theorem 4.2. Consider a continuous random variable X with domain R_X, and let $Y = g(X)$. Suppose that we can partition R_X into a finite number of intervals such that $g(x)$ is strictly monotone and differentiable on each partition. Then the PDF of Y is given by

$$f_Y(y) = \sum_{i=1}^{n} \frac{f_X(x_i)}{|g'(x_i)|} = \sum_{i=1}^{n} f_X(x_i) \cdot \left| \frac{dx_i}{dy} \right| \qquad (4.6)$$

where $x_1, x_2, ..., x_n$ are real solutions to $g(x) = y$.

Let us look at an example to see how we can use Theorem 4.2.

Example 4.10. Let X be a continuous random variable with PDF

$$f_X(x) = \frac{1}{\sqrt{2\pi}} e^{-\frac{x^2}{2}}, \qquad \text{for all } x \in \mathbb{R},$$

and let $Y = X^2$. Find $f_Y(y)$.

Solution: We note that the function $g(x) = x^2$ is strictly decreasing on the interval $(-\infty, 0)$, strictly increasing on the interval $(0, \infty)$, and differentiable on both intervals, $g'(x) = 2x$. Thus, we can use Equation 4.6. First, note that $R_Y = (0, \infty)$. Next, for any $y \in (0, \infty)$ we have two solutions for $y = g(x)$, in particular,[3]

$$x_1 = \sqrt{y}, \quad x_2 = -\sqrt{y}.$$

---

[3]Note that although $0 \in R_X$ it has not been included in our partition of R_X. This is not a problem, since $P(X = 0) = 0$. Indeed, in the statement of Theorem 4.2, we could replace R_X by $R_X - A$, where A is any set for which $P(X \in A) = 0$. In particular, this is convenient when we exclude the endpoints of the intervals.

Thus, we have

$$
\begin{aligned}
f_Y(y) &= \frac{f_X(x_1)}{|g'(x_1)|} + \frac{f_X(x_2)}{|g'(x_2)|} \\
&= \frac{f_X(\sqrt{y})}{|2\sqrt{y}|} + \frac{f_X(-\sqrt{y})}{|-2\sqrt{y}|} \\
&= \frac{1}{2\sqrt{2\pi y}}e^{-\frac{y}{2}} + \frac{1}{2\sqrt{2\pi y}}e^{-\frac{y}{2}} \\
&= \frac{1}{\sqrt{2\pi y}}e^{-\frac{y}{2}}, \text{ for } y \in (0, \infty).
\end{aligned}
$$

4.1.4 Solved Problems

1. Let X be a random variable with PDF given by

$$
f_X(x) = \begin{cases} cx^2 & |x| \leq 1 \\ 0 & \text{otherwise} \end{cases}
$$

(a) Find the constant c.

(b) Find EX and $\text{Var}(X)$.

(c) Find $P(X \geq \frac{1}{2})$.

Solution:

(a) To find c, we can use $\int_{-\infty}^{\infty} f_X(u)du = 1$:

$$
\begin{aligned}
1 = \int_{-\infty}^{\infty} f_X(u)du &= \int_{-1}^{1} cu^2 du \\
&= \frac{2}{3}c.
\end{aligned}
$$

Thus, we must have $c = \frac{3}{2}$.

(b) To find EX, we can write

$$
\begin{aligned}
EX &= \int_{-1}^{1} u f_X(u)du \\
&= \frac{3}{2} \int_{-1}^{1} u^3 du \\
&= 0.
\end{aligned}
$$

In fact, we could have guessed $EX = 0$ because the PDF is symmetric around $x = 0$. To find $\text{Var}(X)$, we have

$$\text{Var}(X) = EX^2 - (EX)^2 = EX^2$$
$$= \int_{-1}^{1} u^2 f_X(u) du$$
$$= \frac{3}{2} \int_{-1}^{1} u^4 du$$
$$= \frac{3}{5}.$$

(c) To find $P(X \geq \frac{1}{2})$, we can write

$$P(X \geq \frac{1}{2}) = \frac{3}{2} \int_{\frac{1}{2}}^{1} x^2 dx = \frac{7}{16}.$$

2. Let X be a continuous random variable with PDF given by

$$f_X(x) = \frac{1}{2} e^{-|x|}, \qquad \text{for all } x \in \mathbb{R}.$$

If $Y = X^2$, find the CDF of Y.

Solution: First, we note that $R_Y = [0, \infty)$. For $y \in [0, \infty)$, we have

$$F_Y(y) = P(Y \leq y)$$
$$= P(X^2 \leq y)$$
$$= P(-\sqrt{y} \leq X \leq \sqrt{y})$$
$$= \int_{-\sqrt{y}}^{\sqrt{y}} \frac{1}{2} e^{-|x|} dx$$
$$= \int_{0}^{\sqrt{y}} e^{-x} dx$$
$$= 1 - e^{-\sqrt{y}}.$$

Thus,

$$F_Y(y) = \begin{cases} 1 - e^{-\sqrt{y}} & y \geq 0 \\ 0 & \text{otherwise} \end{cases}$$

3. Let X be a continuous random variable with PDF

$$f_X(x) = \begin{cases} 4x^3 & 0 < x \leq 1 \\ 0 & \text{otherwise} \end{cases}$$

Find $P(X \leq \frac{2}{3} | X > \frac{1}{3})$.

Solution: We have

$$P(X \leq \frac{2}{3} | X > \frac{1}{3}) = \frac{P(\frac{1}{3} < X \leq \frac{2}{3})}{P(X > \frac{1}{3})}$$

$$= \frac{\int_{\frac{1}{3}}^{\frac{2}{3}} 4x^3 dx}{\int_{\frac{1}{3}}^{1} 4x^3 dx}$$

$$= \frac{3}{16}.$$

4. Let X be a continuous random variable with PDF

$$f_X(x) = \begin{cases} x^2 \left(2x + \frac{3}{2}\right) & 0 < x \leq 1 \\ 0 & \text{otherwise} \end{cases}$$

If $Y = \frac{2}{X} + 3$, find $\text{Var}(Y)$.

Solution: First, note that

$$\text{Var}(Y) = \text{Var}\left(\frac{2}{X} + 3\right) = 4\text{Var}\left(\frac{1}{X}\right), \qquad \text{using Equation 4.4}$$

Thus, it suffices to find $\text{Var}(\frac{1}{X}) = E[\frac{1}{X^2}] - (E[\frac{1}{X}])^2$. Using LOTUS, we have

$$E\left[\frac{1}{X}\right] = \int_0^1 x\left(2x + \frac{3}{2}\right) dx = \frac{17}{12}$$

$$E\left[\frac{1}{X^2}\right] = \int_0^1 \left(2x + \frac{3}{2}\right) dx = \frac{5}{2}.$$

Thus, $\text{Var}(\frac{1}{X}) = E[\frac{1}{X^2}] - (E[\frac{1}{X}])^2 = \frac{71}{144}$. So, we obtain

$$\text{Var}(Y) = 4\text{Var}\left(\frac{1}{X}\right) = \frac{71}{36}.$$

5. Let X be a <u>positive</u> continuous random variable. Prove that $EX = \int_0^\infty P(X \geq x) dx$.

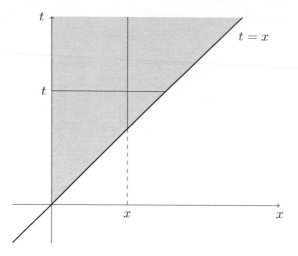

Figure 4.4: The shaded area shows the region of the double integral of Problem 5.

Solution: We have

$$P(X \geq x) = \int_x^\infty f_X(t)dt.$$

Thus, we need to show that

$$\int_0^\infty \int_x^\infty f_X(t)dtdx = EX.$$

The left hand side is a double integral. In particular, it is the integral of $f_X(t)$ over the shaded region in Figure 4.4.

We can take the integral with respect to x or t. Thus, we can write

$$
\begin{aligned}
\int_0^\infty \int_x^\infty f_X(t)dtdx &= \int_0^\infty \int_0^t f_X(t)dxdt \\
&= \int_0^\infty f_X(t) \left(\int_0^t 1dx \right) dt \\
&= \int_0^\infty tf_X(t)dt = EX \qquad \text{since } X \text{ is a positive random variable.}
\end{aligned}
$$

6. Let $X \sim Uniform(-\frac{\pi}{2}, \pi)$ and $Y = \sin(X)$. Find $f_Y(y)$.

Solution: Here $Y = g(X)$, where g is a differentiable function. Although g is not monotone, it can be divided to a finite number of regions in which it is monotone. Thus, we can use Equation 4.6. We note that since $R_X = [-\frac{\pi}{2}, \pi]$, $R_Y = [-1, 1]$.

By looking at the plot of $g(x) = \sin(x)$ over $[-\frac{\pi}{2}, \pi]$, we notice that for $y \in (0, 1)$ there are two solutions to $y = g(x)$, while for $y \in (-1, 0)$, there is only one solution. In particular, if $y \in (0, 1)$, we have two solutions: $x_1 = \arcsin(y)$, and $x_2 = \pi - \arcsin(y)$. If $y \in (-1, 0)$ we have one solution, $x_1 = \arcsin(y)$. Thus, for $y \in (-1, 0)$, we have

$$
\begin{aligned}
f_Y(y) &= \frac{f_X(x_1)}{|g'(x_1)|} \\
&= \frac{f_X(\arcsin(y))}{|\cos(\arcsin(y))|} \\
&= \frac{\frac{2}{3\pi}}{\sqrt{1 - y^2}}.
\end{aligned}
$$

For $y \in (0, 1)$, we have

$$
\begin{aligned}
f_Y(y) &= \frac{f_X(x_1)}{|g'(x_1)|} + \frac{f_X(x_2)}{|g'(x_2)|} \\
&= \frac{f_X(\arcsin(y))}{|\cos(\arcsin(y))|} + \frac{f_X(\pi - \arcsin(y))}{|\cos(\pi - \arcsin(y))|} \\
&= \frac{\frac{2}{3\pi}}{\sqrt{1 - y^2}} + \frac{\frac{2}{3\pi}}{\sqrt{1 - y^2}} \\
&= \frac{4}{3\pi\sqrt{1 - y^2}}.
\end{aligned}
$$

To summarize, we can write

$$
f_Y(y) = \begin{cases} \frac{2}{3\pi\sqrt{1-y^2}} & -1 < y < 0 \\ \frac{4}{3\pi\sqrt{1-y^2}} & 0 < y < 1 \\ 0 & \text{otherwise} \end{cases}
$$

4.2 Special Distributions

Similar to the case of discrete random variables, there are some continuous random variables that are used frequently in practice, so here we would like to introduce some of the well-known continuous random variables.

4.2.1   Uniform Distribution

We have already seen the uniform distribution. In particular, we have the following definition:

A continuous random variable X is said to have a *Uniform* distribution over the interval $[a, b]$, shown as $X \sim Uniform(a, b)$, if its PDF is given by

$$f_X(x) = \begin{cases} \frac{1}{b-a} & a < x < b \\ 0 & x < a \text{ or } x > b \end{cases}$$

We have already found the CDF and the expected value of the uniform distribution. In particular, we know that if $X \sim Uniform(a, b)$, then its CDF is given by Equation 4.1, and its mean is given by

$$EX = \frac{a+b}{2}.$$

To find the variance, we can find EX^2 using LOTUS:

$$EX^2 = \int_{-\infty}^{\infty} x^2 f_X(x) dx$$

$$= \int_a^b x^2 \left(\frac{1}{b-a} \right) dx$$

$$= \frac{a^2 + ab + b^2}{3}.$$

Therefore,

$$Var(X) = EX^2 - (EX)^2$$

$$= \frac{(b-a)^2}{12}.$$

4.2.2   Exponential Distribution

The exponential distribution is one of the widely used continuous distributions. It is often used to model the time elapsed between events. We will now mathematically define the exponential distribution, and derive its mean and expected value. Then we will develop the intuition for the distribution and discuss several interesting properties that it has.

A continuous random variable X is said to have an *exponential* distribution with parameter $\lambda > 0$, shown as $X \sim Exponential(\lambda)$, if its PDF is given by

$$f_X(x) = \begin{cases} \lambda e^{-\lambda x} & x > 0 \\ 0 & \text{otherwise} \end{cases}$$

Figure 4.5 shows the PDF of exponential distribution for several values of λ.

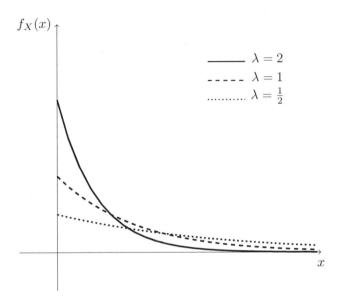

Figure 4.5: PDF of the exponential random variable

It is convenient to use the unit step function defined as

$$u(x) = \begin{cases} 1 & x \geq 0 \\ 0 & \text{otherwise} \end{cases}$$

so we can write the PDF of an $Exponential(\lambda)$ random variable as

$$f_X(x) = \lambda e^{-\lambda x} u(x).$$

Let us find its CDF, mean and variance. For $x > 0$, we have

$$F_X(x) = \int_0^x \lambda e^{-\lambda t} dt = 1 - e^{-\lambda x}.$$

So we can express the CDF as

$$F_X(x) = \left(1 - e^{-\lambda x}\right) u(x).$$

Let $X \sim Exponential(\lambda)$. We can find its expected value as follows, using integration by parts:

$$
\begin{aligned}
EX &= \int_0^\infty x\lambda e^{-\lambda x}dx \\
&= \frac{1}{\lambda}\int_0^\infty ye^{-y}dy \qquad \text{(choosing } y = \lambda x\text{)} \\
&= \frac{1}{\lambda}\left[-e^{-y} - ye^{-y}\right]_0^\infty \\
&= \frac{1}{\lambda}.
\end{aligned}
$$

Now let's find $\text{Var}(X)$. We have

$$
\begin{aligned}
EX^2 &= \int_0^\infty x^2\lambda e^{-\lambda x}dx \\
&= \frac{1}{\lambda^2}\int_0^\infty y^2e^{-y}dy \\
&= \frac{1}{\lambda^2}\left[-2e^{-y} - 2ye^{-y} - y^2e^{-y}\right]_0^\infty \\
&= \frac{2}{\lambda^2}.
\end{aligned}
$$

Thus, we obtain

$$
\text{Var}(X) = EX^2 - (EX)^2 = \frac{2}{\lambda^2} - \frac{1}{\lambda^2} = \frac{1}{\lambda^2}.
$$

If $X \sim Exponential(\lambda)$, then $EX = \frac{1}{\lambda}$ and $\text{Var}(X) = \frac{1}{\lambda^2}$.

An interesting property of the exponential distribution is that it can be viewed as a continuous analogue of the geometric distribution. To see this, recall the random experiment behind the geometric distribution: you toss a coin (repeat a Bernoulli experiment) until you observe the first heads (success). Now, suppose that the coin tosses are Δ seconds apart and in each toss the probability of success is $p = \Delta\lambda$. Also suppose that Δ is very small, so the coin tosses are very close together in time and the probability of success in each trial is very low. Let X be the time you observe the first success. We will show in the Solved Problems section that the distribution of X converges to $Exponential(\lambda)$ as Δ approaches zero.

To get some intuition for this interpretation of the exponential distribution, suppose you are waiting for an event to happen. For example, you are at a store and are waiting for the next customer. In each millisecond, the probability that a new customer enters the store is very small. You can imagine that, in each millisecond, a coin (with a very small $P(H)$) is tossed, and if it lands heads a new customers enters. If you toss a coin every millisecond, the time until a new customer arrives approximately follows an exponential distribution.

The above interpretation of the exponential is useful in better understanding the properties of the exponential distribution. The most important of these properties is that the exponential distribution is **memoryless**. To see this, think of an exponential random variable in the sense of tossing a lot of coins until observing the first heads. If we toss the coin several times and do not observe a heads, from now on it is like we start all over again. In other words, the failed coin tosses do not impact the distribution of waiting time from now on. The reason for this is that the coin tosses are independent. We can state this formally as follows:

If X is exponential with parameter $\lambda > 0$, then X is a *memoryless* random variable, that is

$$P(X > x + a \mid X > a) = P(X > x), \qquad \text{for } a, x \geq 0.$$

From the point of view of waiting time until arrival of a customer, the memoryless property means that it does not matter how long you have waited so far. If you have not observed a customer until time a, the distribution of waiting time (from time a) until the next customer is the same as when you started at time zero. Let us prove the memoryless property of the exponential distribution.

$$
\begin{aligned}
P(X > x + a | X > a) &= \frac{P(X > x + a, X > a)}{P(X > a)} \\
&= \frac{P(X > x + a)}{P(X > a)} \\
&= \frac{1 - F_X(x + a)}{1 - F_X(a)} \\
&= \frac{e^{-\lambda(x+a)}}{e^{-\lambda a}} \\
&= e^{-\lambda x} \\
&= P(X > x).
\end{aligned}
$$

4.2.3   Normal (Gaussian) Distribution

The normal distribution is by far the most important probability distribution. One of the main reasons for that is the *Central Limit Theorem* (CLT) that we will discuss later in the book. To give you an idea, the CLT states that if you add a large number of random variables, the distribution of the sum will be approximately normal under certain conditions. The importance of this result comes from the fact that many random variables in real life can be expressed as the sum of a large number of random variables and, by the CLT, we can argue that distribution of the sum should be normal. The CLT is one of the most important results in probability and we will discuss it later on. Here, we will introduce normal random variables.

We first define the **standard normal random variable**. We will then see that we can obtain other normal random variables by *scaling* and *shifting* a standard normal random variable.

A continuous random variable Z is said to be a *standard normal (standard Gaussian)* random variable, shown as $Z \sim N(0,1)$, if its PDF is given by

$$f_Z(z) = \frac{1}{\sqrt{2\pi}} e^{-\frac{z^2}{2}}, \qquad \text{for all } z \in \mathbb{R}.$$

The $\frac{1}{\sqrt{2\pi}}$ is there to make sure that the area under the PDF is equal to one. We will verify that this holds in the Solved Problems section. Figure 4.6 shows the PDF of the standard normal random variable.

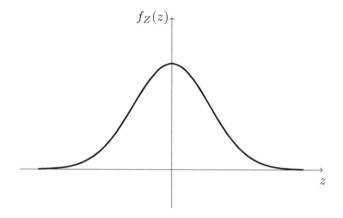

Figure 4.6: PDF of the standard normal random variable

Let us find the mean and variance of the standard normal distribution. To do that, we will use a simple useful fact. Consider a function $g(u) : \mathbb{R} \to \mathbb{R}$. If $g(u)$ is an odd function, i.e., $g(-u) = -g(u)$, and $|\int_0^\infty g(u)du| < \infty$, then

$$\int_{-\infty}^{\infty} g(u)du = 0.$$

For our purpose, let

$$g(u) = u^{2k+1} e^{-\frac{u^2}{2}},$$

where $k = 0, 1, 2, \ldots$ Then $g(u)$ is an odd function. Also $|\int_0^\infty g(u)du| < \infty$. One way to see this is to note that $g(u)$ decays faster than the function e^{-u} and since $|\int_0^\infty e^{-u}du| < \infty$, we conclude that $|\int_0^\infty g(u)du| < \infty$. Now, let Z be a standard normal random variable. Then, we have

$$EZ^{2k+1} = \frac{1}{\sqrt{2\pi}} \int_{-\infty}^{\infty} u^{2k+1} e^{-\frac{u^2}{2}} du = 0,$$

for all $k \in \{0, 1, 2, .., \}$. Thus, we have shown that for a standard normal random variable Z, we have

$$EZ = EZ^3 = EZ^5 = ... = 0.$$

In particular, the standard normal distribution has zero mean. This is not surprising as we can see from Figure 4.6 that the PDF is symmetric around the origin, so we expect that $EZ = 0$. Next, let's find EZ^2.

$$
\begin{aligned}
EZ^2 &= \frac{1}{\sqrt{2\pi}} \int_{-\infty}^{\infty} u^2 e^{-\frac{u^2}{2}} \, du \\
&= \frac{1}{\sqrt{2\pi}} \left[-ue^{-\frac{u^2}{2}} \right]_{-\infty}^{\infty} + \frac{1}{\sqrt{2\pi}} \int_{-\infty}^{\infty} e^{-\frac{u^2}{2}} \, du \qquad \text{(integration by parts)} \\
&= \int_{-\infty}^{\infty} \frac{1}{\sqrt{2\pi}} e^{-\frac{u^2}{2}} \, du \\
&= 1.
\end{aligned}
$$

The last equality holds because we are integrating the standard normal PDF from $-\infty$ to ∞. Thus, we conclude that for a standard normal random variable Z, we have

$$\text{Var}(Z) = 1.$$

So far we have shown the following:

If $Z \sim N(0, 1)$, then $EZ = 0$ and $\text{Var}(Z) = 1$.

CDF of the standard normal: To find the CDF of the standard normal distribution, we need to integrate the PDF function. In particular, we have

$$F_Z(z) = \frac{1}{\sqrt{2\pi}} \int_{-\infty}^{z} e^{-\frac{u^2}{2}} \, du.$$

This integral does not have a closed form solution. Nevertheless, because of the importance of the normal distribution, the values of $F_Z(z)$ have been tabulated and many calculators and software packages have this function. We usually denote the standard normal CDF by Φ.

The CDF of the standard normal distribution is denoted by the Φ function:

$$\Phi(x) = P(Z \le x) = \frac{1}{\sqrt{2\pi}} \int_{-\infty}^{x} e^{-\frac{u^2}{2}} \, du.$$

As we will see in a moment, the CDF of any normal random variable can be written in terms of the Φ function, so the Φ function is widely used in probability. Figure 4.7 shows the Φ function.

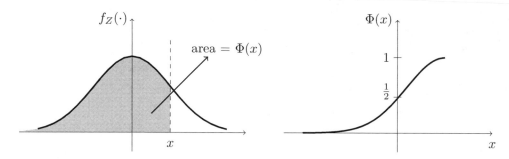

Figure 4.7: The Φ function (CDF of standard normal).

Here are some properties of the Φ function that can be shown from its definition.

1. $\lim\limits_{x \to \infty} \Phi(x) = 1, \quad \lim\limits_{x \to -\infty} \Phi(x) = 0$;

2. $\Phi(0) = \frac{1}{2}$;

3. $\Phi(-x) = 1 - \Phi(x)$, for all $x \in \mathbb{R}$.

Also, since the Φ function does not have a closed form, it is sometimes useful to use upper or lower bounds. In particular we can state the following bounds (see Problem 7 in the Solved Problems section). For all $x \geq 0$,

$$\frac{1}{\sqrt{2\pi}} \frac{x}{x^2 + 1} e^{-\frac{x^2}{2}} \leq 1 - \Phi(x) \leq \frac{1}{\sqrt{2\pi}} \frac{1}{x} e^{-\frac{x^2}{2}} \tag{4.7}$$

As we mentioned earlier, because of the importance of the normal distribution, the values of the Φ function have been tabulated and many calculators and software packages have this function. For example, you can use the normcdf command in MATLAB to compute $\Phi(x)$ for a given number x. More specifically, normcdf(x) returns $\Phi(x)$. Also, the function norminv returns $\Phi^{-1}(x)$. That is, if you run x=norminv(y), then x will be the real number for which $\Phi(x) = y$.

Normal random variables: Now that we have seen the standard normal random variable, we can obtain any normal random variable by shifting and scaling a standard normal random variable. In particular, define

$$X = \sigma Z + \mu, \qquad \text{where } \sigma > 0.$$

Then

$$EX = \sigma EZ + \mu = \mu,$$

$$\text{Var}(X) = \sigma^2 \text{Var}(Z) = \sigma^2.$$

We say that X is a normal random variable with mean μ and variance σ^2. We write $X \sim N(\mu, \sigma^2)$.

If Z is a standard normal random variable and $X = \sigma Z + \mu$, then X is a normal random variable with mean μ and variance σ^2, i.e,

$$X \sim N(\mu, \sigma^2).$$

Conversely, if $X \sim N(\mu, \sigma^2)$, the random variable defined by $Z = \frac{X-\mu}{\sigma}$ is a standard normal random variable, i.e., $Z \sim N(0, 1)$. To find the CDF of $X \sim N(\mu, \sigma^2)$, we can write

$$\begin{aligned}
F_X(x) &= P(X \le x) \\
&= P(\sigma Z + \mu \le x) && \text{(where } Z \sim N(0,1)) \\
&= P\left(Z \le \frac{x-\mu}{\sigma}\right) \\
&= \Phi\left(\frac{x-\mu}{\sigma}\right).
\end{aligned}$$

To find the PDF, we can take the derivative of F_X,

$$\begin{aligned}
f_X(x) &= \frac{d}{dx} F_X(x) \\
&= \frac{d}{dx} \Phi\left(\frac{x-\mu}{\sigma}\right) \\
&= \frac{1}{\sigma} \Phi'\left(\frac{x-\mu}{\sigma}\right) && \text{(chain rule for derivative)} \\
&= \frac{1}{\sigma} f_Z\left(\frac{x-\mu}{\sigma}\right) \\
&= \frac{1}{\sigma\sqrt{2\pi}} e^{-\frac{(x-\mu)^2}{2\sigma^2}}.
\end{aligned}$$

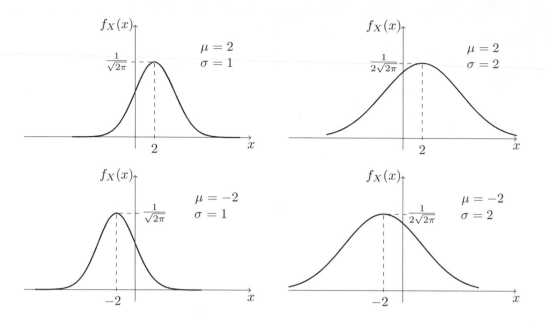

Figure 4.8: PDF for normal distribution

If X is a normal random variable with mean μ and variance σ^2, i.e, $X \sim N(\mu, \sigma^2)$, then

$$f_X(x) = \frac{1}{\sigma\sqrt{2\pi}}e^{-\frac{(x-\mu)^2}{2\sigma^2}},$$

$$F_X(x) = P(X \leq x) = \Phi\left(\frac{x-\mu}{\sigma}\right),$$

$$P(a < X \leq b) = \Phi\left(\frac{b-\mu}{\sigma}\right) - \Phi\left(\frac{a-\mu}{\sigma}\right).$$

Figure 4.8 shows the PDF of the normal distribution for several values of μ and σ.

Example 4.11. Let $X \sim N(-5, 4)$.

(a) Find $P(X < 0)$.

(b) Find $P(-7 < X < -3)$.

(c) Find $P(X > -3 | X > -5)$.

Solution: X is a normal random variable with $\mu = -5$ and $\sigma = \sqrt{4} = 2$, thus we have

(a) Find $P(X < 0)$:

$$P(X < 0) = F_X(0)$$
$$= \Phi\left(\frac{0 - (-5)}{2}\right)$$
$$= \Phi(2.5) \approx 0.99$$

(b) Find $P(-7 < X < -3)$:

$$P(-7 < X < -3) = F_X(-3) - F_X(-7)$$
$$= \Phi\left(\frac{(-3) - (-5)}{2}\right) - \Phi\left(\frac{(-7) - (-5)}{2}\right)$$
$$= \Phi(1) - \Phi(-1)$$
$$= 2\Phi(1) - 1 \qquad \left(\text{since } \Phi(-x) = 1 - \Phi(x)\right)$$
$$\approx 0.68$$

(c) Find $P(X > -3 | X > -5)$:

$$P(X > -3 | X > -5) = \frac{P(X > -3, X > -5)}{P(X > -5)}$$
$$= \frac{P(X > -3)}{P(X > -5)}$$
$$= \frac{1 - \Phi\left(\frac{(-3)-(-5)}{2}\right)}{1 - \Phi\left(\frac{(-5)-(-5)}{2}\right)}$$
$$= \frac{1 - \Phi(1)}{1 - \Phi(0)}$$
$$\approx \frac{0.1587}{0.5} \approx 0.32$$

An important and useful property of the normal distribution is that a linear transformation of a normal random variable is itself a normal random variable. In particular, we have the following theorem:

Theorem 4.3. If $X \sim N(\mu_X, \sigma_X^2)$, and $Y = aX + b$, where $a, b \in \mathbb{R}$, then $Y \sim N(\mu_Y, \sigma_Y^2)$ where

$$\mu_Y = a\mu_X + b, \quad \sigma_Y^2 = a^2 \sigma_X^2.$$

Proof. We can write

$$X = \sigma_X Z + \mu_X \qquad \text{where } Z \sim N(0,1).$$

Thus,

$$
\begin{aligned}
Y &= aX + b \\
&= a(\sigma_X Z + \mu_X) + b \\
&= (a\sigma_X)Z + (a\mu_X + b).
\end{aligned}
$$

Therefore,

$$Y \sim N(a\mu_X + b, a^2 \sigma_X^2).$$

$\blacksquare$

4.2.4   Gamma Distribution

The gamma distribution is another widely used distribution. Its importance is largely due to its relation to exponential and normal distributions. Here, we will provide an introduction to the gamma distribution. In Chapters 6 and 11, we will discuss more properties of the gamma random variables. Before introducing the gamma random variable, we need to introduce the gamma function.

Gamma function: The gamma function [10], shown by $\Gamma(x)$, is an extension of the factorial function to real (and complex) numbers. Specifically, if $n \in \{1, 2, 3, ...\}$, then

$$\Gamma(n) = (n-1)!$$

More generally, for any positive real number α, $\Gamma(\alpha)$ is defined as

$$\Gamma(\alpha) = \int_0^\infty x^{\alpha-1} e^{-x} \mathrm{d}x, \qquad \text{for } \alpha > 0.$$

Figure 4.9 shows the gamma function for some positive real values of α.

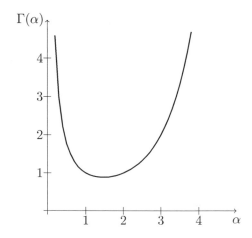

Figure 4.9: The gamma function for some positive real values of α.

Note that for $\alpha = 1$, we can write

$$\Gamma(1) = \int_0^\infty e^{-x}\mathrm{d}x$$
$$= 1.$$

Using the change of variable $x = \lambda y$, we can show the following equation that is often useful when working with the gamma distribution:

$$\Gamma(\alpha) = \lambda^\alpha \int_0^\infty y^{\alpha-1}e^{-\lambda y}\mathrm{d}y, \qquad \text{for } \alpha, \lambda > 0.$$

Also, using integration by parts it can be shown that

$$\Gamma(\alpha + 1) = \alpha\Gamma(\alpha), \qquad \text{for } \alpha > 0.$$

Note that if $\alpha = n$, where n is a positive integer, the above equation reduces to

$$n! = n \cdot (n-1)!$$

Properties of the gamma function

For any positive real number α:

1. $\Gamma(\alpha) = \int_0^\infty x^{\alpha-1}e^{-x}\mathrm{d}x$.

2. $\int_0^\infty x^{\alpha-1}e^{-\lambda x}\mathrm{d}x = \frac{\Gamma(\alpha)}{\lambda^\alpha}$, \qquad for $\lambda > 0$.

3. $\Gamma(\alpha + 1) = \alpha\Gamma(\alpha)$.

4. $\Gamma(n) = (n-1)!$, for $n = 1, 2, 3, \cdots$.

5. $\Gamma(\frac{1}{2}) = \sqrt{\pi}$.

Example 4.12. Answer the following questions:

1. Find $\Gamma(\frac{7}{2})$.

2. Find the value of the following integral:

$$I = \int_0^\infty x^6 e^{-5x}\mathrm{d}x.$$

Solution:

1. To find $\Gamma(\frac{7}{2})$, we can write

$$
\begin{aligned}
\Gamma\left(\frac{7}{2}\right) &= \frac{5}{2} \cdot \Gamma\left(\frac{5}{2}\right) && \text{(using Property 3)} \\
&= \frac{5}{2} \cdot \frac{3}{2} \cdot \Gamma\left(\frac{3}{2}\right) && \text{(using Property 3)} \\
&= \frac{5}{2} \cdot \frac{3}{2} \cdot \frac{1}{2} \cdot \Gamma\left(\frac{1}{2}\right) && \text{(using Property 3)} \\
&= \frac{5}{2} \cdot \frac{3}{2} \cdot \frac{1}{2} \cdot \sqrt{\pi} && \text{(using Property 5)} \\
&= \frac{15}{8} \sqrt{\pi}.
\end{aligned}
$$

2. Using Property 2 with $\alpha = 7$ and $\lambda = 5$, we obtain

$$
\begin{aligned}
I &= \int_0^\infty x^6 e^{-5x} \mathrm{d}x \\
&= \frac{\Gamma(7)}{5^7} \\
&= \frac{6!}{5^7} && \text{(using Property 4)} \\
&\approx 0.0092
\end{aligned}
$$

Gamma Distribution:

We now define the gamma distribution by providing its PDF:

A continuous random variable X is said to have a *gamma* distribution with parameters $\alpha > 0$ and $\lambda > 0$, shown as $X \sim Gamma(\alpha, \lambda)$, if its PDF is given by

$$
f_X(x) = \begin{cases} \frac{\lambda^\alpha x^{\alpha-1} e^{-\lambda x}}{\Gamma(\alpha)} & x > 0 \\ 0 & \text{otherwise} \end{cases}
$$

If we let $\alpha = 1$, we obtain

$$
f_X(x) = \begin{cases} \lambda e^{-\lambda x} & x > 0 \\ 0 & \text{otherwise} \end{cases}
$$

Thus, we conclude $Gamma(1, \lambda) = Exponential(\lambda)$. More generally, if you sum n independent $Exponential(\lambda)$ random variables, then you will get a $Gamma(n, \lambda)$ random variable. We will prove this later on using the moment generating function. The gamma distribution is also related to the normal distribution as will be discussed later. Figure 4.10 shows the PDF of the gamma distribution for several values of α.

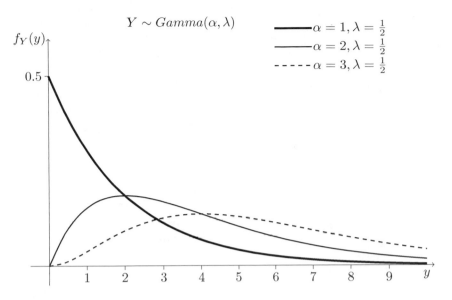

Figure 4.10: PDF of the gamma distribution for some values of α and λ.

Example 4.13. Using the properties of the gamma function, show that the gamma PDF integrates to 1, i.e., show that for $\alpha, \lambda > 0$, we have

$$\int_0^\infty \frac{\lambda^\alpha x^{\alpha-1} e^{-\lambda x}}{\Gamma(\alpha)} \mathrm{d}x = 1.$$

Solution: We can write

$$\int_0^\infty \frac{\lambda^\alpha x^{\alpha-1} e^{-\lambda x}}{\Gamma(\alpha)} \mathrm{d}x = \frac{\lambda^\alpha}{\Gamma(\alpha)} \int_0^\infty x^{\alpha-1} e^{-\lambda x} \mathrm{d}x$$

$$= \frac{\lambda^\alpha}{\Gamma(\alpha)} \cdot \frac{\Gamma(\alpha)}{\lambda^\alpha} \qquad \text{(using Property 2 of the gamma function)}$$

$$= 1.$$

In the Solved Problems section, we calculate the mean and variance for the gamma distribution. In particular, we find out that if $X \sim Gamma(\alpha, \lambda)$, then

$$EX = \frac{\alpha}{\lambda}, \qquad \mathrm{Var}(X) = \frac{\alpha}{\lambda^2}.$$

4.2.5 Other Distributions

In addition to the special distributions that we discussed above, there are many other continuous random variables that are used in practice. Depending on the applications you are interested in you might need to deal with some of them. We have provided a list of important distributions in the appendix. In the next chapters, we will discuss some of them in more detail. There are also some problems at the end of this chapter that discuss some of these distributions. There is no need to try to memorize these distributions. When you understand the general theory behind random variables, you can essentially work with any distribution.

4.2.6   Solved Problems

1. Suppose the number of customers arriving at a store obeys a Poisson distribution with an average of λ customers per unit time. That is, if Y is the number of customers arriving in an interval of length t, then $Y \sim Poisson(\lambda t)$. Suppose that the store opens at time $t = 0$. Let X be the arrival time of the first customer. Show that $X \sim Exponential(\lambda)$.

   Solution: We first find $P(X > t)$:

   $$P(X > t) = P(\text{No arrival in } [0, t])$$
   $$= e^{-\lambda t} \frac{(\lambda t)^0}{0!}$$
   $$= e^{-\lambda t}.$$

   Thus, the CDF of X for $x > 0$ is given by

   $$F_X(x) = 1 - P(X > x) = 1 - e^{-\lambda x},$$

   which is the CDF of $Exponential(\lambda)$. Note that by the same argument, the time between the first and second customer also has $Exponential(\lambda)$ distribution. In general, the time between the kth and $k + 1$th customer is $Exponential(\lambda)$.

2. (Exponential as the limit of Geometric) Let $Y \sim Geometric(p)$, where $p = \lambda \Delta$. Define $X = Y \Delta$, where $\lambda, \Delta > 0$. Prove that for any $x \in (0, \infty)$, we have

   $$\lim_{\Delta \to 0} F_X(x) = 1 - e^{-\lambda x}.$$

   Solution: If $Y \sim Geometric(p)$ and $q = 1 - p$, then

   $$P(Y \leq n) = \sum_{k=1}^{n} pq^{k-1}$$
   $$= p \cdot \frac{1 - q^n}{1 - q} = 1 - (1 - p)^n.$$

   Then for any $y \in (0, \infty)$, we can write

   $$P(Y \leq y) = 1 - (1 - p)^{\lfloor y \rfloor},$$

where $\lfloor y \rfloor$ is the largest integer less than or equal to y. Now, since $X = Y\Delta$, we have

$$F_X(x) = P(X \leq x)$$
$$= P\left(Y \leq \frac{x}{\Delta}\right)$$
$$= 1 - (1-p)^{\lfloor \frac{x}{\Delta} \rfloor} = 1 - (1 - \lambda\Delta)^{\lfloor \frac{x}{\Delta} \rfloor}.$$

Now, we have

$$\lim_{\Delta \to 0} F_X(x) = \lim_{\Delta \to 0} 1 - (1 - \lambda\Delta)^{\lfloor \frac{x}{\Delta} \rfloor}$$
$$= 1 - \lim_{\Delta \to 0} (1 - \lambda\Delta)^{\lfloor \frac{x}{\Delta} \rfloor}$$
$$= 1 - e^{-\lambda x}.$$

The last equality holds because $\frac{x}{\Delta} - 1 \leq \lfloor \frac{x}{\Delta} \rfloor \leq \frac{x}{\Delta}$, and we know

$$\lim_{\Delta \to 0^+} (1 - \lambda\Delta)^{\frac{1}{\Delta}} = e^{-\lambda}.$$

3. Let $U \sim Uniform(0,1)$ and $X = -\ln(1-U)$. Show that $X \sim Exponential(1)$.

Solution: First note that since $R_U = (0,1)$, $R_X = (0,\infty)$. We will find the CDF of X. For $x \in (0, \infty)$, we have

$$F_X(x) = P(X \leq x)$$
$$= P(-\ln(1-U) \leq x)$$
$$= P\left(\frac{1}{1-U} \leq e^x\right)$$
$$= P(U \leq 1 - e^{-x}) = 1 - e^{-x},$$

which is the CDF of an *Exponential*(1) random variable.

4. Let $X \sim N(2,4)$ and $Y = 3 - 2X$.

   (a) Find $P(X > 1)$.
   (b) Find $P(-2 < Y < 1)$.
   (c) Find $P(X > 2|Y < 1)$.

Solution:

(a) Find $P(X > 1)$: We have $\mu_X = 2$ and $\sigma_X = 2$. Thus,

$$P(X > 1) = 1 - \Phi\left(\frac{1-2}{2}\right)$$
$$= 1 - \Phi(-0.5) = \Phi(0.5) = 0.6915$$

(b) Find $P(-2 < Y < 1)$: Since $Y = 3 - 2X$, using Theorem 4.3, we have $Y \sim N(-1, 16)$. Therefore,

$$P(-2 < Y < 1) = \Phi\left(\frac{1-(-1)}{4}\right) - \Phi\left(\frac{(-2)-(-1)}{4}\right)$$
$$= \Phi(0.5) - \Phi(-0.25) = 0.29$$

(c) Find $P(X > 2|Y < 1)$:

$$P(X > 2|Y < 1) = P(X > 2|3 - 2X < 1)$$
$$= P(X > 2|X > 1)$$
$$= \frac{P(X > 2, X > 1)}{P(X > 1)}$$
$$= \frac{P(X > 2)}{P(X > 1)}$$
$$= \frac{1 - \Phi(\frac{2-2}{2})}{1 - \Phi(\frac{1-2}{2})}$$
$$= \frac{1 - \Phi(0)}{1 - \Phi(-0.5)}$$
$$\approx 0.72$$

5. Let $X \sim N(0, \sigma^2)$. Find $E|X|$.

Solution: We can write $X = \sigma Z$, where $Z \sim N(0, 1)$. Thus, $E|X| = \sigma E|Z|$. We have

$$E|Z| = \frac{1}{\sqrt{2\pi}} \int_{-\infty}^{\infty} |t| e^{-\frac{t^2}{2}} \, dt$$
$$= \frac{2}{\sqrt{2\pi}} \int_{0}^{\infty} |t| e^{-\frac{t^2}{2}} \, dt \qquad \text{(integral of an even function)}$$
$$= \sqrt{\frac{2}{\pi}} \int_{0}^{\infty} t e^{-\frac{t^2}{2}} \, dt$$
$$= \sqrt{\frac{2}{\pi}} \left[-e^{-\frac{t^2}{2}} \right]_{0}^{\infty} = \sqrt{\frac{2}{\pi}}.$$

Thus, we conclude $E|X| = \sigma E|Z| = \sigma \sqrt{\frac{2}{\pi}}$.

6. Show that the constant in the normal distribution must be $\frac{1}{\sqrt{2\pi}}$. That is, show that

$$I = \int_{-\infty}^{\infty} e^{-\frac{x^2}{2}} dx = \sqrt{2\pi}.$$

Hint: Write I^2 as a double integral in polar coordinates.

Solution: Let $I = \int_{-\infty}^{\infty} e^{-\frac{x^2}{2}} dx$. We show that $I^2 = 2\pi$. To see this, note

$$I^2 = \int_{-\infty}^{\infty} e^{-\frac{x^2}{2}} dx \int_{-\infty}^{\infty} e^{-\frac{y^2}{2}} dy$$

$$= \int_{-\infty}^{\infty} \int_{-\infty}^{\infty} e^{-\frac{x^2+y^2}{2}} dx dy.$$

To evaluate this double integral we can switch to polar coordinates. This can be done by change of variables $x = r\cos\theta$, $y = r\sin\theta$, and $dxdy = rdrd\theta$. In particular, we have

$$I^2 = \int_{-\infty}^{\infty} \int_{-\infty}^{\infty} e^{-\frac{x^2+y^2}{2}} dx dy$$

$$= \int_{0}^{\infty} \int_{0}^{2\pi} e^{-\frac{r^2}{2}} r d\theta dr$$

$$= 2\pi \int_{0}^{\infty} r e^{-\frac{r^2}{2}} dr$$

$$= 2\pi \left[-e^{-\frac{r^2}{2}} \right]_{0}^{\infty} = 2\pi.$$

7. Let $Z \sim N(0,1)$. Prove for all $x \geq 0$,

$$\frac{1}{\sqrt{2\pi}} \frac{x}{x^2+1} e^{-\frac{x^2}{2}} \leq P(Z \geq x) \leq \frac{1}{\sqrt{2\pi}} \frac{1}{x} e^{-\frac{x^2}{2}}.$$

Solution: To show the upper bound, we can write

$$P(Z \geq x) = \frac{1}{\sqrt{2\pi}} \int_{x}^{\infty} e^{-\frac{u^2}{2}} du$$

$$\leq \frac{1}{\sqrt{2\pi}} \int_{x}^{\infty} \frac{u}{x} e^{-\frac{u^2}{2}} du \qquad \text{(since } u \geq x > 0\text{)}$$

$$= \frac{1}{\sqrt{2\pi}} \frac{1}{x} \left[-e^{-\frac{u^2}{2}} \right]_{x}^{\infty}$$

$$= \frac{1}{\sqrt{2\pi}} \frac{1}{x} e^{-\frac{x^2}{2}}.$$

To show the lower bound, let $Q(x) = P(Z \geq x)$, and

$$h(x) = Q(x) - \frac{1}{\sqrt{2\pi}} \frac{x}{x^2 + 1} e^{-\frac{x^2}{2}}, \qquad \text{for all } x \geq 0.$$

It suffices to show that

$$h(x) \geq 0, \qquad \text{for all } x \geq 0.$$

To see this, note that the function h has the following properties

- $h(0) = \frac{1}{2}$;

- $\lim_{x \to \infty} h(x) = 0$;

- $h'(x) = -\frac{2}{\sqrt{2\pi}} \left(\frac{e^{-\frac{x^2}{2}}}{(x^2+1)^2} \right) < 0$, for all $x \geq 0$.

Therefore, $h(x)$ is a strictly decreasing function that starts at $h(0) = \frac{1}{2}$ and decreases as x increases. It approaches 0 as x goes to infinity. We conclude that $h(x) \geq 0$, for all $x \geq 0$.

8. Let $X \sim Gamma(\alpha, \lambda)$, where $\alpha, \lambda > 0$. Find EX and $\text{Var}(X)$.

Solution: To find EX, we can write

$$
\begin{aligned}
EX &= \int_0^\infty x f_X(x) dx \\
&= \int_0^\infty x \cdot \frac{\lambda^\alpha}{\Gamma(\alpha)} x^{\alpha-1} e^{-\lambda x} dx \\
&= \frac{\lambda^\alpha}{\Gamma(\alpha)} \int_0^\infty x \cdot x^{\alpha-1} e^{-\lambda x} dx \\
&= \frac{\lambda^\alpha}{\Gamma(\alpha)} \int_0^\infty x^\alpha e^{-\lambda x} dx \\
&= \frac{\lambda^\alpha}{\Gamma(\alpha)} \frac{\Gamma(\alpha+1)}{\lambda^{\alpha+1}} \qquad &&\text{(using Property 2 of the gamma function)} \\
&= \frac{\alpha \Gamma(\alpha)}{\lambda \Gamma(\alpha)} \qquad &&\text{(using Property 3 of the gamma function)} \\
&= \frac{\alpha}{\lambda}.
\end{aligned}
$$

Similarly, we can find EX^2:

$$
\begin{aligned}
EX^2 &= \int_0^\infty x^2 f_X(x)dx \\
&= \int_0^\infty x^2 \cdot \frac{\lambda^\alpha}{\Gamma(\alpha)} x^{\alpha-1} e^{-\lambda x} \mathrm{d}x \\
&= \frac{\lambda^\alpha}{\Gamma(\alpha)} \int_0^\infty x^2 \cdot x^{\alpha-1} e^{-\lambda x} \mathrm{d}x \\
&= \frac{\lambda^\alpha}{\Gamma(\alpha)} \int_0^\infty x^{\alpha+1} e^{-\lambda x} \mathrm{d}x \\
&= \frac{\lambda^\alpha}{\Gamma(\alpha)} \frac{\Gamma(\alpha+2)}{\lambda^{\alpha+2}} && \text{(using Property 2 of the gamma function)} \\
&= \frac{(\alpha+1)\Gamma(\alpha+1)}{\lambda^2 \Gamma(\alpha)} && \text{(using Property 3 of the gamma function)} \\
&= \frac{(\alpha+1)\alpha\Gamma(\alpha)}{\lambda^2 \Gamma(\alpha)} && \text{(using Property 3 of the gamma function)} \\
&= \frac{\alpha(\alpha+1)}{\lambda^2}.
\end{aligned}
$$

So, we conclude

$$
\begin{aligned}
\mathrm{Var}(X) &= EX^2 - (EX)^2 \\
&= \frac{\alpha(\alpha+1)}{\lambda^2} - \frac{\alpha^2}{\lambda^2} \\
&= \frac{\alpha}{\lambda^2}.
\end{aligned}
$$

4.3 Mixed Random Variables

Here, we will discuss *mixed* random variables. These are random variables that are neither discrete nor continuous, but are a mixture of both. In particular, a mixed random variable has a continuous part and a discrete part. Thus, we can use our tools from previous chapters to analyze them. In Section 4.3.1, we will provide some examples on how we can do this. Then in Section 4.3.2, we will revisit the concept of mixed random variables using the *delta "function."*

4.3.1 Mixed Random Variables

Example 4.14. Let X be a continuous random variable with the following PDF:

$$
f_X(x) = \begin{cases} 2x & 0 \le x \le 1 \\ 0 & \text{otherwise} \end{cases}
$$

Let also

$$
Y = g(X) = \begin{cases} X & 0 \le X \le \frac{1}{2} \\ \frac{1}{2} & X > \frac{1}{2} \end{cases}
$$

Find the CDF of Y.

Solution: First we note that $R_X = [0, 1]$. For $x \in [0, 1]$, $0 \leq g(x) \leq \frac{1}{2}$. Thus, $R_Y = [0, \frac{1}{2}]$, and therefore

$$F_Y(y) = 0, \qquad \text{for } y < 0,$$
$$F_Y(y) = 1, \qquad \text{for } y > \frac{1}{2}.$$

Now note that

$$P\left(Y = \frac{1}{2}\right) = P\left(X > \frac{1}{2}\right)$$
$$= \int_{\frac{1}{2}}^{1} 2x\,dx = \frac{3}{4}.$$

Also, for $0 < y < \frac{1}{2}$,

$$F_Y(y) = P(Y \leq y)$$
$$= P(X \leq y) = \int_0^y 2x\,dx$$
$$= y^2.$$

Thus, the CDF of Y is given by

$$F_Y(y) = \begin{cases} 1 & y \geq \frac{1}{2} \\ y^2 & 0 \leq y < \frac{1}{2} \\ 0 & \text{otherwise} \end{cases}$$

Figure 4.11 shows the CDF of Y. We note that the CDF is not continuous, so Y is not a continuous random variable. On the other hand, the CDF is not in the staircase form, so it is not a discrete random variable either. It is indeed a *mixed* random variable. There is a jump at $y = \frac{1}{2}$, and the amount of jump is $1 - \frac{1}{4} = \frac{3}{4}$, which is the probability that $Y = \frac{1}{2}$. The CDF is continuous at other points.

The CDF of Y has a continuous part and a discrete part. In particular, we can write

$$F_Y(y) = C(y) + D(y),$$

where $C(y)$ is the continuous part of $F_Y(y)$, i.e.,

$$C(y) = \begin{cases} \frac{1}{4} & y \geq \frac{1}{2} \\ y^2 & 0 \leq y < \frac{1}{2} \\ 0 & y < 0 \end{cases}$$

The discrete part of $F_Y(y)$ is $D(y)$, given by

$$D(y) = \begin{cases} \frac{3}{4} & y \geq \frac{1}{2} \\ 0 & y < \frac{1}{2} \end{cases}$$

In general, the CDF of a mixed random variable Y can be written as the sum of a continuous function and a staircase function:

$$F_Y(y) = C(y) + D(y).$$

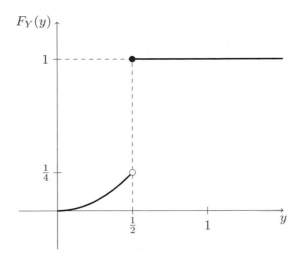

Figure 4.11: CDF of a Mixed random variable, Example 4.14.

We differentiate the continuous part of the CDF. In particular, let's define

$$c(y) = \frac{dC(y)}{dy}, \quad \text{wherever } C(y) \text{ is differentiable.}$$

Note that this is not a valid PDF as it does not integrate to one. Also, let $\{y_1, y_2, y_3, ...\}$ be the set of jump points of $D(y)$, i.e., the points for which $P(Y = y_k) > 0$. We then have

$$\int_{-\infty}^{\infty} c(y)dy + \sum_{y_k} P(Y = y_k) = 1.$$

The expected value of Y can be obtained as

$$EY = \int_{-\infty}^{\infty} yc(y)dy + \sum_{y_k} y_k P(Y = y_k).$$

Example 4.15. Let Y be the mixed random variable defined in Example 4.14.

(a) Find $P(\frac{1}{4} \leq Y \leq \frac{3}{8})$.

(b) Find $P(Y \geq \frac{1}{4})$.

(c) Find EY.

Solution: Since we have the CDF of Y, we can find the probability that Y is in any given interval. We should pay special attention if the interval includes any jump points.

(a) Find $P(\frac{1}{4} \leq Y \leq \frac{3}{8})$: We can write

$$P\left(\frac{1}{4} \leq Y \leq \frac{3}{8}\right) = F_Y\left(\frac{3}{8}\right) - F_Y\left(\frac{1}{4}\right) + P\left(Y = \frac{1}{4}\right)$$

$$= \left(\frac{3}{8}\right)^2 - \left(\frac{1}{4}\right)^2 + 0 = \frac{5}{64}.$$

(b) Find $P(Y \geq \frac{1}{4})$: We have

$$P\left(Y \geq \frac{1}{4}\right) = 1 - F_Y\left(\frac{1}{4}\right) + P\left(Y = \frac{1}{4}\right)$$

$$= 1 - \left(\frac{1}{4}\right)^2 = \frac{15}{16}.$$

(c) Find EY: Here, we can differentiate the continuous part of the CDF to obtain

$$c(y) = \frac{dC(y)}{dy} = \begin{cases} 2y & 0 \leq y \leq \frac{1}{2} \\ 0 & \text{otherwise} \end{cases}$$

So, we can find EY as

$$EY = \int_0^{\frac{1}{2}} y(2y)dy + \frac{1}{2}P\left(Y = \frac{1}{2}\right)$$

$$= \frac{1}{12} + \frac{3}{8} = \frac{11}{24}.$$

4.3.2   Using the Delta Function

In this section, we will use the Dirac delta function [4] to analyze mixed random variables. Remember that any random variable has a CDF. Thus, we can use the CDF to answer questions regarding discrete, continuous, and mixed random variables. On the other hand, the PDF is defined only for continuous random variables, while the PMF is defined only for discrete random variables. Using delta functions will allow us to define the PDF for discrete and mixed random variables. Thus, it allows us to unify the theory of discrete, continuous, and mixed random variables.

Dirac Delta Function:

Remember, we cannot define the PDF for a discrete random variable because its CDF has jumps. If we could somehow differentiate the CDF at jump points, we would be able to define the PDF for discrete random variables as well. This is the idea behind our effort in this section. Here, we will introduce the *Dirac delta function* and discuss its application to probability distributions. If you are less interested in the derivations, you may directly jump to **Definition 4.3** and continue from there. Consider the unit step function $u(x)$ defined by

$$u(x) = \begin{cases} 1 & x \geq 0 \\ 0 & \text{otherwise} \end{cases} \tag{4.8}$$

This function has a jump at $x = 0$. Let us remove the jump and define, for any $\alpha > 0$, the function u_α as

$$u_\alpha(x) = \begin{cases} 1 & x > \frac{\alpha}{2} \\ \frac{1}{\alpha}(x + \frac{\alpha}{2}) & -\frac{\alpha}{2} \leq x \leq \frac{\alpha}{2} \\ 0 & x < -\frac{\alpha}{2} \end{cases}$$

---

[4]Technically speaking, the Dirac delta function is not actually a function. It is what we may call a generalized function. Nevertheless, its definition is intuitive and it simplifies dealing with probability distributions.

The good thing about $u_\alpha(x)$ is that it is a continuous function. Now let us define the function $\delta_\alpha(x)$ as the derivative of $u_\alpha(x)$ wherever it exists.

$$\delta_\alpha(x) = \frac{du_\alpha(x)}{dx} = \begin{cases} \frac{1}{\alpha} & |x| < \frac{\alpha}{2} \\ 0 & |x| > \frac{\alpha}{2} \end{cases}$$

Figure 4.12 shows these functions.

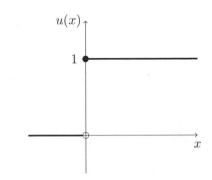

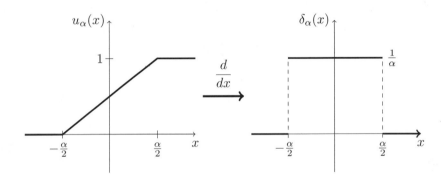

Figure 4.12: Functions $u(x), u_\alpha(x)$, and $\delta_\alpha(x)$

We notice the following relations:

$$\delta_\alpha(x) = \frac{d}{dx}u_\alpha(x), \qquad u(x) = \lim_{\alpha \to 0} u_\alpha(x). \tag{4.9}$$

Now, we would like to define the *delta "function"*, $\delta(x)$, as

$$\delta(x) = \lim_{\alpha \to 0} \delta_\alpha(x). \tag{4.10}$$

Note that as α becomes smaller and smaller, the height of $\delta_\alpha(x)$ becomes larger and larger and its width becomes smaller and smaller. Taking the limit, we obtain

$$\delta(x) = \begin{cases} \infty & x = 0 \\ 0 & \text{otherwise} \end{cases}$$

Combining Equations 4.9 and 4.10, we would like to symbolically write

$$\delta(x) = \frac{d}{dx} u(x).$$

Intuitively, when we are using the delta function, we have in mind $\delta_\alpha(x)$ with extremely small α. In particular, we would like to have the following definitions. Let $g : \mathbb{R} \mapsto \mathbb{R}$ be a continuous function. We <u>define</u>

$$\int_{-\infty}^{\infty} g(x)\delta(x - x_0)dx = \lim_{\alpha \to 0} \left[\int_{-\infty}^{\infty} g(x)\delta_\alpha(x - x_0)dx \right] \tag{4.11}$$

Then, we have the following lemma, which in fact is the most useful property of the delta function.

Lemma 4.1. Let $g : \mathbb{R} \mapsto \mathbb{R}$ be a continuous function. We have

$$\int_{-\infty}^{\infty} g(x)\delta(x - x_0)dx = g(x_0).$$

Proof. Let I be the value of the above integral. Then, we have

$$I = \lim_{\alpha \to 0} \left[\int_{-\infty}^{\infty} g(x)\delta_\alpha(x - x_0)dx \right]$$

$$= \lim_{\alpha \to 0} \left[\int_{x_0 - \frac{\alpha}{2}}^{x_0 + \frac{\alpha}{2}} \frac{g(x)}{\alpha} dx \right].$$

By the mean value theorem in calculus, for any $\alpha > 0$, we have

$$\int_{x_0 - \frac{\alpha}{2}}^{x_0 + \frac{\alpha}{2}} \frac{g(x)}{\alpha} dx = \alpha \frac{g(x_\alpha)}{\alpha} = g(x_\alpha),$$

for some $x_\alpha \in (x_0 - \frac{\alpha}{2}, x_0 + \frac{\alpha}{2})$. Thus, we have

$$I = \lim_{\alpha \to 0} g(x_\alpha) = g(x_0).$$

The last equality holds because $g(x)$ is a continuous function and $\lim_{\alpha \to 0} x_\alpha = x_0$.   ∎

For example, if we let $g(x) = 1$ for all $x \in \mathbb{R}$, we obtain

$$\int_{-\infty}^{\infty} \delta(x)dx = 1.$$

It is worth noting that the Dirac δ function is not strictly speaking a valid function. The reason is that there is no function that can satisfy both of the conditions

$$\delta(x) = 0 \text{ (for } x \neq 0) \quad \text{and} \quad \int_{-\infty}^{\infty} \delta(x)dx = 1.$$

We can think of the delta function as a convenient notation for the integration condition 4.11. The delta function can also be developed formally as a generalized function. Now, let us summarize properties of the delta function.

Properties of the delta function

Definition 4.3. We define the delta function $\delta(x)$ as an object with the following properties:

1. $\delta(x) = \begin{cases} \infty & x = 0 \\ 0 & \text{otherwise} \end{cases}$

2. $\delta(x) = \frac{d}{dx}u(x)$, where $u(x)$ is the unit step function (Equation 4.8);

3. $\int_{-\epsilon}^{\epsilon} \delta(x)dx = 1$, for any $\epsilon > 0$;

4. for any $\epsilon > 0$ and any function $g(x)$ that is continuous over $(x_0 - \epsilon, x_0 + \epsilon)$, we have

$$\int_{-\infty}^{\infty} g(x)\delta(x - x_0)dx = \int_{x_0-\epsilon}^{x_0+\epsilon} g(x)\delta(x - x_0)dx = g(x_0).$$

Figure 4.13 shows how we represent the delta function. The delta function, $\delta(x)$, is shown by an arrow at $x = 0$. The height of the arrow is equal to 1. If we want to represent $2\delta(x)$, the height would be equal to 2. In the figure, we also show the function $\delta(x - x_0)$, which is the shifted version of $\delta(x)$.

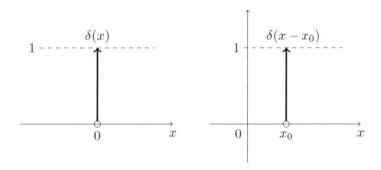

Figure 4.13: Graphical representation of delta function

Using the Delta Function in PDFs of Discrete and Mixed Random Variables:

In this section, we will use the delta function to extend the definition of the PDF to discrete and mixed random variables. Consider a discrete random variable X with range $R_X = \{x_1, x_2, x_3, ...\}$ and PMF $P_X(x_k)$. Note that the CDF for X can be written as

$$F_X(x) = \sum_{x_k \in R_X} P_X(x_k)u(x - x_k).$$

Now that we have symbolically defined the derivative of the step function as the delta function, we can write a PDF for X by "differentiating" the CDF:

$$f_X(x) = \frac{dF_X(x)}{dx}$$

$$= \sum_{x_k \in R_X} P_X(x_k)\frac{d}{dx}u(x - x_k)$$

$$= \sum_{x_k \in R_X} P_X(x_k)\delta(x - x_k).$$

We call this the **generalized** PDF.

For a discrete random variable X with range $R_X = \{x_1, x_2, x_3, ...\}$ and PMF $P_X(x_k)$, we define the (generalized) probability density function (PDF) as

$$f_X(x) = \sum_{x_k \in R_X} P_X(x_k)\delta(x - x_k).$$

Note that for any $x_k \in R_X$, the probability of $X = x_k$ is given by the coefficient of the corresponding δ function, $\delta(x - x_k)$.

It is useful to use the generalized PDF because all random variables have a generalized PDF, so we can use the same formulas for discrete, continuous, and mixed random variables. If the (generalized) PDF of a random variable can be written as the sum of delta functions, then X is a discrete random variable. If the PDF does not include any delta functions, then X is a continuous random variable. Finally, if the PDF has both delta functions and non-delta functions, then X is a mixed random variable. Nevertheless, the formulas for probabilities, expectation and variance are the same for all kinds of random variables.

To see how this works, we will consider the calculation of the expected value of a discrete random variable. Remember that the expected value of a continuous random variable is given by

$$EX = \int_{-\infty}^{\infty} xf_X(x)dx.$$

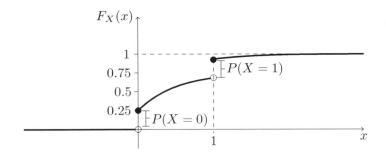

Figure 4.14: The CDF of X in Example 4.16.

Now suppose that I have a discrete random variable X. We can write

$$
\begin{aligned}
EX &= \int_{-\infty}^{\infty} x f_X(x) dx \\
&= \int_{-\infty}^{\infty} x \sum_{x_k \in R_X} P_X(x_k)\delta(x - x_k) dx \\
&= \sum_{x_k \in R_X} P_X(x_k) \int_{-\infty}^{\infty} x\delta(x - x_k) dx \\
&= \sum_{x_k \in R_X} x_k P_X(x_k) \qquad \text{by the 4th property in Definition 4.3,}
\end{aligned}
$$

which is the same as our original definition of expected value for discrete random variables. Let us practice these concepts by looking at an example.

Example 4.16. Let X be a random variable with the following CDF:

$$
F_X(x) = \begin{cases}
\frac{1}{2} + \frac{1}{2}(1 - e^{-x}) & x \geq 1 \\
\frac{1}{4} + \frac{1}{2}(1 - e^{-x}) & 0 \leq x < 1 \\
0 & x < 0
\end{cases}
$$

(a) What kind of random variable is X (discrete, continuous, or mixed)?

(b) Find the (generalized) PDF of X.

(c) Find $P(X > 0.5)$, both using the CDF and using the PDF.

(d) Find EX and $\text{Var}(X)$.

Solution:

(a) Let us plot $F_X(x)$ to better understand the problem. Figure 4.14 shows $F_X(x)$. We see that the CDF has two jumps, at $x = 0$ and $x = 1$. The CDF increases continuously from $x = 0$ to $x = 1$ and also after $x = 1$. Since the CDF is neither in the form of a staircase function, nor is it continuous, we conclude that X is a mixed random variable.

(b) To find the PDF, we need to differentiate the CDF. We must be careful about the points of discontinuity. In particular, we have two jumps: one at $x = 0$, and one at $x = 1$. The size of the jump for both points is equal to $\frac{1}{4}$. Thus, the CDF has two delta functions: $\frac{1}{4}\delta(x) + \frac{1}{4}\delta(x-1)$. The continuous part of the CDF can be written as $\frac{1}{2}(1 - e^{-x})$, for $x > 0$. Thus, we conclude

$$f_X(x) = \frac{1}{4}\delta(x) + \frac{1}{4}\delta(x-1) + \frac{1}{2}e^{-x}u(x).$$

(c) Using the CDF, we have

$$P(X > 0.5) = 1 - F_X(0.5)$$

$$= 1 - \left[\frac{1}{4} + \frac{1}{2}(1 - e^{-x})\right]$$

$$= \frac{1}{4} + \frac{1}{2}e^{-0.5} = 0.5533$$

Using The PDF, we can write

$$P(X > 0.5) = \int_{0.5}^{\infty} f_X(x)dx$$

$$= \int_{0.5}^{\infty} \left(\frac{1}{4}\delta(x) + \frac{1}{4}\delta(x-1) + \frac{1}{2}e^{-x}u(x)\right)dx$$

$$= 0 + \frac{1}{4} + \frac{1}{2}\int_{0.5}^{\infty} e^{-x}dx \qquad \text{(using property 3 in Definition 4.3)}$$

$$= \frac{1}{4} + \frac{1}{2}e^{-0.5} = 0.5533$$

(d) We have

$$EX = \int_{-\infty}^{\infty} x f_X(x)dx$$

$$= \int_{-\infty}^{\infty} \left(\frac{1}{4}x\delta(x) + \frac{1}{4}x\delta(x-1) + \frac{1}{2}xe^{-x}u(x)\right)dx$$

$$= \frac{1}{4} \times 0 + \frac{1}{4} \times 1 + \frac{1}{2}\int_{0}^{\infty} xe^{-x}dx \qquad \text{(using Property 4 in Definition 4.3)}$$

$$= \frac{1}{4} + \frac{1}{2} \times 1 = \frac{3}{4}.$$

Note that here $\int_{0}^{\infty} xe^{-x}dx$ is just the expected value of an *Exponential*(1) random variable, which we know is equal to 1.

$$EX^2 = \int_{-\infty}^{\infty} x^2 f_X(x) dx$$

$$= \int_{-\infty}^{\infty} \left(\frac{1}{4} x^2 \delta(x) + \frac{1}{4} x^2 \delta(x-1) + \frac{1}{2} x^2 e^{-x} u(x) \right) dx$$

$$= \frac{1}{4} \times 0 + \frac{1}{4} \times 1 + \frac{1}{2} \int_0^{\infty} x^2 e^{-x} dx \qquad \text{(using Property 4 in Definition 4.3)}$$

$$= \frac{1}{4} + \frac{1}{2} \times 2 = \frac{5}{4}.$$

Again, note that $\int_0^{\infty} x^2 e^{-x} dx$ is just EX^2 for an *Exponential*(1) random variable, which we know is equal to 2. Thus,

$$\text{Var}(X) = EX^2 - (EX)^2$$

$$= \frac{5}{4} - \left(\frac{3}{4} \right)^2$$

$$= \frac{11}{16}.$$

In general, we can make the following statement:

The (generalized) PDF of a mixed random variable can be written in the form

$$f_X(x) = \sum_k a_k \delta(x - x_k) + g(x),$$

where $a_k = P(X = x_k)$, and $g(x) \geq 0$ does not contain any delta functions. Furthermore, we have

$$\int_{-\infty}^{\infty} f_X(x) dx = \sum_k a_k + \int_{-\infty}^{\infty} g(x) dx = 1.$$

4.3.3 Solved Problems

1. Here is one way to think about a mixed random variable. Suppose that we have a discrete random variable X_d with (generalized) PDF and CDF $f_d(x)$ and $F_d(x)$, and a continuous random variable X_c with PDF and CDF $f_c(x)$ and $F_c(x)$. Now we create a new random variable X in the following way. We have a coin with $P(H) = p$. We toss the coin once. If it lands heads, then the value of X is determined according to the probability distribution of X_d. If the coin lands tails, the value of X is determined according to the probability distribution of X_c.

(a) Find the CDF of $X, F_X(x)$.

(b) Find the PDF of $X, f_X(x)$.

(c) Find EX.

(d) Find $\text{Var}(X)$.

Solution:

(a) Find the CDF of $X, F_X(x)$: We can write

$$
\begin{aligned}
F_X(x) &= P(X \le x) \\
&= P(X \le x|H)P(H) + P(X \le x|T)P(T) \qquad \text{(law of total probability)} \\
&= pP(X_d \le x) + (1-p)P(X_c \le x) \\
&= pF_d(x) + (1-p)F_c(x).
\end{aligned}
$$

(b) Find the PDF of $X, f_X(x)$: By differentiating $F_X(x)$, we obtain

$$
\begin{aligned}
f_X(x) &= \frac{dF_X(x)}{dx} \\
&= pf_d(x) + (1-p)f_c(x).
\end{aligned}
$$

(c) Find EX: We have

$$
\begin{aligned}
EX &= \int_{-\infty}^{\infty} x f_X(x) dx \\
&= p \int_{-\infty}^{\infty} x f_d(x) dx + (1-p) \int_{-\infty}^{\infty} x f_c(x) dx \\
&= pEX_d + (1-p)EX_c.
\end{aligned}
$$

(d) Find $\text{Var}(X)$:

$$
\begin{aligned}
EX^2 &= \int_{-\infty}^{\infty} x^2 f_X(x) dx \\
&= p \int_{-\infty}^{\infty} x^2 f_d(x) dx + (1-p) \int_{-\infty}^{\infty} x^2 f_c(x) dx \\
&= pEX_d^2 + (1-p)EX_c^2.
\end{aligned}
$$

Thus,

$$
\begin{aligned}
&= \text{Var}(X) = EX^2 - (EX)^2 \\
&= pEX_d^2 + (1-p)EX_c^2 - (pEX_d + (1-p)EX_c)^2 \\
&= pEX_d^2 + (1-p)EX_c^2 - p^2(EX_d)^2 - (1-p)^2(EX_c)^2 - 2p(1-p)EX_dEX_c \\
&= p(EX_d^2 - (EX_d)^2) + (1-p)(EX_c^2 - (EX_c)^2) + p(1-p)(EX_d - EX_c)^2 \\
&= p\text{Var}(X_d) + (1-p)\text{Var}(X_c) + p(1-p)(EX_d - EX_c)^2.
\end{aligned}
$$

2. Let X be a random variable with CDF

$$F_X(x) = \begin{cases} 1 & x \geq 1 \\ \frac{1}{2} + \frac{x}{2} & 0 \leq x < 1 \\ 0 & x < 0 \end{cases}$$

(a) What kind of random variable is X: discrete, continuous, or mixed?
(b) Find the PDF of X, $f_X(x)$.
(c) Find $E(e^X)$.
(d) Find $P(X = 0 | X \leq 0.5)$.

Solution:

(a) What kind of random variable is X: discrete, continuous, or mixed? We note that the CDF has a discontinuity at $x = 0$, and it is continuous at other points. Since $F_X(x)$ is not flat in other locations, we conclude X is a mixed random variable. Indeed, we can write

$$F_X(x) = \frac{1}{2}u(x) + \frac{1}{2}F_Y(x),$$

where Y is a $Uniform(0, 1)$ random variable. If we use the interpretation of Problem 1, we can say the following. We toss a fair coin. If it lands heads then $X = 0$, otherwise X is obtained according the a $Uniform(0, 1)$ distribution.

(b) Find the PDF of X, $f_X(x)$: By differentiating the CDF, we obtain

$$f_X(x) = \frac{1}{2}\delta(x) + \frac{1}{2}f_Y(x),$$

where $f_Y(x)$ is the PDF of $Uniform(0, 1)$, i.e.,

$$f_Y(x) = \begin{cases} 1 & 0 < x < 1 \\ 0 & \text{otherwise} \end{cases}$$

(c) Find $E(e^X)$: We can use LOTUS to write

$$E(e^X) = \int_{-\infty}^{\infty} e^x f_X(x)dx$$

$$= \frac{1}{2}\int_{-\infty}^{\infty} e^x \delta(x)dx + \frac{1}{2}\int_{-\infty}^{\infty} e^x f_Y(x)dx$$

$$= \frac{1}{2}e^0 + \frac{1}{2}\int_0^1 e^x dx$$

$$= \frac{1}{2} + \frac{1}{2}(e - 1)$$

$$= \frac{1}{2}e.$$

Here is another way to think about this part: similar to part (c) of Problem 1, we can write

$$E(e^X) = \frac{1}{2} \times e^0 + \frac{1}{2} E[e^Y]$$

$$= \frac{1}{2} + \frac{1}{2} \int_0^1 e^y dy$$

$$= \frac{1}{2} e.$$

(d) Find $P(X = 0 | X \leq 0.5)$: We have

$$P(X = 0 | X \leq 0.5) = \frac{P(X = 0, X \leq 0.5)}{P(X \leq 0.5)}$$

$$= \frac{P(X = 0)}{P(X \leq 0.5)}$$

$$= \frac{0.5}{\int_0^{0.5} f_X(x) dx}$$

$$= \frac{0.5}{0.75} = \frac{2}{3}.$$

3. Let X be a $Uniform(-2, 2)$ continuous random variable. We define $Y = g(X)$, where the function $g(x)$ is defined as

$$g(x) = \begin{cases} 1 & x > 1 \\ x & 0 \leq x \leq 1 \\ 0 & \text{otherwise} \end{cases}$$

Find the CDF and PDF of Y.

Solution: Note that $R_Y = [0, 1]$. Therefore,

$$F_Y(y) = 0, \quad \text{for } y < 0,$$
$$F_Y(y) = 1 \quad \text{for } y \geq 1.$$

We also note that

$$P(Y = 0) = P(X < 0) = \frac{1}{2},$$

$$P(Y = 1) = P(X > 1) = \frac{1}{4}.$$

Also, for $0 < y < 1$,

$$F_Y(y) = P(Y \le y)$$
$$= P(X \le y)$$
$$= F_X(y)$$
$$= \frac{y+2}{4}.$$

Thus, the CDF of Y is given by

$$F_Y(y) = \begin{cases} 1 & y \ge 1 \\ \frac{y+2}{4} & 0 \le y < 1 \\ 0 & \text{otherwise} \end{cases}$$

In particular, we note that there are two jumps in the CDF, one at $y = 0$ and another at $y = 1$. We can find the generalized PDF of Y by differentiating $F_Y(y)$:

$$f_Y(y) = \frac{1}{2}\delta(y) + \frac{1}{4}\delta(y-1) + \frac{1}{4}\big(u(y) - u(y-1)\big).$$

4.4   End of Chapter Problems

1. I choose a real number uniformly at random in the interval $[2, 6]$, and call it X.

   (a) Find the CDF of X, $F_X(x)$.
   (b) Find EX.

2. Let X be a continuous random variable with the following PDF

$$f_X(x) = \begin{cases} ce^{-4x} & x \geq 0 \\ 0 & \text{otherwise} \end{cases}$$

   where c is a positive constant.

   (a) Find c.
   (b) Find the CDF of X, $F_X(x)$.
   (c) Find $P(2 < X < 5)$.
   (d) Find EX.

3. Let X be a continuous random variable with PDF

$$f_X(x) = \begin{cases} x^2 + \frac{2}{3} & 0 \leq x \leq 1 \\ 0 & \text{otherwise} \end{cases}$$

   (a) Find $E(X^n)$, for $n = 1, 2, 3, \cdots$.
   (b) Find the variance of X.

4. Let X be a $uniform(0, 1)$ random variable, and let $Y = e^{-X}$.

   (a) Find the CDF of Y.
   (b) Find the PDF of Y.
   (c) Find EY.

5. Let X be a continuous random variable with PDF

$$f_X(x) = \begin{cases} \frac{5}{32}x^4 & 0 < x \leq 2 \\ 0 & \text{otherwise} \end{cases}$$

   and let $Y = X^2$.

   (a) Find the CDF of Y.
   (b) Find the PDF of Y.
   (c) Find EY.

6. Let $X \sim Exponential(\lambda)$, and $Y = aX$, where a is a positive real number. Show that

$$Y \sim Exponential\left(\frac{\lambda}{a}\right).$$

7. Let $X \sim Exponential(\lambda)$. Show that

   (a) $EX^n = \frac{n}{\lambda}EX^{n-1}$, for $n = 1, 2, 3, \cdots$;

   (b) $EX^n = \frac{n!}{\lambda^n}$, for $n = 1, 2, 3, \cdots$.

8. Let $X \sim N(3, 9)$.

   (a) Find $P(X > 0)$.

   (b) Find $P(-3 < X < 8)$.

   (c) Find $P(X > 5|X > 3)$.

9. Let $X \sim N(3, 9)$ and $Y = 5 - X$.

   (a) Find $P(X > 2)$.

   (b) Find $P(-1 < Y < 3)$.

   (c) Find $P(X > 4|Y < 2)$.

10. Let X be a continuous random variable with PDF

$$f_X(x) = \frac{1}{\sqrt{2\pi}}e^{-\frac{x^2}{2}}, \qquad \text{for all } x \in \mathbb{R}.$$

   and let $Y = \sqrt{|X|}$. Find $f_Y(y)$.

11. Let $X \sim Exponential(2)$ and $Y = 2 + 3X$.

   (a) Find $P(X > 2)$.

   (b) Find EY and $\text{Var}(Y)$.

   (c) Find $P(X > 2|Y < 11)$.

12. The **median** of a continuous random variable X can be defined as the unique real number m that satisfies

$$P(X \geq m) = P(X < m) = \frac{1}{2}.$$

   Find the median of the following random variables:

   (a) $X \sim Uniform(a, b)$.

   (b) $Y \sim Exponential(\lambda)$.

   (c) $W \sim N(\mu, \sigma^2)$.

13. Let X be a random variable with the following CDF:

$$F_X(x) = \begin{cases} 0 & \text{for } x < 0 \\[2mm] x & \text{for } 0 \leq x < \frac{1}{4} \\[2mm] x + \frac{1}{2} & \text{for } \frac{1}{4} \leq x < \frac{1}{2} \\[2mm] 1 & \text{for } x \geq \frac{1}{2} \end{cases}$$

(a) Plot $F_X(x)$ and explain why X is a mixed random variable.

(b) Find $P(X \leq \frac{1}{3})$.

(c) Find $P(X \geq \frac{1}{4})$.

(d) Write CDF of X in the form of

$$F_X(x) = C(x) + D(x),$$

where $C(x)$ is a continuous function and $D(x)$ is in the form of a staircase function, i.e.,

$$D(x) = \sum_k a_k u(x - x_k).$$

(e) Find $c(x) = \frac{d}{dx} C(x)$.

(f) Find EX using $EX = \int_{-\infty}^{\infty} x c(x) dx + \sum_k x_k a_k$

14. Let X be a random variable with the following CDF:

$$F_X(x) = \begin{cases} 0 & \text{for } x < 0 \\[2mm] x & \text{for } 0 \leq x < \frac{1}{4} \\[2mm] x + \frac{1}{2} & \text{for } \frac{1}{4} \leq x < \frac{1}{2} \\[2mm] 1 & \text{for } x \geq \frac{1}{2} \end{cases}$$

(a) Find the generalized PDF of X, $f_X(x)$.

(b) Find EX using $f_X(x)$.

(c) Find $\text{Var}(X)$ using $f_X(x)$.

15. Let X be a mixed random variable with the following generalized PDF:

$$f_X(x) = \frac{1}{3}\delta(x + 2) + \frac{1}{6}\delta(x - 1) + \frac{1}{2} \cdot \frac{1}{\sqrt{2\pi}} e^{-\frac{x^2}{2}}.$$

(a) Find $P(X = 1)$ and $P(X = -2)$.

(b) Find $P(X \geq 1)$.

(c) Find $P(X = 1 | X \geq 1)$.

(d) Find EX and $\text{Var}(X)$.

16. A company makes a certain device. We are interested in the lifetime of the device. It is estimated that around 2% of the devices are defected from the start so they have a lifetime of 0 years. If a device is not defected, then the lifetime of the device is exponentially distributed with parameter $\lambda = 2$ years. Let X be the lifetime of a randomly chosen device:

   (a) Find the generalized PDF of X.

   (b) Find $P(X \geq 1)$.

   (c) Find $P(X > 2|X \geq 1)$.

   (d) Find EX and $\text{Var}(X)$.

17. A continuous random variable is said to have a *Laplace*(μ, b) distribution [14] if its PDF is given by

$$f_X(x) = \frac{1}{2b} \exp\left(-\frac{|x - \mu|}{b}\right)$$

$$= \begin{cases} \frac{1}{2b} \exp\left(\frac{x-\mu}{b}\right) & \text{if } x < \mu \\ \frac{1}{2b} \exp\left(-\frac{x-\mu}{b}\right) & \text{if } x \geq \mu \end{cases}$$

where $\mu \in \mathbb{R}$ and $b > 0$.

   (a) If $X \sim Laplace(0,1)$, find EX and $\text{Var}(X)$.

   (b) If $X \sim Laplace(0,1)$ and $Y = bX + \mu$, show that $Y \sim Laplace(\mu, b)$.

   (c) Let $Y \sim Laplace(\mu, b)$, where $\mu \in \mathbb{R}$ and $b > 0$. Find EY and $\text{Var}(Y)$.

18. Let $X \sim Laplace(0, b)$, i.e.,

$$f_X(x) = \frac{1}{2b} \exp\left(-\frac{|x|}{b}\right),$$

(4.12)

where $b > 0$. Define $Y = |X|$. Show that $Y \sim Exponential(\frac{1}{b})$.

19. A continuous random variable is said to have the **standard Cauchy** distribution if its PDF is given by

$$f_X(x) = \frac{1}{\pi(1 + x^2)}.$$

If X has a standard Cauchy distribution, show that EX is not well-defined. Also, show $EX^2 = \infty$.

20. A continuous random variable is said to have a Rayleigh distribution with parameter σ if its PDF is given by

$$f_X(x) = \frac{x}{\sigma^2} e^{-x^2/2\sigma^2} u(x)$$

$$= \begin{cases} \frac{x}{\sigma^2} e^{-x^2/2\sigma^2} & \text{if } x \geq 0 \\ 0 & \text{if } x < 0 \end{cases}$$

where $\sigma > 0$.

   (a) If $X \sim Rayleigh(\sigma)$, find EX.

(b) If $X \sim Rayleigh(\sigma)$, find the CDF of X, $F_X(x)$.

(c) If $X \sim Exponential(1)$ and $Y = \sqrt{2\sigma^2 X}$, show that $Y \sim Rayleigh(\sigma)$.

21. A continuous random variable is said to have a $Pareto(x_m, \alpha)$ distribution [15] if its PDF is given by

$$f_X(x) = \begin{cases} \alpha \dfrac{x_m^{\alpha}}{x^{\alpha+1}} & \text{for } x \geq x_m \\[2mm] 0 & \text{for } x < x_m \end{cases}$$

where $x_m, \alpha > 0$. Let $X \sim Pareto(x_m, \alpha)$.

(a) Find the CDF of X, $F_X(x)$.

(b) Find $P(X > 3x_m | X > 2x_m)$.

(c) If $\alpha > 2$, find EX and $\text{Var}(X)$.

22. Let $Z \sim N(0, 1)$. If we define $X = e^{\sigma Z + \mu}$, then we say that X has a log-normal distribution with parameters μ and σ, and we write $X \sim LogNormal(\mu, \sigma)$.

(a) If $X \sim LogNormal(\mu, \sigma)$, find the CDF of X in terms of the Φ function.

(b) Find EX and $\text{Var}(X)$.

23. Let $X_1, X_2, \cdots, X_n$ be independent random variables with $X_i \sim Exponential(\lambda)$. Define

$$Y = X_1 + X_2 + \cdots + X_n.$$

As we will see later, Y has a **Gamma** distribution with parameters n and λ, i.e., $Y \sim Gamma(n, \lambda)$. Using this, show that if $Y \sim Gamma(n, \lambda)$, then $EY = \frac{n}{\lambda}$ and $\text{Var}(Y) = \frac{n}{\lambda^2}$.

Chapter 5

Joint Distributions: Two Random Variables

In real life, we are often interested in several random variables that are related to each other. For example, suppose that we choose a random family, and we would like to study the number of people in the family, the household income, the ages of the family members, etc. Each of these is a random variable, and we suspect that they are dependent. In this chapter, we develop tools to study joint distributions of random variables. The concepts are similar to what we have seen so far. The only difference is that instead of one random variable, we consider two or more. In this chapter, we will focus on two random variables, but once you understand the theory for two random variables, the extension to n random variables is straightforward. We will first discuss joint distributions of discrete random variables and then extend the results to continuous random variables.

5.1 Two Discrete Random Variables

5.1.1 Joint Probability Mass Function (PMF)

Remember that for a discrete random variable X, we define the PMF as $P_X(x) = P(X = x)$. Now, if we have two random variables X and Y, and we would like to study them jointly, we define the **joint probability mass function** as follows:

The **joint probability mass function** of two discrete random variables X and Y is defined as

$$P_{XY}(x, y) = P(X = x, Y = y).$$

Note that as usual, the comma means "and," so we can write

$$P_{XY}(x, y) = P(X = x, Y = y)$$
$$= P\big((X = x) \text{ and } (Y = y)\big).$$

We can define the joint range for X and Y as

$$R_{XY} = \{(x, y) | P_{XY}(x, y) > 0\}.$$

In particular, if $R_X = \{x_1, x_2, ...\}$ and $R_Y = \{y_1, y_2, ...\}$, then we can always write

$$R_{XY} \subset R_X \times R_Y$$
$$= \{(x_i, y_j) | x_i \in R_X, y_j \in R_Y\}.$$

In fact, sometimes we define $R_{XY} = R_X \times R_Y$ to simplify the analysis. In this case, for some pairs (x_i, y_j) in $R_X \times R_Y$, $P_{XY}(x_i, y_j)$ might be zero. For two discrete random variables X and Y, we have

$$\sum_{(x_i, y_j) \in R_{XY}} P_{XY}(x_i, y_j) = 1$$

We can use the joint PMF to find $P\big((X, Y) \in A\big)$ for any set $A \subset \mathbb{R}^2$. Specifically, we have

$$P\big((X, Y) \in A\big) = \sum_{(x_i, y_j) \in (A \cap R_{XY})} P_{XY}(x_i, y_j)$$

Note that the event $X = x$ can be written as $\{(x_i, y_j) : x_i = x, \ y_j \in R_Y\}$. Also, the event $Y = y$ can be written as $\{(x_i, y_j) : x_i \in R_X, \ y_j = y\}$. Thus, we can write

$$P_{XY}(x, y) = P(X = x, Y = y)$$
$$= P\big((X = x) \cap (Y = y)\big).$$

Marginal PMFs:

The joint PMF contains all the information regarding the distributions of X and Y. This means that, for example, we can obtain PMF of X from its joint PMF with Y. Indeed, we can write

$$P_X(x) = P(X = x)$$
$$= \sum_{y_j \in R_Y} P(X = x, Y = y_j) \qquad \text{law of total probablity}$$
$$= \sum_{y_j \in R_Y} P_{XY}(x, y_j).$$

Here, we call $P_X(x)$ the **marginal PMF** of X. Similarly, we can find the marginal PMF of Y as

$$P_Y(Y) = \sum_{x_i \in R_X} P_{XY}(x_i, y).$$

Marginal PMFs of X and Y:

$$P_X(x) = \sum_{y_j \in R_Y} P_{XY}(x, y_j), \qquad \text{for any } x \in R_X$$

$$P_Y(y) = \sum_{x_i \in R_X} P_{XY}(x_i, y), \qquad \text{for any } y \in R_Y \tag{5.1}$$

Let's practice these concepts by looking at an example.

Example 5.1. Consider two random variables X and Y with joint PMF given in Table 5.1.

Table 5.1: Joint PMF of X and Y in Example 5.1

	$Y = 0$	$Y = 1$	$Y = 2$
$X = 0$	$\frac{1}{6}$	$\frac{1}{4}$	$\frac{1}{8}$
$X = 1$	$\frac{1}{8}$	$\frac{1}{6}$	$\frac{1}{6}$

Figure 5.1 shows $P_{XY}(x, y)$.

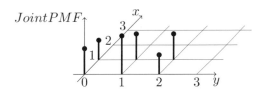

Figure 5.1: Joint PMF of X and Y (Example 5.1).

(a) Find $P(X = 0, Y \leq 1)$.

(b) Find the marginal PMFs of X and Y.

(c) Find $P(Y = 1 | X = 0)$.

(d) Are X and Y independent?

Solution:

(a) To find $P(X = 0, Y \leq 1)$, we can write

$$P(X = 0, Y \leq 1) = P_{XY}(0,0) + P_{XY}(0,1) = \frac{1}{6} + \frac{1}{4} = \frac{5}{12}.$$

(b) Note that from the table,

$$R_X = \{0, 1\} \qquad \text{and} \qquad R_Y = \{0, 1, 2\}.$$

Now we can use Equation 5.1 to find the marginal PMFs. For example, to find $P_X(0)$, we can write

$$P_X(0) = P_{XY}(0,0) + P_{XY}(0,1) + P_{XY}(0,2)$$
$$= \frac{1}{6} + \frac{1}{4} + \frac{1}{8}$$
$$= \frac{13}{24}.$$

We obtain

$$P_X(x) = \begin{cases} \frac{13}{24} & x = 0 \\ \frac{11}{24} & x = 1 \\ 0 & \text{otherwise} \end{cases}$$

$$P_Y(y) = \begin{cases} \frac{7}{24} & y = 0 \\ \frac{5}{12} & y = 1 \\ \frac{7}{24} & y = 2 \\ 0 & \text{otherwise} \end{cases}$$

(c) Find $P(Y = 1 | X = 0)$: Using the formula for conditional probability, we have

$$P(Y = 1 | X = 0) = \frac{P(X = 0, Y = 1)}{P(X = 0)}$$
$$= \frac{P_{XY}(0,1)}{P_X(0)}$$
$$= \frac{\frac{1}{4}}{\frac{13}{24}} = \frac{6}{13}.$$

(d) Are X and Y independent? X and Y are not independent, because as we just found out

$$P(Y = 1|X = 0) = \frac{6}{13} \neq P(Y = 1) = \frac{5}{12}.$$

Caution: If we want to show that X and Y are independent, we need to check that $P(X = x_i, Y = y_j) = P(X = x_i)P(Y = y_j)$, for all $x_i \in R_X$ and all $y_j \in R_Y$. Thus, even if in the above calculation we had found $P(Y = 1|X = 0) = P(Y = 1)$, we would not yet have been able to conclude that X and Y are independent. For that, we would need to check the independence condition for all $x_i \in R_X$ and all $y_j \in R_Y$.

5.1.2 Joint Cumulative Distribution Function (CDF)

Remember that, for a random variable X, we define the CDF as $F_X(x) = P(X \leq x)$. Now, if we have two random variables X and Y and we would like to study them jointly, we can define the **joint cumulative function** as follows:

The **joint cumulative distribution function** of two random variables X and Y is defined as

$$F_{XY}(x, y) = P(X \leq x, Y \leq y).$$

As usual, comma means "and", so we can write

$$F_{XY}(x, y) = P(X \leq x, Y \leq y)$$
$$= P\big((X \leq x) \text{ and } (Y \leq y)\big) = P\big((X \leq x) \cap (Y \leq y)\big).$$

Figure 5.2 shows the region associated with $F_{XY}(x, y)$ in the two-dimensional plane. Note that the above definition of joint CDF is a general definition and is applicable to discrete, continuous, and mixed random variables. Since the joint CDF refers to the probability of an event, we must have $0 \leq F_{XY}(x, y) \leq 1$.

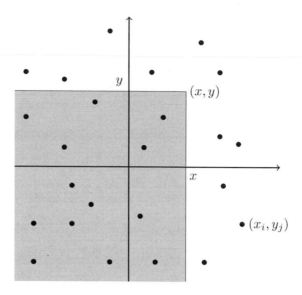

Figure 5.2: $F_{XY}(x, y)$ is the probability that (X, Y) belongs to the shaded region. The dots are the pairs (x_i, y_j) in R_{XY}.

If we know the joint CDF of X and Y, we can find the *marginal* CDFs, $F_X(x)$ and $F_Y(y)$. Specifically, for any $x \in \mathbb{R}$, we have

$$F_{XY}(x, \infty) = P(X \leq x, Y \leq \infty)$$
$$= P(X \leq x) = F_X(x).$$

Here, by $F_{XY}(x, \infty)$, we mean $\lim_{y \to \infty} F_{XY}(x, y)$. Similarly, for any $y \in \mathbb{R}$, we have

$$F_Y(y) = F_{XY}(\infty, y).$$

Marginal CDFs of X and Y:

$$F_X(x) = F_{XY}(x, \infty) = \lim_{y \to \infty} F_{XY}(x, y), \qquad \text{for any } x,$$
$$F_Y(y) = F_{XY}(\infty, y) = \lim_{x \to \infty} F_{XY}(x, y), \qquad \text{for any } y \tag{5.2}$$

Also, note that we must have

$$F_{XY}(\infty, \infty) = 1,$$
$$F_{XY}(-\infty, y) = 0, \qquad \text{for any } y,$$
$$F_{XY}(x, -\infty) = 0, \qquad \text{for any } x.$$

Example 5.2. Let $X \sim Bernoulli(p)$ and $Y \sim Bernoulli(q)$ be independent, where $0 < p, q < 1$. Find the joint PMF and joint CDF for X and Y.

Solution: First note that the joint range of X and Y is given by

$$R_{XY} = \{(0,0), (0,1), (1,0), (1,1)\}.$$

Since X and Y are independent, we have

$$P_{XY}(i,j) = P_X(i)P_Y(j), \qquad \text{for } i, j = 0, 1.$$

Thus, we conclude

$$\begin{aligned}
P_{XY}(0,0) &= P_X(0)P_Y(0) = (1-p)(1-q), \\
P_{XY}(0,1) &= P_X(0)P_Y(1) = (1-p)q, \\
P_{XY}(1,0) &= P_X(1)P_Y(0) = p(1-q), \\
P_{XY}(1,1) &= P_X(1)P_Y(1) = pq.
\end{aligned}$$

Now that we have the joint PMF, we can find the joint CDF

$$F_{XY}(x,y) = P(X \leq x, Y \leq y).$$

Specifically, since $0 \leq X, Y \leq 1$, we conclude

$$\begin{aligned}
F_{XY}(x,y) &= 0, &&\text{if } x < 0, \\
F_{XY}(x,y) &= 0, &&\text{if } y < 0, \\
F_{XY}(x,y) &= 1, &&\text{if } x \geq 1 \text{ and } y \geq 1.
\end{aligned}$$

Now, for $0 \leq x < 1$ and $y \geq 1$, we have

$$\begin{aligned}
F_{XY}(x,y) &= P(X \leq x, Y \leq y) \\
&= P(X = 0, y \leq 1) \\
&= P(X = 0) = 1 - p.
\end{aligned}$$

Similarly, for $0 \leq y < 1$ and $x \geq 1$, we have

$$\begin{aligned}
F_{XY}(x,y) &= P(X \leq x, Y \leq y) \\
&= P(X \leq 1, y = 0) \\
&= P(Y = 0) = 1 - q.
\end{aligned}$$

Finally, for $0 \leq x < 1$ and $0 \leq y < 1$, we have

$$\begin{aligned}
F_{XY}(x,y) &= P(X \leq x, Y \leq y) \\
&= P(X = 0, y = 0) \\
&= P(X = 0)P(Y = 0) = (1-p)(1-q).
\end{aligned}$$

Figure 5.3 shows the values of $F_{XY}(x,y)$ in different regions of the two-dimensional plane. Note that, in general, we actually need a three-dimensional graph to show a joint CDF of two random variables, i.e., we need three axes: x, y, and $z = F_{XY}(x,y)$. However, because the random variables of this example are simple, and can take only two values, a two-dimensional figure suffices.

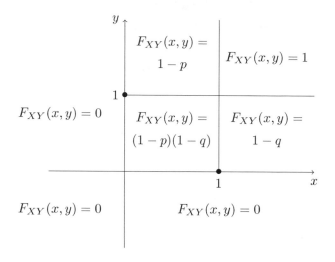

Figure 5.3: Joint CDF for X and Y in Example 5.2.

Here is a useful lemma:

Lemma 5.1. For two random variables X and Y, and real numbers $x_1 \leq x_2$, $y_1 \leq y_2$, we have

$$P(x_1 < X \leq x_2, \; y_1 < Y \leq y_2) =$$
$$F_{XY}(x_2, y_2) - F_{XY}(x_1, y_2) - F_{XY}(x_2, y_1) + F_{XY}(x_1, y_1).$$

To see why the above formula is true, you can look at the region associated with $F_{XY}(x,y)$ (as shown in Figure 5.2) for each of the pairs $(x_2, y_2), (x_1, y_2), (x_2, y_1), (x_1, y_1)$. You can see, as we subtract and add regions, the part that is left is the region $\{x_1 < X \leq x_2, \; y_1 < Y \leq y_2\}$.

5.1.3　Conditioning and Independence

We have discussed conditional probability before, and you have already seen some problems regarding random variables and conditional probability. Here, we will discuss conditioning for random variables more in detail and introduce the conditional PMF, conditional CDF, and conditional expectation. We would like to emphasize that there is only one main formula regarding conditional probability which is

$$P(A|B) = \frac{P(A \cap B)}{P(B)}, \text{ when } P(B) > 0.$$

Any other formula regarding conditional probability can be derived from the above formula. Specifically, if you have two random variables X and Y, you can write

$$P(X \in C | Y \in D) = \frac{P(X \in C, Y \in D)}{P(Y \in D)}, \text{ where } C, D \subset \mathbb{R}.$$

Conditional PMF and CDF:

Remember that the PMF is by definition a probability measure, i.e., it is $P(X = x_k)$. Thus, we can talk about the **conditional PMF**. Specifically, the conditional PMF of X given event A, is defined as

$$P_{X|A}(x_i) = P(X = x_i|A)$$
$$= \frac{P(X = x_i \text{ and } A)}{P(A)}.$$

Example 5.3. I roll a fair die. Let X be the observed number. Find the conditional PMF of X given that we know the observed number was less than 5.

Solution: Here, we condition on the event $A = \{X < 5\}$, where $P(A) = \frac{4}{6}$. Thus,

$$P_{X|A}(1) = P(X = 1|X < 5)$$
$$= \frac{P(X = 1 \text{ and } X < 5)}{P(X < 5)}$$
$$= \frac{P(X = 1)}{P(X < 5)} = \frac{1}{4}.$$

Similarly, we have

$$P_{X|A}(2) = P_{X|A}(3) = P_{X|A}(4) = \frac{1}{4}.$$

Also,

$$P_{X|A}(5) = P_{X|A}(6) = 0.$$

For a discrete random variable X and event A, the **conditional PMF** of X given A is defined as

$$P_{X|A}(x_i) = P(X = x_i|A)$$
$$= \frac{P(X = x_i \text{ and } A)}{P(A)}, \quad \text{for any } x_i \in R_X.$$

Similarly, we define the **conditional CDF** of X given A as

$$F_{X|A}(x) = P(X \leq x|A).$$

Conditional PMF of X Given Y:

In some problems, we have observed the value of a random variable Y, and we need to update the PMF of another random variable X whose value has not yet been observed. In these problems, we use the **conditional PMF** of X given Y. The conditional PMF of X given Y is defined as

$$
\begin{aligned}
P_{X|Y}(x_i|y_j) &= P(X = x_i|Y = y_j) \\
&= \frac{P(X = x_i, Y = y_j)}{P(Y = y_j)} \\
&= \frac{P_{XY}(x_i, y_j)}{P_Y(y_j)}.
\end{aligned}
$$

Similarly, we can define the conditional probability of Y given X:

$$
\begin{aligned}
P_{Y|X}(y_j|x_i) &= P(Y = y_j|X = x_i) \\
&= \frac{P_{XY}(x_i, y_j)}{P_X(x_i)}.
\end{aligned}
$$

For discrete random variables X and Y, the **conditional PMFs** of X given Y and vice versa are defined as

$$
P_{X|Y}(x_i|y_j) = \frac{P_{XY}(x_i, y_j)}{P_Y(y_j)},
$$

$$
P_{Y|X}(y_j|x_i) = \frac{P_{XY}(x_i, y_j)}{P_X(x_i)}
$$

for any $x_i \in R_X$ and $y_j \in R_Y$.

Independent Random Variables:

We have defined independent random variables previously. Now that we have seen joint PMFs and CDFs, we can restate the independence definition.

Two discrete random variables X and Y are independent if

$$
P_{XY}(x, y) = P_X(x)P_Y(y), \quad \text{for all } x, y.
$$

Equivalently, X and Y are independent if

$$
F_{XY}(x, y) = F_X(x)F_Y(y), \quad \text{for all } x, y.
$$

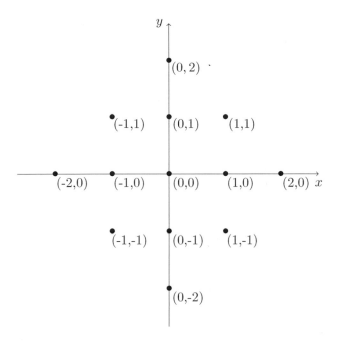

Figure 5.4: Grid for Example 5.4.

So, if X and Y are independent, we have

$$P_{X|Y}(x_i|y_j) = P(X = x_i|Y = y_j)$$
$$= \frac{P_{XY}(x_i, y_j)}{P_Y(y_j)}$$
$$= \frac{P_X(x_i)P_Y(y_j)}{P_Y(y_j)}$$
$$= P_X(x_i).$$

As we expect, for independent random variables, the conditional PMF is equal to the marginal PMF. In other words, knowing the value of Y does not provide any information about X.

Example 5.4. Consider the set of points in the grid shown in Figure 5.4. These are the points in set G defined as

$$G = \{(x, y)|x, y \in \mathbb{Z}, |x| + |y| \le 2\}.$$

Suppose that we pick a point (X, Y) from this grid completely at random. Thus, each point has a probability of $\frac{1}{13}$ of being chosen.

(a) Find the joint and marginal PMFs of X and Y.

(b) Find the conditional PMF of X given $Y = 1$.

(c) Are X and Y independent?

Solution:

(a) Here, note that

$$R_{XY} = G = \{(x,y)|x,y \in \mathbb{Z}, |x| + |y| \leq 2\}.$$

Thus, the joint PMF is given by

$$P_{XY}(x,y) = \begin{cases} \frac{1}{13} & (x,y) \in G \\ 0 & \text{otherwise} \end{cases}$$

To find the marginal PMF of X, $P_X(i)$, we use Equation 5.1. Thus,

$$P_X(-2) = P_{XY}(-2,0) = \frac{1}{13},$$

$$P_X(-1) = P_{XY}(-1,-1) + P_{XY}(-1,0) + P_{XY}(-1,1) = \frac{3}{13},$$

$$P_X(0) = P_{XY}(0,-2) + P_{XY}(0,-1) + P_{XY}(0,0)$$
$$+ P_{XY}(0,1) + P_{XY}(0,2) = \frac{5}{13},$$

$$P_X(1) = P_{XY}(1,-1) + P_{XY}(1,0) + P_{XY}(1,-1) = \frac{3}{13},$$

$$P_X(2) = P_{XY}(2,0) = \frac{1}{13}.$$

Similarly, we can find

$$P_Y(j) = \begin{cases} \frac{1}{13} & \text{for } j = 2,-2 \\ \frac{3}{13} & \text{for } j = -1,1 \\ \frac{5}{13} & \text{for } j = 0 \\ 0 & \text{otherwise} \end{cases}$$

We can write this in a more compact form as

$$P_X(k) = P_Y(k) = \frac{5 - 2|k|}{13}, \quad \text{for } k = -2,-1,0,1,2.$$

(b) For $i = -1,0,1$, we can write

$$P_{X|Y}(i|1) = \frac{P_{XY}(i,1)}{P_Y(1)}$$

$$= \frac{\frac{1}{13}}{\frac{3}{13}} = \frac{1}{3}, \quad \text{for } i = -1,0,1.$$

Thus, we conclude

$$P_{X|Y}(i|1) = \begin{cases} \frac{1}{3} & \text{for } i = -1,0,1 \\ 0 & \text{otherwise} \end{cases}$$

By looking at the above conditional PMF, we conclude that, given $Y = 1$, X is uniformly distributed over the set $\{-1,0,1\}$.

(c) X and Y are **not** independent. We can see this as the conditional PMF of X given $Y = 1$ (calculated above) is not the same as marginal PMF of X, $P_X(x)$.

Conditional Expectation:

Given that we know event A has occurred, we can compute the conditional expectation of a random variable X, $E[X|A]$. Conditional expectation is similar to ordinary expectation. The only difference is that we replace the PMF by the conditional PMF. Specifically, we have

$$E[X|A] = \sum_{x_i \in R_X} x_i P_{X|A}(x_i).$$

Similarly, given that we have observed the value of random variable Y, we can compute the conditional expectation of X. Specifically, the conditional expectation of X given that $Y = y$ is

$$E[X|Y = y] = \sum_{x_i \in R_X} x_i P_{X|Y}(x_i|y).$$

Conditional Expectation of X:

$$E[X|A] = \sum_{x_i \in R_X} x_i P_{X|A}(x_i)$$

$$E[X|Y = y_j] = \sum_{x_i \in R_X} x_i P_{X|Y}(x_i|y_j)$$

Example 5.5. Let X and Y be the same as in Example 5.4.

(a) Find $E[X|Y = 1]$.

(b) Find $E[X| - 1 < Y < 2]$.

(c) Find $E\big[|X|| - 1 < Y < 2\big]$.

Solution:

(a) To find $E[X|Y = 1]$, we have

$$E[X|Y = 1] = \sum_{x_i \in R_X} x_i P_{X|Y}(x_i|1).$$

We found in Example 5.4 that given $Y = 1$, X is uniformly distributed over the set $\{-1, 0, 1\}$. Thus, we conclude that

$$E[X|Y = 1] = \frac{1}{3}(-1 + 0 + 1) = 0.$$

(b) To find $E[X| -1 < Y < 2]$, let A be the event that $-1 < Y < 2$, i.e., $Y \in \{0, 1\}$. To find $E[X|A]$, we need to find the conditional PMF, $P_{X|A}(k)$, for $k = -2, -1, 0, 1, 2$. First, note that

$$P(A) = P_Y(0) + P_Y(1) = \frac{5}{13} + \frac{3}{13} = \frac{8}{13}.$$

Thus, for $k = -2, 1, 0, 1, 2$, we have

$$P_{X|A}(k) = \frac{13}{8} P(X = k, A).$$

So, we can write

$$P_{X|A}(-2) = \frac{13}{8} P(X = -2, A)$$
$$= \frac{13}{8} P_{XY}(-2, 0) = \frac{1}{8},$$
$$P_{X|A}(-1) = \frac{13}{8} P(X = -1, A)$$
$$= \frac{13}{8} \big[P_{XY}(-1, 0) + P_{XY}(-1, 1) \big] = \frac{2}{8} = \frac{1}{4},$$
$$P_{X|A}(\ 0\ ) = \frac{13}{8} P(X = 0, A)$$
$$= \frac{13}{8} \big[P_{XY}(0, 0) + P_{XY}(0, 1) \big] = \frac{2}{8} = \frac{1}{4},$$
$$P_{X|A}(\ 1\ ) = \frac{13}{8} P(X = 1, A)$$
$$= \frac{13}{8} \big[P_{XY}(1, 0) + P_{XY}(1, 1) \big] = \frac{2}{8} = \frac{1}{4},$$
$$P_{X|A}(\ 2\ ) = \frac{13}{8} P(X = 2, A)$$
$$= \frac{13}{8} P_{XY}(2, 0) = \frac{1}{8}.$$

Thus, we have

$$E[X|A] = \sum_{x_i \in R_X} x_i P_{X|A}(x_i)$$
$$= (-2)\frac{1}{8} + (-1)\frac{1}{4} + (0)\frac{1}{4} + (1)\frac{1}{4} + (2)\frac{1}{8} = 0.$$

(c) To find $E\big[|X|| -1 < Y < 2\big]$, we use the conditional PMF and LOTUS. We have

$$E[|X||A] = \sum_{x_i \in R_X} |x_i| P_{X|A}(x_i)$$
$$= |-2| \cdot \frac{1}{8} + |-1| \cdot \frac{1}{4} + 0 \cdot \frac{1}{4} + 1 \cdot \frac{1}{4} + 2 \cdot \frac{1}{8} = 1.$$

Conditional expectation has some interesting properties that are used commonly in practice. Thus, we will revisit conditional expectation in Section 5.1.5, where we discuss properties of conditional expectation, conditional variance, and their applications.

Law of Total Probability:

Remember the law of total probability: If $B_1, B_2, B_3, ...$ is a partition of the sample space S, then for any event A we have

$$P(A) = \sum_i P(A \cap B_i) = \sum_i P(A|B_i)P(B_i).$$

If Y is a discrete random variable with range $R_Y = \{y_1, y_2, ...\}$, then the events $\{Y = y_1\}$, $\{Y = y_2\}$, $\{Y = y_3\}$, $\cdots$ form a partition of the sample space. Thus, we can use the law of total probability. In fact, we have already used the law of total probability to find the marginal PMFs:

$$P_X(x) = \sum_{y_j \in R_Y} P_{XY}(x, y_j) = \sum_{y_j \in R_Y} P_{X|Y}(x|y_j)P_Y(y_j).$$

We can write this more generally as

$$P(X \in A) = \sum_{y_j \in R_Y} P(X \in A|Y = y_j)P_Y(y_j), \quad \text{for any set } A.$$

We can write a similar formula for expectation as well. Indeed, if $B_1, B_2, B_3, ...$ is a partition of the sample space S, then

$$EX = \sum_i E[X|B_i]P(B_i).$$

To see this, just write the definition of $E[X|B_i]$ and apply the law of total probability. The above equation is sometimes called the law of total expectation [2].

Law of Total Probability:

$$P(X \in A) = \sum_{y_j \in R_Y} P(X \in A|Y = y_j)P_Y(y_j), \quad \text{for any set } A.$$

Law of Total Expectation:

1. If $B_1, B_2, B_3, ...$ is a partition of the sample space S,

$$EX = \sum_i E[X|B_i]P(B_i). \tag{5.3}$$

2. For a random variable X and a discrete random variable Y,

$$EX = \sum_{y_j \in R_Y} E[X|Y = y_j]P_Y(y_j). \tag{5.4}$$

Example 5.6. Let $X \sim Geometric(p)$. Find EX by conditioning on the result of the first "coin toss".

Solution: Remember that the random experiment behind $Geometric(p)$ is that we have a coin with $P(H) = p$. We toss the coin repeatedly until we observe the first heads. X is the total number of coin tosses. Now, there are two possible outcomes for the first coin toss: H or T. Thus, we can use the law of total expectation (Equation 5.3):

$$EX = E[X|H]P(H) + E[X|T]P(T)$$
$$= pE[X|H] + (1-p)E[X|T]$$
$$= p \cdot 1 + (1-p)(EX + 1).$$

In this equation, $E[X|T] = 1 + EX$, because the tosses are independent, so if the first toss is tails, it is like starting over on the second toss. Solving for EX, we obtain

$$EX = \frac{1}{p}.$$

Example 5.7. Suppose that the number of customers visiting a fast food restaurant in a given day is $N \sim Poisson(\lambda)$. Assume that each customer purchases a drink with probability p, independently from other customers and independently from the value of N. Let X be the number of customers who purchase drinks. Find EX.

Solution: By the above information, we conclude that given $N = n$, then X is a sum of n independent $Bernoulli(p)$ random variables. Thus, given $N = n$, X has a binomial distribution with parameters n and p. We write

$$X|N = n \sim Binomial(n, p).$$

That is,

$$P_{X|N}(k|n) = \binom{n}{k}p^k(1-p)^{n-k}.$$

Thus, we conclude

$$E[X|N = n] = np.$$

Thus, using the law of total probability, we have

$$E[X] = \sum_{n=0}^{\infty} E[X|N = n]P_N(n)$$
$$= \sum_{n=0}^{\infty} npP_N(n)$$
$$= p\sum_{n=0}^{\infty} nP_N(n) = pE[N] = p\lambda.$$

5.1.4 Functions of two random variables

Analysis of a function of two random variables is pretty much the same as for a function of a single random variable. Suppose that you have two discrete random variables X and Y, and suppose that $Z = g(X, Y)$, where $g : \mathbb{R}^2 \mapsto \mathbb{R}$. Then, if we are interested in the PMF of Z, we can write

$$P_Z(z) = P(g(X, Y) = z)$$
$$= \sum_{(x_i, y_j) \in A_z} P_{XY}(x_i, y_j), \quad \text{where } A_z = \{(x_i, y_j) \in R_{XY} : g(x_i, y_j) = z\}.$$

Note that if we are only interested in $E[g(X, Y)]$, we can directly use LOTUS, without finding $P_Z(z)$:

Law of the unconscious statistician (LOTUS) for two discrete random variables:

$$E[g(X, Y)] = \sum_{(x_i, y_j) \in R_{XY}} g(x_i, y_j) P_{XY}(x_i, y_j) \tag{5.5}$$

Example 5.8. Linearity of Expectation: For two discrete random variables X and Y, show that $E[X + Y] = EX + EY$.

Solution: Let $g(X, Y) = X + Y$. Using LOTUS, we have

$$E[X + Y] = \sum_{(x_i, y_j) \in R_{XY}} (x_i + y_j) P_{XY}(x_i, y_j)$$
$$= \sum_{(x_i, y_j) \in R_{XY}} x_i P_{XY}(x_i, y_j) + \sum_{(x_i, y_j) \in R_{XY}} y_j P_{XY}(x_i, y_j)$$
$$= \sum_{x_i \in R_X} \sum_{y_j \in R_Y} x_i P_{XY}(x_i, y_j) + \sum_{x_i \in R_X} \sum_{y_j \in R_Y} y_j P_{XY}(x_i, y_j)$$
$$= \sum_{x_i \in R_X} x_i \sum_{y_j \in R_Y} P_{XY}(x_i, y_j) + \sum_{y_j \in R_Y} y_j \sum_{x_i \in R_X} P_{XY}(x_i, y_j)$$
$$= \sum_{x_i \in R_X} x_i P_X(x_i) + \sum_{y_j \in R_Y} y_j P_Y(y_j) \quad \text{(marginal PMF (Equation 5.1))}$$
$$= EX + EY.$$

Example 5.9. Let X and Y be two independent *Geometric(p)* random variables. Also let $Z = X - Y$. Find the PMF of Z.

Solution: First note that since $R_X = R_Y = \mathbb{N} = \{1, 2, 3, ...\}$, we have $R_Z = \mathbb{Z} = \{..., -3, -2, -1, 0, 1, 2, 3, ...\}$. Since $X, Y \sim Geometric(p)$, we have

$$P_X(k) = P_Y(k) = pq^{k-1}, \quad \text{for } k = 1, 2, 3, ...,$$

where $q = 1 - p$. We can write for any $k \in \mathbb{Z}$

$$
\begin{aligned}
P_Z(k) &= P(Z = k) \\
&= P(X - Y = k) \\
&= P(X = Y + k) \\
&= \sum_{j=1}^{\infty} P(X = Y + k | Y = j) P(Y = j) \qquad \text{(law of total probability)} \\
&= \sum_{j=1}^{\infty} P(X = j + k | Y = j) P(Y = j) \\
&= \sum_{j=1}^{\infty} P(X = j + k) P(Y = j) \qquad \text{(since } X, Y \text{ are independent)} \\
&= \sum_{j=1}^{\infty} P_X(j + k) P_Y(j).
\end{aligned}
$$

Now, consider two cases: $k \geq 0$ and $k < 0$. If $k \geq 0$, then

$$
\begin{aligned}
P_Z(k) &= \sum_{j=1}^{\infty} P_X(j + k) P_Y(j) \\
&= \sum_{j=1}^{\infty} pq^{j+k-1} pq^{j-1} \\
&= p^2 q^k \sum_{j=1}^{\infty} q^{2(j-1)} \\
&= p^2 q^k \frac{1}{1 - q^2} \qquad \text{(geometric sum (Equation 1.4))} \\
&= \frac{p(1-p)^k}{2 - p}.
\end{aligned}
$$

For $k < 0$, we have

$$
\begin{aligned}
P_Z(k) &= \sum_{j=1}^{\infty} P_X(j + k) P_Y(j) \\
&= \sum_{j=-k+1}^{\infty} pq^{j+k-1} pq^{j-1} \qquad \text{(since } P_X(j+k) = 0 \text{ for } j < -k+1) \\
&= p^2 \sum_{j=-k+1}^{\infty} q^{k+2(j-1)} \\
&= p^2 \left[q^{-k} + q^{-k+2} + q^{-k+4} + \ldots \right] \\
&= p^2 q^{-k} \left[1 + q^2 + q^4 + \ldots \right] \\
&= \frac{p}{(1-p)^k (2 - p)} \qquad \text{(geometric sum (Equation 1.4)).}
\end{aligned}
$$

To summarize, we conclude

$$P_Z(k) = \begin{cases} \frac{p(1-p)^{|k|}}{2-p} & k \in \mathbb{Z} \\ 0 & \text{otherwise} \end{cases}$$

5.1.5 Conditional Expectation (Revisited) and Conditional Variance

In Section 5.1.3, we briefly discussed conditional expectation. Here, we will discuss the properties of conditional expectation in more detail as they are quite useful in practice. We will also discuss conditional variance. An important concept here is that we interpret the conditional expectation as a random variable.

Conditional Expectation as a Function of a Random Variable:

Remember that the conditional expectation of X given that $Y = y$ is given by

$$E[X|Y = y] = \sum_{x_i \in R_X} x_i P_{X|Y}(x_i|y).$$

Note that $E[X|Y = y]$ depends on the value of y. In other words, by changing y, $E[X|Y = y]$ can also change. Thus, we can say $E[X|Y = y]$ is a function of y, so let's write

$$g(y) = E[X|Y = y].$$

Thus, we can think of $g(y) = E[X|Y = y]$ as a function of the value of random variable Y. We then write

$$g(Y) = E[X|Y].$$

We use this notation to indicate that $E[X|Y]$ is a random variable whose value equals $g(y) = E[X|Y = y]$ when $Y = y$. Thus, if Y is a random variable with range $R_Y = \{y_1, y_2, \cdots\}$, then $E[X|Y]$ is also a random variable with

$$E[X|Y] = \begin{cases} E[X|Y = y_1] & \text{with probability } P(Y = y_1) \\ E[X|Y = y_2] & \text{with probability } P(Y = y_2) \\ \cdot & \cdot \\ \cdot & \cdot \\ \cdot & \cdot \end{cases}$$

Let's look at an example.

Example 5.10. Let $X = aY + b$. Then $E[X|Y = y] = E[aY + b|Y = y] = ay + b$. Here, we have $g(y) = ay + b$, and therefore,

$$E[X|Y] = aY + b,$$

which is a function of the random variable Y.

Since $E[X|Y]$ is a random variable, we can find its PMF, CDF, variance, etc. Let's look at an example to better understand $E[X|Y]$.

Example 5.11. Consider two random variables X and Y with joint PMF given in Table 5.2. Let $Z = E[X|Y]$.

(a) Find the Marginal PMFs of X and Y.

(b) Find the conditional PMF of X given $Y = 0$ and $Y = 1$, i.e., find $P_{X|Y}(x|0)$ and $P_{X|Y}(x|1)$.

(c) Find the *PMF* of Z.

(d) Find EZ, and check that $EZ = EX$.

(e) Find $\text{Var}(Z)$.

Table 5.2: Joint PMF of X and Y in Example 5.11

	$Y = 0$	$Y = 1$
$X = 0$	$\frac{1}{5}$	$\frac{2}{5}$
$X = 1$	$\frac{2}{5}$	0

Solution:

(a) Using the table we find out

$$P_X(0) = \frac{1}{5} + \frac{2}{5} = \frac{3}{5},$$
$$P_X(1) = \frac{2}{5} + 0 = \frac{2}{5},$$
$$P_Y(0) = \frac{1}{5} + \frac{2}{5} = \frac{3}{5},$$
$$P_Y(1) = \frac{2}{5} + 0 = \frac{2}{5}.$$

Thus, the marginal distributions of X and Y are both $Bernoulli(\frac{2}{5})$. However, note that X and Y are not independent.

(b) We have

$$P_{X|Y}(0|0) = \frac{P_{XY}(0,0)}{P_Y(0)}$$
$$= \frac{\frac{1}{5}}{\frac{3}{5}} = \frac{1}{3}.$$

Thus,

$$P_{X|Y}(1|0) = 1 - \frac{1}{3} = \frac{2}{3}.$$

We conclude

$$X|Y = 0 \ \sim \ Bernoulli\left(\frac{2}{3}\right).$$

Similarly, we find

$$P_{X|Y}(0|1) = 1$$
$$P_{X|Y}(1|1) = 0.$$

Thus, given $Y = 1$, we have always $X = 0$.

(c) We note that the random variable Y can take two values: 0 and 1. Thus, the random variable $Z = E[X|Y]$ can take two values as it is a function of Y. Specifically,

$$Z = E[X|Y] = \begin{cases} E[X|Y = 0] & \text{if } Y = 0 \\ \\ E[X|Y = 1] & \text{if } Y = 1 \end{cases}$$

Now, using the previous part, we have

$$E[X|Y = 0] = \frac{2}{3}, \quad E[X|Y = 1] = 0,$$

and since $P(y = 0) = \frac{3}{5}$, and $P(y = 1) = \frac{2}{5}$, we conclude that

$$Z = E[X|Y] = \begin{cases} \frac{2}{3} & \text{with probability } \frac{3}{5} \\ \\ 0 & \text{with probability } \frac{2}{5} \end{cases}$$

So we can write

$$P_Z(z) = \begin{cases} \frac{3}{5} & \text{if } z = \frac{2}{3} \\ \\ \frac{2}{5} & \text{if } z = 0 \\ \\ 0 & \text{otherwise} \end{cases}$$

(d) Now that we have found the PMF of Z, we can find its mean and variance. Specifically,

$$E[Z] = \frac{2}{3} \cdot \frac{3}{5} + 0 \cdot \frac{2}{5} = \frac{2}{5}.$$

We also note that $EX = \frac{2}{5}$. Thus, here we have

$$E[X] = E[Z] = E[E[X|Y]].$$

In fact, as we will prove shortly, the above equality always holds. It is called the law of iterated expectations.

(e) To find $\text{Var}(Z)$, we write

$$\text{Var}(Z) = E[Z^2] - (EZ)^2$$
$$= E[Z^2] - \frac{4}{25},$$

where

$$E[Z^2] = \frac{4}{9} \cdot \frac{3}{5} + 0 \cdot \frac{2}{5} = \frac{4}{15}.$$

Thus,

$$\text{Var}(Z) = \frac{4}{15} - \frac{4}{25}$$
$$= \frac{8}{75}.$$

Example 5.12. Let X and Y be two random variables and g and h be two functions. Show that

$$E[g(X)h(Y)|X] = g(X)E[h(Y)|X].$$

Solution: Note that $E[g(X)h(Y)|X]$ is a random variable that is a function of X. In particular, if $X = x$, then $E[g(X)h(Y)|X] = E[g(X)h(Y)|X = x]$. Now, we can write

$$E[g(X)h(Y)|X = x] = E[g(x)h(Y)|X = x]$$
$$= g(x)E[h(Y)|X = x] \qquad \text{since } g(x) \text{ is a constant.}$$

Thinking of this as a function of the random variable X, it can be rewritten as $E[g(X)h(Y)|X] = g(X)E[h(Y)|X]$. This rule is sometimes called "taking out what is known." The idea is that, given X, $g(X)$ is a known quantity, so it can be taken out of the conditional expectation.

$$\boxed{E[g(X)h(Y)|X] = g(X)E[h(Y)|X] \qquad (5.6)}$$

Iterated Expectations:

Let us look again at the law of total probability for expectation. Assuming $g(Y) = E[X|Y]$, we have

$$E[X] = \sum_{y_j \in R_Y} E[X|Y = y_j]P_Y(y_j)$$
$$= \sum_{y_j \in R_Y} g(y_j)P_Y(y_j)$$
$$= E[g(Y)] \qquad \text{by LOTUS (Equation 3.2)}$$
$$= E[E[X|Y]].$$

Thus, we conclude

$$E[X] = E[E[X|Y]]. \tag{5.7}$$

This equation might look a little confusing at first, but it is just another way of writing the law of total expectation (Equation 5.4). To better understand it, let's solve Example 5.7 using this terminology. In that example, we want to find EX. We can write

$$\begin{aligned} E[X] &= E[E[X|N]] \\ &= E[Np] \qquad \big(\text{since } X|N \sim Binomial(N,p)\big) \\ &= pE[N] = p\lambda. \end{aligned}$$

Equation 5.7 is called the *law of iterated expectations*. Since it is basically the same as Equation 5.4, it is also called the law of total expectation [2].

Law of Iterated Expectations: $E[X] = E[E[X|Y]]$

Expectation for Independent Random Variables:

Note that if two random variables X and Y are independent, then the conditional PMF of X given Y will be the same as the marginal PMF of X, i.e., for any $x \in R_X$, we have

$$P_{X|Y}(x|y) = P_X(x).$$

Thus, for independent random variables, we have

$$\begin{aligned} E[X|Y=y] &= \sum_{x \in R_X} x P_{X|Y}(x|y) \\ &= \sum_{x \in R_X} x P_X(x) \\ &= E[X]. \end{aligned}$$

Again, thinking of this as a random variable depending on Y, we obtain

$$E[X|Y] = E[X], \qquad \text{when } X \text{ and } Y \text{ are independent.}$$

More generally, if X and Y are independent then any function of X, say $g(X)$, and Y are independent, thus

$$E[g(X)|Y] = E[g(X)].$$

Remember that for independent random variables, $P_{XY}(x,y) = P_X(x)P_Y(y)$. From this, we can show that $E[XY] = EXEY$.

Lemma 5.2. If X and Y are independent, then $E[XY] = EXEY$.

Proof. Using LOTUS, we have

$$E[XY] = \sum_{x \in R_x} \sum_{y \in R_y} xy P_{XY}(x,y)$$

$$= \sum_{x \in R_x} \sum_{y \in R_y} xy P_X(x) P_Y(y)$$

$$= \left(\sum_{x \in R_x} x P_X(x) \right) \left(\sum_{y \in R_y} y P_Y(y) \right)$$

$$= EXEY.$$

∎

Note that the converse is **not** true. That is, if the only thing that we know about X and Y is that $E[XY] = EXEY$, then X and Y may or may not be independent. Using essentially the same proof as above, we can show if X and Y are independent, then $E[g(X)h(Y)] = E[g(X)]E[h(Y)]$ for any functions $g : \mathbb{R} \mapsto \mathbb{R}$ and $h : \mathbb{R} \mapsto \mathbb{R}$.

If X and Y are independent random variables, then

1. $E[X|Y] = EX$;

2. $E[g(X)|Y] = E[g(X)]$;

3. $E[XY] = EXEY$;

4. $E[g(X)h(Y)] = E[g(X)]E[h(Y)]$.

Conditional Variance:

Similar to the conditional expectation, we can define the conditional variance of X, $\text{Var}(X|Y = y)$, which is the variance of X in the conditional space where we know $Y = y$. If we let $\mu_{X|Y}(y) = E[X|Y = y]$, then

$$\text{Var}(X|Y = y) = E\big[(X - \mu_{X|Y}(y))^2 | Y = y\big]$$

$$= \sum_{x_i \in R_X} \big(x_i - \mu_{X|Y}(y)\big)^2 P_{X|Y}(x_i)$$

$$= E\big[X^2 | Y = y\big] - \mu_{X|Y}(y)^2.$$

Note that $\text{Var}(X|Y = y)$ is a function of y. Similar to our discussion on $E[X|Y = y]$ and $E[X|Y]$, we define $\text{Var}(X|Y)$ as a function of the random variable Y. That is, $\text{Var}(X|Y)$ is a random variable whose value equals $\text{Var}(X|Y = y)$ whenever $Y = y$. Let us look at an example.

Example 5.13. Let X, Y, and $Z = E[X|Y]$ be as in Example 5.11. Let also $V = \text{Var}(X|Y)$.

(a) Find the PMF of V.

(b) Find EV.

(c) Check that $\text{Var}(X) = E(V) + \text{Var}(Z)$.

Solution: In Example 5.11, we found out that $X, Y \sim Bernoulli(\frac{2}{5})$. We also obtained

$$X|Y = 0 \sim Bernoulli\left(\frac{2}{3}\right),$$

$$P(X = 0|Y = 1) = 1,$$

$$\text{Var}(Z) = \frac{8}{75}.$$

(a) To find the PMF of V, we note that V is a function of Y. Specifically,

$$V = \text{Var}(X|Y) = \begin{cases} \text{Var}(X|Y = 0) & \text{if } Y = 0 \\ \text{Var}(X|Y = 1) & \text{if } Y = 1 \end{cases}$$

Therefore,

$$V = \text{Var}(X|Y) = \begin{cases} \text{Var}(X|Y = 0) & \text{with probability } \frac{3}{5} \\ \text{Var}(X|Y = 1) & \text{with probability } \frac{2}{5} \end{cases}$$

Now, since $X|Y = 0 \sim Bernoulli\left(\frac{2}{3}\right)$, we have

$$\text{Var}(X|Y = 0) = \frac{2}{3} \cdot \frac{1}{3} = \frac{2}{9},$$

and since given $Y = 1$, $X = 0$, we have

$$\text{Var}(X|Y = 1) = 0.$$

Thus,

$$V = \text{Var}(X|Y) = \begin{cases} \frac{2}{9} & \text{with probability } \frac{3}{5} \\ 0 & \text{with probability } \frac{2}{5} \end{cases}$$

So we can write

$$P_V(v) = \begin{cases} \frac{3}{5} & \text{if } v = \frac{2}{9} \\ \frac{2}{5} & \text{if } v = 0 \\ 0 & \text{otherwise} \end{cases}$$

(b) To find EV, we write

$$EV = \frac{2}{9} \cdot \frac{3}{5} + 0 \cdot \frac{2}{5} = \frac{2}{15}.$$

(c) To check that $\text{Var}(X) = E(V) + \text{Var}(Z)$, we just note that

$$\text{Var}(X) = \frac{2}{5} \cdot \frac{3}{5} = \frac{6}{25},$$
$$EV = \frac{2}{15},$$
$$\text{Var}(Z) = \frac{8}{75}.$$

In the above example, we checked that $\text{Var}(X) = E(V) + \text{Var}(Z)$, which says

$$\text{Var}(X) = E(\text{Var}(X|Y)) + \text{Var}(E[X|Y]).$$

It turns out this is true in general and it is called *the law of total variance*, or *variance decomposition formula* [3]. Let us first prove the law of total variance, and then we explain it intuitively. Note that if $V = \text{Var}(X|Y)$, and $Z = E[X|Y]$, then

$$V = E[X^2|Y] - (E[X|Y])^2$$
$$= E[X^2|Y] - Z^2.$$

Thus,

$$EV = E[E[X^2|Y]] - E[Z^2]$$
$$= E[X^2] - E[Z^2] \qquad \text{(law of iterated expectations(Equation 5.7))} \qquad (5.8)$$

Next, we have

$$\text{Var}(Z) = E[Z^2] - (EZ)^2$$
$$= E[Z^2] - (EX)^2 \qquad \text{(law of iterated expectations)} \qquad (5.9)$$

Combining Equations 5.8 and 5.9, we obtain the law of total variance.

Law of Total Variance:

$$\text{Var}(X) = E[\text{Var}(X|Y)] + \text{Var}(E[X|Y]) \qquad (5.10)$$

There are several ways that we can look at the law of total variance to get some intuition. Let us first note that all the terms in Equation 5.10 are positive (since variance is always positive). Thus, we conclude

$$\text{Var}(X) \geq E(\text{Var}(X|Y)) \qquad (5.11)$$

This states that when we condition on Y, the variance of X reduces on average. To describe this intuitively, we can say that variance of a random variable is a measure of our uncertainty about that random variable. For example, if $\text{Var}(X) = 0$, we do not have any uncertainty about X. Now, the above inequality simply states that if we obtain some extra information, i.e., we know the value of Y, our uncertainty about the value of the random variable X reduces on average. So, the above inequality makes sense. Now, how do we explain the whole law of total variance?

To describe the law of total variance intuitively, it is often useful to look at a population divided into several groups. In particular, suppose that we have this random experiment: We pick a person in the world at random and look at his/her height. Let's call the resulting value X. Define another random variable Y whose value depends on the country of the chosen person, where $Y = 1, 2, 3, ..., n$, and n is the number of countries in the world. Then, let's look at the two terms in the law of total variance.

$$\text{Var}(X) = E(\text{Var}(X|Y)) + \text{Var}(E[X|Y]).$$

Note that $\text{Var}(X|Y = i)$ is the variance of X in country i. Thus, $E(\text{Var}(X|Y))$ is the average of variances in each country. On the other hand, $E[X|Y = i]$ is the average height in country i. Thus, $\text{Var}(E[X|Y])$ is the variance between countries. So, we can interpret the law of total variance in the following way. Variance of X can be decomposed into two parts: the first is the average of variances in each individual country, while the second is the variance between height averages in each country.

Example 5.14. Let N be the number of customers that visit a certain store in a given day. Suppose that we know $E[N]$ and $\text{Var}(N)$. Let X_i be the amount that the ith customer spends on average. We assume X_i's are independent of each other and also independent of N. We further assume they have the same mean and variance

$$EX_i = EX,$$
$$\text{Var}(X_i) = \text{Var}(X).$$

Let Y be the store's total sales, i.e.,

$$Y = \sum_{i=1}^{N} X_i.$$

Find EY and $\text{Var}(Y)$.

Solution: To find EY, we cannot directly use the linearity of expectation because N is random. But, conditioned on $N = n$, we can use linearity and find $E[Y|N = n]$; so, we use the law of iterated expectations:

$$EY = E[E[Y|N]] \qquad \text{(law of iterated expectations)}$$

$$= E\left[E\left[\sum_{i=1}^{N} X_i|N\right]\right]$$

$$= E\left[\sum_{i=1}^{N} E[X_i|N]\right] \qquad \text{(linearity of expectation)}$$

$$= E\left[\sum_{i=1}^{N} E[X_i]\right] \qquad (X_i\text{'s and } N \text{ are indpendent)}$$

$$= E[NE[X]] \qquad (\text{since } EX_i = EX\text{s})$$

$$= E[X]E[N] \qquad (\text{since } EX \text{ is not random}).$$

To find $\text{Var}(Y)$, we use the law of total variance:

$$\text{Var}(Y) = E(\text{Var}(Y|N)) + \text{Var}(E[Y|N])$$
$$= E(\text{Var}(Y|N)) + \text{Var}(NEX) \qquad \text{(as above)}$$
$$= E(\text{Var}(Y|N)) + (EX)^2\text{Var}(N) \qquad (5.12)$$

To find $E(\text{Var}(Y|N))$, note that, given $N = n$, Y is a sum of n independent random variables. As we discussed before, for n independent random variables, the variance of the sum is equal to sum of the variances. This fact is officially proved in Section 5.3 and also in Chapter 6, but we have occasionally used it as it simplifies the analysis. Thus, we can write

$$\text{Var}(Y|N) = \sum_{i=1}^{N} \text{Var}(X_i|N)$$

$$= \sum_{i=1}^{N} \text{Var}(X_i) \qquad (\text{since } X_i\text{'s are independent of } N)$$

$$= N\text{Var}(X).$$

Thus, we have

$$E(\text{Var}(Y|N)) = EN\text{Var}(X) \qquad (5.13)$$

Combining Equations 5.12 and 5.13, we obtain

$$\text{Var}(Y) = EN\text{Var}(X) + (EX)^2\text{Var}(N).$$

5.1.6   Solved Problems

1. Consider two random variables X and Y with joint PMF given in Table 5.3.

    (a) Find $P(X \leq 2, Y \leq 4)$.

    (b) Find the marginal PMFs of X and Y.

Table 5.3: Joint PMF of X and Y in Example 1

	$Y = 2$	$Y = 4$	$Y = 5$
$X = 1$	$\frac{1}{12}$	$\frac{1}{24}$	$\frac{1}{24}$
$X = 2$	$\frac{1}{6}$	$\frac{1}{12}$	$\frac{1}{8}$
$X = 3$	$\frac{1}{4}$	$\frac{1}{8}$	$\frac{1}{12}$

(c) Find $P(Y = 2|X = 1)$.

(d) Are X and Y independent?

Solution:

(a) To find $P(X \leq 2, Y \leq 4)$, we can write

$$P(X \leq 2, Y \leq 4) = P_{XY}(1,2) + P_{XY}(1,4) + P_{XY}(2,2) + P_{XY}(2,4)$$
$$= \frac{1}{12} + \frac{1}{24} + \frac{1}{6} + \frac{1}{12} = \frac{3}{8}.$$

(b) Note from the table that

$$R_X = \{1,2,3\} \quad \text{and} \quad R_Y = \{2,4,5\}.$$

Now we can use Equation 5.1 to find the marginal PMFs:

$$P_X(x) = \begin{cases} \frac{1}{6} & x = 1 \\ \frac{3}{8} & x = 2 \\ \frac{11}{24} & x = 3 \\ 0 & \text{otherwise} \end{cases}$$

$$P_Y(y) = \begin{cases} \frac{1}{2} & y = 2 \\ \frac{1}{4} & y = 4 \\ \frac{1}{4} & y = 5 \\ 0 & \text{otherwise} \end{cases}$$

(c) Using the formula for conditional probability, we have

$$P(Y = 2|X = 1) = \frac{P(X = 1, Y = 2)}{P(X = 1)}$$
$$= \frac{P_{XY}(1, 2)}{P_X(1)}$$
$$= \frac{\frac{1}{12}}{\frac{1}{6}} = \frac{1}{2}.$$

(d) Are X and Y independent? To check whether X and Y are independent, we need to check that $P(X = x_i, Y = y_j) = P(X = x_i)P(Y = y_j)$, for all $x_i \in R_X$ and all $y_j \in R_Y$. Looking at the table and the results from previous parts, we find

$$P(X = 2, Y = 2) = \frac{1}{6} \neq P(X = 2)P(Y = 2) = \frac{3}{16}.$$

Thus, we conclude that X and Y are not independent.

2. I have a bag containing 40 blue marbles and 60 red marbles. I choose 10 marbles (without replacement) at random. Let X be the number of blue marbles and y be the number of red marbles. Find the joint PMF of X and Y.

Solution: This is, in fact, a hypergeometric distribution. First, note that we must have $X + Y = 10$, so

$$R_{XY} = \{(i, j)|i + j = 10, i, j \in \mathbb{Z}, i, j \geq 0\}$$
$$= \{(0, 10), (1, 9), (2, 8), ..., (10, 0)\}.$$

Then, we can write

$$P_{XY}(i, j) = \begin{cases} \frac{\binom{40}{i}\binom{60}{j}}{\binom{100}{10}} & i + j = 10, i, j \in \mathbb{Z}, i, j \geq 0 \\ 0 & \text{otherwise} \end{cases}$$

3. Let X and Y be two independent discrete random variables with the same CDFs F_X and F_Y . Define

$$Z = \max(X, Y),$$
$$W = \min(X, Y).$$

Find the CDFs of Z and W.

Solution: To find the CDF of Z, we can write

$$
\begin{aligned}
F_Z(z) &= P(Z \leq z) \\
&= P(\max(X, Y) \leq z) \\
&= P\Big((X \leq z) \text{ and } (Y \leq z)\Big) \\
&= P(X \leq z)P(Y \leq z) \qquad \text{(since } X \text{ and } Y \text{ are independent)} \\
&= F_X(z)F_Y(z).
\end{aligned}
$$

To find the CDF of W, we can write

$$
\begin{aligned}
F_W(w) &= P(W \leq w) \\
&= P(\min(X, Y) \leq w) \\
&= 1 - P(\min(X, Y) > w) \\
&= 1 - P\Big((X > w) \text{ and } (Y > w)\Big) \\
&= 1 - P(X > w)P(Y > w) \qquad \text{(since } X \text{ and } Y \text{ are independent)} \\
&= 1 - (1 - F_X(w))(1 - F_Y(w)) \\
&= F_X(w) + F_Y(w) - F_X(w)F_Y(w).
\end{aligned}
$$

4. Let X and Y be two discrete random variables, with range

$$
R_{XY} = \{(i, j) \in \mathbb{Z}^2 | i, j \geq 0, |i - j| \leq 1\},
$$

and joint PMF given by

$$
P_{XY}(i, j) = \frac{1}{6 \cdot 2^{\min(i,j)}}, \qquad \text{for } (i, j) \in R_{XY}.
$$

(a) Pictorially show R_{XY} in the $x - y$ plane.
(b) Find the marginal PMFs $P_X(i)$, $P_Y(j)$.
(c) Find $P(X = Y | X < 2)$.
(d) Find $P(1 \leq X^2 + Y^2 \leq 5)$.
(e) Find $P(X = Y)$.
(f) Find $E[X | Y = 2]$.
(g) Find $\text{Var}(X | Y = 2)$.

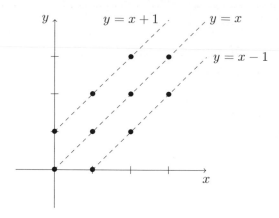

Figure 5.5: Figure shows R_{XY} for X and Y in Problem 4.

Solution:

(a) Figure 5.5 shows the R_{XY} in the $x - y$ plane.

(b) First, by symmetry we note that X and Y have the same PMF. Next, we can write

$$P_X(0) = P_{XY}(0,0) + P_{XY}(0,1) = \frac{1}{6} + \frac{1}{6} = \frac{1}{3},$$

$$P_X(1) = P_{XY}(1,0) + P_{XY}(1,1) + P_{XY}(1,2) = \frac{1}{6}\left(1 + \frac{1}{2} + \frac{1}{2}\right) = \frac{1}{3},$$

$$P_X(2) = P_{XY}(2,1) + P_{XY}(2,2) + P_{XY}(2,3) = \frac{1}{6}\left(\frac{1}{2} + \frac{1}{4} + \frac{1}{4}\right) = \frac{1}{6},$$

$$P_X(3) = P_{XY}(3,2) + P_{XY}(3,3) + P_{XY}(3,4) = \frac{1}{6}\left(\frac{1}{4} + \frac{1}{8} + \frac{1}{8}\right) = \frac{1}{12}.$$

In general, we obtain

$$P_X(k) = P_Y(k) = \begin{cases} \frac{1}{3} & k = 0 \\ \frac{1}{3 \cdot 2^{k-1}} & k = 1, 2, 3, \dots \\ 0 & \text{otherwise} \end{cases}$$

(c) Find $P(X = Y | X < 2)$: We have

$$P(X = Y | X < 2) = \frac{P(X = Y, X < 2)}{P(X < 2)}$$

$$= \frac{P_{XY}(0,0) + P_{XY}(1,1)}{P_X(0) + P_X(1)}$$

$$= \frac{\frac{1}{6} + \frac{1}{12}}{\frac{1}{3} + \frac{1}{3}}$$

$$= \frac{3}{8}.$$

(d) Find $P(1 \le X^2 + Y^2 \le 5)$: We have

$$P(1 \le X^2 + Y^2 \le 5) = P_{XY}(0,1) + P_{XY}(1,0) + P_{XY}(1,1) + P_{XY}(1,2) + P_{XY}(2,1)$$

$$= \frac{1}{6} + \frac{1}{6} + \frac{1}{12} + \frac{1}{12} + \frac{1}{12}$$

$$= \frac{7}{12}.$$

(e) By symmetry, we can argue that $P(X = Y) = \frac{1}{3}$. The reason is that R_{XY} consists of three lines with points with the same probabilities. We can also find $P(X = Y)$ by

$$P(X = Y) = \sum_{i=0}^{\infty} P_{XY}(i,i)$$

$$= \sum_{i=0}^{\infty} \frac{1}{6.2^i}$$

$$= \frac{1}{3}.$$

(f) To find $E[X|Y = 2]$, we first need the conditional PMF of X given $Y = 2$. We have

$$P_{X|Y}(k|2) = \frac{P_{XY}(k,2)}{P(Y = 2)}$$

$$= 6P_{XY}(k,2),$$

so we obtain

$$P_{X|Y}(k|2) = \begin{cases} \frac{1}{2} & k = 1 \\ \frac{1}{4} & k = 2,3 \\ 0 & \text{otherwise} \end{cases}$$

Thus,

$$E[X|Y = 2] = 1 \cdot \frac{1}{2} + 2 \cdot \frac{1}{4} + 3 \cdot \frac{1}{4}$$

$$= \frac{7}{4}.$$

(g) Find $\mathrm{Var}(X|Y=2)$: we have

$$E[X^2|Y=2] = 1 \cdot \frac{1}{2} + 4 \cdot \frac{1}{4} + 9 \cdot \frac{1}{4}$$
$$= \frac{15}{4}.$$

Thus,

$$\mathrm{Var}(X|Y=2) = E[X^2|Y=2] - \big(E[X|Y=2]\big)^2$$
$$= \frac{15}{4} - \frac{49}{16}$$
$$= \frac{11}{16}.$$

5. Suppose that the number of customers visiting a fast food restaurant in a given day is $N \sim Poisson(\lambda)$. Assume that each customer purchases a drink with probability p, independently from other customers, and independently from the value of N. Let X be the number of customers who purchase drinks. Let Y be the number of customers that do not purchase drinks; so $X + Y = N$.

(a) Find the marginal PMFs of X and Y.

(b) Find the joint PMF of X and Y.

(c) Are X and Y independent?

(d) Find $E[X^2Y^2]$.

Solution:

(a) First note that $R_X = R_Y = \{0, 1, 2, ...\}$. Also, given $N = n$, X is a sum of n independent $Bernoulli(p)$ random variables. Thus, given $N = n$, X has a binomial distribution with parameters n and p, so

$$X|N = n \quad \sim \quad Binomial(n, p),$$
$$Y|N = n \quad \sim \quad Binomial(n, q = 1 - p).$$

We have

$$P_X(k) = \sum_{n=0}^{\infty} P(X=k|N=n)P_N(n) \qquad \text{(law of total probability)}$$

$$= \sum_{n=k}^{\infty} \binom{n}{k} p^k q^{n-k} e^{-\lambda} \frac{\lambda^n}{n!}$$

$$= \sum_{n=k}^{\infty} \frac{p^k q^{n-k} e^{-\lambda} \lambda^n}{k!(n-k)!}$$

$$= \frac{e^{-\lambda}(\lambda p)^k}{k!} \sum_{n=k}^{\infty} \frac{(\lambda q)^{n-k}}{(n-k)!}$$

$$= \frac{e^{-\lambda}(\lambda p)^k}{k!} e^{\lambda q} \qquad \text{(Taylor series for } e^x)$$

$$= \frac{e^{-\lambda p}(\lambda p)^k}{k!}, \qquad \text{for } k = 0, 1, 2, \dots$$

Thus, we conclude that

$$X \quad \sim \quad Poisson(\lambda p),$$
$$Y \quad \sim \quad Poisson(\lambda q).$$

(b) To find the joint PMF of X and Y, we can also use the law of total probability:

$$P_{XY}(i,j) = \sum_{n=0}^{\infty} P(X=i, Y=j|N=n)P_N(n) \qquad \text{(law of total probability)}.$$

But note that $P(X=i, Y=j|N=n) = 0$ if $N \neq i+j$, thus

$$P_{XY}(i,j) = P(X=i, Y=j|N=i+j)P_N(i+j)$$

$$= P(X=i|N=i+j)P_N(i+j)$$

$$= \binom{i+j}{i} p^i q^j e^{-\lambda} \frac{\lambda^{i+j}}{(i+j)!}$$

$$= \frac{e^{-\lambda}(\lambda p)^i (\lambda q)^j}{i!j!}$$

$$= \frac{e^{-\lambda p}(\lambda p)^i}{i!} \cdot \frac{e^{-\lambda q}(\lambda q)^j}{j!}$$

$$= P_X(i)P_Y(j).$$

(c) X and Y are independent, since as we saw above

$$P_{XY}(i,j) = P_X(i)P_Y(j).$$

(d) Since X and Y are independent, we have

$$E[X^2 Y^2] = E[X^2]E[Y^2].$$

Also, note that for a Poisson random variable W with parameter λ,

$$E[W^2] = \text{Var}(W) + (EW)^2 = \lambda + \lambda^2.$$

Thus,

$$\begin{aligned}
E[X^2Y^2] &= E[X^2]E[Y^2] \\
&= (\lambda p + \lambda^2 p^2)(\lambda q + \lambda^2 q^2) \\
&= \lambda^2 pq(\lambda^2 pq + \lambda + 1).
\end{aligned}$$

6. I have a coin with $P(H) = p$. I toss the coin repeatedly until I observe two consecutive heads. Let X be the total number of coin tosses. Find EX.

Solution: We solve this problem using a similar approach as in Example 5.6. Let $\mu = EX$. We first condition on the result of the first coin toss. Specifically,

$$\begin{aligned}
\mu = EX &= E[X|H]P(H) + E[X|T]P(T) \\
&= E[X|H]p + (1 + \mu)(1 - p).
\end{aligned}$$

In this equation, $E[X|T] = 1 + EX$, because the tosses are independent, so if the first toss is tails, it is like starting over on the second toss. Thus,

$$p\mu = pE[X|H] + (1 - p). \tag{5.14}$$

We still need to find $E[X|H]$ so we condition on the second coin toss

$$\begin{aligned}
E[X|H] &= E[X|HH]p + E[X|HT](1 - p) \\
&= 2p + (2 + \mu)(1 - p) \\
&= 2 + (1 - p)\mu.
\end{aligned}$$

Here, $E[X|HT] = 2 + EX$ because, if the first two tosses are HT, we have wasted two coin tosses and we start over at the third toss. By letting $E[X|H] = 2 + (1 - p)\mu$ in Equation 5.14, we obtain

$$\mu = EX = \frac{1 + p}{p^2}.$$

7. Let $X, Y \sim Geometric(p)$ be independent, and let $Z = \frac{X}{Y}$.

   (a) Find the range of Z.

(b) Find the PMF of Z.

(c) Find EZ.

Solution:

(a) The range of Z is given by

$$R_Z = \left\{ \frac{m}{n} | m, n \in \mathbb{N} \right\},$$

which is the set of all positive rational numbers.

(b) To find the PMF of Z, let $m, n \in \mathbb{N}$ such that $(m, n) = 1$, where (m, n) is the largest divisor of m and n. Then

$$P_Z \left(\frac{m}{n} \right) = \sum_{k=1}^{\infty} P(X = mk, Y = nk)$$

$$= \sum_{k=1}^{\infty} P(X = mk) P(Y = nk) \qquad \text{(since } X \text{ and } Y \text{ are independent)}$$

$$= \sum_{k=1}^{\infty} pq^{mk-1} pq^{nk-1} \qquad \text{(where } q = 1 - p)$$

$$= p^2 q^{-2} \sum_{k=1}^{\infty} q^{(m+n)k}$$

$$= \frac{p^2 q^{m+n-2}}{1 - q^{m+n}}$$

$$= \frac{p^2 (1 - p)^{m+n-2}}{1 - (1 - p)^{m+n}}.$$

(c) Find EZ: We can use LOTUS to find EZ. Let us first remember the following useful identities:

$$\sum_{k=1}^{\infty} k x^{k-1} = \frac{1}{(1 - x)^2}, \qquad \text{for } |x| < 1,$$

$$-\ln(1 - x) = \sum_{k=1}^{\infty} \frac{x^k}{k}, \qquad \text{for } |x| < 1.$$

The first one is obtained by taking derivative of the geometric sum formula, and the second one is a Taylor series. Now, let's apply LOTUS.

$$E\left[\frac{X}{Y}\right] = \sum_{n=1}^{\infty} \sum_{m=1}^{\infty} \frac{m}{n} P(X = m, Y = n)$$

$$= \sum_{n=1}^{\infty} \sum_{m=1}^{\infty} \frac{m}{n} p^2 q^{m-1} q^{n-1}$$

$$= \sum_{n=1}^{\infty} \frac{1}{n} p^2 q^{n-1} \sum_{m=1}^{\infty} m q^{m-1}$$

$$= \sum_{n=1}^{\infty} \frac{1}{n} p^2 q^{n-1} \frac{1}{(1-q)^2}$$

$$= \sum_{n=1}^{\infty} \frac{1}{n} q^{n-1}$$

$$= \frac{1}{q} \sum_{n=1}^{\infty} \frac{q^n}{n}$$

$$= \frac{1}{1-p} \ln \frac{1}{p}.$$

5.2   Two Continuous Random Variables

In Chapter 4, we introduced continuous random variables. As a simplified view of things, we mentioned that when we move from discrete random variables to continuous random variables, two things happen: sums become integrals, and PMFs become PDFs. The same statement can be repeated when we talk about joint distributions: (double) sums become (double) integrals, and joint PMFs become joint PDFs. Note that the CDF has the same definition for all kinds of random variables. Experience shows that students usually can learn the concepts behind joint continuous random variables without much difficulty; however, they sometimes run into issues when dealing with double integrals. That is, in the discussion of joint continuous distributions, students' problems often relate to multivariate calculus rather than their lack of understanding of probability concepts. The good news is that in practice we do not often need to evaluate multiple integrals anyway. Nevertheless, since this part of the book relies on familiarity with multivariate calculus, we recommend a quick review of double integrals and partial derivatives in case you have not dealt with them recently. We will only need the calculus concepts very lightly and our goal here is to focus on probability. In this section, we discuss joint continuous distributions. Since the ideas behind the theory is very analogous to joint discrete random variables, we will provide a quick introduction to main concepts and then focus on examples.

5.2.1   Joint Probability Density Function (PDF)

Here, we will define jointly continuous random variables. Basically, two random variables are jointly continuous if they have a joint probability density function as defined below.

Definition 5.1. Two random variables X and Y are **jointly continuous** if there exists a nonnegative function $f_{XY} : \mathbb{R}^2 \to \mathbb{R}$, such that, for any set $A \in \mathbb{R}^2$, we have

$$P\big((X,Y) \in A\big) = \iint\limits_{A} f_{XY}(x,y)dxdy \tag{5.15}$$

The function $f_{XY}(x,y)$ is called the **joint probability density function (PDF)** of X and Y.

In the above definition, the domain of $f_{XY}(x,y)$ is the entire $\mathbb{R}^2$. We may define the range of (X,Y) as

$$R_{XY} = \{(x,y) | f_{X,Y}(x,y) > 0\}.$$

The above double integral (Equation 5.15) exists for all sets A of practical interest. If we choose $A = \mathbb{R}^2$, then the probability of $(X,Y) \in A$ must be one, so we must have

$$\int_{-\infty}^{\infty} \int_{-\infty}^{\infty} f_{XY}(x,y)dxdy = 1$$

The intuition behind the joint density $f_{XY}(x,y)$ is similar to that of the PDF of a single random variable. In particular, remember that for a random variable X and small positive δ, we have

$$P(x < X \le x + \delta) \approx f_X(x)\delta.$$

Similarly, for small positive δ_x and δ_y, we can write

$$P(x < X \le x + \delta_x, y \le Y \le y + \delta_y) \approx f_{XY}(x,y)\delta_x\delta_y.$$

Example 5.15. Let X and Y be two jointly continuous random variables with joint PDF

$$f_{XY}(x,y) = \begin{cases} x + cy^2 & 0 \le x \le 1, 0 \le y \le 1 \\ \\ 0 & \text{otherwise} \end{cases}$$

(a) Find the constant c.

(b) Find $P(0 \le X \le \frac{1}{2}, 0 \le Y \le \frac{1}{2})$.

Solution:

(a) To find c, we use

$$\int_{-\infty}^{\infty} \int_{-\infty}^{\infty} f_{XY}(x,y)dxdy = 1.$$

Thus, we have

$$1 = \int_{-\infty}^{\infty} \int_{-\infty}^{\infty} f_{XY}(x,y)dxdy$$

$$= \int_0^1 \int_0^1 x + cy^2 \ dxdy$$

$$= \int_0^1 \left[\frac{1}{2}x^2 + cy^2 x \right]_{x=0}^{x=1} dy$$

$$= \int_0^1 \frac{1}{2} + cy^2 \ dy$$

$$= \left[\frac{1}{2}y + \frac{1}{3}cy^3 \right]_{y=0}^{y=1}$$

$$= \frac{1}{2} + \frac{1}{3}c.$$

Therefore, we obtain $c = \frac{3}{2}$.

(b) To find $P(0 \le X \le \frac{1}{2}, 0 \le Y \le \frac{1}{2})$, we can write

$$P\big((X,Y) \in A\big) = \iint_A f_{XY}(x,y)dxdy, \quad \text{for } A = \{(x,y)|0 \le x,y \le 1\}.$$

Thus,

$$P(0 \le X \le \frac{1}{2}, 0 \le Y \le \frac{1}{2}) = \int_0^{\frac{1}{2}} \int_0^{\frac{1}{2}} \left(x + \frac{3}{2}y^2 \right) dxdy$$

$$= \int_0^{\frac{1}{2}} \left[\frac{1}{2}x^2 + \frac{3}{2}y^2 x \right]_0^{\frac{1}{2}} dy$$

$$= \int_0^{\frac{1}{2}} \left(\frac{1}{8} + \frac{3}{4}y^2 \right) dy$$

$$= \frac{3}{32}.$$

We can find marginal PDFs of X and Y from their joint PDF. This is exactly analogous to what we saw in the discrete case. In particular, by integrating over all y's, we obtain $f_X(x)$. We have

$$\boxed{\begin{array}{c} \text{Marginal PDFs} \\[2mm] f_X(x) = \int_{-\infty}^{\infty} f_{XY}(x,y)dy, \quad \text{for all } x, \\[3mm] f_Y(y) = \int_{-\infty}^{\infty} f_{XY}(x,y)dx, \quad \text{for all } y. \end{array}}$$

Example 5.16. In Example 5.15 find the marginal PDFs $f_X(x)$ and $f_Y(y)$.

Solution: For $0 \le x \le 1$, we have

$$\begin{aligned} f_X(x) &= \int_{-\infty}^{\infty} f_{XY}(x,y)dy \\ &= \int_0^1 \left(x + \frac{3}{2}y^2 \right) dy \\ &= \left[xy + \frac{1}{2}y^3 \right]_0^1 \\ &= x + \frac{1}{2}. \end{aligned}$$

Thus,

$$f_X(x) = \begin{cases} x + \frac{1}{2} & 0 \le x \le 1 \\ 0 & \text{otherwise} \end{cases}$$

Similarly, for $0 \le y \le 1$, we have

$$\begin{aligned} f_Y(y) &= \int_{-\infty}^{\infty} f_{XY}(x,y)dx \\ &= \int_0^1 \left(x + \frac{3}{2}y^2 \right) dx \\ &= \left[\frac{1}{2}x^2 + \frac{3}{2}y^2 x \right]_0^1 \\ &= \frac{3}{2}y^2 + \frac{1}{2}. \end{aligned}$$

Thus,

$$f_Y(y) = \begin{cases} \frac{3}{2}y^2 + \frac{1}{2} & 0 \le y \le 1 \\ 0 & \text{otherwise} \end{cases}$$

Example 5.17. Let X and Y be two jointly continuous random variables with joint PDF

$$f_{XY}(x,y) = \begin{cases} cx^2y & 0 \le y \le x \le 1 \\ 0 & \text{otherwise} \end{cases}$$

(a) Find R_{XY} and show it in the $x - y$ plane.

(b) Find the constant c.

(c) Find marginal PDFs, $f_X(x)$ and $f_Y(y)$.

(d) Find $P(Y \leq \frac{X}{2})$.

(e) Find $P(Y \leq \frac{X}{4} | Y \leq \frac{X}{2})$.

Solution:

(a) From the joint PDF, we find that

$$R_{XY} = \{(x, y) \in \mathbb{R}^2 | 0 \leq y \leq x \leq 1\}.$$

Figure 5.6 shows R_{XY} in the $x - y$ plane.

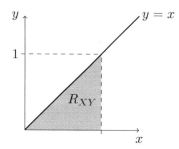 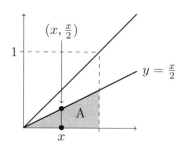

Figure 5.6: Figure shows R_{XY} as well as integration region for finding $P(Y \leq \frac{X}{2})$.

(b) To find the constant c, we can write

$$1 = \int_{-\infty}^{\infty} \int_{-\infty}^{\infty} f_{XY}(x, y) dx dy$$

$$= \int_0^1 \int_0^x cx^2 y \; dy dx$$

$$= \int_0^1 \frac{c}{2} x^4 dx$$

$$= \frac{c}{10}.$$

Thus, $c = 10$.

(c) To find the marginal PDFs, first note that $R_X = R_Y = [0, 1]$. For $0 \leq x \leq 1$, we can write

$$f_X(x) = \int_{-\infty}^{\infty} f_{XY}(x, y) dy$$

$$= \int_0^x 10x^2 y dy$$

$$= 5x^4.$$

Thus,

$$f_X(x) = \begin{cases} 5x^4 & 0 \le x \le 1 \\ 0 & \text{otherwise} \end{cases}$$

For $0 \le y \le 1$, we can write

$$f_Y(y) = \int_{-\infty}^{\infty} f_{XY}(x, y)dx$$

$$= \int_y^1 10x^2 y \, dx$$

$$= \frac{10}{3}y(1 - y^3).$$

Thus,

$$f_Y(y) = \begin{cases} \frac{10}{3}y(1 - y^3) & 0 \le y \le 1 \\ 0 & \text{otherwise} \end{cases}$$

(d) To find $P(Y \le \frac{X}{2})$, we need to integrate $f_{XY}(x, y)$ over region A shown in Figure 5.6. In particular, we have

$$P\left(Y \le \frac{X}{2}\right) = \int_{-\infty}^{\infty} \int_0^{\frac{x}{2}} f_{XY}(x, y)dydx$$

$$= \int_0^1 \int_0^{\frac{x}{2}} 10x^2 y \, dydx$$

$$= \int_0^1 \frac{5}{4}x^4 dx$$

$$= \frac{1}{4}.$$

(e) To find $P(Y \le \frac{X}{4} | Y \le \frac{X}{2})$, we have

$$P\left(Y \le \frac{X}{4} \Big| Y \le \frac{X}{2}\right) = \frac{P\left(Y \le \frac{X}{4}, Y \le \frac{X}{2}\right)}{P\left(Y \le \frac{X}{2}\right)}$$

$$= 4P\left(Y \le \frac{X}{4}\right)$$

$$= 4 \int_0^1 \int_0^{\frac{x}{4}} 10x^2 y \, dydx$$

$$= 4 \int_0^1 \frac{5}{16}x^4 dx$$

$$= \frac{1}{4}.$$

5.2.2   Joint Cumulative Distribution Function (CDF)

We have already seen the joint CDF for discrete random variables. The joint CDF has the same definition for continuous random variables. It also satisfies the same properties.

The **joint cumulative function** of two random variables X and Y is defined as

$$F_{XY}(x, y) = P(X \leq x, Y \leq y).$$

The joint CDF satisfies the following properties:

1. $F_X(x) = F_{XY}(x, \infty)$, for any x (marginal CDF of X);

2. $F_Y(y) = F_{XY}(\infty, y)$, for any y (marginal CDF of Y);

3. $F_{XY}(\infty, \infty) = 1$;

4. $F_{XY}(-\infty, y) = F_{XY}(x, -\infty) = 0$;

5. $P(x_1 < X \leq x_2, \ y_1 < Y \leq y_2) =$
   $F_{XY}(x_2, y_2) - F_{XY}(x_1, y_2) - F_{XY}(x_2, y_1) + F_{XY}(x_1, y_1)$;

6. if X and Y are independent, then $F_{XY}(x, y) = F_X(x)F_Y(y)$.

Example 5.18. Let X and Y be two independent $Uniform(0, 1)$ random variables. Find $F_{XY}(x, y)$.

Solution: Since $X, Y \sim Uniform(0, 1)$, we have

$$F_X(x) = \begin{cases} 0 & \text{for } x < 0 \\ x & \text{for } 0 \leq x \leq 1 \\ 1 & \text{for } x > 1 \end{cases}$$

$$F_Y(y) = \begin{cases} 0 & \text{for } y < 0 \\ y & \text{for } 0 \leq y \leq 1 \\ 1 & \text{for } y > 1 \end{cases}$$

Since X and Y are independent, we obtain

$$F_{XY}(x, y) = F_X(x)F_Y(y) = \begin{cases} 0 & \text{for } y < 0 \text{ or } x < 0 \\ xy & \text{for } 0 \leq x \leq 1, 0 \leq y \leq 1 \\ y & \text{for } x > 1, 0 \leq y \leq 1 \\ x & \text{for } y > 1, 0 \leq x \leq 1 \\ 1 & \text{for } x > 1, y > 1 \end{cases}$$

Figure 5.7 shows the values of $F_{XY}(x, y)$ in the $x-y$ plane. Note that $F_{XY}(x, y)$ is a continuous function in both arguments. This is always true for jointly continuous random variables. This fact sometimes simplifies finding $F_{XY}(x, y)$. The next example (Example 5.19) shows how we can use this fact.

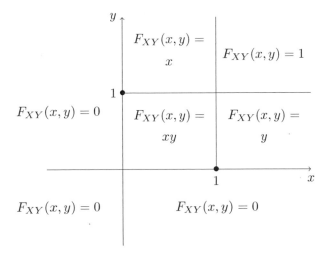

Figure 5.7: The joint CDF of two independent $Uniform(0, 1)$ random variables X and Y.

Remember that, for a single random variable, we have the following relationship between the PDF and CDF:

$$F_X(x) = \int_{-\infty}^{x} f_X(u)du,$$

$$f_X(x) = \frac{dF_X(x)}{dx}.$$

Similar formulas hold for jointly continuous random variables. In particular, we have the following:

$$F_{XY}(x, y) = \int_{-\infty}^{y} \int_{-\infty}^{x} f_{XY}(u, v)dudv$$

$$f_{XY}(x, y) = \frac{\partial^2}{\partial x \partial y} F_{XY}(x, y)$$

Example 5.19. Find the joint CDF for X and Y in Example 5.15.

Solution: In Example 5.15, we found

$$f_{XY}(x, y) = \begin{cases} x + \frac{3}{2}y^2 & 0 \le x, y \le 1 \\ \\ 0 & \text{otherwise} \end{cases}$$

First, note that since $R_{XY} = \{(x, y) | 0 \le x, y \le 1\}$, we find that

$$F_{XY}(x, y) = 0, \qquad \text{for } x < 0 \text{ or } y < 0,$$
$$F_{XY}(x, y) = 1, \qquad \text{for } x \ge 1 \text{ and } y \ge 1.$$

To find the joint CDF for $x > 0$ and $y > 0$, we need to integrate the joint PDF:

$$F_{XY}(x, y) = \int_{-\infty}^{y} \int_{-\infty}^{x} f_{XY}(u, v) du dv$$
$$= \int_{0}^{y} \int_{0}^{x} f_{XY}(u, v) du dv$$
$$= \int_{0}^{\min(y,1)} \int_{0}^{\min(x,1)} \left(u + \frac{3}{2}v^2 \right) du dv.$$

For $0 \le x, y \le 1$, we obtain

$$F_{XY}(x, y) = \int_{0}^{y} \int_{0}^{x} \left(u + \frac{3}{2}v^2 \right) du dv$$
$$= \int_{0}^{y} \left[\frac{1}{2}u^2 + \frac{3}{2}v^2 u \right]_{0}^{x} dv$$
$$= \int_{0}^{y} \left(\frac{1}{2}x^2 + \frac{3}{2}xv^2 \right) dv$$
$$= \frac{1}{2}x^2 y + \frac{1}{2}xy^3.$$

For $0 \le x \le 1$ and $y \ge 1$, we use the fact that F_{XY} is continuous to obtain

$$F_{XY}(x, y) = F_{XY}(x, 1)$$
$$= \frac{1}{2}x^2 + \frac{1}{2}x.$$

Similarly, for $0 \le y \le 1$ and $x \ge 1$, we obtain

$$F_{XY}(x, y) = F_{XY}(1, y)$$
$$= \frac{1}{2}y + \frac{1}{2}y^3.$$

5.2.3   Conditioning and Independence

Here, we will discuss conditioning for continuous random variables. In particular, we will discuss the conditional PDF, conditional CDF, and conditional expectation. We have discussed conditional probability for discrete random variables before. The ideas behind conditional probability for continuous random variables are very similar to the discrete case. The difference lies in the fact that we need to work with probability density in the case of continuous random variables. Nevertheless, we would like to emphasize again that there is only one main formula regarding conditional probability which is

$$P(A|B) = \frac{P(A \cap B)}{P(B)}, \text{ when } P(B) > 0.$$

Any other formula regarding conditional probability can be derived from the above formula. In fact, for some problems we only need to apply the above formula. You have already used this in Example 5.17. As another example, if you have two random variables X and Y, you can write

$$P(X \in C | Y \in D) = \frac{P(X \in C, Y \in D)}{P(Y \in D)}, \text{ where } C, D \subset \mathbb{R}.$$

However, sometimes we need to use the concepts of conditional PDFs and CDFs. The formulas for conditional PDFs and CDFs of continuous random variables are very similar to those of discrete random variables. Since there are no new fundamental ideas in this section, we usually provide the main formulas and guidelines, and then work on examples. Specifically, we do not spend much time deriving formulas. Nevertheless, to give you the basic idea of how to derive these formulas, we start by deriving a formula for the conditional CDF and PDF of a random variable X given that $X \in I = [a, b]$. Consider a continuous random variable X. Suppose that we know that the event $X \in I = [a, b]$ has occurred. Call this event A. The conditional CDF of X given A, denoted by $F_{X|A}(x)$ or $F_{X|a \leq X \leq b}(x)$, is

$$\begin{aligned} F_{X|A}(x) &= P(X \leq x | A) \\ &= P(X \leq x | a \leq X \leq b) \\ &= \frac{P(X \leq x, a \leq X \leq b)}{P(A)}. \end{aligned}$$

Now if $x < a$, then $F_{X|A}(x) = 0$. On the other hand, if $a \leq x \leq b$, we have

$$\begin{aligned} F_{X|A}(x) &= \frac{P(X \leq x, a \leq X \leq b)}{P(A)} \\ &= \frac{P(a \leq X \leq x)}{P(A)} \\ &= \frac{F_X(x) - F_X(a)}{F_X(b) - F_X(a)}. \end{aligned}$$

Finally, if $x > b$, then $F_{X|A}(x) = 1$. Thus, we obtain

$$F_{X|A}(x) = \begin{cases} 1 & x > b \\ \frac{F_X(x) - F_X(a)}{F_X(b) - F_X(a)} & a \leq x < b \\ 0 & \text{otherwise} \end{cases}$$

Note that since X is a continuous random variable, we do not need to be careful about end points, i.e., changing $x > b$ to $x \geq b$ does not make a difference in the above formula. To obtain the conditional PDF of X, denoted by $f_{X|A}(x)$, we can differentiate $F_{X|A}(x)$. We obtain

$$f_{X|A}(x) = \begin{cases} \frac{f_X(x)}{P(A)} & a \le x < b \\ \\ 0 & \text{otherwise} \end{cases}$$

It is insightful if we derive the above formula for $f_{X|A}(x)$ directly from the definition of the PDF for continuous random variables. Recall that the PDF of X can be defined as

$$f_X(x) = \lim_{\Delta \to 0^+} \frac{P(x < X \le x + \Delta)}{\Delta}.$$

Now, the conditional PDF of X given A, denoted by $f_{X|A}(x)$, is

$$\begin{aligned} f_{X|A}(x) &= \lim_{\Delta \to 0^+} \frac{P(x < X \le x + \Delta | A)}{\Delta} \\ &= \lim_{\Delta \to 0^+} \frac{P(x < X \le x + \Delta, A)}{\Delta P(A)} \\ &= \lim_{\Delta \to 0^+} \frac{P(x < X \le x + \Delta, a \le X \le b)}{\Delta P(A)}. \end{aligned}$$

Now consider two cases. If $a \le x < b$, then

$$\begin{aligned} f_{X|A}(x) &= \lim_{\Delta \to 0^+} \frac{P(x < X \le x + \Delta, a \le X \le b)}{\Delta P(A)} \\ &= \frac{1}{P(A)} \lim_{\Delta \to 0^+} \frac{P(x < X \le x + \Delta)}{\Delta} \\ &= \frac{f_X(x)}{P(A)}. \end{aligned}$$

On the other hand, if $x < a$ or $x \ge b$, then

$$\begin{aligned} f_{X|A}(x) &= \lim_{\Delta \to 0^+} \frac{P(x < X \le x + \Delta, a \le X \le b)}{\Delta P(A)} \\ &= 0. \end{aligned}$$

If X is a continuous random variable, and A is the event that $a < X < b$ (where possibly $b = \infty$ or $a = -\infty$), then

$$F_{X|A}(x) = \begin{cases} 1 & x > b \\ \\ \frac{F_X(x) - F_X(a)}{F_X(b) - F_X(a)} & a \le x < b \\ \\ 0 & x < a \end{cases}$$

$$f_{X|A}(x) = \begin{cases} \frac{f_X(x)}{P(A)} & a \le x < b \\ \\ 0 & \text{otherwise} \end{cases}$$

The conditional expectation and variance are defined by replacing the PDF by conditional PDF in the definitions of expectation and variance. In general, for a random variable X and an event A, we have the following:

$$E[X|A] = \int_{-\infty}^{\infty} x f_{X|A}(x) dx$$

$$E[g(X)|A] = \int_{-\infty}^{\infty} g(x) f_{X|A}(x) dx$$

$$\text{Var}(X|A) = E[X^2|A] - (E[X|A])^2$$

Example 5.20. Let $X \sim Exponential(1)$.

(a) Find the conditional PDF and CDF of X given $X > 1$.

(b) Find $E[X|X > 1]$.

(c) Find $\text{Var}(X|X > 1)$.

Solution:

(a) Let A be the event that $X > 1$. Then

$$P(A) = \int_{1}^{\infty} e^{-x} dx$$
$$= \frac{1}{e}.$$

Thus,

$$f_{X|X>1}(x) = \begin{cases} e^{-x+1} & x > 1 \\ \\ 0 & \text{otherwise} \end{cases}$$

For $x > 1$, we have

$$F_{X|A}(A) = \frac{F_X(x) - F_X(1)}{P(A)}$$
$$= 1 - e^{-x+1}.$$

Thus,

$$F_{X|A}(x) = \begin{cases} 1 - e^{-x+1} & x > 1 \\ \\ 0 & \text{otherwise} \end{cases}$$

(b) We have

$$E[X|X > 1] = \int_1^\infty x f_{X|X>1}(x) dx$$

$$= \int_1^\infty x e^{-x+1} dx$$

$$= e \int_1^\infty x e^{-x} dx$$

$$= e \left[-e^{-x} - x e^{-x} \right]_1^\infty$$

$$= e \frac{2}{e}$$

$$= 2.$$

(c) We have

$$E[X^2|X > 1] = \int_1^\infty x^2 f_{X|X>1}(x) dx$$

$$= \int_1^\infty x^2 e^{-x+1} dx$$

$$= e \int_1^\infty x^2 e^{-x} dx$$

$$= e \left[-2e^{-x} - 2x e^{-x} - x^2 e^{-x} \right]_1^\infty$$

$$= e \frac{5}{e}$$

$$= 5.$$

Thus,

$$\text{Var}(X|X > 1) = E[X^2|X > 1] - (E[X|X > 1])^2$$

$$= 5 - 4 = 1.$$

Conditioning by Another Random Variable:

If X and Y are two jointly continuous random variables, and we obtain some information regarding Y, we should update the PDF and CDF of X based on the new information. In particular, if we get to observe the value of the random variable Y, then how do we need to update the PDF and CDF of X? Remember for the discrete case, the conditional PMF of X given $Y = y$ is given by

$$P_{X|Y}(x_i|y_j) = \frac{P_{XY}(x_i, y_j)}{P_Y(y_j)}.$$

Now, if X and Y are jointly continuous, the conditional PDF of X given Y is given by

$$f_{X|Y}(x|y) = \frac{f_{XY}(x, y)}{f_Y(y)}.$$

This means that if we get to observe $Y = y$, then we need to use the above conditional density for the random variable X. To get an intuition about the formula, note that by definition, for small Δ_x and Δ_y we should have

$$
\begin{aligned}
f_{X|Y}(x|y) &\approx \frac{P(x \leq X \leq x + \Delta_x | y \leq Y \leq y + \Delta_y)}{\Delta_x} \quad \text{(definition of PDF)} \\
&= \frac{P(x \leq X \leq x + \Delta_x, y \leq Y \leq y + \Delta_y)}{P(y \leq Y \leq y + \Delta_y)\Delta_x} \\
&\approx \frac{f_{XY}(x,y)\Delta_x\Delta_y}{f_Y(y)\Delta_y\Delta_x} \\
&= \frac{f_{XY}(x,y)}{f_Y(y)}.
\end{aligned}
$$

Similarly, we can write the conditional PDF of Y, given $X = x$, as

$$
f_{Y|X}(y|x) = \frac{f_{XY}(x,y)}{f_X(x)}.
$$

For two jointly continuous random variables X and Y, we can define the following conditional concepts:

1. The conditional PDF of X given $Y = y$:

$$
f_{X|Y}(x|y) = \frac{f_{XY}(x,y)}{f_Y(y)}
$$

2. The conditional probability that $X \in A$ given $Y = y$:

$$
P(X \in A|Y = y) = \int_A f_{X|Y}(x|y)dx
$$

3. The conditional CDF of X given $Y = y$:

$$
F_{X|Y}(x|y) = P(X \leq x|Y = y) = \int_{-\infty}^{x} f_{X|Y}(x|y)dx
$$

Example 5.21. Let X and Y be two jointly continuous random variables with joint PDF

$$
f_{XY}(x,y) = \begin{cases} \frac{x^2}{4} + \frac{y^2}{4} + \frac{xy}{6} & 0 \leq x \leq 1, 0 \leq y \leq 2 \\ 0 & \text{otherwise} \end{cases}
$$

For $0 \leq y \leq 2$, find

(a) the conditional PDF of X given $Y = y$;

(b) $P(X < \frac{1}{2}|Y = y)$.

Solution:

(a) Let us first find the marginal PDF of Y. We have

$$f_Y(y) = \int_0^1 \frac{x^2}{4} + \frac{y^2}{4} + \frac{xy}{6} \ dx$$
$$= \frac{3y^2 + y + 1}{12}, \qquad \text{for } 0 \le y \le 2.$$

Thus, for $0 \le y \le 2$, we obtain

$$f_{X|Y}(x|y) = \frac{f_{XY}(x, y)}{f_Y(y)}$$
$$= \frac{3x^2 + 3y^2 + 2xy}{3y^2 + y + 1}, \qquad \text{for } 0 \le x \le 1.$$

Thus, for $0 \le y \le 2$, we have

$$f_{X|Y}(x|y) = \begin{cases} \frac{3x^2 + 3y^2 + 2xy}{3y^2 + y + 1} & 0 \le x \le 1 \\ \\ 0 & \text{otherwise} \end{cases}$$

(b) We have

$$P\left(X < \frac{1}{2}|Y = y\right) = \int_0^{\frac{1}{2}} \frac{3x^2 + 3y^2 + 2xy}{3y^2 + y + 1} \ dx$$
$$= \frac{1}{3y^2 + y + 1}\left[x^3 + yx^2 + 3y^2x\right]_0^{\frac{1}{2}}$$
$$= \frac{\frac{3}{2}y^2 + \frac{y}{4} + \frac{1}{8}}{3y^2 + y + 1}.$$

Note that, as we expect, $P\left(X < \frac{1}{2}|Y = y\right)$ depends on y.

Conditional expectation and variance are similarly defined. Given $Y = y$, we need to replace $f_X(x)$ by $f_{X|Y}(x|y)$ in the formulas for expectation:

For two jointly continuous random variables X and Y, we have:

1. Expected value of X given $Y = y$:

$$E[X|Y = y] = \int_{-\infty}^{\infty} x f_{X|Y}(x|y) dx$$

2. Conditional LOTUS:

$$E[g(X)|Y = y] = \int_{-\infty}^{\infty} g(x) f_{X|Y}(x|y) dx$$

3. Conditional variance of X given $Y = y$:

$$\text{Var}(X|Y = y) = E[X^2|Y = y] - (E[X|Y = y])^2$$

Example 5.22. Let X and Y be as in Example 5.21. Find $E[X|Y = 1]$ and $\text{Var}(X|Y = 1)$.

Solution: We have

$$
\begin{aligned}
E[X|Y = 1] &= \int_{-\infty}^{\infty} x f_{X|Y}(x|1) dx \\
&= \int_{0}^{1} x \frac{3x^2 + 3y^2 + 2xy}{3y^2 + y + 1} \Big|_{y=1} dx \\
&= \int_{0}^{1} x \frac{3x^2 + 3 + 2x}{3 + 1 + 1} dx \qquad (y = 1) \\
&= \frac{1}{5} \int_{0}^{1} 3x^3 + 2x^2 + 3x \ dx \\
&= \frac{7}{12}
\end{aligned}
$$

$$
\begin{aligned}
E[X^2|Y = 1] &= \int_{-\infty}^{\infty} x^2 f_{X|Y}(x|1) dx \\
&= \frac{1}{5} \int_{0}^{1} 3x^4 + 2x^3 + 3x^2 \ dx \\
&= \frac{21}{50}.
\end{aligned}
$$

So we have

$$\text{Var}(X|Y=1) = E[X^2|Y=1] - (E[X|Y=1])^2$$
$$= \frac{21}{50} - \left(\frac{7}{12}\right)^2$$
$$= \frac{287}{3600}.$$

Independent Random Variables:

When two jointly continuous random variables are independent, we must have

$$f_{X|Y}(x|y) = f_X(x).$$

That is, knowing the value of Y does not change the PDF of X. Since $f_{X|Y}(x|y) = \frac{f_{XY}(x,y)}{f_Y(y)}$, we conclude that for two independent continuous random variables we must have

$$f_{XY}(x,y) = f_X(x)f_Y(y).$$

- Two continuous random variables X and Y are independent if

$$f_{XY}(x,y) = f_X(x)f_Y(y), \quad \text{for all } x, y.$$

Equivalently, X and Y are independent if

$$F_{XY}(x,y) = F_X(x)F_Y(y), \quad \text{for all } x, y.$$

- If X and Y are independent, we have

$$E[XY] = EXEY,$$
$$E[g(X)h(Y)] = E[g(X)]E[h(Y)].$$

Suppose that we are given the joint PDF $f_{XY}(x,y)$ of two random variables X and Y. If we can write

$$f_{XY}(x,y) = f_1(x)f_2(y),$$

then X and Y are independent.

Example 5.23. Determine whether X and Y are independent:

(a) $f_{XY}(x,y) = \begin{cases} 2e^{-x-2y} & x, y > 0 \\ \\ 0 & \text{otherwise} \end{cases}$

(b) $f_{XY}(x, y) = \begin{cases} 8xy & 0 < x < y < 1 \\ 0 & \text{otherwise} \end{cases}$

Solution:

(a) We can write

$$f_{XY}(x, y) = \left[e^{-x}u(x)\right]\left[2e^{-2y}u(y)\right],$$

where $u(x)$ is the unit step function:

$$u(x) = \begin{cases} 1 & x \geq 0 \\ 0 & \text{otherwise} \end{cases}$$

Thus, we conclude that X and Y are independent.

(b) For this case, it does not seem that we can write $f_{XY}(x, y)$ as a product of some $f_1(x)$ and $f_2(y)$. Note that the given region $0 < x < y < 1$ enforces that $x < y$. That is, we always have $X < Y$. Thus, we conclude that X and Y are not independent. To show this, we can obtain the marginal PDFs of X and Y and show that $f_{XY}(x, y) \neq f_X(x)f_Y(y)$, for some x, y. We have, for $0 \leq x \leq 1$,

$$f_X(x) = \int_x^1 8xy \, dy$$
$$= 4x(1 - x^2).$$

Thus,

$$f_X(x) = \begin{cases} 4x(1 - x^2) & 0 < x < 1 \\ 0 & \text{otherwise} \end{cases}$$

Similarly, we obtain

$$f_Y(y) = \begin{cases} 4y^3 & 0 < y < 1 \\ 0 & \text{otherwise} \end{cases}$$

As we see, $f_{XY}(x, y) \neq f_X(x)f_Y(y)$, thus X and Y are NOT independent.

Example 5.24. Consider the unit disc

$$D = \{(x, y) | x^2 + y^2 \leq 1\}.$$

Suppose that we choose a point (X, Y) uniformly at random in D. That is, the joint PDF of X and Y is given by

$$f_{XY}(x, y) = \begin{cases} c & (x, y) \in D \\ 0 & \text{otherwise} \end{cases}$$

(a) Find the constant c.

(b) Find the marginal PDFs $f_X(x)$ and $f_Y(y)$.

(c) Find the conditional PDF of X given $Y = y$, where $-1 \le y \le 1$.

(d) Are X and Y independent?

Solution:

(a) We have

$$
\begin{aligned}
1 &= \int_{-\infty}^{\infty} \int_{-\infty}^{\infty} f_{XY}(x, y) dx dy \\
&= \iint_{D} c \; dx dy \\
&= c(\text{area of } D) \\
&= c(\pi).
\end{aligned}
$$

Thus, $c = \frac{1}{\pi}$.

(b) For $-1 \le x \le 1$, we have

$$
\begin{aligned}
f_X(x) &= \int_{-\infty}^{\infty} f_{XY}(x, y) dy \\
&= \int_{-\sqrt{1-x^2}}^{\sqrt{1-x^2}} \frac{1}{\pi} \; dy \\
&= \frac{2}{\pi} \sqrt{1 - x^2}.
\end{aligned}
$$

Thus,

$$
f_X(x) = \begin{cases} \frac{2}{\pi} \sqrt{1 - x^2} & -1 \le x \le 1 \\ 0 & \text{otherwise} \end{cases}
$$

Similarly,

$$
f_Y(y) = \begin{cases} \frac{2}{\pi} \sqrt{1 - y^2} & -1 \le y \le 1 \\ 0 & \text{otherwise} \end{cases}
$$

(c) We have

$$
\begin{aligned}
f_{X|Y}(x|y) &= \frac{f_{XY}(x, y)}{f_Y(y)} \\
&= \begin{cases} \frac{1}{2\sqrt{1-y^2}} & -\sqrt{1 - y^2} \le x \le \sqrt{1 - y^2} \\ 0 & \text{otherwise} \end{cases}
\end{aligned}
$$

Note that the above equation indicates that, given $Y = y$, X is uniformly distributed on $[-\sqrt{1 - y^2}, \sqrt{1 - y^2}]$. We write

$$
X|Y = y \; \sim \; Uniform(-\sqrt{1 - y^2}, \sqrt{1 - y^2}).
$$

(d) Are X and Y independent? No, because $f_{XY}(x, y) \ne f_X(x) f_Y(y)$.

Law of Total Probability:

Now, we'll discuss the law of total probability for continuous random variables. This is completely analogous to the discrete case. In particular, the law of total probability, the law of total expectation (law of iterated expectations), and the law of total variance can be stated as follows:

Law of Total Probability:

$$P(A) = \int_{-\infty}^{\infty} P(A|X = x)f_X(x) \; dx \qquad (5.16)$$

Law of Total Expectation:

$$E[Y] = \int_{-\infty}^{\infty} E[Y|X = x]f_X(x) \; dx$$
$$= E[E[Y|X]] \qquad (5.17)$$

Law of Total Variance:

$$\mathrm{Var}(Y) = E[\mathrm{Var}(Y|X)] + \mathrm{Var}(E[Y|X]) \qquad (5.18)$$

Let's look at some examples.

Example 5.25. Let X and Y be two independent $Uniform(0,1)$ random variables. Find $P(X^3 + Y > 1)$.

Solution: Using the law of total probability (Equation 5.16), we can write

$$P(X^3 + Y > 1) = \int_{-\infty}^{\infty} P(X^3 + Y > 1 | X = x) f_X(x) \; dx$$

$$= \int_0^1 P(x^3 + Y > 1 | X = x) \; dx$$

$$= \int_0^1 P(Y > 1 - x^3) \; dx \qquad \text{(since } X \text{ and } Y \text{ are independent)}$$

$$= \int_0^1 x^3 \; dx \qquad \qquad \text{(since } Y \sim Uniform(0,1))$$

$$= \frac{1}{4}.$$

Example 5.26. Suppose $X \sim Uniform(1,2)$ and given $X = x$, Y is an exponential random variable with parameter $\lambda = x$, so we can write

$$Y | X = x \quad \sim \quad Exponential(x).$$

We sometimes write this as

$$Y | X \quad \sim \quad Exponential(X).$$

(a) Find EY.

(b) Find $Var(Y)$.

Solution:

(a) We use the law of total expectation (Equation 5.17) to find EY. Remember that if $Y \sim Exponential(\lambda)$, then $EY = \frac{1}{\lambda}$. Thus we conclude

$$E[Y | X = x] = \frac{1}{x}.$$

Using the law of total expectation, we have

$$EY = \int_{-\infty}^{\infty} E[Y | X = x] f_X(x) dx$$

$$= \int_1^2 E[Y | X = x] \cdot 1 \; dx$$

$$= \int_1^2 \frac{1}{x} dx$$

$$= \ln 2.$$

Another way to write the above calculation is

$$EY = E[E[Y|X]] \qquad \text{(law of total expectation)}$$

$$= E\left[\frac{1}{X}\right] \qquad \text{(since } E[Y|X] = \frac{1}{X})$$

$$= \int_1^2 \frac{1}{x} dx$$

$$= \ln 2.$$

(b) To find $\text{Var}(Y)$, we can write

$$\text{Var}(Y) = E[Y^2] - (E[Y])^2$$
$$= E[Y^2] - (\ln 2)^2$$
$$= E[E[Y^2|X]] - (\ln 2)^2 \qquad \text{(law of total expectation)}$$
$$= E\left[\frac{2}{X^2}\right] - (\ln 2)^2 \qquad \left(\text{since } Y|X \sim Exponential(X)\right)$$
$$= \int_1^2 \frac{2}{x^2} dx - (\ln 2)^2$$
$$= 1 - (\ln 2)^2.$$

Another way to find $\text{Var}(Y)$ is to apply the law of total variance:

$$\text{Var}(Y) = E[\text{Var}(Y|X)] + \text{Var}(E[Y|X]).$$

Since $Y|X \sim Exponential(X)$, we conclude

$$E[Y|X] = \frac{1}{X},$$
$$\text{Var}(Y|X) = \frac{1}{X^2}.$$

Therefore

$$\text{Var}(Y) = E\left[\frac{1}{X^2}\right] + \text{Var}\left(\frac{1}{X}\right)$$

$$= E\left[\frac{1}{X^2}\right] + E\left[\frac{1}{X^2}\right] - \left(E\left[\frac{1}{X}\right]\right)^2$$

$$= E\left[\frac{2}{X^2}\right] - (\ln 2)^2$$

$$= 1 - (\ln 2)^2.$$

5.2.4 Functions of Two Continuous Random Variables

So far, we have seen several examples involving functions of random variables. When we have two continuous random variables $g(X, Y)$, the ideas are still the same. First, if we are just interested in $E[g(X, Y)]$, we can use LOTUS:

LOTUS for two continuous random variables:

$$E[g(X,Y)] = \int_{-\infty}^{\infty} \int_{-\infty}^{\infty} g(x,y) f_{XY}(x,y) \; dxdy \qquad (5.19)$$

Example 5.27. Let X and Y be two jointly continuous random variables with joint PDF

$$f_{XY}(x,y) = \begin{cases} x+y & 0 \le x \le 1, 0 \le y \le 1 \\ \\ 0 & \text{otherwise} \end{cases}$$

Find $E[XY^2]$.

Solution: We have

$$\begin{aligned} E[XY^2] &= \int_{-\infty}^{\infty} \int_{-\infty}^{\infty} (xy^2) f_{XY}(x,y) \; dxdy \\ &= \int_0^1 \int_0^1 xy^2(x+y) \; dxdy \\ &= \int_0^1 \int_0^1 x^2y^2 + xy^3 \; dxdy \\ &= \int_0^1 \left(\frac{1}{3}y^2 + \frac{1}{2}y^3 \right) \; dy \\ &= \frac{17}{72}. \end{aligned}$$

If $Z = g(X,Y)$ and we are interested in its distribution, we can start by writing

$$\begin{aligned} F_Z(z) &= P(Z \le z) \\ &= P(g(X,Y) \le z) \\ &= \iint_D f_{XY}(x,y) \; dxdy, \end{aligned}$$

where $D = \{(x,y) | g(x,y) < z\}$. To find the PDF of Z, we differentiate $F_Z(z)$.

Example 5.28. Let X and Y be two independent $Uniform(0,1)$ random variables, and $Z = XY$. Find the CDF and PDF of Z.

Solution: First note that $R_Z = [0,1]$. Thus,

$$\begin{aligned} F_Z(z) &= 0, & \text{for } z \le 0, \\ F_Z(z) &= 1, & \text{for } z \ge 1. \end{aligned}$$

For $0 < z < 1$, we have

$$F_Z(z) = P(Z \le z)$$
$$= P(XY \le z)$$
$$= P\left(X \le \frac{z}{Y}\right).$$

Just to get some practice, we will show you two ways to calculate $P(X \le \frac{z}{Y})$ for $0 < z < 1$. The first way is just integrating $f_{XY}(x,y)$ in the region $x \le \frac{z}{y}$. We have

$$P\left(X \le \frac{z}{Y}\right) = \int_0^1 \int_0^{\frac{z}{y}} f_{XY}(x,y) \; dxdy$$
$$= \int_0^1 \int_0^{\min(1, \frac{z}{y})} 1 \; dxdy$$
$$= \int_0^1 \min\left(1, \frac{z}{y}\right) \; dy.$$

Note that if we let $g(y) = \min\left(1, \frac{z}{y}\right)$, then

$$g(y) = \begin{cases} 1 & \text{for } 0 < y < z \\ \frac{z}{y} & \text{for } z \le y \le 1 \end{cases}$$

Therefore,

$$P\left(X \le \frac{z}{Y}\right) = \int_0^1 g(y) \; dy$$
$$= \int_0^z 1 \; dy + \int_z^1 \frac{z}{y} \; dy$$
$$= z - z \ln z.$$

The second way to find $P(X \le \frac{z}{Y})$ is to use the law of total probability. We have

$$P\left(X \le \frac{z}{Y}\right) = \int_0^1 P\left(X \le \frac{z}{Y} \Big| Y = y\right) f_Y(y) \; dy$$
$$= \int_0^1 P\left(X \le \frac{z}{y}\right) f_Y(y) \; dy \qquad \text{since } X \text{ and } Y \text{ are independent.}$$

$$(5.20)$$

Note that

$$P\left(X \le \frac{z}{y}\right) = \begin{cases} 1 & \text{for } 0 < y < z \\ \frac{z}{y} & \text{for } z \le y \le 1 \end{cases}$$

Therefore,

$$P\left(X \leq \frac{z}{Y}\right) = \int_0^1 P\left(X \leq \frac{z}{y}\right) f_Y(y) \ dy$$

$$= \int_0^z 1 \ dy + \int_z^1 \frac{z}{y} \ dy$$

$$= z - z \ln z.$$

Thus, in the end we obtain

$$F_Z(z) = \begin{cases} 0 & z \leq 0 \\ z - z \ln z & 0 < z < 1 \\ 1 & z \geq 1 \end{cases}$$

You can check that $F_Z(z)$ is a continuous function. To find the PDF, we differentiate the CDF. We have

$$f_Z(z) = \begin{cases} -\ln z & 0 < z < 1 \\ 0 & \text{otherwise} \end{cases}$$

The Method of Transformations:

When we have functions of two or more jointly continuous random variables, we may be able to use a method similar to Theorems 4.1 and 4.2 to find the resulting PDFs. In particular, we can state the following theorem. While the statement of the theorem might look a little confusing, its application is quite straightforward and we will see a few examples to illustrate the methodology.

Theorem 5.1. Let X and Y be two jointly continuous random variables. Let $(Z, W) = g(X, Y) = (g_1(X, Y), g_2(X, Y))$, where $g : \mathbb{R}^2 \mapsto \mathbb{R}^2$ is a continuous one-to-one (invertible) function with continuous partial derivatives. Let $h = g^{-1}$, i.e., $(X, Y) = h(Z, W) = (h_1(Z, W), h_2(Z, W))$. Then Z and W are jointly continuous and their joint PDF, $f_{ZW}(z, w)$, for $(z, w) \in R_{ZW}$ is given by

$$f_{ZW}(z, w) = f_{XY}(h_1(z, w), h_2(z, w))|J|,$$

where J is the Jacobian of h defined by

$$J = \det \begin{bmatrix} \frac{\partial h_1}{\partial z} & \frac{\partial h_1}{\partial w} \\ \frac{\partial h_2}{\partial z} & \frac{\partial h_2}{\partial w} \end{bmatrix} = \frac{\partial h_1}{\partial z} \cdot \frac{\partial h_2}{\partial w} - \frac{\partial h_2}{\partial z} \cdot \frac{\partial h_1}{\partial w}.$$

The following examples show how to apply the above theorem.

Example 5.29. Let X and Y be two independent standard normal random variables. Let also

$$\begin{cases} Z = 2X - Y \\ W = -X + Y \end{cases}$$

Find $f_{ZW}(z, w)$.

Solution: X and Y are jointly continuous and their joint PDF is given by

$$f_{XY}(x, y) = f_X(x) f_Y(y) = \frac{1}{2\pi} e^{-\frac{x^2 + y^2}{2}}, \qquad \text{for all } x, y \in \mathbb{R}.$$

Here, the function g is defined by $(z, w) = g(x, y) = (g_1(x, y), g_2(x, y)) = (2x - y, -x + y)$. Solving for x and y, we obtain the inverse function h:

$$\begin{cases} x = z + w = h_1(z, w) \\ y = z + 2w = h_2(z, w) \end{cases}$$

We have

$$f_{ZW}(z, w) = f_{XY}(h_1(z, w), h_2(z, w))|J|$$
$$= f_{XY}(z + w, z + 2w)|J|,$$

where

$$J = \det \begin{bmatrix} \frac{\partial h_1}{\partial z} & \frac{\partial h_1}{\partial w} \\ \frac{\partial h_2}{\partial z} & \frac{\partial h_2}{\partial w} \end{bmatrix} = \det \begin{bmatrix} 1 & 1 \\ 1 & 2 \end{bmatrix} = 1.$$

Thus, we conclude that

$$f_{ZW}(z, w) = f_{XY}(z + w, z + 2w)|J|$$
$$= \frac{1}{2\pi} e^{-\frac{(z+w)^2 + (z+2w)^2}{2}}$$
$$= \frac{1}{2\pi} e^{-\frac{2z^2 + 5w^2 + 6zw}{2}}.$$

Example 5.30. Let X and Y be two random variables with joint PDF $f_{XY}(x, y)$. Let $Z = X + Y$. Find $f_Z(z)$.

Solution: To apply Theorem 5.1, we need two random variables Z and W. We can simply define $W = X$. Thus, the function g is given by

$$\begin{cases} z = x + y \\ w = x \end{cases}$$

Then, we can find the inverse transform:

$$\begin{cases} x = w \\ y = z - w \end{cases}$$

Then, we have

$$|J| = \left| \det \begin{bmatrix} 0 & 1 \\ 1 & -1 \end{bmatrix} \right| = |-1| = 1.$$

Thus,

$$f_{ZW}(z, w) = f_{XY}(w, z - w).$$

But since we are interested in the marginal PDF, $f_Z(z)$, we have

$$f_Z(z) = \int_{-\infty}^{\infty} f_{XY}(w, z - w) dw.$$

Note that, if X and Y are independent, then $f_{XY}(x, y) = f_X(x) f_Y(y)$ and we conclude that

$$f_Z(z) = \int_{-\infty}^{\infty} f_X(w) f_Y(z - w) dw.$$

The above integral is called the *convolution* of f_X and f_Y, and we write

$$f_Z(z) = f_X(z) * f_Y(z)$$
$$= \int_{-\infty}^{\infty} f_X(w) f_Y(z - w) dw = \int_{-\infty}^{\infty} f_Y(w) f_X(z - w) dw.$$

If X and Y are two jointly continuous random variables and $Z = X + Y$, then

$$f_Z(z) = \int_{-\infty}^{\infty} f_{XY}(w, z - w) dw = \int_{-\infty}^{\infty} f_{XY}(z - w, w) dw.$$

If X and Y are also independent, then

$$f_Z(z) = f_X(z) * f_Y(z)$$
$$= \int_{-\infty}^{\infty} f_X(w) f_Y(z - w) dw = \int_{-\infty}^{\infty} f_Y(w) f_X(z - w) dw.$$

Example 5.31. Let X and Y be two independent standard normal random variables, and let $Z = X + Y$. Find the PDF of Z.

Solution: We have

$$f_Z(z) = f_X(z) * f_Y(z)$$

$$= \int_{-\infty}^{\infty} f_X(w) f_Y(z-w) dw$$

$$= \int_{-\infty}^{\infty} \frac{1}{2\pi} e^{-\frac{w^2}{2}} e^{-\frac{(z-w)^2}{2}} dw$$

$$= \frac{1}{\sqrt{4\pi}} e^{\frac{-z^2}{4}} \int_{-\infty}^{\infty} \frac{1}{\sqrt{\pi}} e^{-(w-\frac{z}{2})^2} dw$$

$$= \frac{1}{\sqrt{4\pi}} e^{\frac{-z^2}{4}},$$

where $\int_{-\infty}^{\infty} \frac{1}{\sqrt{\pi}} e^{-(w-\frac{z}{2})^2} dw = 1$ because it is the integral of the PDF of a normal random variable with mean $\frac{z}{2}$ and variance $\frac{1}{2}$. Thus, we conclude that $Z \sim N(0,2)$. In fact, this is one of the interesting properties of the normal distribution: the sum of two independent normal random variables is also normal. In particular, similar to our calculation above, we can show the following:

Theorem 5.2. If $X \sim N(\mu_X, \sigma_X^2)$ and $Y \sim N(\mu_Y, \sigma_Y^2)$ are independent, then

$$X + Y \sim N\left(\mu_X + \mu_Y, \sigma_X^2 + \sigma_Y^2\right).$$

We will see an easier proof of Theorem 5.2 when we discuss *moment generating functions*.

5.2.5 Solved Problems

1. Let X and Y be jointly continuous random variables with joint PDF

$$f_{X,Y}(x,y) = \begin{cases} cx+1 & x,y \geq 0, x+y < 1 \\ 0 & \text{otherwise} \end{cases}$$

(a) Show the range of (X,Y), R_{XY}, in the $x-y$ plane.

(b) Find the constant c.

(c) Find the marginal PDFs $f_X(x)$ and $f_Y(y)$.

(d) Find $P(Y < 2X^2)$.

Solution:

(a) Figure 5.8(a) shows R_{XY} in the $x - y$ plane.

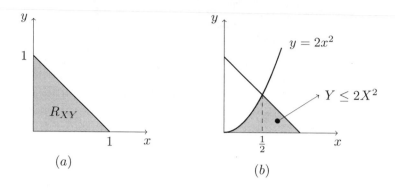

Figure 5.8: The figure shows (a) R_{XY} as well as (b) the integration region for finding $P(Y < 2X^2)$ for Solved Problem 1.

(b) To find the constant c, we write

$$1 = \int_{-\infty}^{\infty} \int_{-\infty}^{\infty} f_{XY}(x,y)dxdy$$

$$= \int_{0}^{1} \int_{0}^{1-x} cx + 1 \ dydx$$

$$= \int_{0}^{1} (cx+1)(1-x) \ dx$$

$$= \frac{1}{2} + \frac{1}{6}c.$$

Thus, we conclude $c = 3$.

(c) We first note that $R_X = R_Y = [0,1]$.

$$f_X(x) = \int_{-\infty}^{\infty} f_{XY}(x,y)dy$$

$$= \int_{0}^{1-x} 3x + 1 \ dy$$

$$= (3x+1)(1-x), \quad \text{for } x \in [0,1].$$

Thus, we have

$$f_X(x) = \begin{cases} (3x+1)(1-x) & 0 \le x \le 1 \\ 0 & \text{otherwise} \end{cases}$$

Similarly, we obtain

$$f_Y(y) = \int_{-\infty}^{\infty} f_{XY}(x,y)dx$$

$$= \int_0^{1-y} 3x + 1 \ dx$$

$$= \frac{1}{2}(1-y)(5-3y), \quad \text{for } y \in [0,1].$$

Thus, we have

$$f_Y(y) = \begin{cases} \frac{1}{2}(1-y)(5-3y) & 0 \le y \le 1 \\ \\ 0 & \text{otherwise} \end{cases}$$

(d) To find $P(Y < 2X^2)$, we need to integrate $f_{XY}(x,y)$ over the region shown in Figure 5.8(b). We have

$$P(Y < 2X^2) = \int_{-\infty}^{\infty} \int_{-\infty}^{2x^2} f_{XY}(x,y)dydx$$

$$= \int_0^1 \int_0^{\min(2x^2, 1-x)} 3x + 1 \ dydx$$

$$= \int_0^1 (3x+1)\min(2x^2, 1-x) \ dx$$

$$= \int_0^{\frac{1}{2}} 2x^2(3x+1) \ dx + \int_{\frac{1}{2}}^1 (3x+1)(1-x) \ dx$$

$$= \frac{53}{96}.$$

2. Let X and Y be jointly continuous random variables with joint PDF

$$f_{X,Y}(x,y) = \begin{cases} 6e^{-(2x+3y)} & x,y \ge 0 \\ \\ 0 & \text{otherwise} \end{cases}$$

(a) Are X and Y independent?
(b) Find $E[Y|X > 2]$.
(c) Find $P(X > Y)$.

Solution:

(a) We can write

$$f_{X,Y}(x,y) = f_X(x)f_Y(y),$$

where

$$f_X(x) = 2e^{-2x}u(x), \qquad f_Y(y) = 3e^{-3y}u(y).$$

Thus, X and Y are independent.

(b) Since X and Y are independent, we have $E[Y|X > 2] = E[Y]$. Note that $Y \sim$ $Exponential(3)$, thus $EY = \frac{1}{3}$.

(c) We have

$$P(X > Y) = \int_0^\infty \int_y^\infty 6e^{-(2x+3y)}\,dx\,dy$$

$$= \int_0^\infty 3e^{-5y}\,dy$$

$$= \frac{3}{5}.$$

3. Let X be a continuous random variable with PDF

$$f_X(x) = \begin{cases} 2x & 0 \le x \le 1 \\ 0 & \text{otherwise} \end{cases}$$

We know that given $X = x$, the random variable Y is uniformly distributed on $[-x, x]$.

(a) Find the joint PDF $f_{XY}(x, y)$.

(b) Find $f_Y(y)$.

(c) Find $P(|Y| < X^3)$.

Solution:

(a) First note that, by the assumption

$$f_{Y|X}(y|x) = \begin{cases} \frac{1}{2x} & -x \le y \le x \\ 0 & \text{otherwise} \end{cases}$$

Thus, we have

$$f_{XY}(x, y) = f_{Y|X}(y|x)f_X(x) = \begin{cases} 1 & 0 \le x \le 1, -x \le y \le x \\ 0 & \text{otherwise} \end{cases}$$

Thus,

$$f_{XY}(x, y) = \begin{cases} 1 & |y| \le x \le 1 \\ 0 & \text{otherwise} \end{cases}$$

(b) First, note that $R_Y = [-1, 1]$. To find $f_Y(y)$, we can write

$$f_Y(y) = \int_{-\infty}^{\infty} f_{XY}(x, y)dx$$

$$= \int_{|y|}^{1} 1 \ dx$$

$$= 1 - |y|.$$

Thus,

$$f_Y(y) = \begin{cases} 1 - |y| & |y| \leq 1 \\ 0 & \text{otherwise} \end{cases}$$

(c) To find $P(|Y| < X^3)$, we can use the law of total probability (Equation 5.16):

$$P(|Y| < X^3) = \int_0^1 P(|Y| < X^3 | X = x) f_X(x) \ dx$$

$$= \int_0^1 P(|Y| < x^3 | X = x) 2x \ dx$$

$$= \int_0^1 \left(\frac{2x^3}{2x} \right) 2x dx \qquad \text{since } Y | X = x \sim Uniform(-x, x)$$

$$= \frac{1}{2}.$$

4. Let X and Y be two jointly continuous random variables with joint PDF

$$f_{X,Y}(x, y) = \begin{cases} 6xy & 0 \leq x \leq 1, 0 \leq y \leq \sqrt{x} \\ 0 & \text{otherwise} \end{cases}$$

(a) Show R_{XY} in the $x - y$ plane.

(b) Find $f_X(x)$ and $f_Y(y)$.

(c) Are X and Y independent?

(d) Find the conditional PDF of X given $Y = y$, $f_{X|Y}(x|y)$.

(e) Find $E[X|Y = y]$, for $0 \leq y \leq 1$.

(f) Find $\text{Var}(X|Y = y)$, for $0 \leq y \leq 1$.

Solution:

(a) Figure 5.9 shows R_{XY} in the $x - y$ plane.

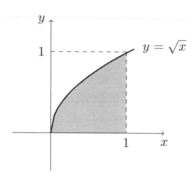

Figure 5.9: The figure shows R_{XY} for Solved Problem 4.

(b) First, note that $R_X = R_Y = [0, 1]$. To find $f_X(x)$ for $0 \leq x \leq 1$, we can write

$$f_X(x) = \int_{-\infty}^{\infty} f_{XY}(x, y) \ dy$$

$$= \int_0^{\sqrt{x}} 6xy \ dy$$

$$= 3x^2.$$

Thus,

$$f_X(x) = \begin{cases} 3x^2 & 0 \leq x \leq 1 \\ \\ 0 & \text{otherwise} \end{cases}$$

To find $f_Y(y)$ for $0 \leq y \leq 1$, we can write

$$f_Y(y) = \int_{-\infty}^{\infty} f_{XY}(x, y) \ dx$$

$$= \int_{y^2}^1 6xy \ dx$$

$$= 3y(1 - y^4).$$

$$f_Y(y) = \begin{cases} 3y(1 - y^4) & 0 \leq y \leq 1 \\ \\ 0 & \text{otherwise} \end{cases}$$

(c) X and Y are not independent, since $f_{XY}(x, y) \neq f_X(x) f_Y(y)$.

(d) We have

$$f_{X|Y}(x|y) = \frac{f_{XY}(x, y)}{f_Y(y)}$$

$$= \begin{cases} \frac{2x}{1 - y^4} & y^2 \leq x \leq 1 \\ \\ 0 & \text{otherwise} \end{cases}$$

(e) We have

$$E[X|Y = y] = \int_{-\infty}^{\infty} x f_{X|Y}(x|y) \ dx$$

$$= \int_{y^2}^{1} x \frac{2x}{1 - y^4} \ dx$$

$$= \frac{2(1 - y^6)}{3(1 - y^4)}.$$

(f) We have

$$E[X^2|Y = y] = \int_{-\infty}^{\infty} x^2 f_{X|Y}(x|y) \ dx$$

$$= \int_{y^2}^{1} x^2 \frac{2x}{1 - y^4} \ dx$$

$$= \frac{1 - y^8}{2(1 - y^4)}.$$

Thus,

$$\text{Var}(X|Y = y) = E[X^2|Y = y] - (E[X|Y = y])^2$$

$$= \frac{1 - y^8}{2(1 - y^4)} - \left(\frac{2(1 - y^6)}{3(1 - y^4)} \right)^2.$$

5. Consider the unit disc

$$D = \{(x, y) | x^2 + y^2 \leq 1\}.$$

Suppose that we choose a point (X, Y) uniformly at random in D. That is, the joint PDF of X and Y is given by

$$f_{XY}(x, y) = \begin{cases} \frac{1}{\pi} & (x, y) \in D \\ 0 & \text{otherwise} \end{cases}$$

Let (R, Θ) be the corresponding polar coordinates as shown in Figure 5.10. The inverse transformation is given by

$$\begin{cases} X = R \cos \Theta \\ Y = R \sin \Theta \end{cases}$$

where $R \geq 0$ and $-\pi < \Theta \leq \pi$. Find the joint PDF of R and Θ.

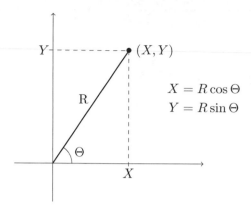

Figure 5.10: Polar coordinates

Solution: Here (X, Y) are jointly continuous and are related to (R, Θ) by a one-to-one relationship. We use the method of transformations (Theorem 5.1). The function $h(r, \theta)$ is given by

$$\begin{cases} x = h_1(r, \theta) = r \cos \theta \\ y = h_2(r, \theta) = r \sin \theta \end{cases}$$

Thus, we have

$$f_{R\Theta}(r, \theta) = f_{XY}(h_1(r, \theta), h_2(r, \theta))|J| \\ = f_{XY}(r \cos \theta, r \sin \theta)|J|.$$

where

$$J = \det \begin{bmatrix} \frac{\partial h_1}{\partial r} & \frac{\partial h_1}{\partial \theta} \\ \frac{\partial h_2}{\partial r} & \frac{\partial h_2}{\partial \theta} \end{bmatrix} = \det \begin{bmatrix} \cos \theta & -r \sin \theta \\ \sin \theta & r \cos \theta \end{bmatrix} = r \cos^2 \theta + r \sin^2 \theta = r.$$

We conclude that

$$f_{R\Theta}(r, \theta) = f_{XY}(r \cos \theta, r \sin \theta)|J| \\ = \begin{cases} \frac{r}{\pi} & r \in [0, 1], \theta \in (-\pi, \pi] \\ 0 & \text{otherwise} \end{cases}$$

Note that from above we can write

$$f_{R\Theta}(r, \theta) = f_R(r) f_\Theta(\theta),$$

where

$$f_R(r) = \begin{cases} 2r & r \in [0, 1] \\ 0 & \text{otherwise} \end{cases}$$

$$f_\Theta(\theta) = \begin{cases} \frac{1}{2\pi} & \theta \in (-\pi, \pi] \\ 0 & \text{otherwise} \end{cases}$$

Thus, we conclude that R and Θ are independent.

5.3 More Topics on Two Random Variables

5.3.1 Covariance and Correlation

Consider two random variables X and Y. Here, we define the **covariance** between X and Y, written $\text{Cov}(X, Y)$. The covariance gives some information about how X and Y are statistically related. Let us provide the definition, then discuss the properties and applications of covariance.

The **covariance** between X and Y is defined as

$$\text{Cov}(X, Y) = E\big[(X - EX)(Y - EY)\big] = E[XY] - (EX)(EY).$$

Note that

$$\begin{aligned} E\big[(X - EX)(Y - EY)\big] &= E\big[XY - X(EY) - (EX)Y + (EX)(EY)\big] \\ &= E[XY] - (EX)(EY) - (EX)(EY) + (EX)(EY) \\ &= E[XY] - (EX)(EY). \end{aligned}$$

Intuitively, the covariance between X and Y indicates how the values of X and Y move relative to each other. If large values of X tend to happen with large values of Y, then $(X-EX)(Y-EY)$ is positive on average. In this case, the covariance is positive and we say X and Y are positively correlated. On the other hand, if X tends to be small when Y is large, then $(X-EX)(Y-EY)$ is negative on average. In this case, the covariance is negative and we say X and Y are negatively correlated.

Example 5.32. Suppose $X \sim Uniform(1, 2)$, and given $X = x$, Y is exponential with parameter $\lambda = x$. Find $\text{Cov}(X, Y)$.

Solution: We can use $\text{Cov}(X, Y) = EXY - EXEY$. We have $EX = \frac{3}{2}$ and

$$\begin{aligned} EY &= E[E[Y|X]] && \big(\text{law of iterated expectations (Equation 5.17)}\big) \\ &= E\left[\frac{1}{X}\right] && \big(\text{since } Y|X \sim Exponential(X)\big) \\ &= \int_1^2 \frac{1}{x}dx \\ &= \ln 2. \end{aligned}$$

We also have

$$EXY = E[E[XY|X]] \qquad \text{(law of iterated expectations)}$$
$$EXY = E[XE[Y|X]] \qquad \left(\text{since} E[X|X=x]=x\right)$$
$$= E\left[X\frac{1}{X}\right] \qquad \left(\text{since } Y|X \sim Exponential(X)\right)$$
$$= 1.$$

Thus,

$$\text{Cov}(X,Y) = E[XY] - (EX)(EY) = 1 - \frac{3}{2}\ln 2.$$

Now we discuss the properties of covariance.

Lemma 5.3. The covariance has the following properties.

1. $\text{Cov}(X,X) = \text{Var}(X)$

2. If X and Y are independent then $\text{Cov}(X,Y) = 0$.

3. $\text{Cov}(X,Y) = \text{Cov}(Y,X)$

4. $\text{Cov}(aX,Y) = a\text{Cov}(X,Y)$

5. $\text{Cov}(X+c,Y) = \text{Cov}(X,Y)$

6. $\text{Cov}(X+Y,Z) = \text{Cov}(X,Z) + \text{Cov}(Y,Z)$

7. More generally,

$$\text{Cov}\left(\sum_{i=1}^{m} a_i X_i, \sum_{j=1}^{n} b_j Y_j\right) = \sum_{i=1}^{m}\sum_{j=1}^{n} a_i b_j \text{Cov}(X_i, Y_j).$$

All of the above results can be proven directly from the definition of covariance. For example, if X and Y are independent, then as we have seen before $E[XY] = EXEY$, so

$$\text{Cov}(X,Y) = E[XY] - EXEY = 0.$$

Note that the converse is not necessarily true. That is, if $\text{Cov}(X,Y) = 0$, X and Y may or may not be independent.

Let us prove Item 6 in Lemma 5.3, $\text{Cov}(X + Y, Z) = \text{Cov}(X, Z) + \text{Cov}(Y, Z)$. We have

$$
\begin{aligned}
\text{Cov}(X + Y, Z) &= E[(X + Y)Z] - E(X + Y)EZ \\
&= E[XZ + YZ] - (EX + EY)EZ \\
&= EXZ - EXEZ + EYZ - EYEZ \\
&= \text{Cov}(X, Z) + \text{Cov}(Y, Z).
\end{aligned}
$$

You can prove the rest of the items in Lemma 5.3 similarly.

Example 5.33. Let X and Y be two independent $N(0, 1)$ random variables and

$$
\begin{aligned}
Z &= 1 + X + XY^2, \\
W &= 1 + X.
\end{aligned}
$$

Find $\text{Cov}(Z, W)$.

Solution: We have

$$
\begin{aligned}
\text{Cov}(Z, W) &= \text{Cov}(1 + X + XY^2, 1 + X) \\
&= \text{Cov}(X + XY^2, X) && \text{(by part 5 of Lemma 5.3)} \\
&= \text{Cov}(X, X) + \text{Cov}(XY^2, X) && \text{(by part 6 of Lemma 5.3)} \\
&= \text{Var}(X) + E[X^2Y^2] - E[XY^2]EX && \text{(by part 1 of Lemma 5.3 and definition of Cov)} \\
&= 1 + E[X^2]E[Y^2] - E[X]^2E[Y^2] && \text{(since X and Y are independent)} \\
&= 1 + 1 - 0 = 2.
\end{aligned}
$$

Variance of a sum:

One of the applications of covariance is finding the variance of a sum of several random variables. In particular, if $Z = X + Y$, then

$$
\begin{aligned}
\text{Var}(Z) &= \text{Cov}(Z, Z) \\
&= \text{Cov}(X + Y, X + Y) \\
&= \text{Cov}(X, X) + \text{Cov}(X, Y) + \text{Cov}(Y, X) + \text{Cov}(Y, Y) \\
&= \text{Var}(X) + \text{Var}(Y) + 2\text{Cov}(X, Y).
\end{aligned}
$$

More generally, for $a, b \in \mathbb{R}$, we conclude:

$$
\text{Var}(aX + bY) = a^2\text{Var}(X) + b^2\text{Var}(Y) + 2ab\text{Cov}(X, Y) \tag{5.21}
$$

Correlation Coefficient:

The **correlation coefficient**, denoted by ρ_{XY} or $\rho(X, Y)$, is obtained by normalizing the covariance. In particular, we define the correlation coefficient of two random variables X and Y as the covariance of the standardized versions of X and Y. Define the standardized versions of X and Y as

$$U = \frac{X - EX}{\sigma_X}, \quad V = \frac{Y - EY}{\sigma_Y} \tag{5.22}$$

Then,

$$
\begin{aligned}
\rho_{XY} = \mathrm{Cov}(U, V) &= \mathrm{Cov}\left(\frac{X - EX}{\sigma_X}, \frac{Y - EY}{\sigma_Y}\right) \\
&= \mathrm{Cov}\left(\frac{X}{\sigma_X}, \frac{Y}{\sigma_Y}\right) \qquad \text{(by Item 5 of Lemma 5.3)} \\
&= \frac{\mathrm{Cov}(X, Y)}{\sigma_X \sigma_Y}.
\end{aligned}
$$

$$\rho_{XY} = \rho(X, Y) = \frac{\mathrm{Cov}(X, Y)}{\sqrt{\mathrm{Var(X)}\ \mathrm{Var(Y)}}} = \frac{\mathrm{Cov}(X, Y)}{\sigma_X \sigma_Y}$$

A nice thing about the correlation coefficient is that it is always between -1 and 1. This is an immediate result of the Cauchy-Schwarz inequality that is discussed in Section 6.2.4. One way to prove that $-1 \leq \rho \leq 1$ is to use the following inequality:

$$\alpha\beta \leq \frac{\alpha^2 + \beta^2}{2}, \qquad \text{for } \alpha, \beta \in \mathbb{R}.$$

This is because $(\alpha - \beta)^2 \geq 0$. The equality holds only if $\alpha = \beta$. From this, we can conclude that for any two random variables U and V,

$$E[UV] \leq \frac{EU^2 + EV^2}{2},$$

with equality only if $U = V$ with probability one. Now, let U and V be the standardized versions of X and Y as defined in Equation 5.22. Then, by definition $\rho_{XY} = \mathrm{Cov}(U, V) = EUV$. But since $EU^2 = EV^2 = 1$, we conclude

$$\rho_{XY} = E[UV] \leq \frac{EU^2 + EV^2}{2} = 1,$$

with equality only if $U = V$. That is,

$$\frac{Y - EY}{\sigma_Y} = \frac{X - EX}{\sigma_X},$$

which implies

$$Y = \frac{\sigma_Y}{\sigma_X} X + \left(EY - \frac{\sigma_Y}{\sigma_X} EX \right)$$

$$= aX + b, \qquad \text{where } a \text{ and } b \text{ are constants.}$$

Replacing X by $-X$, we conclude that

$$\rho(-X, Y) \leq 1.$$

But $\rho(-X, Y) = -\rho(X, Y)$, thus we conclude $\rho(X, Y) \geq -1$. Thus, we can summarize some properties of the correlation coefficient as follows.

Properties of the correlation coefficient:

1. $-1 \leq \rho(X, Y) \leq 1$;

2. If $\rho(X, Y) = 1$, then $Y = aX + b$, where $a > 0$;

3. If $\rho(X, Y) = -1$, then $Y = aX + b$, where $a < 0$;

4. $\rho(aX + b, cY + d) = \rho(X, Y)$ for $a, c > 0$.

Definition 5.2. Consider two random variables X and Y:

- If $\rho(X, Y) = 0$, we say that X and Y are **uncorrelated**.

- If $\rho(X, Y) > 0$, we say that X and Y are **positively** correlated.

- If $\rho(X, Y) < 0$, we say that X and Y are **negatively** correlated.

Note that as we discussed previously, two independent random variables are always uncorrelated, but the converse is not necessarily true. That is, if X and Y are uncorrelated, then X and Y may or may not be independent. Also, note that if X and Y are uncorrelated from Equation 5.21, we conclude that $\text{Var}(X + Y) = \text{Var}(X) + \text{Var}(Y)$.

If X and Y are uncorrelated, then

$$\text{Var}(X + Y) = \text{Var}(X) + \text{Var}(Y).$$

More generally, if $X_1, X_2, ..., X_n$ are pairwise uncorrelated, i.e., $\rho(X_i, X_j) = 0$ when $i \neq j$, then

$$\text{Var}(X_1 + X_2 + ... + X_n) = \text{Var}(X_1) + \text{Var}(X_2) + ... + \text{Var}(X_n).$$

Note that if X and Y are independent, then they are uncorrelated, and so $\text{Var}(X + Y) = \text{Var}(X) + \text{Var}(Y)$. This is a fact that we stated previously in Chapter 3, and now we could easily prove using covariance.

Example 5.34. Let X and Y be as in Example 5.24 in Section 5.2.3, i.e., suppose that we choose a point (X, Y) uniformly at random in the unit disc

$$D = \{(x, y) | x^2 + y^2 \leq 1\}.$$

Are X and Y uncorrelated?

Solution: We need to check whether $\text{Cov}(X, Y) = 0$. First note that, in Example 5.24 of Section 5.2.3, we found out that X and Y are not independent and in fact, we found that

$$X|Y \ \sim \ Uniform(-\sqrt{1 - Y^2}, \sqrt{1 - Y^2}).$$

Now let's find $\text{Cov}(X, Y) = EXY - EXEY$. We have

$$
\begin{aligned}
EX &= E[E[X|Y]] && \text{(law of iterated expectations (Equation 5.17))}\\
&= E[0] = 0 && \left(\text{since } X|Y \ \sim \ Uniform(-\sqrt{1 - Y^2}, \sqrt{1 - Y^2})\right).
\end{aligned}
$$

Also, we have

$$
\begin{aligned}
E[XY] &= E[E[XY|Y]] && \text{(law of iterated expectations (Equation 5.17))}\\
&= E[YE[X|Y]] && \text{(Equation 5.6)}\\
&= E[Y \cdot 0] = 0.
\end{aligned}
$$

Thus,

$$\text{Cov}(X, Y) = E[XY] - EXEY = 0.$$

Thus, X and Y are uncorrelated.

5.3.2   Bivariate Normal Distribution

Remember that the normal distribution is very important in probability theory and it shows up in many different applications. We have discussed a single normal random variable previously; we will now talk about two or more normal random variables. We recently saw in Theorem 5.2 that the sum of two independent normal random variables is also normal. However, if the two normal random variables are not independent, then their sum is not necessarily normal. Here is a simple counterexample:

Example 5.35. Let $X \sim N(0, 1)$ and $W \sim Bernoulli\left(\frac{1}{2}\right)$ be independent random variables. Define the random variable Y as a function of X and W:

$$Y = h(X, W) = \begin{cases} X & \text{if } W = 0 \\ -X & \text{if } W = 1 \end{cases}$$

Find the PDF of Y and $X + Y$.

Solution: Note that by symmetry of $N(0,1)$ around zero, $-X$ is also $N(0,1)$. In particular, we can write

$$
\begin{aligned}
F_Y(y) &= P(Y \le y) \\
&= P(Y \le y | W = 0)P(W = 0) + P(Y \le y | W = 1)P(W = 1) \\
&= \frac{1}{2}P(X \le y | W = 0) + \frac{1}{2}P(-X \le y | W = 1) \\
&= \frac{1}{2}P(X \le y) + \frac{1}{2}P(-X \le y) \qquad \text{(since X and W are independent)} \\
&= \frac{1}{2}\Phi(y) + \frac{1}{2}\Phi(y) \qquad \text{(since X and $-X$ are $N(0,1)$)} \\
&= \Phi(y).
\end{aligned}
$$

Thus, $Y \sim N(0,1)$. Now, note that

$$
Z = X + Y = \begin{cases} 2X & \text{with probability } \frac{1}{2} \\ \\ 0 & \text{with probability } \frac{1}{2} \end{cases}
$$

Thus, Z is a mixed random variable and its PDF is given by

$$
f_Z(z) = \frac{1}{2}\delta(z) + \frac{1}{2}(\text{PDF of } 2X \text{ at } z)
$$

$$
f_Z(z) = \frac{1}{2}\delta(z) + \frac{1}{2}(\text{PDF of a } N(0,4) \text{ at } z)
$$

$$
= \frac{1}{2}\delta(z) + \frac{1}{4\sqrt{2\pi}}e^{-\frac{z^2}{8}}.
$$

In particular, note that X and Y are both normal but their sum is not. Now, we are ready to define **bivariate normal** or **jointly normal** random variables.

Definition 5.3. Two random variables X and Y are said to be **bivariate normal**, or **jointly normal**, if $aX + bY$ has a normal distribution for all $a, b \in \mathbb{R}$.

In the above definition, if we let $a = b = 0$, then $aX + bY = 0$. We agree that the constant zero is a normal random variable with mean and variance 0. From the above definition, we can immediately conclude the following facts:

- If X and Y are bivariate normal, then by letting $a = 1$, $b = 0$, we conclude X must be normal.

- If X and Y are bivariate normal, then by letting $a = 0$, $b = 1$, we conclude Y must be normal.

- If $X \sim N(\mu_X, \sigma_X^2)$ and $Y \sim N(\mu_Y, \sigma_Y^2)$ are independent, then they are jointly normal (Theorem 5.2).

- If $X \sim N(\mu_X, \sigma_X^2)$ and $Y \sim N(\mu_Y, \sigma_Y^2)$ are jointly normal, then $X + Y \sim N\Big(\mu_X +$

$\mu_Y, \sigma_X^2 + \sigma_Y^2 + 2\rho(X, Y)\sigma_X \sigma_Y\Big)$ (Equation 5.21).

But how can we obtain the joint normal PDF in general? Can we provide a simple way to generate jointly normal random variables? The basic idea is that we can start from several independent random variables and by considering their linear combinations, we can obtain bivariate normal random variables. Similar to our discussion on normal random variables, we start by introducing the **standard bivariate normal distribution** and then obtain the general case from the standard one. The following example gives the idea.

Example 5.36. Let Z_1 and Z_2 be two independent $N(0, 1)$ random variables. Define

$$X = Z_1,$$
$$Y = \rho Z_1 + \sqrt{1 - \rho^2} Z_2,$$

where ρ is a real number in $(-1, 1)$.

(a) Show that X and Y are bivariate normal.

(b) Find the joint PDF of X and Y.

(c) Find $\rho(X, Y)$.

   Solution: First, note that since Z_1 and Z_2 are normal and independent, they are jointly normal, with the joint PDF

$$f_{Z_1 Z_2}(z_1, z_2) = f_{Z_1}(z_1) f_{Z_2}(z_2)$$
$$= \frac{1}{2\pi} \exp\left\{ -\frac{1}{2}\left[z_1^2 + z_2^2\right] \right\}.$$

(a) We need to show $aX + bY$ is normal for all $a, b \in \mathbb{R}$. We have

$$aX + bY = aZ_1 + b(\rho Z_1 + \sqrt{1 - \rho^2} Z_2)$$
$$= (a + b\rho)Z_1 + b\sqrt{1 - \rho^2} Z_2,$$

which is a linear combination of Z_1 and Z_2 and thus it is normal.

(b) We can use the method of transformations (Theorem 5.1) to find the joint PDF of X and Y. The inverse transformation is given by

$$Z_1 = X = h_1(X, Y),$$
$$Z_2 = -\frac{\rho}{\sqrt{1 - \rho^2}} X + \frac{1}{\sqrt{1 - \rho^2}} Y = h_2(X, Y).$$

We have

$$f_{XY}(z_1, z_2) = f_{Z_1 Z_2}(h_1(x, y), h_2(x, y))|J|$$
$$= f_{Z_1 Z_2}(x, -\frac{\rho}{\sqrt{1 - \rho^2}} x + \frac{1}{\sqrt{1 - \rho^2}} y)|J|,$$

where

$$J = \det \begin{bmatrix} \frac{\partial h_1}{\partial x} & \frac{\partial h_1}{\partial y} \\ \frac{\partial h_2}{\partial x} & \frac{\partial h_2}{\partial y} \end{bmatrix} = \det \begin{bmatrix} 1 & 0 \\ -\frac{\rho}{\sqrt{1-\rho^2}} & \frac{1}{\sqrt{1-\rho^2}} \end{bmatrix} = \frac{1}{\sqrt{1-\rho^2}}.$$

Thus, we conclude that

$$f_{XY}(x,y) = f_{Z_1 Z_2}\left(x, -\frac{\rho}{\sqrt{1-\rho^2}}x + \frac{1}{\sqrt{1-\rho^2}}y\right)|J|$$

$$= \frac{1}{2\pi} \exp\left\{-\frac{1}{2}\left[x^2 + \frac{1}{1-\rho^2}(-\rho x + y)^2\right]\right\} \cdot \frac{1}{\sqrt{1-\rho^2}}$$

$$= \frac{1}{2\pi\sqrt{1-\rho^2}} \exp\left\{-\frac{1}{2(1-\rho^2)}\left[x^2 - 2\rho xy + y^2\right]\right\}.$$

(c) To find $\rho(X,Y)$, first note

$$\text{Var}(X) = \text{Var}(Z_1) = 1,$$
$$\text{Var}(Y) = \rho^2 \text{Var}(Z_1) + (1-\rho^2)\text{Var}(Z_2) = 1.$$

Therefore,

$$\rho(X,Y) = \text{Cov}(X,Y)$$
$$= \text{Cov}(Z_1, \rho Z_1 + \sqrt{1-\rho^2} Z_2)$$
$$= \rho \text{Cov}(Z_1, Z_1) + \sqrt{1-\rho^2} \text{Cov}(Z_1, Z_2)$$
$$= \rho \cdot 1 + \sqrt{1-\rho^2} \cdot 0$$
$$= \rho.$$

We call the above joint distribution for X and Y the **standard bivariate normal distribution with correlation coefficient** ρ. It is the distribution for two jointly normal random variables when their variances are equal to one and their correlation coefficient is ρ.

Two random variables X and Y are said to have the **standard bivariate normal distribution with correlation coefficient** ρ if their joint PDF is given by

$$f_{XY}(x,y) = \frac{1}{2\pi\sqrt{1-\rho^2}} \exp\left\{-\frac{1}{2(1-\rho^2)}\left[x^2 - 2\rho xy + y^2\right]\right\},$$

where $\rho \in (-1,1)$. If $\rho = 0$, then we just say X and Y have the standard bivariate normal distribution.

Now, if you want two jointly normal random variables X and Y such that $X \sim N(\mu_X, \sigma_X^2)$, $Y \sim N(\mu_Y, \sigma_Y^2)$, and $\rho(X, Y) = \rho$, you can start with two independent $N(0, 1)$ random variables, Z_1 and Z_2, and define

$$\begin{cases} X & = \sigma_X Z_1 + \mu_X \\ Y & = \sigma_Y(\rho Z_1 + \sqrt{1 - \rho^2} Z_2) + \mu_Y \end{cases} \tag{5.23}$$

We can find the joint PDF of X and Y as above. While the joint PDF has a big formula, we usually do not need to use the formula itself. Instead, we usually work with properties of jointly normal random variables such as their mean, variance, and covariance.[1]

Definition 5.4. Two random variables X and Y are said to have a **bivariate normal distribution** with parameters μ_X, σ_X^2, μ_Y, σ_Y^2, and ρ, if their joint PDF is given by

$$f_{XY}(x, y) = \frac{1}{2\pi \sigma_X \sigma_Y \sqrt{1 - \rho^2}} \; \cdot$$
$$\exp\left\{ -\frac{1}{2(1 - \rho^2)} \left[\left(\frac{x - \mu_X}{\sigma_X}\right)^2 + \left(\frac{y - \mu_Y}{\sigma_Y}\right)^2 - 2\rho \frac{(x - \mu_X)(y - \mu_Y)}{\sigma_X \sigma_Y} \right] \right\} \tag{5.24}$$

where $\mu_X, \mu_Y \in \mathbb{R}$, $\sigma_X, \sigma_Y > 0$ and $\rho \in (-1, 1)$ are all constants.

In the above discussion, we introduced bivariate normal distributions by starting from independent normal random variables, Z_1 and Z_2. Another approach would have been to define the bivariate normal distribution using the joint PDF. The two definitions are equivalent mathematically. In particular, we can state the following theorem.

Theorem 5.3. Let X and Y be two bivariate normal random variables, i.e., their joint PDF is given by Equation 5.24. Then there exist independent standard normal random variables Z_1 and Z_2 such that

$$\begin{cases} X & = \sigma_X Z_1 + \mu_X \\ Y & = \sigma_Y(\rho Z_1 + \sqrt{1 - \rho^2} Z_2) + \mu_Y \end{cases}$$

Proof. (Sketch) To prove the theorem, define

$$\begin{cases} Z_1 & = \frac{X - \mu_X}{\sigma_X} \\ Z_2 & = -\frac{\rho}{\sqrt{1 - \rho^2}} \frac{X - \mu_X}{\sigma_X} + \frac{1}{\sqrt{1 - \rho^2}} \frac{Y - \mu_Y}{\sigma_Y} \end{cases}$$

---

[1]Definitions 5.3 and 5.4 are equivalent in the sense that, if X and Y are jointly normal based on one definition, they are jointly normal based on the other definition, too. The proof of their equivalence can be concluded from Problem 10 in Section 6.1.6. In that problem, we show that the two definitions result in the same moment generating functions.

Now find the joint PDF of Z_1 and Z_2 using the method of transformations (Theorem 5.1), similar to what we did above. You will find out that Z_1 and Z_2 are independent and standard normal and by definition satisfy the equations of Theorem 5.3. ∎

The reason we started our discussion on bivariate normal random variables from Z_1 and Z_2 is three fold. First, it is more convenient and insightful than the joint PDF formula. Second, sometimes the construction using Z_1 and Z_2 can be used to solve problems regarding bivariate normal distributions. Third, this method gives us a way to generate samples from the bivariate normal distribution using a computer program. Since most computing packages have a built-in command for independent normal random variable generation, we can simply use this command to generate bivariate normal variables using Equation 5.23.

Example 5.37. Let X and Y be jointly normal random variables with parameters μ_X, σ_X^2, μ_Y, σ_Y^2, and ρ. Find the conditional distribution of Y given $X = x$.

Solution: One way to solve this problem is by using the joint PDF formula (Equation 5.24). In particular, since $X \sim N(\mu_X, \sigma_X^2)$, we can use

$$f_{Y|X}(y|x) = \frac{f_{XY}(x,y)}{f_X(x)}.$$

Another way to solve this problem is to use Theorem 5.3. We can write

$$\begin{cases} X &= \sigma_X Z_1 + \mu_X \\ Y &= \sigma_Y(\rho Z_1 + \sqrt{1-\rho^2}Z_2) + \mu_Y \end{cases}$$

Thus, given $X = x$, we have

$$Z_1 = \frac{x - \mu_X}{\sigma_X},$$

and

$$Y = \sigma_Y \rho \frac{x - \mu_X}{\sigma_X} + \sigma_Y \sqrt{1-\rho^2}Z_2 + \mu_Y.$$

Since Z_1 and Z_2 are independent, knowing Z_1 does not provide any information on Z_2. We have shown that given $X = x$, Y is a linear function of Z_2, thus it is normal. In particular

$$E[Y|X=x] = \sigma_Y \rho \frac{x - \mu_X}{\sigma_X} + \sigma_Y \sqrt{1-\rho^2}E[Z_2] + \mu_Y$$

$$= \mu_Y + \rho\sigma_Y \frac{x - \mu_X}{\sigma_X}$$

$$\text{Var}(Y|X=x) = \sigma_Y^2(1-\rho^2)\text{Var}(Z_2)$$

$$= (1-\rho^2)\sigma_Y^2.$$

We conclude that given $X = x$, Y is normally distributed with mean $\mu_Y + \rho\sigma_Y \frac{x-\mu_X}{\sigma_X}$ and variance $(1-\rho^2)\sigma_Y^2$.

Theorem 5.4. Suppose X and Y are jointly normal random variables with parameters μ_X, σ_X^2, μ_Y, σ_Y^2, and ρ. Then, given $X = x$, Y is normally distributed with

$$E[Y|X = x] = \mu_Y + \rho\sigma_Y \frac{x - \mu_X}{\sigma_X},$$

$$\text{Var}(Y|X = x) = (1 - \rho^2)\sigma_Y^2.$$

Example 5.38. Let X and Y be jointly normal random variables with parameters $\mu_X = 1$, $\sigma_X^2 = 1$, $\mu_Y = 0$, $\sigma_Y^2 = 4$, and $\rho = \frac{1}{2}$.

(a) Find $P(2X + Y \leq 3)$.

(b) Find $\text{Cov}(X + Y, 2X - Y)$.

(c) Find $P(Y > 1|X = 2)$.

Solution:

(a) Since X and Y are jointly normal, the random variable $V = 2X + Y$ is normal. We have

$$EV = 2EX + EY = 2,$$

$$\begin{aligned}
\text{Var}(V) &= 4\text{Var}(X) + \text{Var}(Y) + 4\text{Cov}(X, Y) \\
&= 4 + 4 + 4\sigma_X\sigma_Y\rho(X, Y) \\
&= 8 + 4 \times 1 \times 2 \times \frac{1}{2} \\
&= 12.
\end{aligned}$$

Thus, $V \sim N(2, 12)$. Therefore,

$$P(V \leq 3) = \Phi\left(\frac{3 - 2}{\sqrt{12}}\right) = \Phi\left(\frac{1}{\sqrt{12}}\right) = 0.6136$$

(b) Note that $\text{Cov}(X, Y) = \sigma_X\sigma_Y\rho(X, Y) = 1$. We have

$$\begin{aligned}
\text{Cov}(X + Y, 2X - Y) &= 2\text{Cov}(X, X) - \text{Cov}(X, Y) + 2\text{Cov}(Y, X) - \text{Cov}(Y, Y) \\
&= 2 - 1 + 2 - 4 = -1.
\end{aligned}$$

(c) Using Theorem 5.4, we conclude that given $X = 2$, Y is normally distributed with

$$E[Y|X = 2] = \mu_Y + \rho\sigma_Y \frac{2 - \mu_X}{\sigma_X} = 1,$$

$$\text{Var}(Y|X = x) = (1 - \rho^2)\sigma_Y^2 = 3.$$

Thus

$$P(Y > 1|X = 2) = 1 - \Phi\left(\frac{1 - 1}{\sqrt{3}}\right) = \frac{1}{2}.$$

Remember that if two random variables X and Y are independent, then they are uncorrelated, i.e., $\text{Cov}(X, Y) = 0$. However, the converse is not true in general. In the case of jointly normal random variables, the converse is true. Thus, for jointly normal random variables, being independent and being uncorrelated are equivalent.

Theorem 5.5. If X and Y are bivariate normal and uncorrelated, then they are independent.

Proof. Since X and Y are uncorrelated, we have $\rho(X, Y) = 0$. By Theorem 5.4, given $X = x$, Y is normally distributed with

$$E[Y|X = x] = \mu_Y + \rho\sigma_Y \frac{x - \mu_X}{\sigma_X} = \mu_Y,$$
$$\text{Var}(Y|X = x) = (1 - \rho^2)\sigma_Y^2 = \sigma_Y^2.$$

Thus, $f_{Y|X}(y|x) = f_Y(y)$ for all $x, y \in \mathbb{R}$. Thus X and Y are independent. Another way to prove the theorem is to let $\rho = 0$ in Equation 5.24 and observe that $f_{XY}(x, y) = f_X(x)f_Y(y)$. ∎

5.3.3 Solved Problems

1. Let X and Y be two jointly continuous random variables with joint PDF

$$f_{XY}(x, y) = \begin{cases} 2 & y + x \leq 1, x > 0, y > 0 \\ 0 & \text{otherwise} \end{cases}$$

Find $\text{Cov}(X, Y)$ and $\rho(X, Y)$.

Solution: For $0 \leq x \leq 1$, we have

$$f_X(x) = \int_{-\infty}^{\infty} f_{XY}(x, y)dy$$
$$= \int_0^{1-x} 2dy$$
$$= 2(1 - x).$$

Thus,

$$f_X(x) = \begin{cases} 2(1 - x) & 0 \leq x \leq 1 \\ 0 & \text{otherwise} \end{cases}$$

Similarly, we obtain

$$f_Y(y) = \begin{cases} 2(1-y) & 0 \leq y \leq 1 \\ 0 & \text{otherwise} \end{cases}$$

Thus, we have

$$EX = \int_0^1 2x(1-x)dx$$
$$= \frac{1}{3} = EY.$$

$$EX^2 = \int_0^1 2x^2(1-x)dx$$
$$= \frac{1}{6} = EY^2.$$

Thus,

$$\text{Var}(X) = \text{Var}(Y) = \frac{1}{18}.$$

We also have

$$EXY = \int_0^1 \int_0^{1-x} 2xy\,dy\,dx$$
$$= \int_0^1 x(1-x)^2 dx$$
$$= \frac{1}{12}.$$

Now, we can find $\text{Cov}(X,Y)$ and $\rho(X,Y)$:

$$\text{Cov}(X,Y) = EXY - EXEY$$
$$= \frac{1}{12} - \left(\frac{1}{3}\right)^2$$
$$= -\frac{1}{36},$$

$$\rho(X,Y) = \frac{\text{Cov}(X,Y)}{\sqrt{\text{Var}(X)\text{Var}(Y)}}$$
$$= -\frac{1}{2}.$$

2. I roll a fair die n times. Let X be the number of 1's that I observe and let Y be the number of 2's that I observe. Find $\text{Cov}(X,Y)$ and $\rho(X,Y)$. *Hint:* One way to solve this problem is to look at $\text{Var}(X+Y)$.

Solution: Note that you can look at this as a binomial experiment. In particular, we can say that X and Y are $Binomial(n,\frac{1}{6})$. Also, $X+Y$ is $Binomial(n,\frac{2}{6})$. Remember the variance of a $Binomial(n,p)$ random variable is $np(1-p)$. Thus, we can write

$$n\frac{2}{6}\cdot\frac{4}{6} = \text{Var}(X+Y)$$
$$= \text{Var}(X) + \text{Var}(Y) + 2\text{Cov}(X,Y)$$
$$= n\frac{1}{6}\cdot\frac{5}{6} + n\frac{1}{6}\cdot\frac{5}{6} + 2\text{Cov}(X,Y).$$

Thus,

$$\text{Cov}(X,Y) = -\frac{n}{36}.$$

And,

$$\rho(X,Y) = \frac{\text{Cov}(X,Y)}{\sqrt{\text{Var}(X)\text{Var}(Y)}} = -\frac{1}{5}.$$

3. In this problem, you will provide another proof for the fact that $|\rho(X,Y)| \leq 1$. By definition $\rho_{XY} = \text{Cov}(U,V)$, where U and V are the normalized versions of X and Y as defined in Equation 5.22:

$$U = \frac{X - EX}{\sigma_X}, \quad V = \frac{Y - EY}{\sigma_Y}.$$

Use the fact that $\text{Var}(U+V) \geq 0$ to show that $|\rho(X,Y)| \leq 1$.

Solution: We have

$$\text{Var}(U+V) = \text{Var}(U) + \text{Var}(V) + 2\text{Cov}(U,V)$$
$$= 1 + 1 + 2\rho_{XY}.$$

Since $\text{Var}(U+V) \geq 0$, we conclude $\rho(X,Y) \geq -1$. Also, from this we conclude that

$$\rho(-X,Y) \geq -1.$$

But $\rho(-X,Y) = -\rho(X,Y)$, so we conclude $\rho(X,Y) \leq 1$.

4. Let X and Y be two independent $Uniform(0,1)$ random variables. Let also $Z = \max(X,Y)$ and $W = \min(X,Y)$. Find $\text{Cov}(Z,W)$.

Solution: It is useful to find the distributions of Z and W. To find the CDF of Z, we can write

$$\begin{aligned}
F_Z(z) &= P(Z \leq z) \\
&= P(\max(X,Y) \leq z) \\
&= P\Big((X \leq z) \text{ and } (Y \leq z) \Big) \\
&= P(X \leq z)P(Y \leq z) \qquad \text{(since X and Y are independent)} \\
&= F_X(z)F_Y(z).
\end{aligned}$$

Thus, we conclude

$$F_Z(z) = \begin{cases} 0 & z < 0 \\ z^2 & 0 \leq z \leq 1 \\ 1 & z > 1 \end{cases}$$

Therefore,

$$f_Z(z) = \begin{cases} 2z & 0 \leq z \leq 1 \\ 0 & \text{otherwise} \end{cases}$$

From this we obtain $EZ = \frac{2}{3}$. Note that we can find EW as follows

$$\begin{aligned}
1 = E[X+Y] = E[Z+W] &= EZ + EW \\
&= \frac{2}{3} + EW.
\end{aligned}$$

Thus, $EW = \frac{1}{3}$. Nevertheless, it is a good exercise to find the CDF and PDF of W, too. To find the CDF of W, we can write

$$\begin{aligned}
F_W(w) &= P(W \leq w) \\
&= P(\min(X,Y) \leq w) \\
&= 1 - P(\min(X,Y) > w) \\
&= 1 - P\Big((X > w) \text{ and } (Y > w) \Big) \\
&= 1 - P(X > w)P(Y > w) \qquad \text{(since X and Y are independent)} \\
&= 1 - (1 - F_X(w))(1 - F_Y(w)) \\
&= F_X(w) + F_Y(w) - F_X(w)F_Y(w).
\end{aligned}$$

Thus,

$$F_W(w) = \begin{cases} 0 & w < 0 \\ 2w - w^2 & 0 \leq w \leq 1 \\ 1 & w > 1 \end{cases}$$

Therefore,

$$f_W(w) = \begin{cases} 2 - 2w & 0 \leq w \leq 1 \\ 0 & \text{otherwise} \end{cases}$$

From the above PDF we can verify that $EW = \frac{1}{3}$. Now, to find $\text{Cov}(Z, W)$, we can write

$$\begin{aligned} \text{Cov}(Z, W) &= E[ZW] - EZEW \\ &= E[XY] - EZEW \\ &= E[X]E[Y] - E[Z]E[W] \quad \text{(since X and Y are independent)} \\ &= \frac{1}{2} \cdot \frac{1}{2} - \frac{2}{3} \cdot \frac{1}{3} \\ &= \frac{1}{36}. \end{aligned}$$

Note that $\text{Cov}(Z, W) > 0$ as we expect intuitively.

5. Let X and Y be jointly (bivariate) normal, with $\text{Var}(X) = \text{Var}(Y)$. Show that the two random variables $X + Y$ and $X - Y$ are independent.

Solution: Note that since X and Y are jointly normal, we conclude that the random variables $X + Y$ and $X - Y$ are also jointly normal. We have

$$\begin{aligned} \text{Cov}(X + Y, X - Y) &= \text{Cov}(X, X) - \text{Cov}(X, Y) + \text{Cov}(Y, X) - \text{Cov}(Y, Y) \\ &= \text{Var}(X) - \text{Var}(Y) \\ &= 0. \end{aligned}$$

Since $X + Y$ and $X - Y$ are jointly normal and uncorrelated, they are independent.

6. Let X and Y be jointly normal random variables with parameters $\mu_X = 0$, $\sigma_X^2 = 1$, $\mu_Y = -1$, $\sigma_Y^2 = 4$, and $\rho = -\frac{1}{2}$.

(a) Find $P(X + Y > 0)$.

(b) Find the constant a if we know $aX + Y$ and $X + 2Y$ are independent.

(c) Find $P(X + Y > 0 | 2X - Y = 0)$.

Solution:

(a) Since X and Y are jointly normal, the random variable $U = X + Y$ is normal. We have

$$EU = EX + EY = -1,$$

$$\text{Var}(U) = \text{Var}(X) + \text{Var}(Y) + 2\text{Cov}(X,Y)$$
$$= 1 + 4 + 2\sigma_X\sigma_Y\rho(X,Y)$$
$$= 5 - 2 \times 1 \times 2 \times \frac{1}{2}$$
$$= 3.$$

Thus, $U \sim N(-1,3)$. Therefore,

$$P(U > 0) = 1 - \Phi\left(\frac{0 - (-1)}{\sqrt{3}}\right) = 1 - \Phi\left(\frac{1}{\sqrt{3}}\right) = 0.2819$$

(b) Note that $aX + Y$ and $X + 2Y$ are jointly normal. Thus, for them, independence is equivalent to having $\text{Cov}(aX + Y, X + 2Y) = 0$. Also, note that $\text{Cov}(X,Y) = \sigma_X\sigma_Y\rho(X,Y) = -1$. We have

$$\text{Cov}(aX + Y, X + 2Y) = a\text{Cov}(X,X) + 2a\text{Cov}(X,Y) + \text{Cov}(Y,X) + 2\text{Cov}(Y,Y)$$
$$= a - (2a + 1) + 8$$
$$= -a + 7.$$

Thus, $a = 7$.

(c) If we define $U = X + Y$ and $V = 2X - Y$, then note that U and V are jointly normal. We have

$$EU = -1, \ \text{Var}(U) = 3,$$
$$EV = 1, \ \text{Var}(V) = 12,$$

and

$$\text{Cov}(U,V) = \text{Cov}(X + Y, 2X - Y)$$
$$= 2\text{Cov}(X,X) - \text{Cov}(X,Y) + 2\text{Cov}(Y,X) - \text{Cov}(Y,Y)$$
$$= 2\text{Var}(X) + \text{Cov}(X,Y) - \text{Var}(Y)$$
$$= 2 - 1 - 4$$
$$= -3.$$

Thus,

$$\rho(U,V) = \frac{\text{Cov}(U,V)}{\sqrt{\text{Var}(U)\text{Var}(V)}}$$
$$= -\frac{1}{2}.$$

Using Theorem 5.4, we conclude that given $V = 0$, U is normally distributed with

$$E[U|V = 0] = \mu_U + \rho(U,V)\sigma_U\frac{0 - \mu_V}{\sigma_V} = -\frac{3}{4}$$
$$\text{Var}(U|V = 0) = (1 - \rho_{UV}^2)\sigma_U^2 = \frac{9}{4}.$$

Thus

$$P(X + Y > 0 | 2X - Y = 0) = P(U > 0 | V = 0)$$
$$= 1 - \Phi\left(\frac{0 - (-\frac{3}{4})}{\frac{3}{2}}\right)$$
$$= 1 - \Phi\left(\frac{1}{2}\right) = 0.3085$$

5.4   End of Chapter Problems

1. Consider two random variables X and Y with joint PMF given in Table 5.4.

Table 5.4: Joint PMF of X and Y in Problem 1

	$Y = 1$	$Y = 2$
$X = 1$	$\frac{1}{3}$	$\frac{1}{12}$
$X = 2$	$\frac{1}{6}$	0
$X = 4$	$\frac{1}{12}$	$\frac{1}{3}$

   (a) Find $P(X \leq 2, Y > 1)$.
   (b) Find the marginal PMFs of X and Y.
   (c) Find $P(Y = 2 | X = 1)$.
   (d) Are X and Y independent?

2. Let X and Y be as defined in Problem 1. I define a new random variable $Z = X - 2Y$.

   (a) Find the PMF of Z.
   (b) Find $P(X = 2 | Z = 0)$.

3. A box contains two coins: a regular coin and a biased coin with $P(H) = \frac{2}{3}$. I choose a coin at random and toss it once. I define the random variable X as a Bernoulli random variable associated with this coin toss, i.e., $X = 1$ if the result of the coin toss is heads and $X = 0$ otherwise. Then I take the remaining coin in the box and toss it once. I define the random variable Y as a Bernoulli random variable associated with the second coin toss. Find the joint PMF of X and Y. Are X and Y independent?

4. Consider two random variables X and Y with joint PMF given by

$$P_{XY}(k, l) = \frac{1}{2^{k+l}}, \qquad \text{for } k, l = 1, 2, 3, \ldots$$

   (a) Show that X and Y are independent and find the marginal PMFs of X and Y.
   (b) Find $P(X^2 + Y^2 \leq 10)$.

5. Let X and Y be as defined in Problem 1. Also, suppose that we are given that $Y = 1$.

   (a) Find the conditional PMF of X given $Y = 1$. That is, find $P_{X|Y}(x|1)$.
   (b) Find $E[X|Y = 1]$.

(c) Find $\text{Var}(X|Y = 1)$.

6. The number of customers visiting a store in one hour has a Poisson distribution with mean $\lambda = 10$. Each customer is a female with probability $p = \frac{3}{4}$ independent of other customers. Let X be the total number of customers in a one-hour interval and Y be the total number of female customers in the same interval. Find the joint PMF of X and Y.

7. Let $X \sim Geometric(p)$. Find $\text{Var}(X)$ as follows: Find EX and EX^2 by conditioning on the result of the first "coin toss" and use $\text{Var}(X) = EX^2 - (EX)^2$.

8. Let X and Y be two independent $Geometric(p)$ random variables. Find $E\left[\frac{X^2+Y^2}{XY}\right]$.

9. Consider the set of points in the set C:

$$C = \{(x, y)|x, y \in \mathbb{Z}, x^2 + |y| \leq 2\}.$$

Suppose that we pick a point (X, Y) from this set completely at random. Thus, each point has a probability of $\frac{1}{11}$ of being chosen.

(a) Find the joint and marginal PMFs of X and Y.
(b) Find the conditional PMF of X given $Y = 1$.
(c) Are X and Y independent?
(d) Find $E[XY^2]$.

10. Consider the set of points in the set C:

$$C = \{(x, y)|x, y \in \mathbb{Z}, x^2 + |y| \leq 2\}.$$

Suppose that we pick a point (X, Y) from this set completely at random. Thus, each point has a probability of $\frac{1}{11}$ of being chosen.

(a) Find $E[X|Y = 1]$.
(b) Find $\text{Var}(X|Y = 1)$.
(c) Find $E[X||Y| \leq 1]$.
(d) Find $E[X^2||Y| \leq 1]$.

11. The number of cars being repaired at a small repair shop has the following PMF:

$$P_N(n) = \begin{cases} \frac{1}{8} & \text{for } n = 0 \\ \frac{1}{8} & \text{for } n = 1 \\ \frac{1}{4} & \text{for } n = 2 \\ \frac{1}{2} & \text{for } n = 3 \\ 0 & \text{otherwise} \end{cases}$$

Each car that is being repaired is a four-door car with probability $\frac{3}{4}$ and a two-door car with probability $\frac{1}{4}$, independently from other cars and independently from the number of cars being repaired. Let X be the number of four-door cars and Y be the number of two-door cars currently being repaired.

(a) Find the marginal PMFs of X and Y.

(b) Find the joint PMF of X and Y.

(c) Are X and Y independent?

12. Let X and Y be two independent random variables with PMFs

$$P_X(k) = P_Y(k) = \begin{cases} \frac{1}{5} & \text{for } x = 1, 2, 3, 4, 5 \\ 0 & \text{otherwise} \end{cases}$$

Define $Z = X - Y$. Find the PMF of Z.

13. Consider two random variables X and Y with joint PMF given in Table 5.5.

Table 5.5: Joint PMF of X and Y in Problem 13

	$Y = 0$	$Y = 1$	$Y = 2$
$X = 0$	$\frac{1}{6}$	$\frac{1}{6}$	$\frac{1}{8}$
$X = 1$	$\frac{1}{8}$	$\frac{1}{6}$	$\frac{1}{4}$

Define the random variable Z as $Z = E[X|Y]$.

(a) Find the Marginal PMFs of X and Y.

(b) Find the conditional PMF of X, given $Y = 0$ and $Y = 1$, i.e., find $P_{X|Y}(x|0)$ and $P_{X|Y}(x|1)$.

(c) Find the PMF of Z.

(d) Find EZ, and check that $EZ = EX$.

(e) Find $\text{Var}(Z)$.

14. Let X, Y, and $Z = E[X|Y]$ be as in Problem 13. Define the random variable V as $V = \text{Var}(X|Y)$.

(a) Find the PMF of V.

(b) Find EV.

(c) Check that $\text{Var}(X) = EV + \text{Var}(Z)$.

15. Let N be the number of phone calls made by the customers of a phone company in a given hour. Suppose that $N \sim Poisson(\beta)$, where $\beta > 0$ is known. Let X_i be the length of the i'th phone call, for $i = 1, 2, ..., N$. We assume X_i's are independent of each other and also independent of N. We further assume

$$X_i \sim Exponential(\lambda),$$

where $\lambda > 0$ is known. Let Y be the sum of the lengths of the phone calls, i.e.,

$$Y = \sum_{i=1}^{N} X_i.$$

Find EY and $\text{Var}(Y)$.

16. Let X and Y be two jointly continuous random variables with joint PDF

$$f_{XY}(x, y) = \begin{cases} \frac{1}{2}e^{-x} + \frac{cy}{(1+x)^2} & 0 \leq x, \ 0 \leq y \leq 1 \\ 0 & \text{otherwise} \end{cases}$$

(a) Find the constant c.

(b) Find $P(0 \leq X \leq 1, 0 \leq Y \leq \frac{1}{2})$.

(c) Find $P(0 \leq X \leq 1)$.

17. Let X and Y be two jointly continuous random variables with joint PDF

$$f_{XY}(x, y) = \begin{cases} e^{-xy} & 1 \leq x \leq e, \ y > 0 \\ 0 & \text{otherwise} \end{cases}$$

(a) Find the marginal PDFs, $f_X(x)$ and $f_Y(y)$.

(b) Write an integral to compute $P(0 \leq Y \leq 1, 1 \leq X \leq \sqrt{e})$.

18. Let X and Y be two jointly continuous random variables with joint PDF

$$f_{XY}(x, y) = \begin{cases} \frac{1}{4}x^2 + \frac{1}{6}y & -1 \leq x \leq 1, \ 0 \leq y \leq 2 \\ 0 & \text{otherwise} \end{cases}$$

(a) Find the marginal PDFs, $f_X(x)$ and $f_Y(y)$.

(b) Find $P(X > 0, Y < 1)$.

(c) Find $P(X > 0 \text{ or } Y < 1)$.

(d) Find $P(X > 0 | Y < 1)$.

(e) Find $P(X + Y > 0)$.

19. Let X and Y be two jointly continuous random variables with joint CDF

$$F_{XY}(x, y) = \begin{cases} 1 - e^{-x} - e^{-2y} + e^{-(x+2y)} & x, y > 0 \\ 0 & \text{otherwise} \end{cases}$$

(a) Find the joint PDF, $f_{XY}(x, y)$.

(b) Find $P(X < 2Y)$.

(c) Are X and Y independent?

20. Let $X \sim N(0, 1)$.

    (a) Find the conditional PDF and CDF of X given $X > 0$.
    (b) Find $E[X|X > 0]$.
    (c) Find $\text{Var}(X|X > 0)$.

21. Let X and Y be two jointly continuous random variables with joint PDF

$$f_{XY}(x, y) = \begin{cases} x^2 + \frac{1}{3}y & -1 \leq x \leq 1, 0 \leq y \leq 1 \\ 0 & \text{otherwise} \end{cases}$$

    (a) Find the conditional PDF of X given $Y = y$, for $0 \leq y \leq 1$.
    (b) Find $P(X > 0|Y = y)$, for $0 \leq y \leq 1$. Does this value depend on y?
    (c) Are X and Y independent?

22. Let X and Y be two jointly continuous random variables with joint PDF

$$f_{XY}(x, y) = \begin{cases} \frac{1}{2}x^2 + \frac{2}{3}y & -1 \leq x \leq 1, 0 \leq y \leq 1 \\ 0 & \text{otherwise} \end{cases}$$

    Find $E[Y|X = 0]$ and $\text{Var}(Y|X = 0)$.

23. Consider the set

$$E = \{(x, y) | |x| + |y| \leq 1\}.$$

    Suppose that we choose a point (X, Y) uniformly at random in E. That is, the joint PDF of X and Y is given by

$$f_{XY}(x, y) = \begin{cases} c & (x, y) \in E \\ 0 & \text{otherwise} \end{cases}$$

    (a) Find the constant c.
    (b) Find the marginal PDFs $f_X(x)$ and $f_Y(y)$.
    (c) Find the conditional PDF of X given $Y = y$, where $-1 \leq y \leq 1$.
    (d) Are X and Y independent?

24. Let X and Y be two independent $Uniform(0, 2)$ random variables. Find $P(XY < 1)$.

25. Suppose $X \sim Exponential(1)$ and given $X = x$, Y is a uniform random variable in $[0, x]$, i.e.,

$$Y|X = x \quad \sim \quad Uniform(0, x),$$

    or equivalently

$$Y|X \quad \sim \quad Uniform(0, X).$$

(a) Find EY.

(b) Find $\text{Var}(Y)$.

26. Let X and Y be two independent $Uniform(0,1)$ random variables. Find

    (a) $E[XY]$
    (b) $E[e^{X+Y}]$
    (c) $E[X^2 + Y^2 + XY]$
    (d) $E[Ye^{XY}]$

27. Let X and Y be two independent $Uniform(0,1)$ random variables, and $Z = \frac{X}{Y}$. Find the CDF and PDF of Z.

28. Let X and Y be two independent $N(0,1)$ random variables, and $U = X + Y$.

    (a) Find the conditional PDF of U given $X = x$, $f_{U|X}(u|x)$.
    (b) Find the PDF of U, $f_U(u)$.
    (c) Find the conditional PDF of X given $U = u$, $f_{X|U}(x|u)$.
    (d) Find $E[X|U = u]$, and $\text{Var}(X|U = u)$.

29. Let X and Y be two independent standard normal random variables. Consider the point (X, Y) in the $x - y$ plane. Let (R, Θ) be the corresponding polar coordinates as shown in Figure 5.11. The inverse transformation is given by

$$\begin{cases} X = R\cos\Theta \\ Y = R\sin\Theta \end{cases}$$

where, $R \geq 0$ and $-\pi < \Theta \leq \pi$. Find the joint PDF of R and Θ. Show that R and Θ are independent.

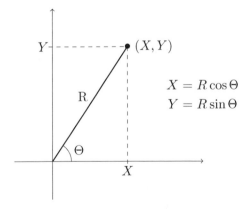

Figure 5.11: Polar coordinates

30. In Problem 29, suppose that X and Y are independent $Uniform(0,1)$ random variables. Find the joint PDF of R and Θ. Are R and Θ independent?

Table 5.6: Joint PMF of X and Y in Problem 31

	$Y = 0$	$Y = 1$	$Y = 2$
$X = 0$	$\frac{1}{6}$	$\frac{1}{4}$	$\frac{1}{8}$
$X = 1$	$\frac{1}{8}$	$\frac{1}{6}$	$\frac{1}{6}$

31. Consider two random variables X and Y with joint PMF given in Table 5.6. Find $\text{Cov}(X, Y)$ and $\rho(X, Y)$.

32. Let X and Y be two independent $N(0, 1)$ random variables and

$$Z = 11 - X + X^2 Y,$$
$$W = 3 - Y.$$

Find $\text{Cov}(Z, W)$.

33. Let X and Y be two random variables. Suppose that $\sigma_X^2 = 4$, and $\sigma_Y^2 = 9$. If we know that the two random variables $Z = 2X - Y$ and $W = X + Y$ are independent, find $\text{Cov}(X, Y)$ and $\rho(X, Y)$.

34. Let $X \sim Uniform(1, 3)$ and $Y|X \sim Exponential(X)$. Find $\text{Cov}(X, Y)$.

35. Let X and Y be two independent $N(0, 1)$ random variables and

$$Z = 7 + X + Y,$$
$$W = 1 + Y.$$

Find $\rho(Z, W)$.

36. Let X and Y be jointly normal random variables with parameters $\mu_X = -1$, $\sigma_X^2 = 4$, $\mu_Y = 1$, $\sigma_Y^2 = 1$, and $\rho = -\frac{1}{2}$.

    (a) Find $P(X + 2Y \leq 3)$.

    (b) Find $\text{Cov}(X - Y, X + 2Y)$.

37. Let X and Y be jointly normal random variables with parameters $\mu_X = 1$, $\sigma_X^2 = 4$, $\mu_Y = 1$, $\sigma_Y^2 = 1$, and $\rho = 0$.

    (a) Find $P(X + 2Y > 4)$.

    (b) Find $E[X^2 Y^2]$.

38. Let X and Y be jointly normal random variables with parameters $\mu_X = 2$, $\sigma_X^2 = 4$, $\mu_Y = 1$, $\sigma_Y^2 = 9$, and $\rho = -\frac{1}{2}$.

(a) Find $E[Y|X = 3]$.

(b) Find $\text{Var}(Y|X = 2)$.

(c) Find $P(X + 2Y \leq 5|X + Y = 3)$.

Chapter 6

Multiple Random Variables

All the concepts that we have seen regarding one and two random variables can be extended to more random variables. In particular, we can define joint PDF, joint PMF, and joint CDF for three or more random variables. We will provide these definitions in Section 6.1.1. However, working with these functions quickly becomes computationally intractable as the number of random variables grows. Thus, in dealing with multiple random variables we usually resort to other techniques. We will discuss some of these techniques, such as moment generating functions and probability bounds, in this chapter.

6.1 Methods for more than Two Random Variables

6.1.1 Joint Distributions and Independence

For three or more random variables, the joint PDF, joint PMF, and joint CDF are defined in a similar way to what we have already seen for the case of two random variables. Let X_1, X_2, $\cdots$, X_n be n discrete random variables. The joint PMF of X_1, X_2, $\cdots$, X_n is defined as

$$P_{X_1, X_2, \ldots, X_n}(x_1, x_2, \ldots, x_n) = P(X_1 = x_1, X_2 = x_2, \ldots, X_n = x_n).$$

For n jointly continuous random variables X_1, X_2, $\cdots$, X_n, the joint PDF is defined to be the function $f_{X_1 X_2 \ldots X_n}(x_1, x_2, \ldots, x_n)$ such that the probability of any set $A \subset \mathbb{R}^n$ is given by the integral of the PDF over the set A. In particular, for a set $A \in \mathbb{R}^n$, we can write

$$P\bigg((X_1, X_2, \cdots, X_n) \in A\bigg) = \int \cdots \int_A \cdots \int f_{X_1 X_2 \cdots X_n}(x_1, x_2, \cdots, x_n) \ dx_1 dx_2 \cdots dx_n.$$

The marginal PDF of X_i can be obtained by integrating all other X_j's. For example,

$$f_{X_1}(x_1) = \int_{-\infty}^{\infty} \cdots \int_{-\infty}^{\infty} f_{X_1 X_2 \ldots X_n}(x_1, x_2, \ldots, x_n) \ dx_2 \cdots dx_n.$$

The joint CDF of n random variables X_1, X_2,...,X_n is defined as

$$F_{X_1, X_2, \ldots, X_n}(x_1, x_2, \ldots, x_n) = P(X_1 \leq x_1, X_2 \leq x_2, \ldots, X_n \leq x_n).$$

Example 6.1. Let X, Y and Z be three jointly continuous random variables with joint PDF

$$f_{XYZ}(x,y,z) = \begin{cases} c(x + 2y + 3z) & 0 \le x, y, z \le 1 \\ 0 & \text{otherwise} \end{cases}$$

1. Find the constant c.

2. Find the marginal PDF of X.

Solution:

1.

$$\begin{aligned}
1 &= \int_{-\infty}^{\infty} \int_{-\infty}^{\infty} \int_{-\infty}^{\infty} f_{XYZ}(x,y,z)\,dxdydz \\
&= \int_0^1 \int_0^1 \int_0^1 c(x + 2y + 3z)\ dxdydz \\
&= \int_0^1 \int_0^1 c\left(\frac{1}{2} + 2y + 3z\right)\ dydz \\
&= \int_0^1 c\left(\frac{3}{2} + 3z\right)\ dz \\
&= 3c.
\end{aligned}$$

Thus, $c = \frac{1}{3}$.

2. To find the marginal PDF of X, we note that $R_X = [0,1]$. For $0 \le x \le 1$, we can write

$$\begin{aligned}
f_X(x) &= \int_{-\infty}^{\infty} \int_{-\infty}^{\infty} f_{XYZ}(x,y,z)dydz \\
&= \int_0^1 \int_0^1 \frac{1}{3}(x + 2y + 3z)\ dydz \\
&= \int_0^1 \frac{1}{3}(x + 1 + 3z)\ dz \\
&= \frac{1}{3}\left(x + \frac{5}{2}\right).
\end{aligned}$$

Thus,

$$f_X(x) = \begin{cases} \frac{1}{3}\left(x + \frac{5}{2}\right) & 0 \le x \le 1 \\ 0 & \text{otherwise} \end{cases}$$

Independence: The idea of independence is exactly the same as what we have seen before. We restate it here in terms of the joint PMF, joint PDF, and joint CDF. Random variables X_1, $X_2, \dots, X_n$ are independent, if for all $(x_1, x_2, \dots, x_n) \in \mathbb{R}^n$,

$$F_{X_1, X_2, \dots, X_n}(x_1, x_2, \dots, x_n) = F_{X_1}(x_1) F_{X_2}(x_2) \cdots F_{X_n}(x_n).$$

Equivalently, if X_1, X_2, ..., X_n are discrete, then they are independent if for all $(x_1, x_2, ..., x_n) \in \mathbb{R}^n$, we have

$$P_{X_1, X_2, ..., X_n}(x_1, x_2, ..., x_n) = P_{X_1}(x_1)P_{X_2}(x_2) \cdots P_{X_n}(x_n).$$

If X_1, X_2, ..., X_n are continuous, then they are independent if for all $(x_1, x_2, ..., x_n) \in \mathbb{R}^n$, we have

$$f_{X_1, X_2, ..., X_n}(x_1, x_2, ..., x_n) = f_{X_1}(x_1)f_{X_2}(x_2) \cdots f_{X_n}(x_n).$$

If random variables X_1, X_2, ..., X_n are independent, then we have

$$E[X_1 X_2 \cdots X_n] = E[X_1]E[X_2] \cdots E[X_n].$$

In some situations, we are dealing with random variables that are independent and are also identically distributed, i.e, they have the same CDFs. It is usually easier to deal with such random variables, since independence and being identically distributed often simplify the analysis. We will see examples of such analyses shortly.

Definition 6.1. Random variables X_1, X_2, ..., X_n are said to be **independent and identically distributed (i.i.d.)** if they are *independent*, and they have the *same marginal distributions*:

$$F_{X_1}(x) = F_{X_2}(x) = ... = F_{X_n}(x), \quad \text{for all } x \in \mathbb{R}.$$

For example, if random variables X_1, X_2, ..., X_n are i.i.d., they will have the same means and variances, so we can write

$$\begin{aligned} E[X_1 X_2 \cdots X_n] &= E[X_1]E[X_2] \cdots E[X_n] \quad && \text{(because } X_i\text{'s are independent)} \\ &= E[X_1]E[X_1] \cdots E[X_1] \quad && \text{(because } X_i\text{'s are identically distributed)} \\ &= E[X_1]^n. \end{aligned}$$

6.1.2 Sums of Random Variables

In many applications, we need to work with a sum of several random variables. In particular, we might need to study a random variable Y given by

$$Y = X_1 + X_2 + \cdots + X_n.$$

The linearity of expectation tells us that

$$EY = EX_1 + EX_2 + \cdots + EX_n.$$

We can also find the variance of Y based on our discussion in Section 5.3. In particular, we saw that the variance of a sum of two random variables is

$$\text{Var}(X_1 + X_2) = \text{Var}(X_1) + \text{Var}(X_2) + 2\text{Cov}(X_1, X_2).$$

For $Y = X_1 + X_2 + \cdots + X_n$, we can obtain a more general version of the above equation. We can write

$$\text{Var}(Y) = \text{Cov}\left(\sum_{i=1}^{n} X_i, \sum_{j=1}^{n} X_j\right)$$

$$= \sum_{i=1}^{n}\sum_{j=1}^{n} \text{Cov}(X_i, X_j) \qquad \text{(using part 7 of Lemma 5.3)}$$

$$= \sum_{i=1}^{n} \text{Var}(X_i) + 2\sum_{i<j} \text{Cov}(X_i, X_j).$$

$$\text{Var}\left(\sum_{i=1}^{n} X_i\right) = \sum_{i=1}^{n} \text{Var}(X_i) + 2\sum_{i<j} \text{Cov}(X_i, X_j)$$

If the X_i's are independent, then $\text{Cov}(X_i, X_j) = 0$ for $i \neq j$. In this case, we can write the following:

$$\text{If } X_1, X_2, \dots, X_n \text{ are independent, } \text{Var}\left(\sum_{i=1}^{n} X_i\right) = \sum_{i=1}^{n} \text{Var}(X_i).$$

Example 6.2. N people sit around a round table, where $N > 5$. Each person tosses a coin. Anyone whose outcome is different from his/her two neighbors will receive a present. Let X be the number of people who receive presents. Find EX and $\text{Var}(X)$.

Solution: Number the N people from 1 to N. Let X_i be the indicator random variable for the ith person, that is, $X_i = 1$ if the ith person receives a present and zero otherwise. Then

$$X = X_1 + X_2 + \dots + X_N.$$

First note that $P(X_i = 1) = \frac{1}{4}$. This is the probability that the person to the right has a different outcome times the probability that the person to the left has a different outcome. In other words, if we define H_i and T_i be the events that the ith person's outcome is heads and tails respectively, then we can write

$$\begin{aligned}
EX_i &= P(X_i = 1) \\
&= P(H_{i-1}, T_i, H_{i+1}) + P(T_{i-1}, H_i, T_{i+1}) \\
&= \frac{1}{8} + \frac{1}{8} = \frac{1}{4}.
\end{aligned}$$

Thus, we find

$$EX = EX_1 + EX_2 + \ldots + EX_N = \frac{N}{4}.$$

Next, we can write

$$\operatorname{Var}(X) = \sum_{i=1}^{N} \operatorname{Var}(X_i) + \sum_{i=1}^{N} \sum_{j \neq i} \operatorname{Cov}(X_i, X_j).$$

Since $X_i \sim Bernoulli(\frac{1}{4})$, we have

$$\operatorname{Var}(X_i) = \frac{1}{4} \cdot \frac{3}{4} = \frac{3}{16}.$$

It remains to find $\operatorname{Cov}(X_i, X_j)$. First note that X_i and X_j are independent if there are at least two people between the ith person and the jth person. In other words, if $2 < |i - j| < N - 2$, then X_i and X_j are independent, so

$$\operatorname{Cov}(X_i, X_j) = 0, \qquad \text{for} \quad 2 < |i - j| < N - 2.$$

Also, note that there is a lot of symmetry in the problem:

$$\operatorname{Cov}(X_1, X_2) = \operatorname{Cov}(X_2, X_3) = \operatorname{Cov}(X_3, X_4) = \ldots = \operatorname{Cov}(X_{N-1}, X_N) = \operatorname{Cov}(X_N, X_1),$$
$$\operatorname{Cov}(X_1, X_3) = \operatorname{Cov}(X_2, X_4) = \operatorname{Cov}(X_3, X_5) = \ldots = \operatorname{Cov}(X_{N-1}, X_1) = \operatorname{Cov}(X_N, X_2).$$

Thus, we can write

$$\begin{aligned}
\operatorname{Var}(X) &= N\operatorname{Var}(X_1) + 2N\operatorname{Cov}(X_1, X_2) + 2N\operatorname{Cov}(X_1, X_3) \\
&= \frac{3N}{16} + 2N\operatorname{Cov}(X_1, X_2) + 2N\operatorname{Cov}(X_1, X_3).
\end{aligned}$$

So we need to find $\operatorname{Cov}(X_1, X_2)$ and $\operatorname{Cov}(X_1, X_3)$. We have

$$\begin{aligned}
E[X_1 X_2] &= P(X_1 = 1, X_2 = 1) \\
&= P(H_N, T_1, H_2, T_3) + P(T_N, H_1, T_2, H_3) \\
&= \frac{1}{16} + \frac{1}{16} = \frac{1}{8}.
\end{aligned}$$

Thus,

$$\begin{aligned}
\operatorname{Cov}(X_1, X_2) &= E[X_1 X_2] - E[X_1]E[X_2] \\
&= \frac{1}{8} - \frac{1}{16} = \frac{1}{16},
\end{aligned}$$

$$E[X_1 X_3] = P(X_1 = 1, X_3 = 1)$$
$$= P(H_N, T_1, H_2, T_3, H_4) + P(T_N, H_1, T_2, H_3, T_4)$$
$$= \frac{1}{32} + \frac{1}{32} = \frac{1}{16}.$$

Thus,

$$\text{Cov}(X_1, X_3) = E[X_1 X_3] - E[X_1]E[X_3]$$
$$= \frac{1}{16} - \frac{1}{16} = 0.$$

Therefore,

$$\text{Var}(X) = \frac{3N}{16} + 2N\text{Cov}(X_1, X_2) + 2N\text{Cov}(X_1, X_3)$$
$$= \frac{3N}{16} + \frac{2N}{16}$$
$$= \frac{5N}{16}.$$

We now know how to find the mean and variance of a sum of n random variables, but we might need to go beyond that. Specifically, what if we need to know the PDF of $Y = X_1 + X_2 + ... + X_n$? In fact, we have addressed that problem for the case where $Y = X_1 + X_2$ and X_1 and X_2 are independent (Example 5.30 in Section 5.2.4). For this case, we found out the PDF is given by convolving the PDF of X_1 and X_2, that is

$$f_Y(y) = f_{X_1}(y) * f_{X_2}(y) = \int_{-\infty}^{\infty} f_{X_1}(x) f_{X_2}(y - x) dx.$$

For $Y = X_1 + X_2 + ... + X_n$, we can use the above formula repeatedly to obtain the PDF of Y:

$$f_Y(y) = f_{X_1}(y) * f_{X_2}(y) * ... * f_{X_n}(y).$$

Nevertheless, this quickly becomes computationally difficult. Thus, we often resort to other methods if we can. One method that is often useful is using moment generating functions, as we will discuss in the next section.

6.1.3   Moment Generating Functions

Here, we will introduce and discuss **moment generating functions (MGFs)**. Moment generating functions are useful for several reasons, one of which is their application to analysis of sums of random variables. Before discussing MGFs, let's define moments.

Definition 6.2. The **nth moment** of a random variable X is defined to be $E[X^n]$. The **nth central moment** of X is defined to be $E[(X - EX)^n]$.

For example, the first moment is the expected value $E[X]$. The second central moment is the variance of X. Similar to mean and variance, other moments give useful information about random variables.

The moment generating function (MGF) of a random variable X is a function $M_X(s)$ defined as

$$M_X(s) = E\left[e^{sX}\right].$$

We say that MGF of X exists, if there exists a positive constant a such that $M_X(s)$ is finite for all $s \in [-a, a]$.

Before going any further, let's look at an example.

Example 6.3. For each of the following random variables, find the MGF.

(a) X is a discrete random variable, with PMF

$$P_X(k) = \begin{cases} \frac{1}{3} & k = 1 \\ \frac{2}{3} & k = 2 \end{cases}$$

(b) Y is a $Uniform(0, 1)$ random variable.

Solution:

(a) For X, we have

$$M_X(s) = E\left[e^{sX}\right]$$
$$= \frac{1}{3}e^s + \frac{2}{3}e^{2s},$$

which is well-defined for all $s \in \mathbb{R}$.

(b) For Y, we can write

$$M_Y(s) = E\left[e^{sY}\right]$$
$$= \int_0^1 e^{sy}\,dy$$
$$= \frac{e^s - 1}{s}.$$

Note that we always have $M_Y(0) = E[e^{0\cdot Y}] = 1$, thus $M_Y(s)$ is also well-defined for all $s \in \mathbb{R}$.

Why is the MGF useful? There are basically two reasons for this. First, the MGF of X gives us all moments of X. That is why it is called the moment generating function. Second, the MGF (if it exists) uniquely determines the distribution. That is, if two random variables have the same MGF, then they must have the same distribution. Thus, if you find the MGF of a random variable, you have indeed determined its distribution. We will see that this method is very useful when we work on sums of several independent random variables. Let's discuss these in detail.

Finding Moments from MGF:

Remember the Taylor series for e^x: for all $x \in \mathbb{R}$, we have

$$e^x = 1 + x + \frac{x^2}{2!} + \frac{x^3}{3!} + \dots = \sum_{k=0}^{\infty} \frac{x^k}{k!}.$$

Now, we can write

$$e^{sX} = \sum_{k=0}^{\infty} \frac{(sX)^k}{k!} = \sum_{k=0}^{\infty} \frac{X^k s^k}{k!}.$$

Thus, we have

$$M_X(s) = E[e^{sX}] = \sum_{k=0}^{\infty} E[X^k] \frac{s^k}{k!}.$$

We conclude that the kth moment of X is the coefficient of $\frac{s^k}{k!}$ in the Taylor series of $M_X(s)$. Thus, if we have the Taylor series of $M_X(s)$, we can obtain all moments of X.

Example 6.4. If $Y \sim Uniform(0, 1)$, find $E[Y^k]$ using $M_Y(s)$.

Solution: We found $M_Y(s)$ in Example 6.3, so we have

$$M_Y(s) = \frac{e^s - 1}{s}$$

$$= \frac{1}{s} \left(\sum_{k=0}^{\infty} \frac{s^k}{k!} - 1 \right)$$

$$= \frac{1}{s} \sum_{k=1}^{\infty} \frac{s^k}{k!}$$

$$= \sum_{k=1}^{\infty} \frac{s^{k-1}}{k!}$$

$$= \sum_{k=0}^{\infty} \frac{1}{k+1} \frac{s^k}{k!}.$$

Thus, the coefficient of $\frac{s^k}{k!}$ in the Taylor series for $M_Y(s)$ is $\frac{1}{k+1}$, so

$$E[X^k] = \frac{1}{k+1}.$$

We remember from calculus that the coefficient of $\frac{s^k}{k!}$ in the Taylor series of $M_X(s)$ is obtained by taking the kth derivative of $M_X(s)$ and evaluating it at $s = 0$. Thus, we can write

$$E[X^k] = \frac{d^k}{ds^k} M_X(s) \Big|_{s=0}$$

We can obtain all moments of X^k from its MGF:

$$M_X(s) = \sum_{k=0}^{\infty} E[X^k] \frac{s^k}{k!}$$

$$E[X^k] = \frac{d^k}{ds^k} M_X(s) \bigg|_{s=0}$$

Example 6.5. Let $X \sim Exponential(\lambda)$. Find the MGF of X, $M_X(s)$, and all of its moments, $E[X^k]$.

Solution: Recall that the PDF of X is

$$f_X(x) = \lambda e^{-\lambda x} u(x),$$

where $u(x)$ is the unit step function. We conclude

$$\begin{aligned}
M_X(s) &= E[e^{sX}] \\
&= \int_0^{\infty} \lambda e^{-\lambda x} e^{sx} dx \\
&= \left[-\frac{\lambda}{\lambda - s} e^{-(\lambda - s)x} \right]_0^{\infty}, \quad \text{for } s < \lambda \\
&= \frac{\lambda}{\lambda - s}, \qquad \text{for } s < \lambda.
\end{aligned}$$

Therefore, $M_X(s)$ exists for all $s < \lambda$. To find the moments of X, we can write

$$\begin{aligned}
M_X(s) &= \frac{\lambda}{\lambda - s} \\
&= \frac{1}{1 - \frac{s}{\lambda}} \\
&= \sum_{k=0}^{\infty} \left(\frac{s}{\lambda} \right)^k, \quad \text{for } \left| \frac{s}{\lambda} \right| < 1 \\
&= \sum_{k=0}^{\infty} \frac{k!}{\lambda^k} \frac{s^k}{k!}.
\end{aligned}$$

We conclude that

$$E[X^k] = \frac{k!}{\lambda^k}, \quad \text{for } k = 0, 1, 2, \dots$$

Example 6.6. Let $X \sim Poisson(\lambda)$. Find the MGF of X, $M_X(s)$.

Solution: We have

$$P_X(k) = e^{-\lambda}\frac{\lambda^k}{k!}, \quad \text{for } k = 0, 1, 2, \ldots$$

Thus,

$$
\begin{aligned}
M_X(s) &= E[e^{sX}] \\
&= \sum_{k=0}^{\infty} e^{sk} e^{-\lambda}\frac{\lambda^k}{k!} \\
&= e^{-\lambda}\sum_{k=0}^{\infty} e^{sk}\frac{\lambda^k}{k!} \\
&= e^{-\lambda}\sum_{k=0}^{\infty}\frac{(\lambda e^s)^k}{k!} \\
&= e^{-\lambda}e^{\lambda e^s} \quad \text{(Taylor series for } e^x) \\
&= e^{\lambda(e^s - 1)}, \quad \text{for all } s \in \mathbb{R}.
\end{aligned}
$$

As we discussed previously, the MGF uniquely determines the distribution. This is a very useful fact. We will see examples of how we use it shortly. Right now let's state this fact more precisely as a theorem. We omit the proof here.

Theorem 6.1. Consider two random variables X and Y. Suppose that there exists a positive constant c such that MGFs of X and Y are finite and identical for all values of s in $[-c, c]$. Then,

$$F_X(t) = F_Y(t), \quad \text{for all } t \in \mathbb{R}.$$

Example 6.7. For a random variable X, we know that

$$M_X(s) = \frac{2}{2 - s}, \quad \text{for } s \in (-2, 2).$$

Find the distribution of X.

Solution: We note that the above MGF is the MGF of an exponential random variable with $\lambda = 2$ (Example 6.5). Thus, we conclude that $X \sim Exponential(2)$.

Sum of Independent Random Variables:

Suppose X_1, X_2, ..., X_n are n independent random variables, and the random variable Y is defined as

$$Y = X_1 + X_2 + \cdots + X_n.$$

Then,

$$
\begin{aligned}
M_Y(s) &= E[e^{sY}] \\
&= E[e^{s(X_1+X_2+\cdots+X_n)}] \\
&= E[e^{sX_1}e^{sX_2}\cdots e^{sX_n}] \\
&= E[e^{sX_1}]E[e^{sX_2}]\cdots E[e^{sX_n}] \qquad \text{(since } X_i\text{'s are independent)} \\
&= M_{X_1}(s)M_{X_2}(s)\cdots M_{X_n}(s).
\end{aligned}
$$

If X_1, X_2, ..., X_n are n <u>independent</u> random variables, then

$$
M_{X_1+X_2+\cdots+X_n}(s) = M_{X_1}(s)M_{X_2}(s)\cdots M_{X_n}(s).
$$

Example 6.8. If $X \sim Binomial(n,p)$ find the MGF of X.

Solution: We can solve this question directly using the definition of MGF, but an easier way to solve it is to use the fact that a binomial random variable can be considered as the sum of n independent and identically distributed (i.i.d.) Bernoulli random variables. Thus, we can write

$$
X = X_1 + X_2 + \cdots + X_n,
$$

where $X_i \sim Bernoulli(p)$. Thus,

$$
\begin{aligned}
M_X(s) &= M_{X_1}(s)M_{X_2}(s)\cdots M_{X_n}(s) \\
&= \big(M_{X_1}(s)\big)^n \qquad \text{(since } X_i\text{'s are i.i.d.).}
\end{aligned}
$$

Also,

$$
M_{X_1}(s) = E[e^{sX_1}] = pe^s + 1 - p.
$$

Thus, we conclude

$$
M_X(s) = \big(pe^s + 1 - p\big)^n.
$$

Example 6.9. Using MGFs prove that if $X \sim Binomial(m,p)$ and $Y \sim Binomial(n,p)$ are independent, then $X + Y \sim Binomial(m+n,p)$.

Solution: We have

$$
\begin{aligned}
M_X(s) &= \big(pe^s + 1 - p\big)^m, \\
M_Y(s) &= \big(pe^s + 1 - p\big)^n.
\end{aligned}
$$

Since X and Y are independent, we conclude that

$$
\begin{aligned}
M_{X+Y}(s) &= M_X(s)M_Y(s) \\
&= \big(pe^s + 1 - p\big)^{m+n},
\end{aligned}
$$

which is the MGF of a $Binomial(m+n,p)$ random variable. Thus, $X+Y \sim Binomial(m+n,p)$.

6.1.4   Characteristic Functions

There are random variables for which the moment generating function does not exist on any real interval with positive length. For example, consider the random variable X that has a *Cauchy* distribution

$$f_X(x) = \frac{\frac{1}{\pi}}{1+x^2}, \qquad \text{for all } x \in \mathbb{R}.$$

You can show that for any nonzero real number s

$$M_X(s) = \int_{-\infty}^{\infty} e^{sx} \frac{\frac{1}{\pi}}{1+x^2} dx = \infty.$$

Therefore, the moment generating function does not exist for this random variable on any real interval with positive length.

   If a random variable does not have a well-defined MGF, we can use the characteristic function defined as

$$\phi_X(\omega) = E[e^{j\omega X}],$$

where $j = \sqrt{-1}$ and ω is a real number.[1] The advantage of the characteristic function is that it is defined for all real-valued random variables. Specifically, if X is a real-valued random variable, we can write

$$|e^{j\omega X}| = 1.$$

Therefore, we conclude

$$\begin{aligned}
|\phi_X(\omega)| &= |E[e^{j\omega X}]| \\
&\leq E[|e^{j\omega X}|] \\
&\leq 1.
\end{aligned}$$

The characteristic function has similar properties to the MGF. For example, if X and Y are independent

$$\begin{aligned}
\phi_{X+Y}(\omega) &= E[e^{j\omega(X+Y)}] \\
&= E[e^{j\omega X} e^{j\omega Y}] \\
&= E[e^{j\omega X}]E[e^{j\omega Y}] \qquad \text{(since X and Y are independent)} \\
&= \phi_X(\omega)\phi_Y(\omega).
\end{aligned}$$

More generally, if X_1, X_2, ..., X_n are n <u>independent</u> random variables, then

$$\phi_{X_1+X_2+\cdots+X_n}(\omega) = \phi_{X_1}(\omega)\phi_{X_2}(\omega)\cdots\phi_{X_n}(\omega).$$

---

[1]It is worth noting that $e^{j\omega X}$ is a complex-valued random variable. We have not discussed complex-valued random variables. Nevertheless, you can imagine that a complex random variable can be written as $X = Y + jZ$, where Y and Z are ordinary real-valued random variables. Thus, working with a complex random variable is like working with two real-valued random variables.

Example 6.10. If $X \sim Exponential(\lambda)$, show that

$$\phi_X(\omega) = \frac{\lambda}{\lambda - j\omega}.$$

Solution: Recall that the PDF of X is

$$f_X(x) = \lambda e^{-\lambda x} u(x),$$

where $u(x)$ is the unit step function. We conclude

$$\phi_X(\omega) = E[e^{j\omega X}]$$
$$= \int_0^\infty \lambda e^{-\lambda x} e^{j\omega x} dx$$
$$= \left[\frac{\lambda}{j\omega - \lambda} e^{(j\omega - \lambda)x} \right]_0^\infty$$
$$= \frac{\lambda}{\lambda - j\omega}.$$

Note that since $\lambda > 0$, the value of $e^{(j\omega - \lambda)x}$, when evaluated at $x = +\infty$, is zero.

6.1.5 Random Vectors

When dealing with multiple random variables, it is sometimes useful to use vector and matrix notations. This makes the formulas more compact and lets us use facts from linear algebra. In this section, we will briefly explore this avenue. The reader should be familiar with matrix algebra before reading this section.

When we have n random variables $X_1, X_2, ..., X_n$ we can put them in a (column) vector $\mathbf{X}$:

$$\mathbf{X} = \begin{bmatrix} X_1 \\ X_2 \\ \vdots \\ X_n \end{bmatrix}.$$

We call $\mathbf{X}$ a **random vector**. Here $\mathbf{X}$ is an n-dimensional vector because it consists of n random variables. In this book, we usually use bold capital letters such as $\mathbf{X}$, $\mathbf{Y}$ and $\mathbf{Z}$ to represent a random vector. To show a possible value of a random vector we usually use bold lowercase letters such as $\mathbf{x}$, $\mathbf{y}$ and $\mathbf{z}$. Thus, we can write the CDF of the random vector $\mathbf{X}$ as

$$F_\mathbf{X}(\boldsymbol{x}) = F_{X_1, X_2, ..., X_n}(x_1, x_2, ..., x_n)$$
$$= P(X_1 \leq x_1, X_2 \leq x_2, ..., X_n \leq x_n).$$

If the X_i's are jointly continuous, the PDF of $\mathbf{X}$ can be written as

$$f_\mathbf{X}(\boldsymbol{x}) = f_{X_1, X_2, ..., X_n}(x_1, x_2, ..., x_n).$$

Expectation:

The **expected value vector** or the **mean vector** of the random vector $\mathbf{X}$ is defined as

$$E\mathbf{X} = \begin{bmatrix} EX_1 \\ EX_2 \\ \vdots \\ EX_n \end{bmatrix}.$$

Similarly, a **random matrix** is a matrix whose elements are random variables. In particular, we can have an m by n random matrix $\mathbf{M}$ as

$$\mathbf{M} = \begin{bmatrix} X_{11} & X_{12} & ... & X_{1n} \\ X_{21} & X_{22} & ... & X_{2n} \\ \vdots & \vdots & \vdots & \vdots \\ X_{m1} & X_{m2} & ... & X_{mn} \end{bmatrix}.$$

We sometimes write this as $\mathbf{M} = [X_{ij}]$, which means that X_{ij} is the element in the ith row and jth column of $\mathbf{M}$. The mean matrix of $\mathbf{M}$ is given by

$$E\mathbf{M} = \begin{bmatrix} EX_{11} & EX_{12} & ... & EX_{1n} \\ EX_{21} & EX_{22} & ... & EX_{2n} \\ \vdots & \vdots & \vdots & \vdots \\ EX_{m1} & EX_{m2} & ... & EX_{mn} \end{bmatrix}.$$

Linearity of expectation is also valid for random vectors and matrices. In particular, let $\mathbf{X}$ be an n-dimensional random vector and the random vector $\mathbf{Y}$ be defined as

$$\mathbf{Y} = \mathbf{A}\mathbf{X} + \mathbf{b},$$

where $\mathbf{A}$ is a fixed (non-random) m by n matrix and $\mathbf{b}$ is a fixed m-dimensional vector. Then we have

$$E\mathbf{Y} = \mathbf{A}E\mathbf{X} + \mathbf{b}.$$

Also, if $\mathbf{X}_1$, $\mathbf{X}_2$, ..., $\mathbf{X}_k$ are n-dimensional random vectors, then we have

$$E[\mathbf{X}_1 + \mathbf{X}_2 + \cdots + \mathbf{X}_k] = E\mathbf{X}_1 + E\mathbf{X}_2 + \cdots + E\mathbf{X}_k.$$

Correlation and Covariance Matrix:

For a random vector $\mathbf{X}$, we define the **correlation matrix**, $\mathbf{R_X}$, as

$$\mathbf{R_X} = E[\mathbf{X}\mathbf{X}^T] = E\begin{bmatrix} X_1^2 & X_1X_2 & ... & X_1X_n \\ X_2X_1 & X_2^2 & ... & X_2X_n \\ \vdots & \vdots & \vdots & \vdots \\ X_nX_1 & X_nX_2 & ... & X_n^2 \end{bmatrix} = \begin{bmatrix} EX_1^2 & E[X_1X_2] & ... & E[X_1X_n] \\ EX_2X_1 & E[X_2^2] & ... & E[X_2X_n] \\ \vdots & \vdots & \vdots & \vdots \\ E[X_nX_1] & E[X_nX_2] & ... & E[X_n^2] \end{bmatrix},$$

where ^{T} shows matrix transposition.

The **covariance matrix**, $\mathbf{C_X}$, is defined as

$$\mathbf{C_X} = E[(\mathbf{X} - E\mathbf{X})(\mathbf{X} - E\mathbf{X})^T]$$

$$= E \begin{bmatrix} (X_1 - EX_1)^2 & (X_1 - EX_1)(X_2 - EX_2) & \dots & (X_1 - EX_1)(X_n - EX_n) \\ (X_2 - EX_2)(X_1 - EX_1) & (X_2 - EX_2)^2 & \dots & (X_2 - EX_2)(X_n - EX_n) \\ \vdots & \vdots & \vdots & \vdots \\ (X_n - EX_n)(X_1 - EX_1) & (X_n - EX_n)(X_2 - EX_2) & \dots & (X_n - EX_n)^2 \end{bmatrix}$$

$$= \begin{bmatrix} \mathrm{Var}(X_1) & \mathrm{Cov}(X_1, X_2) & \dots & \mathrm{Cov}(X_1, X_n) \\ \mathrm{Cov}(X_2, X_1) & \mathrm{Var}(X_2) & \dots & \mathrm{Cov}(X_2, X_n) \\ \vdots & \vdots & \vdots & \vdots \\ \mathrm{Cov}(X_n, X_1) & \mathrm{Cov}(X_n X_2) & \dots & \mathrm{Var}(X_n) \end{bmatrix}.$$

The covariance matrix is a generalization of the variance of a random variable. Remember that for a random variable, we have $\mathrm{Var}(X) = EX^2 - (EX)^2$. The following example extends this formula to random vectors.

Example 6.11. For a random vector $\mathbf{X}$, show

$$\mathbf{C_X} = \mathbf{R_X} - E\mathbf{X}E\mathbf{X}^T.$$

Solution: We have

$$\begin{aligned} \mathbf{C_X} &= E[(\mathbf{X} - E\mathbf{X})(\mathbf{X} - E\mathbf{X})^T] \\ &= E[(\mathbf{X} - E\mathbf{X})(\mathbf{X}^T - E\mathbf{X}^T)] \\ &= E[\mathbf{X}\mathbf{X}^T] - E\mathbf{X}E\mathbf{X}^T - E\mathbf{X}E\mathbf{X}^T + E\mathbf{X}E\mathbf{X}^T \quad \text{(by linearity of expectation)} \\ &= \mathbf{R_X} - E\mathbf{X}E\mathbf{X}^T. \end{aligned}$$

Correlation matrix of $\mathbf{X}$:
$$\mathbf{R_X} = E[\mathbf{X}\mathbf{X}^T]$$
Covariance matrix of $\mathbf{X}$:
$$\mathbf{C_X} = E[(\mathbf{X} - E\mathbf{X})(\mathbf{X} - E\mathbf{X})^T] = \mathbf{R_X} - E\mathbf{X}E\mathbf{X}^T$$

Example 6.12. Let $\mathbf{X}$ be an n-dimensional random vector and the random vector $\mathbf{Y}$ be defined as

$$\mathbf{Y} = \mathbf{AX} + \mathbf{b},$$

where $\mathbf{A}$ is a fixed m by n matrix and $\mathbf{b}$ is a fixed m-dimensional vector. Show that

$$\mathbf{C_Y} = \mathbf{AC_XA}^T.$$

Solution: Note that by linearity of expectation, we have

$$E\mathbf{Y} = \mathbf{A}E\mathbf{X} + \mathbf{b}.$$

By definition, we have

$$
\begin{aligned}
\mathbf{C_Y} &= E[(\mathbf{Y} - E\mathbf{Y})(\mathbf{Y} - E\mathbf{Y})^T] \\
&= E[(\mathbf{AX} + \mathbf{b} - \mathbf{A}E\mathbf{X} - \mathbf{b})(\mathbf{AX} + \mathbf{b} - \mathbf{A}E\mathbf{X} - \mathbf{b})^T] \\
&= E[\mathbf{A}(\mathbf{X} - E\mathbf{X})(\mathbf{X} - E\mathbf{X})^T\mathbf{A}^T] \\
&= \mathbf{A}E[(\mathbf{X} - E\mathbf{X})(\mathbf{X} - E\mathbf{X})^T]\mathbf{A}^T \qquad\qquad \text{(by linearity of expectation)} \\
&= \mathbf{AC_XA}^T.
\end{aligned}
$$

Example 6.13. Let X and Y be two jointly continuous random variables with joint PDF

$$
f_{X,Y}(x,y) = \begin{cases} \frac{3}{2}x^2 + y & 0 < x, y < 1 \\[2mm] 0 & \text{otherwise} \end{cases}
$$

and let the random vector $\mathbf{U}$ be defined as

$$
\mathbf{U} = \begin{bmatrix} X \\ Y \end{bmatrix}.
$$

Find the correlation and covariance matrices of $\mathbf{U}$.

Solution: We first obtain the marginal PDFs of X and Y. Note that $R_X = R_Y = (0,1)$. We have for $x \in R_X$

$$
\begin{aligned}
f_X(x) &= \int_0^1 \frac{3}{2}x^2 + y \ \ dy \\
&= \frac{3}{2}x^2 + \frac{1}{2}, \qquad \text{for } 0 < x < 1.
\end{aligned}
$$

Similarly, for $y \in R_Y$, we have

$$
\begin{aligned}
f_Y(y) &= \int_0^1 \frac{3}{2}x^2 + y \ \ dx \\
&= y + \frac{1}{2}, \qquad \text{for } 0 < y < 1.
\end{aligned}
$$

From these, we obtain $EX = \frac{5}{8}$, $EX^2 = \frac{7}{15}$, $EY = \frac{7}{12}$, and $EY^2 = \frac{5}{12}$. We also need EXY. By LOTUS, we can write

$$EXY = \int_0^1 \int_0^1 xy \left(\frac{3}{2}x^2 + y \right) \; dxdy$$

$$= \int_0^1 \frac{3}{8}y + \frac{1}{2}y^2 \; dy$$

$$= \frac{17}{48}.$$

From this, we also obtain

$$\text{Cov}(X,Y) = EXY - EXEY$$

$$= \frac{17}{48} - \frac{5}{8}\cdot\frac{7}{12}$$

$$= -\frac{1}{96}.$$

The correlation matrix $\mathbf{R_U}$ is given by

$$\mathbf{R_U} = E[\mathbf{UU}^T] = \begin{bmatrix} EX^2 & EXY \\ EYX & EY^2 \end{bmatrix} = \begin{bmatrix} \frac{7}{15} & \frac{17}{48} \\ \frac{17}{48} & \frac{5}{12} \end{bmatrix}.$$

The covariance matrix $\mathbf{C_U}$ is given by

$$\mathbf{C_U} = \begin{bmatrix} \text{Var}(X) & \text{Cov}(X,Y) \\ \text{Cov}(Y,X) & \text{Var}(Y) \end{bmatrix} = \begin{bmatrix} \frac{73}{960} & -\frac{1}{96} \\ -\frac{1}{96} & \frac{11}{144} \end{bmatrix}.$$

Properties of the Covariance Matrix:

The covariance matrix is the generalization of the variance to random vectors. It is an important matrix and is used extensively. Let's take a moment and discuss its properties. Here, we use concepts from linear algebra such as eigenvalues and positive definiteness. First note that, for any random vector $\mathbf{X}$, the covariance matrix $\mathbf{C_X}$ is a symmetric matrix. This is because if $\mathbf{C_X} = [c_{ij}]$, then

$$c_{ij} = \text{Cov}(X_i, X_j) = \text{Cov}(X_j, X_i) = c_{ji}.$$

Thus, the covariance matrix has all the nice properties of symmetric matrices. In particular, $\mathbf{C_X}$ can be diagonalized and all the eigenvalues of $\mathbf{C_X}$ are real.[2] A special important property of the covariance matrix is that it is positive semi-definite (PSD). Remember from linear algebra that a symmetric matrix $\mathbf{M}$ is **positive semi-definite (PSD)** if, for all vectors $\mathbf{b}$, we have

$$\mathbf{b}^T\mathbf{Mb} \geq 0.$$

Also, $\mathbf{M}$ is said to be **positive definite (PD)**, if for all vectors $\mathbf{b} \neq 0$, we have

$$\mathbf{b}^T\mathbf{Mb} > 0.$$

By the above definitions, we note that every PD matrix is also PSD, but the converse is not generally true. Here, we will show that covariance matrices are always PSD.

---

[2] Here, we assume $\mathbf{X}$ is a real random vector, i.e., the X_i's can only take real values.

Theorem 6.2. Let X be a random vector with n elements. Then, its covariance matrix $\mathbf{C_X}$ is positive semi-definite(PSD).

Proof. Let $\mathbf{b}$ be any fixed vector with n elements. Define the random variable Y as

$$Y = \mathbf{b}^T(\mathbf{X} - E\mathbf{X}).$$

We have

$$0 \leq EY^2$$
$$= E(YY^T)$$
$$= \mathbf{b}^T E\left[(\mathbf{X} - E\mathbf{X})(\mathbf{X} - E\mathbf{X})^T\right]\mathbf{b}$$
$$= \mathbf{b}^T \mathbf{C_X}\mathbf{b}.$$

■

Note that the eigenvalues of a PSD matrix are always larger than or equal to zero. If all the eigenvalues are strictly larger than zero, then the matrix is positive definite. From linear algebra, we know that a real symmetric matrix is positive definite if and only if all its eigenvalues are positive. Since $\mathbf{C_X}$ is a real symmetric matrix, we can state the following theorem.

Theorem 6.3. Let X be a random vector with n elements. Then its covariance matrix $\mathbf{C_X}$ is positive definite (PD), if and only if all its eigenvalues are larger than zero. Equivalently, $\mathbf{C_X}$ is positive definite (PD), if and only if $\det(\mathbf{C_X}) > 0$.

Note that the second part of the theorem is implied by the first part. This is because the determinant of a matrix is the product of its eigenvalues, and we already know that all eigenvalues of $\mathbf{C_X}$ are larger than or equal to zero.

Example 6.14. Let X and Y be two independent $Uniform(0,1)$ random variables. Let the random vectors $\mathbf{U}$ and $\mathbf{V}$ be defined as

$$\mathbf{U} = \begin{bmatrix} X \\ X+Y \end{bmatrix}, \qquad \mathbf{V} = \begin{bmatrix} X \\ Y \\ X+Y \end{bmatrix}.$$

Determine whether $\mathbf{C_U}$ and $\mathbf{C_V}$ are positive definite.

Solution: Let us first find $\mathbf{C_U}$. We have

$$\mathbf{C_U} = \begin{bmatrix} \text{Var}(X) & \text{Cov}(X, X+Y) \\ \text{Cov}(X+Y, X) & \text{Var}(X+Y) \end{bmatrix}.$$

Since X and Y are independent $Uniform(0,1)$ random variables, we have

$$\text{Var}(X) = \text{Var}(Y) = \frac{1}{12},$$
$$\text{Cov}(X, X+Y) = \text{Cov}(X, X) + \text{Cov}(X, Y)$$
$$= \frac{1}{12} + 0 = \frac{1}{12},$$
$$\text{Var}(X+Y) = \text{Var}(X) + \text{Var}(Y) = \frac{1}{6}.$$

Thus,

$$\mathbf{C_U} = \begin{bmatrix} \frac{1}{12} & \frac{1}{12} \\ \frac{1}{12} & \frac{1}{6} \end{bmatrix}.$$

So we conclude

$$\det(\mathbf{C_U}) = \frac{1}{12}\cdot\frac{1}{6} - \frac{1}{12}\cdot\frac{1}{12}$$
$$= \frac{1}{144} > 0.$$

Therefore, $\mathbf{C_U}$ is positive definite. For $\mathbf{C_V}$, we have

$$\mathbf{C_V} = \begin{bmatrix} \mathrm{Var}(X) & \mathrm{Cov}(X,Y) & \mathrm{Cov}(X,X+Y) \\ \mathrm{Cov}(Y,X) & \mathrm{Var}(Y) & \mathrm{Cov}(Y,X+Y) \\ \mathrm{Cov}(X+Y,X) & \mathrm{Cov}(X+Y,Y) & \mathrm{Var}(X+Y) \end{bmatrix}$$

$$= \begin{bmatrix} \frac{1}{12} & 0 & \frac{1}{12} \\ 0 & \frac{1}{12} & \frac{1}{12} \\ \frac{1}{12} & \frac{1}{12} & \frac{1}{6} \end{bmatrix}.$$

So we conclude

$$\det(\mathbf{C_V}) = \frac{1}{12}\left(\frac{1}{12}\cdot\frac{1}{6} - \frac{1}{12}\cdot\frac{1}{12}\right) - 0 + \frac{1}{12}\left(0 - \frac{1}{12}\cdot\frac{1}{12}\right)$$
$$= 0.$$

Thus, $\mathbf{C_V}$ is not positive definite (we already know that it is positive semi-definite).

Finally, if we have two random vectors, $\mathbf{X}$ and $\mathbf{Y}$, we can define the **cross correlation matrix** of $\mathbf{X}$ and $\mathbf{Y}$ as

$$\mathbf{R_{XY}} = E[\mathbf{X}\mathbf{Y}^T].$$

Also, the **cross covariance matrix** of $\mathbf{X}$ and $\mathbf{Y}$ is

$$\mathbf{C_{XY}} = E[(\mathbf{X} - E\mathbf{X})(\mathbf{Y} - E\mathbf{Y})^T].$$

Functions of Random Vectors: The Method of Transformations

A function of a random vector is a random vector. Thus, the methods that we discussed regarding functions of two random variables can be used to find distributions of functions of random vectors. For example, we can state a more general form of Theorem 5.1 (method of transformations). Let us first explain the method and then see some examples on how to use it.

Let $\mathbf{X}$ be an n-dimensional random vector with joint PDF $f_{\mathbf{X}}(\boldsymbol{x})$. Let $G : \mathbb{R}^n \mapsto \mathbb{R}^n$ be a continuous and invertible function with continuous partial derivatives and let $H = G^{-1}$. Suppose that the random vector $\mathbf{Y}$ is given by $\mathbf{Y} = G(\mathbf{X})$ and thus $\mathbf{X} = G^{-1}(\mathbf{Y}) = H(\mathbf{Y})$. That is,

$$\mathbf{X} = \begin{bmatrix} X_1 \\ X_2 \\ \vdots \\ X_n \end{bmatrix} = \begin{bmatrix} H_1(Y_1, Y_2, ..., Y_n) \\ H_2(Y_1, Y_2, ..., Y_n) \\ \vdots \\ H_n(Y_1, Y_2, ..., Y_n) \end{bmatrix}.$$

Then, the PDF of $\mathbf{Y}$, $f_{Y_1, Y_2, ..., Y_n}(y_1, y_2, ..., y_n)$, is given by

$$f_{\mathbf{Y}}(\mathbf{y}) = f_{\mathbf{X}}\big(H(\mathbf{y})\big)|J|$$

where J is the Jacobian of H defined by

$$J = \det \begin{bmatrix} \frac{\partial H_1}{\partial y_1} & \frac{\partial H_1}{\partial y_2} & \cdots & \frac{\partial H_1}{\partial y_n} \\ \frac{\partial H_2}{\partial y_1} & \frac{\partial H_2}{\partial y_2} & \cdots & \frac{\partial H_2}{\partial y_n} \\ \vdots & \vdots & \vdots & \vdots \\ \frac{\partial H_n}{\partial y_1} & \frac{\partial H_n}{\partial y_2} & \cdots & \frac{\partial H_n}{\partial y_n} \end{bmatrix},$$

and evaluated at $(y_1, y_2, ..., y_n)$.

Example 6.15. Let $\mathbf{X}$ be an n-dimensional random vector. Let $\mathbf{A}$ be a fixed (non-random) invertible n by n matrix, and $\mathbf{b}$ be a fixed n-dimensional vector. Define the random vector $\mathbf{Y}$ as

$$\mathbf{Y} = \mathbf{A}\mathbf{X} + \mathbf{b}.$$

Find the PDF of $\mathbf{Y}$ in terms of PDF of $\mathbf{X}$.

Solution: Since A is invertible, we can write

$$\mathbf{X} = \mathbf{A}^{-1}(\mathbf{Y} - \mathbf{b}).$$

We can also check that

$$J = \det(\mathbf{A}^{-1}) = \frac{1}{\det \mathbf{A}}.$$

Thus, we conclude that

$$f_{\mathbf{Y}}(\mathbf{y}) = \frac{1}{|\det \mathbf{A}|} f_{\mathbf{X}}\big(\mathbf{A}^{-1}(\mathbf{y} - \mathbf{b})\big).$$

Normal (Gaussian) Random Vectors:

We discussed two jointly normal random variables previously in Section 5.3.2. In particular, two random variables X and Y are said to be **bivariate normal** or **jointly normal**, if $aX + bY$ has normal distribution for all $a, b \in \mathbb{R}$. We can extend this definition to n jointly normal random variables.

Random variables X_1, X_2,..., X_n are said to be **jointly normal** if, for all $a_1, a_2,..., a_n \in \mathbb{R}$, the random variable

$$a_1 X_1 + a_2 X_2 + ... + a_n X_n$$

is a normal random variable.

As before, we agree that the constant zero is a normal random variable with zero mean and variance, i.e., $N(0,0)$. When we have several jointly normal random variables, we often put them in a vector. The resulting random vector is a called a normal (Gaussian) random vector.

A random vector

$$\mathbf{X} = \begin{bmatrix} X_1 \\ X_2 \\ \vdots \\ X_n \end{bmatrix}$$

is said to be **normal** or **Gaussian** if the random variables X_1, X_2,..., X_n are jointly normal.

To find the general form for the PDF of a Gaussian random vector it is convenient to start from the simplest case where the X_i's are independent and identically distributed (i.i.d.), $X_i \sim N(0, 1)$. In this case, we know how to find the joint PDF. It is simply the product of the individual (marginal) PDFs. Let's call such a random vector the **standard normal random vector**. So, let

$$\mathbf{Z} = \begin{bmatrix} Z_1 \\ Z_2 \\ \vdots \\ Z_n \end{bmatrix},$$

where Z_i's are i.i.d. and $Z_i \sim N(0,1)$. Then, we have

$$f_{\mathbf{Z}}(\mathbf{z}) = f_{Z_1, Z_2, \ldots, Z_n}(z_1, z_2, \ldots, z_n)$$

$$= \prod_{i=1}^{n} f_{Z_i}(z_i)$$

$$= \frac{1}{(2\pi)^{\frac{n}{2}}} \exp\left\{ -\frac{1}{2} \sum_{i=1}^{n} z_i^2 \right\}$$

$$= \frac{1}{(2\pi)^{\frac{n}{2}}} \exp\left\{ -\frac{1}{2} \mathbf{z}^T \mathbf{z} \right\}.$$

For a standard normal random vector $\mathbf{Z}$, where the Z_i's are i.i.d. and $Z_i \sim N(0,1)$, the PDF is given by

$$f_{\mathbf{Z}}(\mathbf{z}) = \frac{1}{(2\pi)^{\frac{n}{2}}} \exp\left\{ -\frac{1}{2} \mathbf{z}^T \mathbf{z} \right\}.$$

Now, we need to extend this formula to a general normal random vector $\mathbf{X}$ with mean $\mathbf{m}$ and covariance matrix $\mathbf{C}$. This is very similar to when we defined general normal random variables from the standard normal random variable. We remember that if $Z \sim N(0,1)$, then the random variable $X = \sigma Z + \mu$ has $N(\mu, \sigma^2)$ distribution. We would like to do the same thing for normal random vectors.

Assume that I have a normal random vector $\mathbf{X}$ with mean $\mathbf{m}$ and covariance matrix $\mathbf{C}$. We write $\mathbf{X} \sim N(\mathbf{m}, \mathbf{C})$. Further, assume that $\mathbf{C}$ is a positive definite matrix.[3] Then from linear algebra we know that there exists an n by n matrix $\mathbf{Q}$ such that

$$\mathbf{Q}\mathbf{Q}^T = \mathbf{I} \qquad (\mathbf{I} \text{ is the identity matrix}),$$

$$\mathbf{C} = \mathbf{Q}\mathbf{D}\mathbf{Q}^T,$$

where $\mathbf{D}$ is a diagonal matrix

$$\mathbf{D} = \begin{bmatrix} d_{11} & 0 & \ldots & 0 \\ 0 & d_{22} & \ldots & 0 \\ \vdots & \vdots & \vdots & \vdots \\ 0 & 0 & \ldots & d_{nn} \end{bmatrix}.$$

---

[3]The positive definiteness assumption here does not create any limitations. We already know that $\mathbf{C}$ is positive semi-definite (Theorem 6.2), so $\det(\mathbf{C}) \geq 0$. We also know that $\mathbf{C}$ is positive definite if and only if $\det(\mathbf{C}) > 0$ (Theorem 6.3). So here, we are only excluding the case $\det(\mathbf{C}) = 0$. If $\det(\mathbf{C}) = 0$, then you can show that you can write some X_i's as a linear combination of others, so indeed we can remove them from the vector without losing any information.

The positive definiteness assumption guarantees that all d_{ii}'s are positive. Let's define

$$\mathbf{D}^{\frac{1}{2}} = \begin{bmatrix} \sqrt{d_{11}} & 0 & \cdots & 0 \\ 0 & \sqrt{d_{22}} & \cdots & 0 \\ \vdots & \vdots & \vdots & \vdots \\ 0 & 0 & \cdots & \sqrt{d_{nn}} \end{bmatrix}.$$

We have $\mathbf{D}^{\frac{1}{2}}\mathbf{D}^{\frac{1}{2}} = \mathbf{D}$ and $\mathbf{D}^{\frac{1}{2}} = \mathbf{D}^{\frac{1}{2}T}$. Also define

$$\mathbf{A} = \mathbf{Q}\mathbf{D}^{\frac{1}{2}}\mathbf{Q}^T.$$

Then,

$$\mathbf{A}\mathbf{A}^T = \mathbf{A}^T\mathbf{A} = \mathbf{C}.$$

Now we are ready to define the transformation that converts a standard Gaussian vector to $\mathbf{X} \sim N(\mathbf{m}, \mathbf{C})$. Let $\mathbf{Z}$ be a standard Gaussian vector, i.e., $\mathbf{Z} \sim N(\mathbf{0}, \mathbf{I})$. Define

$$\mathbf{X} = \mathbf{A}\mathbf{Z} + \mathbf{m}.$$

We claim that $\mathbf{X} \sim N(\mathbf{m}, \mathbf{C})$. To see this, first note that $\mathbf{X}$ is a normal random vector. The reason is that any linear combination of components of $\mathbf{X}$ is indeed a linear combination of components of $\mathbf{Z}$ plus a constant. Thus, every linear combination of components of $\mathbf{X}$ is a normal random variable. It remains to show that $E\mathbf{X} = \mathbf{m}$ and $\mathbf{C_X} = \mathbf{C}$. First note that by linearity of expectation we have

$$\begin{aligned} E\mathbf{X} &= E\left[\mathbf{A}\mathbf{Z} + \mathbf{m}\right] \\ &= \mathbf{A}E[\mathbf{Z}] + \mathbf{m} \\ &= \mathbf{m}. \end{aligned}$$

Also, by Example 6.12 we have

$$\begin{aligned} \mathbf{C_X} &= \mathbf{A}\mathbf{C_Z}\mathbf{A}^T \\ &= \mathbf{A}\mathbf{A}^T && \text{(since } \mathbf{C_Z} = \mathbf{I}) \\ &= \mathbf{C}. \end{aligned}$$

Thus, we have shown that $\mathbf{X}$ is a random vector with mean $\mathbf{m}$ and covariance matrix $\mathbf{C}$. Now we can use Example 6.15 to find the PDF of $\mathbf{X}$. We have

$$\begin{aligned} f_{\mathbf{X}}(\mathbf{x}) &= \frac{1}{|\det \mathbf{A}|} f_{\mathbf{Z}}\big(\mathbf{A}^{-1}(\mathbf{x} - \mathbf{m})\big) \\ &= \frac{1}{(2\pi)^{\frac{n}{2}}|\det \mathbf{A}|} \exp\left\{-\frac{1}{2}(\mathbf{A}^{-1}(\mathbf{x} - \mathbf{m}))^T(\mathbf{A}^{-1}(\mathbf{x} - \mathbf{m}))\right\} \\ &= \frac{1}{(2\pi)^{\frac{n}{2}}\sqrt{\det \mathbf{C}}} \exp\left\{-\frac{1}{2}(\mathbf{x} - \mathbf{m})^T\mathbf{A}^{-T}\mathbf{A}^{-1}(\mathbf{x} - \mathbf{m})\right\} \\ &= \frac{1}{(2\pi)^{\frac{n}{2}}\sqrt{\det \mathbf{C}}} \exp\left\{-\frac{1}{2}(\mathbf{x} - \mathbf{m})^T\mathbf{C}^{-1}(\mathbf{x} - \mathbf{m})\right\}. \end{aligned}$$

> For a normal random vector $\mathbf{X}$ with mean $\mathbf{m}$ and covariance matrix $\mathbf{C}$, the PDF is given by
>
> $$f_{\mathbf{X}}(\mathbf{x}) = \frac{1}{(2\pi)^{\frac{n}{2}} \sqrt{\det \mathbf{C}}} \exp\left\{-\frac{1}{2}(\mathbf{x} - \mathbf{m})^T \mathbf{C}^{-1}(\mathbf{x} - \mathbf{m})\right\} \tag{6.1}$$

Example 6.16. Let X and Y be two jointly normal random variables with $X \sim N(\mu_X, \sigma_X)$, $Y \sim N(\mu_Y, \sigma_Y)$, and $\rho(X, Y) = \rho$. Show that the above PDF formula for PDF of $\begin{bmatrix} X \\ Y \end{bmatrix}$ is the same as $f_{X,Y}(x, y)$ given in Definition 5.4 in Section 5.3.2. That is,

$$f_{XY}(x, y) = \frac{1}{2\pi\sigma_X\sigma_Y \sqrt{1 - \rho^2}} \cdot$$

$$\exp\left\{-\frac{1}{2(1 - \rho^2)}\left[\left(\frac{x - \mu_X}{\sigma_X}\right)^2 + \left(\frac{y - \mu_Y}{\sigma_Y}\right)^2 - 2\rho\frac{(x - \mu_X)(y - \mu_Y)}{\sigma_X\sigma_Y}\right]\right\}.$$

Solution: Both formulas are in the form $ae^{-\frac{1}{2}b}$. Thus, it suffices to show that they have the same a and b. Here we have

$$\mathbf{m} = \begin{bmatrix} \mu_X \\ \mu_Y \end{bmatrix}.$$

We also have

$$\mathbf{C} = \begin{bmatrix} \text{Var}(X) & \text{Cov}(X, Y) \\ \text{Cov}(Y, X) & \text{Var}(Y) \end{bmatrix} = \begin{bmatrix} \sigma_X^2 & \rho\sigma_X\sigma_Y \\ \rho\sigma_X\sigma_Y & \sigma_Y^2 \end{bmatrix}.$$

From this, we obtain

$$\det \mathbf{C} = \sigma_X^2\sigma_Y^2(1 - \rho^2).$$

Thus, in both formulas for PDF a is given by

$$a = \frac{1}{2\pi\sigma_X\sigma_Y \sqrt{1 - \rho^2}}.$$

Next, we check b. We have

$$\mathbf{C}^{-1} = \frac{1}{\sigma_X^2\sigma_Y^2(1 - \rho^2)} \begin{bmatrix} \sigma_Y^2 & -\rho\sigma_X\sigma_Y \\ -\rho\sigma_X\sigma_Y & \sigma_X^2 \end{bmatrix}.$$

Now by matrix multiplication we obtain

$$(\mathbf{x} - \mathbf{m})^T \mathbf{C}^{-1}(\mathbf{x} - \mathbf{m}) =$$

$$= \frac{1}{\sigma_X^2\sigma_Y^2(1 - \rho^2)} \begin{bmatrix} x - \mu_X \\ y - \mu_Y \end{bmatrix}^T \begin{bmatrix} \sigma_Y^2 & -\rho\sigma_X\sigma_Y \\ -\rho\sigma_X\sigma_Y & \sigma_X^2 \end{bmatrix} \begin{bmatrix} x - \mu_X \\ y - \mu_Y \end{bmatrix}$$

$$= -\frac{1}{2(1 - \rho^2)}\left[\left(\frac{x - \mu_X}{\sigma_X}\right)^2 + \left(\frac{y - \mu_Y}{\sigma_Y}\right)^2 - 2\rho\frac{(x - \mu_X)(y - \mu_Y)}{\sigma_X\sigma_Y}\right],$$

which agrees with the formula in Definition 5.4.

Remember that two jointly normal random variables X and Y are independent if and only if they are uncorrelated. We can extend this to multiple jointly normal random variables. Thus, if you have a normal random vector whose components are uncorrelated, you can conclude that the components are independent. To show this, note that if the X_i's are uncorrelated, then the covariance matrix $\mathbf{C_X}$ is diagonal, so its inverse $\mathbf{C_X}^{-1}$ is also diagonal. You can see that in this case the PDF (Equation 6.1) becomes the products of marginal PDFs.

If $\mathbf{X} = [X_1, X_2, ..., X_n]^T$ is a normal random vector, and we know $\text{Cov}(X_i, X_j) = 0$ for all $i \neq j$, then $X_1, X_2, ..., X_n$ are independent.

Another important result is that if $\mathbf{X} = [X_1, X_2, ..., X_n]^T$ is a normal random vector then $\mathbf{Y} = \mathbf{AX} + \mathbf{b}$ is also a random vector because any linear combination of components of $\mathbf{Y}$ is also a linear combination of components of $\mathbf{X}$ plus a constant value.

If $\mathbf{X} = [X_1, X_2, ..., X_n]^T$ is a normal random vector, $\mathbf{X} \sim N(\mathbf{m}, \mathbf{C})$, $\mathbf{A}$ is an m by n fixed matrix, and $\mathbf{b}$ is an m-dimensional fixed vector, then the random vector $\mathbf{Y} = \mathbf{AX} + \mathbf{b}$ is a normal random vector with mean $\mathbf{A}E\mathbf{X} + \mathbf{b}$ and covariance matrix $\mathbf{ACA}^T$.

$$\mathbf{Y} \sim N(\mathbf{A}E\mathbf{X} + \mathbf{b}, \mathbf{ACA}^T)$$

6.1.6 Solved Problems

1. Let X, Y and Z be three jointly continuous random variables with joint PDF

$$f_{XYZ}(x, y, z) = \begin{cases} \frac{1}{3}(x + 2y + 3z) & 0 \leq x, y, z \leq 1 \\ 0 & \text{otherwise} \end{cases}$$

Find the joint PDF of X and Y, $f_{XY}(x, y)$.

Solution:

$$f_{XY}(x,y) = \int_{-\infty}^{\infty} f_{XYZ}(x,y,z)dz$$

$$= \int_0^1 \frac{1}{3}(x + 2y + 3z)dz$$

$$= \frac{1}{3}\left[(x + 2y)z + \frac{3}{2}z^2\right]_0^1$$

$$= \frac{1}{3}\left(x + 2y + \frac{3}{2}\right), \qquad\qquad \text{for} \quad 0 \leq x, y \leq 1.$$

Thus,

$$f_{XY}(x,y) = \begin{cases} \frac{1}{3}\left(x + 2y + \frac{3}{2}\right) & 0 \leq x \leq 1, 0 \leq y \leq 1 \\ \\ 0 & \text{otherwise} \end{cases}$$

2. Let X, Y and Z be three independent random variables with $X \sim N(\mu, \sigma^2)$, and $Y, Z \sim Uniform(0,2)$. We also know that

$$E[X^2Y + XYZ] = 13,$$
$$E[XY^2 + ZX^2] = 14.$$

Find μ and σ.

Solution:

$$X, Y, \text{ and } Z \, are \text{ independent} \Rightarrow \begin{cases} EX^2 \cdot EY + EX \cdot EY \cdot EZ = 13 \\ EX \cdot EY^2 + EZ \cdot EX^2 = 14 \end{cases}$$

Since $Y, Z \sim Uniform(0,2)$, we conclude

$$EY = EZ = 1; \; \text{Var}(Y) = \text{Var}(Z) = \frac{(2-0)^2}{12} = \frac{1}{3}.$$

Therefore,

$$EY^2 = \frac{1}{3} + 1 = \frac{4}{3}.$$

Thus,

$$\begin{cases} EX^2 + EX = 13 \\ \frac{4}{3}EX + EX^2 = 14 \end{cases}$$

We conclude $EX = 3$, $EX^2 = 10$. Therefore,

$$\begin{cases} \mu = 3 \\ \mu^2 + \sigma^2 = 10 \end{cases}$$

So, we obtain $\mu = 3, \sigma = 1$.

3. Let X_1, X_2, and X_3 be three i.i.d. *Bernoulli*(p) random variables and

$$Y_1 = \max(X_1, X_2),$$
$$Y_2 = \max(X_1, X_3),$$
$$Y_3 = \max(X_2, X_3),$$
$$Y = Y_1 + Y_2 + Y_3.$$

Find EY and $\text{Var}(Y)$.

Solution: We have

$$EY = EY_1 + EY_2 + EY_3 = 3EY_1, \qquad \text{by symmetry.}$$

Also,

$$\text{Var}(Y) = \text{Var}(Y_1) + \text{Var}(Y_2) + \text{Var}(Y_3) + 2\text{Cov}(Y_1, Y_2) + 2\text{Cov}(Y_1, Y_3) + 2\text{Cov}(Y_2, Y_3)$$

$$= 3\text{Var}(Y_1) + 6\text{Cov}(Y_1, Y_2), \qquad \text{by symmetry.}$$

Note that Y_i's are also Bernoulli random variables (but they are not independent). In particular, we have

$$P(Y_1 = 1) = P\big((X_1 = 1) \text{ or } (X_2 = 1)\big)$$
$$= P(X_1 = 1) + P(X_2 = 1) - P(X_1 = 1, X_2 = 1) \quad \text{(comma means "and")}$$
$$= 2p - p^2.$$

Thus, $Y_1 \sim Bernoulli(2p - p^2)$, and we obtain

$$EY_1 = 2p - p^2 = p(2 - p)$$
$$\text{Var}(Y_1) = (2p - p^2)(1 - 2p + p^2) = p(2 - p)(1 - p)^2.$$

It remains to find $\text{Cov}(Y_1, Y_2)$. We can write

$$\text{Cov}(Y_1, Y_2) = E[Y_1 Y_2] - E[Y_1]E[Y_2]$$
$$= E[Y_1 Y_2] - p^2(2 - p)^2.$$

Note that $Y_1 Y_2$ is also a Bernoulli random variable. We have

$$E[Y_1 Y_2] = P\big(Y_1 = 1, Y_2 = 1\big)$$
$$= P\Big((X_1 = 1) \text{ or } \big(X_2 = 1, X_3 = 1\big)\Big)$$
$$= P(X_1 = 1) + P\big(X_2 = 1, X_3 = 1\big) - P\big(X_1 = 1, X_2 = 1, X_3 = 1\big)$$
$$= p + p^2 - p^3.$$

Thus, we obtain

$$\text{Cov}(Y_1, Y_2) = E[Y_1 Y_2] - p^2(2-p)^2$$
$$= p + p^2 - p^3 - p^2(2-p)^2.$$

Finally, we obtain

$$EY = 3EY_1 = 3p(2-p).$$

Also,

$$\text{Var}(Y) = 3\text{Var}(Y_1) + 6\text{Cov}(Y_1, Y_2)$$
$$= 3p(2-p)(1-p)^2 + 6(p + p^2 - p^3 - p^2(2-p)^2).$$

4. Let $M_X(s)$ be finite for $s \in [-c, c]$, where $c > 0$. Show that MGF of $Y = aX + b$ is given by

$$M_Y(s) = e^{sb} M_X(as),$$

and it is finite in $\left[-\frac{c}{|a|}, \frac{c}{|a|}\right]$.

Solution: We have

$$M_Y(s) = E[e^{sY}]$$
$$= E[e^{saX} e^{sb}]$$
$$= e^{sb} E[e^{(sa)X}]$$
$$= e^{sb} M_X(as).$$

Also, since $M_X(s)$ is finite for $s \in [-c, c]$, $M_X(as)$ is finite for $s \in \left[-\frac{c}{|a|}, \frac{c}{|a|}\right]$.

5. Let $Z \sim N(0, 1)$ Find the MGF of Z. Extend your result to $X \sim N(\mu, \sigma)$.

Solution: We have

$$M_Z(s) = E[e^{sZ}]$$
$$= \frac{1}{\sqrt{2\pi}} \int_{-\infty}^{\infty} e^{sx} e^{-\frac{x^2}{2}} dx$$
$$= \frac{1}{\sqrt{2\pi}} \int_{-\infty}^{\infty} e^{\frac{s^2}{2}} e^{-\frac{(x-s)^2}{2}} dx$$
$$= e^{\frac{s^2}{2}} \frac{1}{\sqrt{2\pi}} \int_{-\infty}^{\infty} e^{-\frac{(x-s)^2}{2}} dx$$
$$= e^{\frac{s^2}{2}} \qquad \text{(PDF of normal integrates to 1).}$$

Using Problem 4, we obtain

$$M_X(s) = e^{s\mu + \frac{\sigma^2 s^2}{2}}, \qquad \text{for all} \quad s \in \mathbb{R}.$$

6. Let $Y = X_1 + X_2 + X_3 + ... + X_n$, where the X_i's are independent and $X_i \sim Poisson(\lambda_i)$. Find the distribution of Y.

Solution: We have

$$M_{X_i}(s) = e^{\lambda_i(e^s - 1)}, \qquad \text{for all } s \in \mathbb{R}.$$

Thus,

$$M_Y(s) = \prod_{i=1}^{n} e^{\lambda_i(e^s - 1)}$$
$$= e^{(\sum_{i=1}^{n} \lambda_i)(e^s - 1)}, \qquad \text{for all } s \in \mathbb{R}.$$

which is the MGF of a Poisson random variable with parameter $\lambda = \sum_{i=1}^{n} \lambda_i$, thus

$$Y \sim Poisson(\sum_{i=1}^{n} \lambda_i).$$

7. **Probability Generating Functions (PGFs):** For many important discrete random variables, the range is a subset of $\{0, 1, 2, ...\}$. For these random variables it is usually useful to work with *probability generating functions (PGFs)* defined as

$$G_X(z) = E[z^X] = \sum_{n=0}^{\infty} P(X = n)z^n,$$

for all $z \in \mathbb{R}$ that $G_X(z)$ is finite.

(a) Show that $G_X(z)$ is always finite for $|z| \leq 1$.

(b) Show that if X and Y are independent, then

$$G_{X+Y}(z) = G_X(z)G_Y(z).$$

(c) Show that

$$\frac{1}{k!} \frac{d^k G_X(z)}{dz^k}\bigg|_{z=0} = P(X = k).$$

(d) Show that

$$\frac{d^k G_X(z)}{dz^k}\bigg|_{z=1} = E[X(X-1)(X-2)...(X-k+1)].$$

Solution:

(a) If $|z| \leq 1$, then $z^n \leq |z| \leq 1$, so we have

$$G_X(z) = \sum_{n=0}^{\infty} P(X = n)z^n$$

$$\leq \sum_{n=0}^{\infty} P(X = n) = 1.$$

(b) If X and Y are independent, then

$$\begin{aligned}
G_{X+Y}(z) &= E[z^{X+Y}] \\
&= E[z^X z^Y] \\
&= E[z^X]E[z^Y] \qquad \text{(since } X \text{ and } Y \text{ are independent)} \\
&= G_X(z)G_Y(z).
\end{aligned}$$

(c) By differentiation we obtain

$$\frac{d^k G_X(z)}{dz^k} = \sum_{n=k}^{\infty} n(n-1)(n-2)...(n-k+1)P(X = n)z^{n-k}.$$

Thus,

$$\frac{d^k G_X(z)}{dz^k} = k!P(X = k) + \sum_{n=k+1}^{\infty} n(n-1)(n-2)...(n-k+1)P(X = n)z^{n-k}.$$

Thus,

$$\frac{1}{k!} \frac{d^k G_X(z)}{dz^k}\bigg|_{z=0} = P(X = k).$$

(d) By letting $Z = 1$ in

$$\frac{d^k G_X(z)}{dz^k} = \sum_{n=k}^{\infty} n(n-1)(n-2)...(n-k+1)P(X = n)z^{n-k},$$

we obtain

$$\frac{d^k G_X(z)}{dz^k}\bigg|_{z=1} = \sum_{n=k}^{\infty} n(n-1)(n-2)...(n-k+1)P(X = n),$$

which by LOTUS is equal to $E[X(X - 1)(X - 2)...(X - k + 1)]$.

8. Let $M_X(s)$ be finite for $s \in [-c, c]$ where $c > 0$. Prove

$$\lim_{n \to \infty} \left[M_X(\frac{s}{n}) \right]^n = e^{sEX}.$$

Solution: Equivalently, we show

$$\lim_{n \to \infty} n \ln \left(M_X(\frac{s}{n}) \right) = sEX.$$

We have

$$\lim_{n \to \infty} n \ln \left(M_X(\frac{s}{n}) \right) = \lim_{n \to \infty} \frac{\ln \left(M_X(\frac{s}{n}) \right)}{\frac{1}{n}}$$

$$= \frac{0}{0}.$$

So, we can use L'Hôpital's rule

$$\lim_{n \to \infty} \frac{\ln \left(M_X(\frac{s}{n}) \right)}{\frac{1}{n}} = \lim_{t \to 0} \frac{\ln \left(M_X(ts) \right)}{t} \quad (\text{let} \quad t = \frac{1}{n})$$

$$= \lim_{t \to 0} \frac{\frac{s M_X'(ts)}{M_X(ts)}}{1} \quad (\text{by L'Hôpital's rule})$$

$$= \frac{s M_X'(0)}{M_X(0)}$$

$$= s\mu \quad (\text{since} \quad M_X'(0) = \mu, M_X(0) = 1).$$

9. Let $M_X(s)$ be finite for $s \in [-c, c]$, where $c > 0$. Assume $EX = 0$, and $\text{Var}(X) = 1$. Prove

$$\lim_{n \to \infty} \left[M_X \left(\frac{s}{\sqrt{n}} \right) \right]^n = e^{\frac{s^2}{2}}.$$

Note: From this, we can prove the Central Limit Theorem (CLT) which is discussed in Section 7.1.

Solution: Equivalently, we show

$$\lim_{n \to \infty} n \ln \left(M_X(\frac{s}{\sqrt{n}}) \right) = \frac{s^2}{2}.$$

We have

$$\lim_{n\to\infty} n \ln\left(M_X(\frac{s}{\sqrt{n}})\right) = \lim_{n\to\infty} \frac{\ln\left(M_X(\frac{s}{\sqrt{n}})\right)}{\frac{1}{n}} \quad (\text{let} \quad t = \frac{1}{\sqrt{n}})$$

$$= \lim_{t\to 0} \frac{\ln\left(M_X(ts)\right)}{t^2}$$

$$= \lim_{t\to 0} \frac{\frac{sM_X'(ts)}{M_X(ts)}}{2t} \quad (\text{by L'Hôpital's rule})$$

$$= \lim_{t\to 0} \frac{sM_X'(ts)}{2t} \quad (\text{again} \quad \frac{0}{0})$$

$$= \lim_{t\to 0} \frac{s^2 M_X''(ts)}{2} \quad (\text{by L'Hôpital's rule})$$

$$= \frac{s^2}{2} \quad (\text{since} \quad M_X''(0) = EX^2 = 1).$$

10. We can define MGF for jointly distributed random variables as well. For example, for two random variables (X, Y), the MGF is defined by

$$M_{XY}(s, t) = E[e^{sX + tY}].$$

Similar to the MGF of a single random variable, the MGF of the joint distributions uniquely determines the joint distribution. Let X and Y be two jointly normal random variables with $EX = \mu_X$, $EY = \mu_Y$, $\text{Var}(X) = \sigma_X^2$, $\text{Var}(Y) = \sigma_Y^2$, $\rho(X, Y) = \rho$. Find $M_{XY}(s, t)$.

Solution:

Note that $U = sX + tY$ is a linear combination of X and Y and thus it is a normal random variable. We have

$$EU = sEX + tEY = s\mu_X + t\mu_Y,$$

$$\text{Var}(U) = s^2\text{Var}(X) + t^2\text{Var}(Y) + 2st\rho(X, Y)\sigma_X\sigma_Y$$

$$= s^2\sigma_X^2 + t^2\sigma_Y^2 + 2st\rho\sigma_X\sigma_Y.$$

Thus

$$U \sim N(s\mu_X + t\mu_Y, s^2\sigma_X^2 + t^2\sigma_Y^2 + 2st\rho\sigma_X\sigma_Y).$$

Note that for a normal random variable with mean μ and variance σ^2, the MGF is given by $e^{s\mu + \frac{\sigma^2 s^2}{2}}$. Thus

$$M_{XY}(s, t) = E[e^U] = M_U(1)$$

$$= e^{\mu_U + \frac{\sigma_U^2}{2}}$$

$$= e^{s\mu_X + t\mu_Y + \frac{1}{2}(s^2\sigma_X^2 + t^2\sigma_Y^2 + 2st\rho\sigma_X\sigma_Y)}.$$

11. Let $\mathbf{X} = \begin{bmatrix} X_1 \\ X_2 \end{bmatrix}$ be a normal random vector with the following mean vector and covariance matrix

$$\mathbf{m} = \begin{bmatrix} 0 \\ 1 \end{bmatrix}, \qquad \mathbf{C} = \begin{bmatrix} 1 & -1 \\ -1 & 2 \end{bmatrix}.$$

Let also

$$\mathbf{A} = \begin{bmatrix} 1 & 2 \\ 2 & 1 \\ 1 & 1 \end{bmatrix}, \qquad \mathbf{b} = \begin{bmatrix} 0 \\ 1 \\ 2 \end{bmatrix}, \qquad \mathbf{Y} = \begin{bmatrix} Y_1 \\ Y_2 \\ Y_3 \end{bmatrix} = \mathbf{A}\mathbf{X} + \mathbf{b}.$$

(a) Find $P(0 \leq X_2 \leq 1)$.

(b) Find the expected value vector of $\mathbf{Y}$, $\mathbf{m_Y} = E\mathbf{Y}$.

(c) Find the covariance matrix of $\mathbf{Y}$, $\mathbf{C_Y}$.

(d) Find $P(Y_3 \leq 4)$.

Solution:

(a) From $\mathbf{m}$ and $\mathbf{c}$ we have $X_2 \sim N(1, 2)$. Thus

$$P(0 \leq X_2 \leq 1) = \Phi\left(\frac{1-1}{\sqrt{2}}\right) - \Phi\left(\frac{0-1}{\sqrt{2}}\right)$$

$$= \Phi(0) - \Phi\left(\frac{-1}{\sqrt{2}}\right) = 0.2602$$

(b)

$$m_Y = E\mathbf{Y} = AE\mathbf{X} + \mathbf{b}$$

$$= \begin{bmatrix} 1 & 2 \\ 2 & 1 \\ 1 & 1 \end{bmatrix} \cdot \begin{bmatrix} 0 \\ 1 \end{bmatrix} + \begin{bmatrix} 0 \\ 1 \\ 2 \end{bmatrix}$$

$$= \begin{bmatrix} 2 \\ 2 \\ 3 \end{bmatrix}.$$

(c)

$$\mathbf{C_Y} = \mathbf{A}\mathbf{C_X}\mathbf{A}^T$$

$$= \begin{bmatrix} 1 & 2 \\ 2 & 1 \\ 1 & 1 \end{bmatrix} \cdot \begin{bmatrix} 1 & -1 \\ -1 & 2 \end{bmatrix} \cdot \begin{bmatrix} 1 & 2 & 1 \\ 2 & 1 & 1 \end{bmatrix}$$

$$= \begin{bmatrix} 5 & 1 & 2 \\ 1 & 2 & 1 \\ 2 & 1 & 1 \end{bmatrix}.$$

(d) From $\mathbf{m_Y}$ and $\mathbf{c_Y}$ we have $Y_3 \sim N(3,1)$ thus

$$P(Y_3 \leq 4) = \Phi\left(\frac{4-3}{1}\right) = \Phi(1) = 0.8413$$

12. (Whitening/decorrelating transformation) Let $\mathbf{X}$ be an n-dimensional zero-mean random vector. Since $\mathbf{C_X}$ is a real symmetric matrix, we conclude that it can be diagonalized. That is, there exists an n by n matrix $\mathbf{Q}$ such that

$$\mathbf{QQ}^T = \mathbf{I} \qquad (\mathbf{I} \text{ is the identity matrix})$$
$$\mathbf{C_X} = \mathbf{QDQ}^T,$$

where $\mathbf{D}$ is a diagonal matrix

$$\mathbf{D} = \begin{bmatrix} d_{11} & 0 & \dots & 0 \\ 0 & d_{22} & \dots & 0 \\ \vdots & \vdots & \vdots & \vdots \\ 0 & 0 & \dots & d_{nn} \end{bmatrix}.$$

Now suppose we define a new random vector $\mathbf{Y}$ as $\mathbf{Y} = \mathbf{Q}^T\mathbf{X}$, thus

$$\mathbf{X} = \mathbf{QY}.$$

Show that $\mathbf{Y}$ has a diagonal covariance matrix, and conclude that components of $\mathbf{Y}$ are uncorrelated, i.e., $\text{Cov}(Y_i, Y_j) = 0$ if $i \neq j$.

Solution:

$$\begin{aligned} \mathbf{C_Y} &= E[(\mathbf{Y} - E\mathbf{Y})(\mathbf{Y} - E\mathbf{Y})^T] \\ &= E[(\mathbf{Q}^T\mathbf{X} - E\mathbf{Q}^T\mathbf{X})(\mathbf{Q}^T\mathbf{X} - E\mathbf{Q}^T\mathbf{X})^T] \\ &= E[\mathbf{Q}^T(\mathbf{X} - E\mathbf{X})(\mathbf{X} - E\mathbf{X})^T]\mathbf{Q}] \\ &= \mathbf{Q}^T\mathbf{C_X}\mathbf{Q} \\ &= \mathbf{Q}^T\mathbf{QDQ}^T\mathbf{Q} \\ &= \mathbf{D} \quad (\text{since} \quad \mathbf{Q}^T\mathbf{Q} = \mathbf{I}). \end{aligned}$$

Therefore, $\mathbf{Y}$ has a diagonal covariance matrix, and $\text{Cov}(Y_i, Y_j) = 0$ if $i \neq j$.

6.2   Probability Bounds

In this section, we will discuss probability bounds. These are inequalities that are usually applicable to a general scenario. There are several scenarios in which we resort to inequalities.

Sometimes we do not have enough information to calculate a desired quantity (such as the probability of an event or the expected value of a random variable). In other situations, the problem might be complicated and exact calculation might be very difficult. In other scenarios, we might want to provide a result that is general and applicable to wide range of problems.

For example, suppose that you are an engineer and you design a communication system. Your company wants to ensure that the error probability in your system be less than a given value, say 10^{-5}. Calculating the exact value of probability might be difficult due to some unknown parameters or simply because the communication system is a complicated one. Here, you do not actually need to find the error probability exactly, but all you need to do is to show that it is less than 10^{-5}.

In this section, we will discuss several inequalities. Depending on the problem you are dealing with, you might decide which one to use.

6.2.1 The Union Bound and its Extensions

The **union bound** or **Boole's inequality** [13] is applicable when you need to show that the probability of union of some events is less than some value. Remember that for any two events A and B we have

$$P(A \cup B) = P(A) + P(B) - P(A \cap B)$$
$$\leq P(A) + P(B).$$

Similarly, for three events A, B, and C, we can write

$$P(A \cup B \cup C) = P\big((A \cup B) \cup C\big)$$
$$\leq P(A \cup B) + P(C)$$
$$\leq P(A) + P(B) + P(C).$$

In general, using induction we prove the following

The Union Bound

For any events $A_1, A_2, ..., A_n$, we have

$$P\left(\bigcup_{i=1}^{n} A_i\right) \leq \sum_{i=1}^{n} P(A_i) \tag{6.2}$$

The union bound is a very simple but useful result. It is used frequently in different applications. Here, we look at one application in the area of *random graphs*. Random graphs are widely used when analyzing social networks, wireless networks, and the internet. A simple model for random graphs is the Erdös-Rényi model $G(n, p)$ [11, 12]. In this model, we have n nodes in the graph. In social networking context, each node might represent a person. Every pair of nodes are connected by an edge with probability p. The occurrence of each edge in the graph is independent from other edges in the graph. Figure 6.1 shows an example of a randomly generated graph using this model. Here, $n = 5$ and p was chosen to be $\frac{1}{2}$.

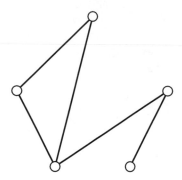

Figure 6.1: An example of a randomly generated graph based on the $G(n, p)$ model. Here $n = 5$ and p was chosen to be $\frac{1}{2}$.

The question we are interested in here is the probability that there exists an isolated node in the graph [11,12]. An isolated node is a node that is not connected to any other nodes in the graph. In a wireless networking context, an isolated node is a node that cannot communicate with any other node in the network.

Example 6.17. Let B_n be the event that a graph randomly generated according to $G(n, p)$ model has at least one isolated node. Show that

$$P(B_n) \leq n(1-p)^{n-1}.$$

And conclude that for any $\epsilon > 0$, if $p = p_n = (1 + \epsilon)\frac{\ln(n)}{n}$, then

$$\lim_{n \to \infty} P(B_n) = 0.$$

Solution: There are n nodes in the network. Let's call them Node 1, Node 2,..., Node n. Let A_i be the event that the ith node is isolated. Then we have

$$B_n = \bigcup_{i=1}^{n} A_i.$$

Thus, using the union bound we conclude that

$$P(B_n) = P(\bigcup_{i=1}^{n} A_i) \leq \sum_{i=1}^{n} P(A_i).$$

By symmetry, for all i, j, we have $P(A_i) = P(A_j)$, so

$$P(B_n) \leq nP(A_1).$$

Thus, we only need to find $P(A_1)$. The event A_1 occurs if Node 1 is not connected to any of the other $n - 1$ nodes. Since the connections are independent, we conclude that

$$P(A_1) = (1-p)^{n-1}.$$

Therefore, we obtain

$$P(B_n) \leq n(1-p)^{n-1},$$

which is the desired result. To prove the limit result, we use

$$\lim_{x \to \infty} \left(1 + \frac{c}{x}\right)^x = e^c, \qquad \text{for any constant } c \in \mathbb{R}.$$

So, we obtain

$$\lim_{n \to \infty} P(B_n) \leq \lim_{n \to \infty} n(1 - p_n)^{n-1}$$

$$= \lim_{n \to \infty} n\left[1 - (1+\epsilon)\frac{\ln n}{n}\right]^{n-1}$$

$$= \lim_{n \to \infty} n\left[1 - \frac{1+\epsilon}{\frac{n}{\ln n}}\right]^{n-1}$$

$$= \lim_{n \to \infty} n\left(\left[1 - \frac{1+\epsilon}{\frac{n}{\ln n}}\right]^{\frac{n}{\ln n}}\right)^{\frac{(n-1)\ln n}{n}}$$

$$= \lim_{n \to \infty} n e^{-(1+\epsilon)\ln n}$$

$$= \lim_{n \to \infty} \frac{1}{n^\epsilon}$$

$$= 0.$$

But since $P(B_n) \geq 0$, we conclude

$$\lim_{n \to \infty} P(B_n) = 0.$$

It is an interesting exercise to calculate $P(B_n)$ exactly using the inclusion-exclusion principle:

$$P\left(\bigcup_{i=1}^{n} A_i\right) = \sum_{i=1}^{n} P(A_i) - \sum_{i<j} P(A_i \cap A_j)$$

$$+ \sum_{i<j<k} P(A_i \cap A_j \cap A_k) - \cdots + (-1)^{n-1} P\left(\bigcap_{i=1}^{n} A_i\right).$$

In fact, the union bound states that the probability of union of some events is smaller than the first term in the inclusion-exclusion formula. We can in fact extend the union bound to obtain lower and upper bounds on the probability of union of events. These bounds are known as **Bonferroni inequalities** [13]. The idea is very simple. Start writing the inclusion-exclusion formula. If you stop at the first term, you obtain an upper bound on the probability of union. If you stop at the second term, you obtain a lower bound. If you stop at the third term, you obtain an upper bound, etc. So in general, if you write an odd number of terms, you get an upper bound and if you write an even number of terms, you get a lower bound.

Generalization of the Union Bound: Bonferroni inequalities

For any events $A_1, A_2, ..., A_n$, we have

$$P\left(\bigcup_{i=1}^{n} A_i\right) \leq \sum_{i=1}^{n} P(A_i)$$

$$P\left(\bigcup_{i=1}^{n} A_i\right) \geq \sum_{i=1}^{n} P(A_i) - \sum_{i<j} P(A_i \cap A_j)$$

$$P\left(\bigcup_{i=1}^{n} A_i\right) \leq \sum_{i=1}^{n} P(A_i) - \sum_{i<j} P(A_i \cap A_j) + \sum_{i<j<k} P(A_i \cap A_j \cap A_k)$$

$$\cdot$$
$$\cdot$$
$$\cdot$$

Example 6.18. Let B_n be the event that a graph randomly generated according to $G(n, p)$ model has at least one isolated node. Show that

$$P(B_n) \geq n(1 - p)^{n-1} - \binom{n}{2}(1 - p)^{2n-3}.$$

Solution: Similar to Example 6.17, let A_i be the event that the ith node is isolated. Then we have

$$B_n = \bigcup_{i=1}^{n} A_i.$$

Thus, using two terms in the inclusion-exclusion principle, we obtain

$$P(B_n) = P(\bigcup_{i=1}^{n} A_i) \geq \sum_{i=1}^{n} P(A_i) - \sum_{i<j} P(A_i \cap A_j).$$

By symmetry, we obtain

$$\sum_{i=1}^{n} P(A_i) = nP(A_1),$$

$$\sum_{i<j} P(A_i \cap A_j) = \binom{n}{2} P(A_1 \cap A_2).$$

Thus, we conclude

$$P(B_n) \geq nP(A_1) - \binom{n}{2} P(A_1 \cap A_2).$$

In Example 6.17 we found

$$P(A_1) = (1-p)^{n-1}.$$

Similarly, we obtain

$$P(A_1 \cap A_2) = (1-p)^{2(n-2)+1} = (1-p)^{2n-3}.$$

The reason for this is that $A_1 \cap A_2$ is the event that Nodes 1 and 2 are isolated. There are $2(n-2)$ potential edges from the rest of the graph to Nodes 1 and 2, and there is also a potential edge from Node 1 to Node 2. These edges exist independently from each other and with probability p_n. We conclude

$$P(B_n) \geq n(1-p)^{n-1} - \binom{n}{2}(1-p)^{2n-3},$$

which is the desired result.

Expected Value of the Number of Events: It is interesting to note that the union bound formula is also equal to the expected value of the number of occurred events. To see this, let $A_1, A_2, ..., A_n$ be any events. Define the indicator random variables $X_1, X_2,...,X_n$ as

$$X_i = \begin{cases} 1 & \text{if } A_i \text{ occurs} \\ 0 & \text{otherwise} \end{cases}$$

If we define $X = X_1 + X_2 + X_3 + ... + X_n$, then X shows the number of events that actually occur. We then have

$$\begin{aligned} EX &= EX_1 + EX_2 + EX_3 + ... + EX_n \quad \text{by linearity of expectation} \\ &= P(A_1) + P(A_2) + ... + P(A_n), \end{aligned}$$

which is indeed the righthand-side of the union bound. For example, from this we can conclude that the expected number of isolated nodes in a graph randomly generated according to $G(n,p)$ is equal to

$$EX = n(1-p)^{n-1}.$$

6.2.2   Markov and Chebyshev Inequalities

Let X be any positive continuous random variable, we can write

$$
\begin{aligned}
EX &= \int_{-\infty}^{\infty} x f_X(x)dx \\
&= \int_0^{\infty} x f_X(x)dx && \text{(since } X \text{ is positive-valued)} \\
&\geq \int_a^{\infty} x f_X(x)dx && \text{(for any } a > 0) \\
&\geq \int_a^{\infty} a f_X(x)dx && \text{(since } x > a \text{ in the integrated region)} \\
&= a \int_a^{\infty} f_X(x)dx \\
&= aP(X \geq a).
\end{aligned}
$$

Thus, we conclude

$$
P(X \geq a) \leq \frac{EX}{a}, \qquad \text{for any } a > 0.
$$

We can prove the above inequality for discrete or mixed random variables similarly (using the generalized PDF), so we have the following result, called **Markov's inequality**.

Markov's Inequality

If X is any nonnegative random variable, then

$$
P(X \geq a) \leq \frac{EX}{a}, \qquad \text{for any } a > 0.
$$

Example 6.19. Prove the union bound using Markov's inequality.

Solution: Similar to the discussion in the previous section, let $A_1, A_2, ..., A_n$ be any events and X be the number events A_i that occur. We saw that

$$
EX = P(A_1) + P(A_2) + ... + P(A_n) = \sum_{i=1}^{n} P(A_i).
$$

Since X is a nonnegative random variable, we can apply Markov's inequality. Choosing $a = 1$, we have

$$
P(X \geq 1) \leq EX = \sum_{i=1}^{n} P(A_i).
$$

But note that $P(X \geq 1) = P\left(\bigcup_{i=1}^{n} A_i\right)$.

Example 6.20. Let $X \sim Binomial(n, p)$. Using Markov's inequality, find an upper bound on $P(X \geq \alpha n)$, where $p < \alpha < 1$. Evaluate the bound for $p = \frac{1}{2}$ and $\alpha = \frac{3}{4}$.

Solution: Note that X is a nonnegative random variable and $EX = np$. Applying Markov's inequality, we obtain

$$P(X \geq \alpha n) \leq \frac{EX}{\alpha n} = \frac{pn}{\alpha n} = \frac{p}{\alpha}.$$

For $p = \frac{1}{2}$ and $\alpha = \frac{3}{4}$, we obtain

$$P\left(X \geq \frac{3n}{4}\right) \leq \frac{2}{3}.$$

Chebyshev's Inequality:

Let X be any random variable. If you define $Y = (X - EX)^2$, then Y is a nonnegative random variable, so we can apply Markov's inequality to Y. In particular, for any positive real number b, we have

$$P(Y \geq b^2) \leq \frac{EY}{b^2}.$$

But note that

$$EY = E(X - EX)^2 = \text{Var}(X)$$
$$P(Y \geq b^2) = P\left((X - EX)^2 \geq b^2\right) = P\left(|X - EX| \geq b\right).$$

Thus, we conclude that

$$P\left(|X - EX| \geq b\right) \leq \frac{\text{Var}(X)}{b^2}.$$

This is **Chebyshev's inequality**.

Chebyshev's Inequality

If X is any random variable, then for any $b > 0$ we have

$$P\left(|X - EX| \geq b\right) \leq \frac{\text{Var}(X)}{b^2}.$$

Chebyshev's inequality states that the difference between X and EX is somehow limited by $\text{Var}(X)$. This is intuitively expected as variance shows on average how far we are from the mean.

Example 6.21. Let $X \sim Binomial(n, p)$. Using Chebyshev's inequality, find an upper bound on $P(X \geq \alpha n)$, where $p < \alpha < 1$. Evaluate the bound for $p = \frac{1}{2}$ and $\alpha = \frac{3}{4}$.

Solution: One way to obtain a bound is to write

$$P(X \geq \alpha n) = P(X - np \geq \alpha n - np)$$
$$\leq P\big(|X - np| \geq n\alpha - np\big)$$
$$\leq \frac{\text{Var}(X)}{(n\alpha - np)^2}$$
$$= \frac{p(1 - p)}{n(\alpha - p)^2}.$$

For $p = \frac{1}{2}$ and $\alpha = \frac{3}{4}$, we obtain

$$P(X \geq \frac{3n}{4}) \leq \frac{4}{n}.$$

6.2.3   Chernoff Bounds

If X is a random variable, then for any $a \in \mathbb{R}$, we can write

$$P(X \geq a) = P(e^{sX} \geq e^{sa}), \qquad \text{for } s > 0,$$
$$P(X \leq a) = P(e^{sX} \geq e^{sa}), \qquad \text{for } s < 0.$$

Now, note that e^{sX} is always a positive random variable for all $s \in \mathbb{R}$. Thus, we can apply Markov's inequality. So for $s > 0$, we can write

$$P(X \geq a) = P(e^{sX} \geq e^{sa})$$
$$\leq \frac{E[e^{sX}]}{e^{sa}}, \qquad \text{by Markov's inequality.}$$

Similarly, for $s < 0$, we can write

$$P(X \leq a) = P(e^{sX} \geq e^{sa})$$
$$\leq \frac{E[e^{sX}]}{e^{sa}}.$$

Note that $E[e^{sX}]$ is in fact the moment generating function, $M_X(s)$. Thus, we conclude

---

Chernoff Bounds:

$$P(X \geq a) \leq e^{-sa} M_X(s), \qquad \text{for all } s > 0,$$
$$P(X \leq a) \leq e^{-sa} M_X(s), \qquad \text{for all } s < 0$$

Since Chernoff bounds are valid for all values of $s > 0$ and $s < 0$, we can choose s in a way to obtain the best bound, that is we can write

$$P(X \geq a) \leq \min_{s>0} e^{-sa} M_X(s),$$
$$P(X \leq a) \leq \min_{s<0} e^{-sa} M_X(s).$$

Let us look at an example to see how we can use Chernoff bounds.

Example 6.22. Let $X \sim Binomial(n, p)$. Using Chernoff bounds, find an upper bound on $P(X \geq \alpha n)$, where $p < \alpha < 1$. Evaluate the bound for $p = \frac{1}{2}$ and $\alpha = \frac{3}{4}$.

Solution: For $X \sim Binomial(n, p)$, we have

$$M_X(s) = (pe^s + q)^n, \qquad \text{where } q = 1 - p.$$

Thus, the Chernoff bound for $P(X \geq a)$ can be written as

$$P(X \geq \alpha n) \leq \min_{s > 0} e^{-sa} M_X(s)$$
$$= \min_{s > 0} e^{-sa} (pe^s + q)^n. \tag{6.3}$$

To find the minimizing value of s, we can write

$$\frac{d}{ds} e^{-sa} (pe^s + q)^n = 0,$$

which results in

$$e^s = \frac{aq}{np(1 - \alpha)}.$$

By using this value of s in Equation 6.3 and some algebra, we obtain

$$P(X \geq \alpha n) \leq \left(\frac{1 - p}{1 - \alpha}\right)^{(1 - \alpha)n} \left(\frac{p}{\alpha}\right)^{\alpha n}.$$

For $p = \frac{1}{2}$ and $\alpha = \frac{3}{4}$, we obtain

$$P(X \geq \frac{3}{4}n) \leq \left(\frac{16}{27}\right)^{\frac{n}{4}}.$$

Comparison between Markov, Chebyshev, and Chernoff Bounds:

Above, we found upper bounds on $P(X \geq \alpha n)$ for $X \sim Binomial(n, p)$. It is interesting to compare them. Here are the results that we obtain for $p = \frac{1}{4}$ and $\alpha = \frac{3}{4}$:

$$P(X \geq \frac{3n}{4}) \leq \frac{2}{3} \qquad \text{Markov,}$$
$$P(X \geq \frac{3n}{4}) \leq \frac{4}{n} \qquad \text{Chebyshev,}$$
$$P(X \geq \frac{3n}{4}) \leq \left(\frac{16}{27}\right)^{\frac{n}{4}} \qquad \text{Chernoff.}$$

The bound given by Markov is the "weakest" one. It is constant and does not change as n increases. The bound given by Chebyshev's inequality is "stronger" than the one given by Markov's inequality. In particular, note that $\frac{4}{n}$ goes to zero as n goes to infinity. The strongest bound is the Chernoff bound. It goes to zero exponentially fast.

6.2.4 Cauchy-Schwarz Inequality

You might have seen the **Cauchy-Schwarz inequality** in your linear algebra course. The same inequality is valid for random variables. Let us state and prove the Cauchy-Schwarz inequality for random variables.

Cauchy-Schwarz Inequality

For any two random variables X and Y, we have

$$|EXY| \leq \sqrt{E[X^2]E[Y^2]},$$

where equality holds if and only if $X = \alpha Y$, for some constant $\alpha \in \mathbb{R}$.

You can prove the Cauchy-Schwarz inequality with the same methods that we used to prove $|\rho(X,Y)| \leq 1$ in Section 5.3.1. Here, we provide another proof. Define the random variable $W = (X - \alpha Y)^2$. Clearly, W is a nonnegative random variable for any value of $\alpha \in \mathbb{R}$. Thus, we obtain

$$\begin{aligned}
0 \leq EW = E(X - \alpha Y)^2 \\
= E[X^2 - 2\alpha XY + \alpha^2 Y^2] \\
= E[X^2] - 2\alpha E[XY] + \alpha^2 E[Y^2].
\end{aligned}$$

So, if we let $f(\alpha) = E[X^2] - 2\alpha E[XY] + \alpha^2 E[Y^2]$, then we know that $f(\alpha) \geq 0$, for all $\alpha \in \mathbb{R}$. Moreover, if $f(\alpha) = 0$ for some α, then we have $EW = E(X - \alpha Y)^2 = 0$, which essentially means $X = \alpha Y$ with probability one. To prove the Cauchy-Schwarz inequality, choose $\alpha = \frac{EXY}{EY^2}$. We obtain

$$\begin{aligned}
0 \leq E[X^2] - 2\alpha E[XY] + \alpha^2 E[Y^2] \\
= E[X^2] - 2\frac{EXY}{EY^2}E[XY] + \frac{(EXY)^2}{(EY^2)^2}E[Y^2] \\
= E[X^2] - \frac{(E[XY])^2}{EY^2}.
\end{aligned}$$

Thus, we conclude

$$(E[XY])^2 \leq E[X^2]E[Y^2],$$

which implies

$$|EXY| \leq \sqrt{E[X^2]E[Y^2]}.$$

Also, if $|EXY| = \sqrt{E[X^2]E[Y^2]}$, we conclude that $f(\frac{EXY}{EY^2}) = 0$, which implies $X = \frac{EXY}{EY^2}Y$ with probability one.

Example 6.23. Using the Cauchy-Schwarz inequality, show that for any two random variables X and Y

$$|\rho(X,Y)| \leq 1.$$

Also, $|\rho(X,Y)| = 1$ if and only if $Y = aX + b$ for some constants $a, b \in \mathbb{R}$.

Solution: Let

$$U = \frac{X - EX}{\sigma_X}, \qquad V = \frac{Y - EY}{\sigma_Y}.$$

Then $EU = EV = 0$, and $\text{Var}(U) = \text{Var}(V) = 1$. Using the Cauchy-Schwarz inequality for U and V, we obtain

$$|EUV| \leq \sqrt{E[U^2]E[V^2]} = 1.$$

But note that $EUV = \rho(X, Y)$, thus we conclude

$$|\rho(X, Y)| \leq 1,$$

where equality holds if and only if $V = \alpha U$ for some constant $\alpha \in \mathbb{R}$. That is

$$\frac{Y - EY}{\sigma_Y} = \alpha \frac{X - EX}{\sigma_X},$$

which implies

$$Y = \frac{\alpha \sigma_Y}{\sigma_X} X + \left(EY - \frac{\alpha \sigma_Y}{\sigma_X} EX \right).$$

In the Solved Problems section, we provide a generalization of the Cauchy-Schwrarz inequality, called *Hölder's inequality*.

6.2.5 Jensen's Inequality

Remember that variance of every random variable X is a positive value, i.e.,

$$\text{Var}(X) = EX^2 - (EX)^2 \geq 0.$$

Thus,

$$EX^2 \geq (EX)^2.$$

If we define $g(x) = x^2$, we can write the above inequality as

$$E[g(X)] \geq g(E[X]).$$

The function $g(x) = x^2$ is an example of **convex** function. **Jensen's inequality** states that, for any convex function g, we have $E[g(X)] \geq g(E[X])$. So what is a convex function? Figure 6.2 depicts a convex function. A function is convex if, when you pick any two points on the graph of the function and draw a line segment between the two points, the entire segment lies above the graph. On the other hand, if the line segment always lies below the graph, the function is said to be **concave**. In other words, $g(x)$ is convex if and only if $-g(x)$ is concave.

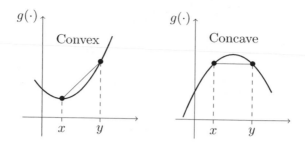

Figure 6.2: Pictorial representation of a convex function and a concave function.

We can state the definition for convex and concave functions in the following way:

Definition 6.3. Consider a function $g : I \to \mathbb{R}$, where I is an interval in $\mathbb{R}$. We say that g is a **convex** function if for any two points x and y in I and any $\alpha \in [0, 1]$, we have

$$g(\alpha x + (1 - \alpha)y) \leq \alpha g(x) + (1 - \alpha)g(y).$$

We say that g is **concave** if

$$g(\alpha x + (1 - \alpha)y) \geq \alpha g(x) + (1 - \alpha)g(y).$$

Note that in the above definition the term $\alpha x + (1 - \alpha)y$ is the weighted average of x and y. Also, $\alpha g(x) + (1 - \alpha)g(y)$ is the weighted average of $g(x)$ and $g(y)$. More generally, for a convex function $g : I \to \mathbb{R}$, and $x_1, x_2,...,x_n$ in I and nonnegative real numbers α_i such that $\alpha_1 + \alpha_2 + ... + \alpha_n = 1$, we have

$$g(\alpha_1 x_1 + \alpha_2 x_2 + ... + \alpha_n x_n) \leq \alpha_1 g(x_1) + \alpha_2 g(x_2) + ... + \alpha_n g(x_n) \tag{6.4}$$

If $n = 2$, the above statement is the definition of convex functions. You can extend it to higher values of n by induction.

Now, consider a discrete random variable X with n possible values $x_1, x_2,...,x_n$. In Equation 6.4, we can choose $\alpha_i = P(X = x_i) = P_X(x_i)$. Then, the left-hand side of 6.4 becomes $g(EX)$ and the right-hand side becomes $E[g(X)]$ (by LOTUS). So we can prove the Jensen's inequality in this case. Using limiting arguments, this result can be extended to other types of random variables.

Jensen's Inequality:

If $g(x)$ is a convex function on R_X, and $E[g(X)]$ and $g(E[X])$ are finite, then

$$E[g(X)] \geq g(E[X]).$$

To use Jensen's inequality, we need to determine if a function g is convex. A useful method is the second derivative.

A twice-differentiable function $g : I \to \mathbb{R}$ is convex if and only if $g''(x) \geq 0$ for all $x \in I$.

For example, if $g(x) = x^2$, then $g''(x) = 2 \geq 0$, thus $g(x) = x^2$ is convex over $\mathbb{R}$.

Example 6.24. Let X be a positive random variable. Compare $E[X^a]$ with $(E[X])^a$ for all values of $a \in \mathbb{R}$.

Solution: First note

$$E[X^a] = 1 = (E[X])^a, \qquad \text{if } a = 0,$$
$$E[X^a] = EX = (E[X])^a, \qquad \text{if } a = 1.$$

So let's assume $a \neq 0, 1$. Letting $g(x) = x^a$, we have

$$g''(x) = a(a-1)x^{a-2}.$$

On $(0, \infty)$, we can say $g''(x)$ is positive, if $a < 0$ or $a > 1$. It is negative, if $0 < a < 1$. Therefore we conclude that $g(x)$ is convex, if $a < 0$ or $a > 1$. It is concave, if $0 < a < 1$. Using Jensen's inequality we conclude

$$E[X^a] \geq (E[X])^a, \qquad \text{if } a < 0 \text{ or } a > 1,$$
$$E[X^a] \leq (E[X])^a, \qquad \text{if } 0 < a < 1.$$

6.2.6 Solved Problems

1. Your friend tells you that he had four job interviews last week. He says that based on how the interviews went, he thinks he has a 20% chance of receiving an offer from each of the companies he interviewed with. Nevertheless, since he interviewed with four companies, he is 90% sure that he will receive at least one offer. Is he right?

   Solution:

Let A_i be the event that your friend receives an offer from the ith company, i=1,2,3,4. Then, by the union bound,

$$P\left(\bigcup_{i=1}^{4} A_i\right) \leq \sum P(A_i)$$

$$= 0.2 + 0.2 + 0.2 + 0.2$$

$$= 0.8$$

Thus the probability of receiving at least one offer is less than or equal to 80%.

2. An isolated edge in a network is an edge that connects two nodes in the network such that neither of the two nodes is connected to any other nodes in the network. Let C_n be the event that a graph randomly generated according to $G(n, p)$ model has at least one isolated edge.

   (a) Show that

   $$P(C_n) \leq \binom{n}{2} p(1-p)^{2(n-2)}.$$

   (b) Show that, for any constant $b > \frac{1}{2}$, if $p = p_n = b\frac{\ln(n)}{n}$ then

   $$\lim_{n \to \infty} P(C_n) = 0.$$

Solution:

There are $\binom{n}{2}$ possible edges in the graph. Let E_i be the event that the ith edge is an isolated edge, then

$$P(E_i) = p(1-p)^{2(n-2)},$$

where p in the above equation is the probability that the ith edge is present and $(1-p)^{2(n-2)}$ is the probability that no other nodes are connected to this edge. By the union bound, we have

$$P(C_n) = P\left(\bigcup E_i\right)$$

$$\leq \sum_i P(E_i)$$

$$= \binom{n}{2} p(1-p)^{2(n-2)},$$

which is the desired result. Now, let $p = b\frac{\ln n}{n}$, where $b > \frac{1}{2}$.

Here, it is convenient to use the following inequality:

$$1 - x \leq e^{-x}, \quad \text{for all} \quad x \in \mathbb{R}.$$

You can prove it by differentiating $f(x) = e^{-x} + x - 1$, and showing that the minimum occurs at $x = 0$.

Now, we can write

$$P(C_n) \leq \binom{n}{2} p(1-p)^{2(n-2)}$$

$$= \frac{n(n-1)}{2} \frac{b \ln n}{n} (1-p)^{2(n-2)}$$

$$\leq \frac{(n-1)b \ln n}{2} e^{-2p(n-2)} \quad (\text{using} \quad 1 - x \leq e^{-x})$$

$$= \frac{(n-1)}{2} b \ln n \, e^{-2 \frac{b \ln n}{n}(n-2)}.$$

Thus,

$$\lim_{n \to \infty} P(C_n) \leq \lim_{n \to \infty} \frac{(n-1)}{2} b \ln n \, e^{-2 \frac{b \ln n}{n}(n-2)}$$

$$= \lim_{n \to \infty} \frac{(n-1)}{2} b \ln n \, n^{-2b}$$

$$= \frac{b}{2} \lim_{n \to \infty} (n^{1-2b} \, \ln n)$$

$$= 0 \quad (\text{since} \quad b > \frac{1}{2}).$$

3. Let $X \sim Exponential(\lambda)$. Using Markov's inequality find an upper bound for $P(X \geq a)$, where $a > 0$. Compare the upper bound with the actual value of $P(X \geq a)$.

   Solution:

   If $X \sim Exponential(\lambda)$, then $EX = \frac{1}{\lambda}$, using Markov's inequality

   $$P(X \geq a) \leq \frac{EX}{a} = \frac{1}{\lambda a}.$$

   The actual value of $P(X \geq a)$ is $e^{-\lambda a}$, and we always have $\frac{1}{\lambda a} \geq e^{-\lambda a}$.

4. Let $X \sim Exponential(\lambda)$. Using Chebyshev's inequality find an upper bound for $P(|X - EX| \geq b)$, where $b > 0$.

   Solution:

   We have $EX = \frac{1}{\lambda}$ and $VarX = \frac{1}{\lambda^2}$. Using Chebyshev's inequality, we have

$$P\left(|X - EX| \geq b\right) \leq \frac{\mathrm{Var}(X)}{b^2}$$

$$= \frac{1}{\lambda^2 b^2}.$$

5. Let $X \sim Exponential(\lambda)$. Using Chernoff bounds find an upper bound for $P(X \geq a)$, where $a > EX$. Compare the upper bound with the actual value of $P(X \geq a)$.

Solution:

If $X \sim Exponential(\lambda)$, then

$$M_X(s) = \frac{\lambda}{\lambda - s}, \quad \text{for} \quad s < \lambda.$$

Using Chernoff bounds, we have

$$P\left(X \geq a\right) \leq \min_{s>0} \left[e^{-sa} M_X(s)\right]$$

$$= \min_{s>0} \left[e^{-sa} \frac{\lambda}{\lambda - s}\right].$$

If $f(s) = e^{-sa} \frac{\lambda}{\lambda - s}$, to find $\min_{s>0} f(s)$ we write

$$\frac{d}{ds} f(s) = 0.$$

Therefore,

$$s^* = \lambda - \frac{1}{a}.$$

Note since $a > EX = \frac{1}{\lambda}$, then $\lambda - \frac{1}{a} > 0$. Thus,

$$P\left(X \geq a\right) \leq e^{-s^* a} \frac{\lambda}{\lambda - s^*} = a\lambda e^{1 - \lambda a}.$$

The real value of $P\left(X \geq a\right)$ is $e^{-\lambda a}$ and we have $e^{-\lambda a} \leq a\lambda e^{1 - \lambda a}$, or equivalently, $a\lambda e \geq 1$, which is true since $a > \frac{1}{\lambda}$.

6. Let X and Y be two random variables with $EX = 1, \mathrm{Var}(X) = 4$, and $EY = 2, \mathrm{Var}(Y) = 1$. Find the maximum possible value for $E[XY]$.

Solution:

Using $\rho(X, Y) \le 1$ and $\rho(X, Y) = \frac{\text{Cov}(X,Y)}{\sigma_X \sigma_Y}$, we conclude

$$\frac{EXY - EXEY}{\sigma_X \sigma_Y} \le 1.$$

Thus

$$EXY \le \sigma_X \sigma_Y + EXEY$$
$$= 2 \times 1 + 2 \times 1$$
$$= 4.$$

In fact, we can achieve $EXY = 4$, if we choose $Y = aX + b$.

$$Y = aX + b \quad \Rightarrow \quad \begin{cases} 2 = a + b \\ \\ 1 = (a^2)(4) \end{cases}$$

Solving for a and b, we obtain

$$a = \frac{1}{2}, \quad b = \frac{3}{2}.$$

Note that if you use the Cauchy-Schwarz inequality directly, you obtain

$$|EXY|^2 \le EX^2 \cdot EY^2$$
$$= 5 \times 5.$$

Thus
$$EXY \le 5.$$

But $EXY = 5$ cannot be achieved because equality in the Cauchy-Schwarz is obtained only when $Y = \alpha X$. But here this is not possible.

7. **(Hölder's Inequality)** Prove

$$E\left[|XY|\right] \le E\left[|X|^p\right]^{\frac{1}{p}} E\left[|Y|^q\right]^{\frac{1}{q}},$$

where $1 < p, q < \infty$ and $\frac{1}{p} + \frac{1}{q} = 1$. Note that, for $p = q = \frac{1}{2}$, Hölder's ineqality becomes the Cauchy-Schwarz inequality. *Hint:* You can use Young's inequality [4] which states that for nonnegative real numbers α and β and integers p and q such that $1 < p, q < \infty$ and $\frac{1}{p} + \frac{1}{q} = 1$, we have

$$\alpha\beta \le \frac{\alpha^p}{p} + \frac{\beta^q}{q},$$

with quality only if $\alpha^p = \beta^q$.

Solution:

Using Young's inequality, we conclude that for random variables U and V we have

$$E|UV| \le \frac{E|U|^p}{p} + \frac{E|V|^q}{q}.$$

Choose $U = \dfrac{|X|}{(E|X|^p)^{\frac{1}{p}}}$ and $V = \dfrac{|Y|}{(E|Y|^q)^{\frac{1}{q}}}$. We obtain

$$\frac{E|XY|}{(E|X|^p)^{\frac{1}{p}}(E|Y|^q)^{\frac{1}{q}}} \le \frac{E|X|^p}{pE|X|^p} + \frac{E|Y|^q}{qE|Y|^q}$$

$$= \frac{1}{p} + \frac{1}{q}$$

$$= 1.$$

8. Show that if $h : \mathbb{R} \mapsto \mathbb{R}$ is convex and non-decreasing, and $g : \mathbb{R} \mapsto \mathbb{R}$ is convex, then $h(g(x))$ is a convex function.

Solution:

Since g is convex, we have

$$g(\alpha x + (1 - \alpha)y) \le \alpha g(x) + (1 - \alpha)g(y), \quad \text{for all } \alpha \in [0, 1].$$

Therefore, we have

$$h(g(\alpha x + (1 - \alpha)y)) \le h(\alpha g(x) + (1 - \alpha)g(y)) \quad \text{(h is non-decreasing)}$$
$$\le \alpha h(g(x)) + (1 - \alpha)h(g(y)) \quad \text{(h is convex)}.$$

9. Let X be a positive random variable with $EX = 10$. What can you say about the following quantities?

   (a) $E\left[\frac{1}{X+1}\right]$

   (b) $E\left[e^{\frac{1}{X+1}}\right]$

   (c) $E[\ln \sqrt{X}]$

Solution:

(a)

$$g(x) = \frac{1}{x+1},$$

$$g''(x) = \frac{2}{(1+x)^3} > 0, \quad \text{for} \quad x > 0.$$

Thus g is convex on $(0, \infty)$. We conclude

$$E\left[\frac{1}{X+1}\right] \geq \frac{1}{1+EX} \quad \text{(Jensen's inequality)}$$

$$= \frac{1}{1+10}$$

$$= \frac{1}{11}.$$

(b) If we let $h(x) = e^x, g(x) = \frac{1}{1+x}$ then h is convex and non-decreasing, and g is convex; thus by Problem 8, $e^{\frac{1}{x+1}}$ is a convex function. Therefore,

$$E\left[e^{\frac{1}{1+X}}\right] \geq e^{\frac{1}{1+EX}} \quad \text{(by Jensen's inequality)}$$

$$= e^{\frac{1}{11}}.$$

(c) If $g(x) = \ln\sqrt{x} = \frac{1}{2}\ln x$, then $g'(x) = \frac{1}{2x}$ for $x > 0$, and $g''(x) = -\frac{1}{2x^2}$. Thus g is concave on $(0, \infty)$. We conclude

$$E\left[\ln\sqrt{X}\right] = E\left[\frac{1}{2}\ln X\right]$$

$$\leq \frac{1}{2}\ln EX \quad \text{(by Jensen's inequality)}$$

$$= \frac{1}{2}\ln 10.$$

6.3   End of Chapter Problems

1. Let X, Y and Z be three jointly continuous random variables with joint PDF

$$f_{XYZ}(x,y,z) = \begin{cases} x+y & 0 \le x,y,z \le 1 \\ 0 & \text{otherwise} \end{cases}$$

   (a) Find the joint PDF of X and Y.
   (b) Find the marginal PDF of X.
   (c) Find the conditional PDF of $f_{XY|Z}(x,y|z)$ using

$$f_{XY|Z}(x,y|z) = \frac{f_{XYZ}(x,y,z)}{f_Z(z)}.$$

   (d) Are X and Y independent of Z?

2. Suppose that X, Y, and Z are three independent random variables. If $X, Y \sim N(0,1)$ and $Z \sim Exponential(1)$, find

   (a) $E[XY|Z=1]$.
   (b) $E[X^2Y^2Z^2|Z=1]$.

3. Let X, Y, and Z be three independent $N(1,1)$ random variables. Find $E[XY|Y+Z=1]$.

4. Let $X_1, X_2, \cdots, X_n$ be i.i.d. random variables, where $X_i \sim Bernoulli(p)$. Define

$$Y_1 = X_1 X_2,$$
$$Y_2 = X_2 X_3,$$
$$\vdots$$
$$Y_{n-1} = X_{n-1} X_n,$$
$$Y_n = X_n X_1.$$

   If $Y = Y_1 + Y_2 + \cdots + Y_n$, find

   (a) $E[Y]$.
   (b) $Var(Y)$.

5. In this problem, our goal is to find the variance of the hypergeometric distribution. Let's remember the random experiment behind the hypergeometric distribution. You have a bag that contains b blue marbles and r red marbles. You choose $k \le b+r$ marbles at random (without replacement) and let X be the number of blue marbles in your sample. Then $X \sim Hypergeometric(b,r,k)$. Now let us define the indicator random variables X_i as follows.

$$X_i = \begin{cases} 1 & \text{if the } i\text{th chosen marble is blue} \\ 0 & \text{otherwise} \end{cases}$$

   Then, we can write

$$X = X_1 + X_2 + \cdots + X_k$$

   Using the above equation, show

(a) $EX = \frac{kb}{b+r}$.

(b) $\text{Var}(X) = \frac{kbr}{(b+r)^2} \frac{b+r-k}{b+r-1}$.

6. (MGF of the geometric distribution) If $X \sim Geometric(p)$, find the MGF of X.

7. If $M_X(s) = \frac{1}{4} + \frac{1}{2}e^s + \frac{1}{4}e^{2s}$, find EX and $\text{Var}(X)$.

8. Using MGFs show that if $X \sim N(\mu_X, \sigma_X^2)$ and $Y \sim N(\mu_Y, \sigma_Y^2)$ are independent, then

$$X + Y \sim N\left(\mu_X + \mu_Y, \sigma_X^2 + \sigma_Y^2\right).$$

9. (MGF of the Laplace distribution) Let X be a continuous random variable with the following PDF

$$f_X(x) = \frac{\lambda}{2}e^{-\lambda|x|}.$$

Find the MGF of X, $M_X(s)$.

10. (MGF of Gamma distribution) Remember that a continuous random variable X is said to have a *Gamma* distribution with parameters $\alpha > 0$ and $\lambda > 0$, shown as $X \sim Gamma(\alpha, \lambda)$, if its PDF is given by

$$f_X(x) = \begin{cases} \frac{\lambda^\alpha x^{\alpha-1}e^{-\lambda x}}{\Gamma(\alpha)} & x > 0 \\ 0 & \text{otherwise} \end{cases}$$

If $X \sim Gamma(\alpha, \lambda)$, find the MGF of X. *Hint:* Remember that $\int_0^\infty x^{\alpha-1}e^{-\lambda x}dx = \frac{\Gamma(\alpha)}{\lambda^\alpha}$, for $\alpha, \lambda > 0$.

11. Using the MGFs show that if $Y = X_1 + X_2 + \cdots + X_n$, where X_i's are independent $Exponential(\lambda)$ random variables, then $Y \sim Gamma(n, \lambda)$.

12. Let X be a random variable with characteristic function $\phi_X(\omega)$. If $Y = aX + b$, show that

$$\phi_Y(\omega) = e^{j\omega b}\phi_X(a\omega).$$

13. Let X and Y be two jointly continuous random variables with joint PDF

$$f_{X,Y}(x,y) = \begin{cases} \frac{1}{2}(3x+y) & 0 \le x, y \le 1 \\ 0 & \text{otherwise} \end{cases}$$

and let the random vector $\mathbf{U}$ be defined as

$$\mathbf{U} = \begin{bmatrix} X \\ Y \end{bmatrix}.$$

(a) Find the mean vector of $\mathbf{U}$, $E\mathbf{U}$.

(b) Find the correlation matrix of $\mathbf{U}$, $\mathbf{R_U}$.

(c) Find the covariance matrix of $\mathbf{U}$, $\mathbf{C_U}$.

14. Let $X \sim Uniform(0,1)$. Suppose that given $X = x$, Y and Z are independent and $Y|X = x \sim Uniform(0,x)$ and $Z|X = x \sim Uniform(0,2x)$. Define the random vector $\mathbf{U}$ as

$$\mathbf{U} = \begin{bmatrix} X \\ Y \\ Z \end{bmatrix}.$$

(a) Find the PDFs of Y and Z.

(b) Find the PDF of $\mathbf{U}$, $f_{\mathbf{U}}(\mathbf{u})$, by using

$$f_{\mathbf{U}}(\mathbf{u}) = f_{XYZ}(x,y,z)$$
$$= f_X(x)f_{Y|X}(y|x)f_{Z|X,Y}(z|x,y).$$

15. Let $\mathbf{X} = \begin{bmatrix} X_1 \\ X_2 \end{bmatrix}$ be a normal random vector with the following mean and covariance matrices

$$\mathbf{m} = \begin{bmatrix} 1 \\ 2 \end{bmatrix}, \qquad \mathbf{C} = \begin{bmatrix} 4 & 1 \\ 1 & 1 \end{bmatrix}.$$

Let also

$$\mathbf{A} = \begin{bmatrix} 2 & 1 \\ -1 & 1 \\ 1 & 3 \end{bmatrix}, \qquad \mathbf{b} = \begin{bmatrix} -1 \\ 0 \\ 1 \end{bmatrix}, \qquad \mathbf{Y} = \begin{bmatrix} Y_1 \\ Y_2 \\ Y_3 \end{bmatrix} = \mathbf{AX} + \mathbf{b}.$$

(a) Find $P(X_2 > 0)$.

(b) Find expected value vector of $\mathbf{Y}$, $\mathbf{m_Y} = E\mathbf{Y}$.

(c) Find the covariance matrix of $\mathbf{Y}$, $\mathbf{C_Y}$.

(d) Find $P(Y_2 \le 2)$.

16. Let $\mathbf{X} = \begin{bmatrix} X_1 \\ X_2 \\ X_3 \end{bmatrix}$ be a normal random vector with the following mean and covariance

$$\mathbf{m} = \begin{bmatrix} 1 \\ 2 \\ 0 \end{bmatrix}, \qquad \mathbf{C} = \begin{bmatrix} 9 & 1 & -1 \\ 1 & 4 & 2 \\ -1 & 2 & 4 \end{bmatrix}.$$

Find the MGF of $\mathbf{X}$ defined as

$$M_{\mathbf{X}}(s,t,r) = E\left[e^{sX_1 + tX_2 + rX_3}\right].$$

17. A system consists of 4 components in a series, so the system works properly if all of the components are functional. In other words, the system fails if and only if at least one of its components fails. Suppose the probability that the component i fails is less than or equal to $p_f = \frac{1}{100}$, for $i = 1,2,3,4$. Find an upper bound on the probability that the system fails.

18. A sensor network consists of n sensors that are distributed randomly on the unit square. Each node's location is uniform over the unit square and is independent of the locations of the other node. A node is isolated if there are no nodes that are within distance r of that node, where $0 < r < 1$.

    (a) Show that the probability that a given node is isolated is less than or equal to
    $$p_d = (1 - \frac{\pi r^2}{4})^{(n-1)}.$$

    (b) Using the union bound, find an upper bound on the probability that the sensor network contains at least one isolated node.

19. Let $X \sim Geometric(p)$. Using Markov's inequality find an upper bound for $P(X \geq a)$, for a positive integer a. Compare the upper bound with the real value of $P(X \geq a)$.

20. Let $X \sim Geometric(p)$. Using Chebyshev's inequality find an upper bound for $P(|X - EX| \geq b)$.

21. (Cantelli's inequality [16]) Let X be a random variable with $EX = 0$ and $\text{Var}(X) = \sigma^2$. We would like to prove that for any $a > 0$, we have

    $$P(X \geq a) \leq \frac{\sigma^2}{\sigma^2 + a^2}.$$

    This inequality is sometimes called the one-sided Chebyshev inequality.

    Hint: One way to show this is to use $P(X \geq a) = P(X + c \geq a + c)$ for any constant $c \in \mathbb{R}$.

22. The number of customers visiting a store during a day is a random variable with mean $EX = 100$ and variance $Var(X) = 225$.

    (a) Using Chebyshev's inequality, find an upper bound for having more than 120 or less than 80 customers in a day. That is, find an upper bound on

    $$P(X \leq 80 \text{ or } X \geq 120).$$

    (b) Using the one-sided Chebyshev inequality (Problem 21), find an upper bound for having more than 120 customers in a day.

23. Let X_i be i.i.d. and $X_i \sim Exponential(\lambda)$. Using Chernoff bounds find an upper bound for $P(X_1 + X_2 + \cdots + X_n \geq an)$, where $a > \frac{1}{\lambda}$. Show that the bound goes to zero exponentially fast as a function of n.

24. (Minkowski's inequality [17]) Prove for two random variables X and Y with finite moments, and $1 \leq p < \infty$, we have

    $$E\left[|X + Y|^p\right]^{\frac{1}{p}} \leq E\left[|X|^p\right]^{\frac{1}{p}} + E\left[|Y|^p\right]^{\frac{1}{p}}.$$

    Hint: Note that
    $$|X + Y|^p = |X + Y|^{p-1}|X + Y|$$
    $$\leq |X + Y|^{p-1}(|X| + |Y|)$$
    $$\leq |X + Y|^{p-1}|X| + |X + Y|^{p-1}|Y|.$$

Therefore

$$E|X + Y|^p \le E\left[|X + Y|^{p-1}|X|\right] + E\left[|X + Y|^{p-1}|Y|\right].$$

Now, apply Hölder's inequality.

25. Let X be a positive random variable with $EX = 10$. What can you say about the following quantities?

   (a) $E[X - X^3]$

   (b) $E[X \ln \sqrt{X}]$

   (c) $E\left[|2 - X|\right]$

26. Let X be a random variable with $EX = 1$ and $R_X = (0, 2)$. If $Y = X^3 - 6X^2$, show that $EY \le -5$.

Chapter 7

Limit Theorems and Convergence of Random Variables

In this chapter, we will discuss limit theorems and convergence modes for random variables. Limit theorems are among the most fundamental results in probability theory. We will discuss two important limit theorems in Section 7.1: *the law of large numbers (LLN)* and *the central limit theorem(CLT)*. We will also talk about the importance of these theorems as applied in practice. In Section 7.2, we will discuss the convergence of sequences of random variables.

7.1 Limit Theorems

In this section, we will discuss two important theorems in probability, *the law of large numbers(LLN)* and *the central limit theorem(CLT)*. The LLN basically states that the average of a large number of i.i.d. random variables converges to the expected value. The CLT states that, under some conditions, the sum of a large number of random variables has an approximately normal distribution.

7.1.1 Law of Large Numbers

The **law of large numbers** has a very central role in probability and statistics. It states that if you repeat an experiment independently a large number of times and average the result, what you obtain should be close to the expected value. There are two main versions of the law of large numbers. They are called the **weak** and the **strong** laws of large numbers. The difference between them is mostly theoretical. In this section, we state and prove the weak law of large numbers (WLLN). The strong law of large numbers is discussed in Section 7.2. Before discussing the WLLN, let us define the *sample mean*.

377

Definition 7.1. For i.i.d. random variables $X_1, X_2, ..., X_n$, the **sample mean**, denoted by $\overline{X}$, is defined as

$$\overline{X} = \frac{X_1 + X_2 + ... + X_n}{n}.$$

Another common notation for the sample mean is M_n. If the X_i's have CDF $F_X(x)$, we might show the sample mean by $M_n(X)$ to indicate the distribution of the X_i's.

Note that since the X_i's are random variables, the sample mean, $\overline{X} = M_n(X)$, is also a random variable. In particular, we have

$$E[\overline{X}] = \frac{EX_1 + EX_2 + ... + EX_n}{n} \qquad \text{(by linearity of expectation)}$$

$$= \frac{nEX}{n} \qquad\qquad\qquad \text{(since } EX_i = EX)$$

$$= EX.$$

Also, the variance of $\overline{X}$ is given by

$$\text{Var}(\overline{X}) = \frac{\text{Var}(X_1 + X_2 + ... + X_n)}{n^2} \qquad \text{(since Var}(aX) = a^2\text{Var}(X))$$

$$= \frac{\text{Var}(X_1) + \text{Var}(X_2) + ... + \text{Var}(X_n)}{n^2} \qquad \text{(since } X_i\text{'s are independent)}$$

$$= \frac{n\text{Var}(X)}{n^2} \qquad\qquad\qquad \text{(since Var}(X_i) = \text{Var}(X))$$

$$= \frac{\text{Var}(X)}{n}.$$

Now let us state and prove the **weak law of large numbers (WLLN)**.

The weak law of large numbers (WLLN)

Let $X_1, X_2, ..., X_n$ be i.i.d. random variables with a finite expected value $EX_i = \mu < \infty$. Then, for any $\epsilon > 0$,

$$\lim_{n \to \infty} P(|\overline{X} - \mu| \geq \epsilon) = 0.$$

Proof. The proof of the weak law of large number is easier if we assume $\text{Var}(X) = \sigma^2$ is finite.

In this case we can use Chebyshev's inequality to write

$$P(|\overline{X} - \mu| \geq \epsilon) \leq \frac{\text{Var}(\overline{X})}{\epsilon^2}$$

$$= \frac{\text{Var}(X)}{n\epsilon^2},$$

which goes to zero as $n \to \infty$. ∎

7.1.2 Central Limit Theorem (CLT)

The **central limit theorem (CLT)** is one of the most important results in probability theory. It states that, under certain conditions, the sum of a large number of random variables is approximately normal. Here, we state a version of the CLT that applies to i.i.d. random variables. Suppose that $X_1, X_2, ..., X_n$ are i.i.d. random variables with expected values $EX_i = \mu < \infty$ and variance $\text{Var}(X_i) = \sigma^2 < \infty$. Then as we saw above, the sample mean $\overline{X} = \frac{X_1 + X_2 + ... + X_n}{n}$ has mean $E\overline{X} = \mu$ and variance $\text{Var}(\overline{X}) = \frac{\sigma^2}{n}$. Thus, the normalized random variable

$$Z_n = \frac{\overline{X} - \mu}{\sigma/\sqrt{n}} = \frac{X_1 + X_2 + ... + X_n - n\mu}{\sqrt{n}\sigma}$$

has mean $EZ_n = 0$ and variance $\text{Var}(Z_n) = 1$. The central limit theorem states that the CDF of Z_n converges to the standard normal CDF.

The Central Limit Theorem (CLT)

Let $X_1, X_2, ..., X_n$ be i.i.d. random variables with expected value $EX_i = \mu < \infty$ and variance $0 < \text{Var}(X_i) = \sigma^2 < \infty$. Then, the random variable

$$Z_n = \frac{\overline{X} - \mu}{\sigma/\sqrt{n}} = \frac{X_1 + X_2 + ... + X_n - n\mu}{\sqrt{n}\sigma}$$

converges in distribution to the standard normal random variable as n goes to infinity, that is

$$\lim_{n \to \infty} P(Z_n \leq x) = \Phi(x), \qquad \text{for all } x \in \mathbb{R},$$

where $\Phi(x)$ is the standard normal CDF.

An interesting thing about the CLT is that it does not matter what the distribution of the X_i's is. The X_i's can be discrete, continuous, or mixed random variables. To get a feeling for the CLT, let us look at some examples. Let's assume that X_i's are *Bernoulli(p)*. Then

$EX_i = p$, $\text{Var}(X_i) = p(1-p)$. Also, $Y_n = X_1 + X_2 + ... + X_n$ has $Binomial(n,p)$ distribution. Thus,

$$Z_n = \frac{Y_n - np}{\sqrt{np(1-p)}},$$

where $Y_n \sim Binomial(n,p)$. Figure 7.1 shows the PMF of Z_n for different values of n. As you see, the shape of the PMF gets closer to a normal PDF curve as n increases. Here, Z_n is a discrete random variable, so mathematically speaking it has a PMF not a PDF. That is why the CLT states that the CDF (not the PDF) of Z_n converges to the standard normal CDF. Nevertheless, since PMF and PDF are conceptually similar, the figure is useful in visualizing the convergence to normal distribution.

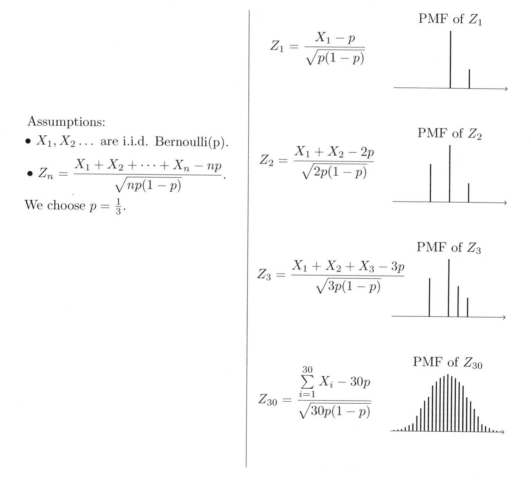

Figure 7.1: Z_n is the normalized sum of n independent $Bernoulli(p)$ random variables. The shape of its PMF, $P_{Z_n}(z)$, resembles the normal curve as n increases.

As another example, let's assume that the X_i's are $Uniform(0,1)$. Then $EX_i = \frac{1}{2}$, $\mathrm{Var}(X_i) = \frac{1}{12}$. In this case,

$$Z_n = \frac{X_1 + X_2 + \dots + X_n - \frac{n}{2}}{\sqrt{n/12}}.$$

Figure 7.2 shows the PDF of Z_n for different values of n. As you see, the shape of the PDF gets closer to the normal PDF as n increases.

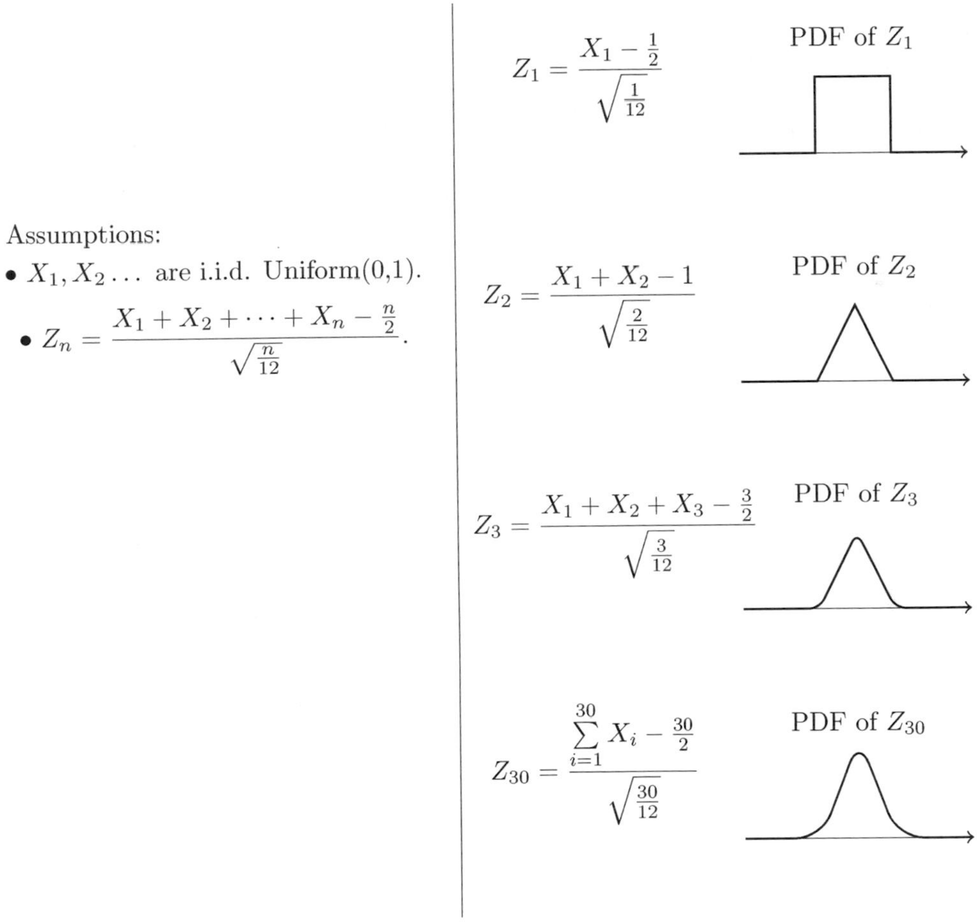

Assumptions:

- $X_1, X_2 \dots$ are i.i.d. Uniform(0,1).
- $Z_n = \dfrac{X_1 + X_2 + \dots + X_n - \frac{n}{2}}{\sqrt{\frac{n}{12}}}.$

$Z_1 = \dfrac{X_1 - \frac{1}{2}}{\sqrt{\frac{1}{12}}}$ PDF of Z_1

$Z_2 = \dfrac{X_1 + X_2 - 1}{\sqrt{\frac{2}{12}}}$ PDF of Z_2

$Z_3 = \dfrac{X_1 + X_2 + X_3 - \frac{3}{2}}{\sqrt{\frac{3}{12}}}$ PDF of Z_3

$Z_{30} = \dfrac{\sum_{i=1}^{30} X_i - \frac{30}{2}}{\sqrt{\frac{30}{12}}}$ PDF of Z_{30}

Figure 7.2: Z_n is the normalized sum of n independent $Uniform(0,1)$ random variables. The shape of its PDF, $f_{Z_n}(z)$, gets closer to the normal curve as n increases.

We could have directly looked at $Y_n = X_1 + X_2 + \dots + X_n$, so why do we normalize it first and say that the normalized version (Z_n) becomes approximately normal? This is because $EY_n = nEX_i$ and $\mathrm{Var}(Y_n) = n\sigma^2$ go to infinity as n goes to infinity. We normalize Y_n in order to have a finite mean and variance ($EZ_n = 0$, $\mathrm{Var}(Z_n) = 1$). Nevertheless, for any fixed n, the

CDF of Z_n is obtained by scaling and shifting the CDF of Y_n. Thus, the two CDFs have similar shapes.

The importance of the central limit theorem stems from the fact that, in many real applications, a certain random variable of interest is a sum of a large number of independent random variables. In these situations, we are often able to use the CLT to justify using the normal distribution. Examples of such random variables are found in almost every discipline. Here are a few:

- Laboratory measurement errors are usually modeled by normal random variables.

- In communication and signal processing, Gaussian noise is the most frequently used model for noise.

- In finance, the percentage changes in the prices of some assets are sometimes modeled by normal random variables.

- When we do random sampling from a population to obtain statistical knowledge about the population, we often model the resulting quantity as a normal random variable.

The CLT is also very useful in the sense that it can simplify our computations significantly. If you have a problem in which you are interested in a sum of one thousand i.i.d. random variables, it might be extremely difficult, if not impossible, to find the distribution of the sum by direct calculation. Using the CLT we can immediately write the distribution, if we know the mean and variance of the X_i's.

Another question that comes to mind is how large n should be so that we can use the normal approximation. The answer generally depends on the distribution of the X_is. Nevertheless, as a rule of thumb it is often stated that if n is larger than or equal to 30, then the normal approximation is very good.

Let's summarize how we use the CLT to solve problems:

How to Apply The Central Limit Theorem (CLT)

Here are the steps that we need in order to apply the CLT:

1. Write the random variable of interest, Y, as the sum of n i.i.d. random variable X_i's:

$$Y = X_1 + X_2 + ... + X_n.$$

2. Find EY and $\text{Var}(Y)$ by noting that

$$EY = n\mu, \qquad \text{Var}(Y) = n\sigma^2,$$

where $\mu = EX_i$ and $\sigma^2 = \text{Var}(X_i)$.

3. According to the CLT, conclude that $\frac{Y-EY}{\sqrt{\text{Var}(Y)}} = \frac{Y-n\mu}{\sqrt{n}\sigma}$ is approximately standard

---

normal; thus, to find $P(y_1 \leq Y \leq y_2)$, we can write

$$P(y_1 \leq Y \leq y_2) = P\left(\frac{y_1 - n\mu}{\sqrt{n}\sigma} \leq \frac{Y - n\mu}{\sqrt{n}\sigma} \leq \frac{y_2 - n\mu}{\sqrt{n}\sigma}\right)$$
$$\approx \Phi\left(\frac{y_2 - n\mu}{\sqrt{n}\sigma}\right) - \Phi\left(\frac{y_1 - n\mu}{\sqrt{n}\sigma}\right).$$

Let us look at some examples to see how we can use the central limit theorem.

Example 7.1. A bank teller serves customers standing in the queue one by one. Suppose that the service time X_i for customer i has mean $EX_i = 2$ (minutes) and $\text{Var}(X_i) = 1$. We assume that service times for different bank customers are independent. Let Y be the total time the bank teller spends serving 50 customers. Find $P(90 < Y < 110)$.

Solution: We have

$$Y = X_1 + X_2 + \ldots + X_n,$$

where $n = 50$, $EX_i = \mu = 2$, and $\text{Var}(X_i) = \sigma^2 = 1$. Thus, we can write

$$P(90 < Y \leq 110) = P\left(\frac{90 - n\mu}{\sqrt{n}\sigma} < \frac{Y - n\mu}{\sqrt{n}\sigma} < \frac{110 - n\mu}{\sqrt{n}\sigma}\right)$$
$$= P\left(\frac{90 - 100}{\sqrt{50}} < \frac{Y - n\mu}{\sqrt{n}\sigma} < \frac{110 - 100}{\sqrt{50}}\right)$$
$$= P\left(-\sqrt{2} < \frac{Y - n\mu}{\sqrt{n}\sigma} < \sqrt{2}\right).$$

By the CLT, $\frac{Y - n\mu}{\sqrt{n}\sigma}$ is approximately standard normal, so we can write

$$P(90 < Y \leq 110) \approx \Phi(\sqrt{2}) - \Phi(-\sqrt{2})$$
$$= 0.8427$$

Example 7.2. In a communication system each data packet consists of 1000 bits. Due to the noise, each bit may be received in error with probability 0.1. It is assumed bit errors occur independently. Find the probability that there are more than 120 errors in a certain data packet.

Solution: Let us define X_i as the indicator random variable for the ith bit in the packet. That is, $X_i = 1$ if the ith bit is received in error, and $X_i = 0$ otherwise. Then the X_i's are i.i.d. and $X_i \sim Bernoulli(p = 0.1)$. If Y is the total number of bit errors in the packet, we have

$$Y = X_1 + X_2 + \ldots + X_n.$$

Since $X_i \sim Bernoulli(p = 0.1)$, we have

$$EX_i = \mu = p = 0.1, \qquad \text{Var}(X_i) = \sigma^2 = p(1-p) = 0.09$$

Using the CLT, we have

$$P(Y > 120) = P\left(\frac{Y - n\mu}{\sqrt{n}\sigma} > \frac{120 - n\mu}{\sqrt{n}\sigma}\right)$$

$$= P\left(\frac{Y - n\mu}{\sqrt{n}\sigma} > \frac{120 - 100}{\sqrt{90}}\right)$$

$$\approx 1 - \Phi\left(\frac{20}{\sqrt{90}}\right)$$

$$= 0.0175$$

Continuity Correction:

Let us assume that $Y \sim Binomial(n = 20, p = \frac{1}{2})$, and suppose that we are interested in $P(8 \leq Y \leq 10)$. We know that a $Binomial(n = 20, p = \frac{1}{2})$ can be written as the sum of n i.i.d. $Bernoulli(p)$ random variables:

$$Y = X_1 + X_2 + ... + X_n.$$

Since $X_i \sim Bernoulli(p = \frac{1}{2})$, we have

$$EX_i = \mu = p = \frac{1}{2}, \qquad \text{Var}(X_i) = \sigma^2 = p(1 - p) = \frac{1}{4}.$$

Thus, we may want to apply the CLT to write

$$P(8 \leq Y \leq 10) = P\left(\frac{8 - n\mu}{\sqrt{n}\sigma} < \frac{Y - n\mu}{\sqrt{n}\sigma} < \frac{10 - n\mu}{\sqrt{n}\sigma}\right)$$

$$= P\left(\frac{8 - 10}{\sqrt{5}} < \frac{Y - n\mu}{\sqrt{n}\sigma} < \frac{10 - 10}{\sqrt{5}}\right)$$

$$\approx \Phi(0) - \Phi\left(\frac{-2}{\sqrt{5}}\right)$$

$$= 0.3145$$

Since, here, $n = 20$ is relatively small, we can actually find $P(8 \leq Y \leq 10)$ accurately. We have

$$P(8 \leq Y \leq 10) = \sum_{k=8}^{10} \binom{n}{k} p^k (1 - p)^{n-k}$$

$$= \left[\binom{20}{8} + \binom{20}{9} + \binom{20}{10}\right] \left(\frac{1}{2}\right)^{20}$$

$$= 0.4565$$

We notice that our approximation is not so good. Part of the error is due to the fact that Y is a discrete random variable and we are using a continuous distribution to find $P(8 \leq Y \leq 10)$. Here is a trick to get a better approximation, called **continuity correction**. Since Y can only take integer values, we can write

$$P(8 \leq Y \leq 10) = P(7.5 < Y < 10.5)$$
$$= P\left(\frac{7.5 - n\mu}{\sqrt{n}\sigma} < \frac{Y - n\mu}{\sqrt{n}\sigma} < \frac{10.5 - n\mu}{\sqrt{n}\sigma}\right)$$
$$= P\left(\frac{7.5 - 10}{\sqrt{5}} < \frac{Y - n\mu}{\sqrt{n}\sigma} < \frac{10.5 - 10}{\sqrt{5}}\right)$$
$$\approx \Phi\left(\frac{0.5}{\sqrt{5}}\right) - \Phi\left(\frac{-2.5}{\sqrt{5}}\right)$$
$$= 0.4567$$

As we see, using continuity correction, our approximation improved significantly. The continuity correction is particularly useful when we would like to find $P(y_1 \leq Y \leq y_2)$, where Y is binomial and y_1 and y_2 are close to each other.

Continuity Correction for Discrete Random Variables

Let $X_1, X_2, \cdots, X_n$ be independent discrete random variables and let

$$Y = X_1 + X_2 + \cdots + X_n.$$

Suppose that we are interested in finding $P(A) = P(l \leq Y \leq u)$ using the CLT, where l and u are integers. Since Y is an integer-valued random variable, we can write

$$P(A) = P(l - \frac{1}{2} \leq Y \leq u + \frac{1}{2}).$$

It turns out that the above expression sometimes provides a better approximation for $P(A)$ when applying the CLT. This is called the continuity correction and it is particularly useful when X_i's are Bernoulli (i.e., Y is binomial).

7.1.3 Solved Problems

1. There are 100 men on a plane. Let X_i be the weight (in pounds) of the ith man on the plane. Suppose that the X_i's are i.i.d., and $EX_i = \mu = 170$ and $\sigma_{X_i} = \sigma = 30$. Find the probability that the total weight of the men on the plane exceeds $18,000$ pounds.

Solution: If W is the total weight, then $W = X_1 + X_2 + \cdots + X_n$, where $n = 100$. We have

$$EW = n\mu$$
$$= (100)(170)$$
$$= 17000,$$
$$\text{Var}(W) = 100\text{Var}(X_i)$$
$$= (100)(30)^2$$
$$= 90000.$$

Thus, $\sigma_W = 300$. We have

$$P(W > 18000) = P\left(\frac{W - 17000}{300} > \frac{18000 - 17000}{300}\right)$$
$$= P\left(\frac{W - 17000}{300} > \frac{10}{3}\right)$$
$$= 1 - \Phi\left(\frac{10}{3}\right) \quad \text{(by CLT)}$$
$$\approx 4.3 \times 10^{-4}.$$

2. Let $X_1, X_2, \cdots, X_{25}$ be i.i.d. with the following PMF

$$P_X(k) = \begin{cases} 0.6 & k = 1 \\ 0.4 & k = -1 \\ 0 & \text{otherwise} \end{cases}$$

And let

$$Y = X_1 + X_2 + \cdots + X_n.$$

Using the CLT and continuity correction, estimate $P(4 \leq Y \leq 6)$.

Solution: We have

$$EX_i = (0.6)(1) + (0.4)(-1)$$
$$= \frac{1}{5},$$

$$EX_i^2 = 0.6 + 0.4$$
$$= 1.$$

Therefore,

$$\text{Var}(X_i) = 1 - \frac{1}{25}$$
$$= \frac{24}{25};$$

$$\text{thus,} \quad \sigma_{X_i} = \frac{2\sqrt{6}}{5}.$$

Therefore,

$$EY = 25 \times \frac{1}{5}$$
$$= 5,$$

$$\text{Var}(Y) = 25 \times \frac{24}{25}$$
$$= 24;$$

$$\text{thus,} \quad \sigma_Y = 2\sqrt{6}.$$

$$P(4 \leq Y \leq 6) = P(3.5 \leq Y \leq 6.5) \quad \text{(continuity correction)}$$
$$= P\left(\frac{3.5 - 5}{2\sqrt{6}} \leq \frac{Y - 5}{2\sqrt{6}} \leq \frac{6.5 - 5}{2\sqrt{6}}\right)$$
$$= P\left(-0.3062 \leq \frac{Y - 5}{2\sqrt{6}} \leq +0.3062\right)$$
$$\approx \Phi(0.3062) - \Phi(-0.3062) \quad \text{(by the CLT)}$$
$$= 2\Phi(0.3062) - 1$$
$$\approx 0.2405$$

3. You have invited 64 guests to a party. You need to make sandwiches for the guests. You believe that a guest might need 0, 1 or 2 sandwiches with probabilities $\frac{1}{4}$, $\frac{1}{2}$, and $\frac{1}{4}$ respectively. You assume that the number of sandwiches each guest needs is independent from other guests. How many sandwiches should you make so that you are 95% sure that there is no shortage?

Solution:

Let X_i be the number of sandwiches that the ith person needs, and let

$$Y = X_1 + X_2 + \cdots + X_{64}.$$

The goal is to find y such that

$$P(Y \leq y) \geq 0.95$$

First note that

$$EX_i = \frac{1}{4}(0) + \frac{1}{2}(1) + \frac{1}{4}(2)$$
$$= 1,$$

$$EX_i^2 = \frac{1}{4}(0^2) + \frac{1}{2}(1^2) + \frac{1}{4}(2^2)$$
$$= \frac{3}{2}.$$

Thus,

$$\text{Var}(X_i) = EX_i^2 - (EX_i)^2$$
$$= \frac{3}{2} - 1$$
$$= \frac{1}{2} \quad \rightarrow \quad \sigma_{X_i} = \frac{1}{\sqrt{2}}.$$

Thus,

$$EY = 64 \times 1$$
$$= 64,$$

$$\text{Var}(Y) = 64 \times \frac{1}{2}$$
$$= 32 \rightarrow \sigma_Y = 4\sqrt{2}.$$

Now, we can use the CLT to find y

$$P(Y \leq y) = P\left(\frac{Y - 64}{4\sqrt{2}} \leq \frac{y - 64}{4\sqrt{2}}\right)$$
$$= \Phi\left(\frac{y - 64}{4\sqrt{2}}\right) \quad \text{(by CLT)}$$

We can write

$$\Phi\left(\frac{y - 64}{4\sqrt{2}}\right) = 0.95$$

Therefore,

$$\frac{y - 64}{4\sqrt{2}} = \Phi^{-1}(0.95)$$
$$\approx 1.6449$$

Thus, $y = 73.3$.

Therefore, if you make 74 sandwiches, you are 95% sure that there is no shortage. Note that you can find the numerical value of $\Phi^{-1}(0.95)$ by running the *norminv(0.95)* command in MATLAB.

4. Let $X_1, X_2, \cdots, X_n$ be i.i.d. *Exponential*(λ) random variables with $\lambda = 1$. Let

$$\overline{X} = \frac{X_1 + X_2 + \cdots + X_n}{n}.$$

How large should n be such that

$$P\left(0.9 \leq \overline{X} \leq 1.1\right) \geq 0.95 \ ?$$

Solution: Let $Y = X_1 + X_2 + \cdots + X_n$, so $\overline{X} = \frac{Y}{n}$. Since $X_i \sim Exponential(1)$, we have

$$E(X_i) = \frac{1}{\lambda} = 1, \quad \text{Var}(X_i) = \frac{1}{\lambda^2} = 1.$$

Therefore,

$$E(Y) = nEX_i = n, \quad \text{Var}(Y) = n\text{Var}(X_i) = n,$$

$$\begin{aligned} P(0.9 \leq \overline{X} \leq 1.1) &= P\left(0.9 \leq \frac{Y}{n} \leq 1.1\right) \\ &= P\left(0.9n \leq Y \leq 1.1n\right) \\ &= P\left(\frac{0.9n - n}{\sqrt{n}} \leq \frac{Y - n}{\sqrt{n}} \leq \frac{1.1n - n}{\sqrt{n}}\right) \\ &= P\left(-0.1\sqrt{n} \leq \frac{Y - n}{\sqrt{n}} \leq 0.1\sqrt{n}\right). \end{aligned}$$

By the CLT $\frac{Y-n}{\sqrt{n}}$ is approximately $N(0,1)$, so

$$\begin{aligned} P(0.9 \leq \overline{X} \leq 1.1) &\approx \Phi\left(0.1\sqrt{n}\right) - \Phi(-0.1\sqrt{n}) \\ &= 2\Phi\left(0.1\sqrt{n}\right) - 1 \quad (\text{since} \quad \Phi(-x) = 1 - \Phi(x)). \end{aligned}$$

We need to have

$$2\Phi\left(0.1\sqrt{n}\right) - 1 \geq 0.95, \quad \text{so} \quad \Phi\left(0.1\sqrt{n}\right) \geq 0.975.$$

Thus,

$$0.1\sqrt{n} \geq \Phi^{-1}(0.975) = 1.96$$
$$\sqrt{n} \geq 19.6$$
$$n \geq 384.16$$

Since n is an integer, we conclude $n \geq 385$.

5. For this problem and the next, you will need to be familiar with moment generating functions (Section 6.1.3). The goal here is to prove the (weak) law of large numbers using MGFs. [1] In particular, let $X_1, X_2, \ldots, X_n$ be i.i.d. random variables with expected value $EX_i = \mu < \infty$ and MGF $M_X(s)$ that is finite on some interval $[-c, c]$ where $c > 0$ is a constant. As usual, let

$$\overline{X} = \frac{X_1 + X_2 + \cdots + X_n}{n}.$$

Prove

$$\lim_{n \to \infty} M_{\overline{X}}(s) = e^{s\mu}, \qquad \text{for all } s \in [-c, c].$$

Since this is the MGF of constant random variable μ, we conclude that the distribution of $\overline{X}$ converges to μ. *Hint:* Use the result of Problem 8 in Section 6.1.6: for a random variable X with a well-defined MGF, $M_X(s)$, we have

$$\lim_{n \to \infty} \left[M_X\left(\frac{s}{n}\right) \right]^n = e^{sEX}.$$

Solution: We have

$$M_{\overline{X}}(s) = E[e^{s\overline{X}}]$$
$$= E[e^{s\frac{X_1 + X_2 + \cdots + X_n}{n}}]$$
$$= E[e^{s\frac{X_1}{n}} e^{s\frac{X_2}{n}} \cdots e^{s\frac{X_n}{n}}]$$
$$= E[e^{\frac{sX_1}{n}}] \cdot E[e^{\frac{sX_2}{n}}] \cdots E[e^{\frac{sX_n}{n}}] \quad \text{(since X_i's are independent)}$$
$$= \left[M_X\left(\frac{s}{n}\right) \right]^n \quad \text{(since X_i's are identically distributed)}$$

Therefore,

$$\lim_{n \to \infty} M_{\overline{X}}(s) = \lim_{n \to \infty} [M_X\left(\frac{s}{n}\right)]^n$$
$$= e^{sEX} \quad \text{(by the hint)}$$
$$= e^{s\mu}.$$

---

[1] Technically, here we prove that $\overline{X}$ converges *in distribution* to μ. To prove the weak law of large numbers, we need to show that $\overline{X}$ converges *in probability* to μ. Luckily, as it is discussed in Section 7.2, if a sequence of random variables converges in distribution to a constant number, then it also converges to that constant in probability.

Note that $e^{s\mu}$ is the MGF of a constant random variable Y, with value $Y = \mu$. This means that the random variable $\overline{X}$ converges to μ (in distribution).

6. The goal in this problem is to prove the central limit theorem using MGFs. In particular, let $X_1, X_2, ..., X_n$ be i.i.d. random variables with expected value $EX_i = \mu < \infty$, $\text{Var}(X_i) = \sigma^2 < \infty$, and MGF $M_X(s)$ that is finite on some interval $[-c, c]$, where $c > 0$ is a constant. As usual, let

$$Z_n = \frac{\overline{X} - \mu}{\sigma/\sqrt{n}} = \frac{X_1 + X_2 + \cdots + X_n - n\mu}{\sqrt{n}\sigma}.$$

Prove

$$\lim_{n \to \infty} M_{Z_n}(s) = e^{\frac{s^2}{2}}, \qquad \text{for all } s \in [-c, c].$$

Since this is the MGF of a standard normal random variable, we conclude that the distribution of Z_n converges to the standard normal random variable. *Hint:* Use the result of Problem 9 in Section 6.1.6: for a random variable Y with a well-defined MGF, $M_Y(s)$, and $EY = 0$, $\text{Var}(Y) = 1$, we have

$$\lim_{n \to \infty} \left[M_Y \left(\frac{s}{\sqrt{n}} \right) \right]^n = e^{\frac{s^2}{2}}.$$

Solution: Let Y_i's be the normalized versions of the X_i's, i.e.,

$$Y_i = \frac{X_i - \mu}{\sigma}.$$

Then, Y_i's are i.i.d. and

$$EY_i = 0,$$
$$\text{Var}(Y_i) = 1.$$

We also have

$$Z_n = \frac{\overline{X} - \mu}{\frac{\sigma}{\sqrt{n}}}$$
$$= \frac{Y_1 + Y_2 + \cdots + Y_n}{\sqrt{n}}.$$

Thus, we have

$$M_{Z_n}(s) = E[e^{s\frac{Y_1 + Y_2 + \cdots + Y_n}{\sqrt{n}}}]$$
$$= E[e^{\frac{sY_1}{\sqrt{n}}}] \cdot E[e^{\frac{sY_2}{\sqrt{n}}}] \cdots E[e^{\frac{sY_n}{\sqrt{n}}}] \quad \text{(since } Y_i\text{'s are independent)}$$
$$= M_{Y_1}(\frac{s}{\sqrt{n}})^n \quad (Y_i\text{'s are identically distributed)}.$$

Thus, we conclude

$$\lim_{n\to\infty} M_{Z_n}(s) = \lim_{n\to\infty} M_{Y_1}\left(\frac{s}{\sqrt{n}}\right)^n$$
$$= e^{\frac{s^2}{2}} \quad \text{(by the hint)}.$$

Since this is the MGF of a standard normal random variable, we conclude the CDF of Z_n converges to the standard normal CDF.

7.2 Convergence of Random Variables

In some situations, we would like to see if a sequence of random variables X_1, X_2, X_3, $\cdots$ "converges" to a random variable X. That is, we would like to see if X_n gets closer and closer to X in some sense as n increases. For example, suppose that we are interested in knowing the value of a random variable X, but we are not able to observe X directly. Instead, you can do some measurements and come up with an estimate of X: call it X_1. You then perform more measurements and update your estimate of X and call it X_2. You continue this process to obtain X_1, X_2, X_3, $\cdots$. Your hope is that as n increases, your estimate gets better and better. That is, you hope that as n increases, X_n gets closer and closer to X. In other words, you hope that X_n converges to X.

In fact, we have already seen the concept of convergence in Section 7.1 when we discussed limit theorems (the weak law of large numbers (WLLN) and the central limit theorem (CLT)). The WLLN states that the average of a large number of i.i.d. random variables converges in probability to the expected value. The CLT states that the normalized average of a sequence of i.i.d. random variables converges in distribution to a standard normal distribution. In this section, we will develop the theoretical background to study the convergence of a sequence of random variables in more detail. In particular, we will define different types of convergence. When we say that the sequence X_n converges to X, it means that X_n's are getting "closer and closer" to X. Different types of convergence refer to different ways of defining what "closer" means. We also discuss how different types of convergence are related.

7.2.1 Convergence of a Sequence of Numbers

Before discussing convergence for a sequence of random variables, let us remember what convergence means for a sequence of real numbers. If we have a sequence of real numbers $a_1, a_2, a_3, \cdots$, we can ask whether the sequence converges. For example, the sequence

$$\frac{1}{2}, \frac{2}{3}, \frac{3}{4}, \cdots, \frac{n}{n+1}, \cdots$$

is defined as

$$a_n = \frac{n}{n+1}, \qquad \text{for } n = 1, 2, 3, \cdots$$

This sequence converges to 1. We say that a sequence a_1, a_2, a_3, $\cdots$ converges to a limit L if a_n approaches L as n goes to infinity.

Definition 7.2. A sequence a_1, a_2, a_3, $\cdots$ converges to a limit L if

$$\lim_{n \to \infty} a_n = L.$$

That is, for any $\epsilon > 0$, there exists an $N \in \mathbb{N}$ such that

$$|a_n - L| < \epsilon, \quad \text{for all } n > N.$$

7.2.2 Sequence of Random Variables

Here, we would like to discuss what we precisely mean by a sequence of random variables. Remember that, in any probability model, we have a sample space S and a probability measure P. For simplicity, suppose that our sample space consists of a finite number of elements, i.e.,

$$S = \{s_1, s_2, \cdots, s_k\}.$$

Then, a random variable X is a mapping that assigns a real number to any of the possible outcomes s_i, $i = 1, 2, \cdots, k$. Thus, we may write

$$X(s_i) = x_i, \quad \text{for } i = 1, 2, \cdots, k.$$

When we have a sequence of random variables X_1, X_2, X_3, $\cdots$, it is also useful to remember that we have an underlying sample space S. In particular, each X_n is a function from S to real numbers. Thus, we may write

$$X_n(s_i) = x_{ni}, \quad \text{for } i = 1, 2, \cdots, k.$$

In sum, a sequence of random variables is in fact a sequence of functions $X_n : S \to \mathbb{R}$.

Example 7.3. Consider the following random experiment: A fair coin is tossed once. Here, the sample space has only two elements $S = \{H, T\}$. We define a sequence of random variables X_1, X_2, X_3, $\cdots$ on this sample space as follows:

$$X_n(s) = \begin{cases} \frac{1}{n+1} & \text{if } s = H \\ \\ 1 & \text{if } s = T \end{cases}$$

(a) Are the X_i's independent?

(b) Find the PMF and CDF of X_n, $F_{X_n}(x)$ for $n = 1, 2, 3, \cdots$.

(c) As n goes to infinity, what does $F_{X_n}(x)$ look like?

Solution:

(a) The X_i's are not independent because their values are determined by the same coin toss. In particular, to show that X_1 and X_2 are not independent, we can write

$$P(X_1 = 1, X_2 = 1) = P(T)$$
$$= \frac{1}{2},$$

which is different from

$$P(X_1 = 1) \cdot P(X_2 = 1) = P(T) \cdot P(T)$$
$$= \frac{1}{4}.$$

(b) Each X_i can take only two possible values that are equally likely. Thus, the PMF of X_n is given by

$$P_{X_n}(x) = P(X_n = x) = \begin{cases} \frac{1}{2} & \text{if } x = \frac{1}{n+1} \\ \\ \frac{1}{2} & \text{if } x = 1 \end{cases}$$

From this we can obtain the CDF of X_n

$$F_{X_n}(x) = P(X_n \leq x) = \begin{cases} 1 & \text{if } x \geq 1 \\ \\ \frac{1}{2} & \text{if } \frac{1}{n+1} \leq x < 1 \\ \\ 0 & \text{if } x < \frac{1}{n+1} \end{cases}$$

(c) Figure 7.3 shows the CDF of X_n for different values of n. We see in the figure that the CDF of X_n approaches the CDF of a *Bernoulli* $\left(\frac{1}{2}\right)$ random variable as $n \to \infty$. As we will discuss in the next sections, this means that the sequence X_1, X_2, X_3, $\cdots$ converges *in distribution* to a *Bernoulli* $\left(\frac{1}{2}\right)$ random variable as $n \to \infty$.

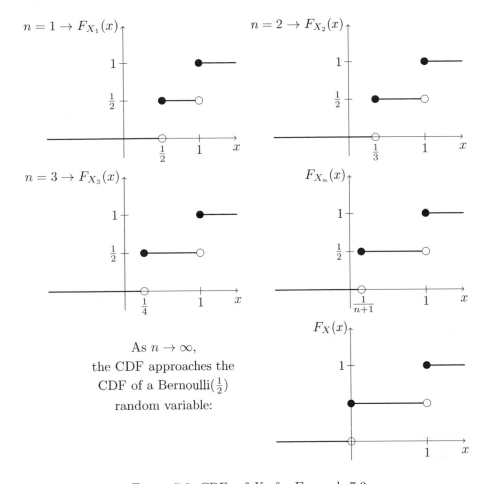

Figure 7.3: CDFs of X_n for Example 7.3

The previous example was defined on a very simple sample space $S = \{H, T\}$. Let us look at an example that is defined on a more interesting sample space.

Example 7.4. Consider the following random experiment: A fair coin is tossed repeatedly forever. Here, the sample space S consists of all possible sequences of heads and tails. We define the sequence of random variables $X_1, X_2, X_3, \cdots$ as follows:

$$X_n = \begin{cases} 0 & \text{if the } n\text{th coin toss results in a heads} \\ 1 & \text{if the } n\text{th coin toss results in a tails} \end{cases}$$

In this example, the X_i's are independent because each X_i is a result of a different coin toss. In fact, the X_i's are i.i.d. *Bernoulli* $\left(\frac{1}{2}\right)$ random variables. Thus, when we would like to refer to such a sequence, we usually say, "Let $X_1, X_2, X_3, \cdots$ be a sequence of i.i.d. *Bernoulli* $\left(\frac{1}{2}\right)$ random variables." We usually do not state the sample space because it is implied that the sample space S consists of all possible sequences of heads and tails.

7.2.3 Different Types of Convergence for Sequences of Random Variables

Here, we would like to provide definitions of different types of convergence and discuss how they are related. Consider a sequence of random variables X_1, X_2, X_3, $\cdots$, i.e, $\{X_n, n \in \mathbb{N}\}$. This sequence might "converge" to a random variable X. There are four types of convergence that we will discuss in this section:

1. Convergence in distribution,

2. Convergence in probability,

3. Convergence in mean,

4. Almost sure convergence.

These are all different kinds of convergence. A sequence might converge in one sense but not another. Some of these convergence types are "stronger" than others and some are "weaker." By this, we mean the following: If Type A convergence is stronger than Type B convergence, it means that Type A convergence implies Type B convergence. Figure 7.4 summarizes how these types of convergence are related. In this figure, the stronger types of convergence are on top and, as we move to the bottom, the convergence becomes weaker. For example, using the figure, we conclude that if a sequence of random variables converges *in probability* to a random variable X, then the sequence converges *in distribution* to X as well.

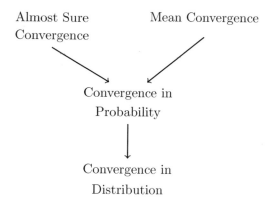

Figure 7.4: Relations between different types of convergence

7.2.4 Convergence in Distribution

Convergence in distribution is in some sense the weakest type of convergence. All it says is that the CDF of X_n's converges to the CDF of X as n goes to infinity. It does not require any dependence between the X_n's and X. We saw this type of convergence before when we discussed the central limit theorem. To say that X_n converges in distribution to X, we write

$$X_n \xrightarrow{d} X.$$

Here is a formal definition of convergence in distribution:

Convergence in Distribution

A sequence of random variables X_1, X_2, X_3, $\cdots$ converges **in distribution** to a random variable X, shown by $X_n \overset{d}{\to} X$, if

$$\lim_{n \to \infty} F_{X_n}(x) = F_X(x),$$

for all x at which $F_X(x)$ is continuous.

Example 7.5. If X_1, X_2, X_3, $\cdots$ is a sequence of i.i.d. random variables with CDF $F_X(x)$, then $X_n \overset{d}{\to} X$. This is because

$$F_{X_n}(x) = F_X(x), \qquad \text{for all } x.$$

Therefore,

$$\lim_{n \to \infty} F_{X_n}(x) = F_X(x), \qquad \text{for all } x.$$

Example 7.6. Let X_2, X_3, X_4, $\cdots$ be a sequence of random variable such that

$$F_{X_n}(x) = \begin{cases} 1 - \left(1 - \frac{1}{n}\right)^{nx} & x > 0 \\ \\ 0 & \text{otherwise} \end{cases}$$

Show that X_n converges in distribution to $Exponential(1)$.

Solution: Let $X \sim Exponential(1)$. For $x \leq 0$, we have

$$F_{X_n}(x) = F_X(x) = 0, \qquad \text{for } n = 2, 3, 4, \cdots.$$

For $x \geq 0$, we have

$$\begin{aligned} \lim_{n \to \infty} F_{X_n}(x) &= \lim_{n \to \infty} \left(1 - \left(1 - \frac{1}{n}\right)^{nx}\right) \\ &= 1 - \lim_{n \to \infty} \left(1 - \frac{1}{n}\right)^{nx} \\ &= 1 - e^{-x} \\ &= F_X(x), \qquad \text{for all } x. \end{aligned}$$

Thus, we conclude that $X_n \overset{d}{\to} X$.

When working with integer-valued random variables, the following theorem is often useful.

Theorem 7.1. Consider the sequence X_1, X_2, X_3, $\cdots$ and the random variable X. Assume that X and X_n (for all n) are non-negative and integer-valued, i.e.,

$$R_X \subset \{0, 1, 2, \cdots\},$$
$$R_{X_n} \subset \{0, 1, 2, \cdots\}, \qquad \text{for } n = 1, 2, 3, \cdots.$$

Then $X_n \xrightarrow{d} X$ if and only if

$$\lim_{n\to\infty} P_{X_n}(k) = P_X(k), \qquad \text{for } k = 0, 1, 2, \cdots.$$

Proof. Since X is integer-valued, its CDF, $F_X(x)$, is continuous at all $x \in \mathbb{R} - \{0, 1, 2, ...\}$. If $X_n \xrightarrow{d} X$, then

$$\lim_{n\to\infty} F_{X_n}(x) = F_X(x), \qquad \text{for all } x \in \mathbb{R} - \{0, 1, 2, ...\}.$$

Thus, for $k = 0, 1, 2, \cdots$, we have

$$\lim_{n\to\infty} P_{X_n}(k) = \lim_{n\to\infty} \left[F_{X_n}\left(k + \frac{1}{2}\right) - F_{X_n}\left(k - \frac{1}{2}\right) \right] \qquad \text{(since } X_n\text{'s are integer-valued)}$$

$$= \lim_{n\to\infty} F_{X_n}\left(k + \frac{1}{2}\right) - \lim_{n\to\infty} F_{X_n}\left(k - \frac{1}{2}\right)$$

$$= F_X\left(k + \frac{1}{2}\right) - F_X\left(k - \frac{1}{2}\right) \qquad \text{(since } X_n \xrightarrow{d} X\text{)}$$

$$= P_X(k) \qquad \text{(since } X \text{ is integer-valued)}.$$

To prove the converse, assume that we know

$$\lim_{n\to\infty} P_{X_n}(k) = P_X(k), \qquad \text{for } k = 0, 1, 2, \cdots.$$

Then, for all $x \in \mathbb{R}$, we have

$$\lim_{n\to\infty} F_{X_n}(x) = \lim_{n\to\infty} P(X_n \leq x)$$

$$= \lim_{n\to\infty} \sum_{k=0}^{\lfloor x \rfloor} P_{X_n}(k),$$

where $\lfloor x \rfloor$ shows the largest integer less than or equal to x. Since for any fixed x, the set $\{0, 1, \cdots, \lfloor x \rfloor\}$ is a finite set, we can change the order of the limit and the sum, so we obtain

$$\lim_{n\to\infty} F_{X_n}(x) = \sum_{k=0}^{\lfloor x \rfloor} \lim_{n\to\infty} P_{X_n}(k)$$

$$= \sum_{k=0}^{\lfloor x \rfloor} P_X(k) \qquad \text{(by assumption)}$$

$$= P(X \le x) = F_X(x).$$

$\blacksquare$

Example 7.7. Let $X_1, X_2, X_3, \cdots$ be a sequence of random variable such that

$$X_n \sim Binomial\left(n, \frac{\lambda}{n}\right), \qquad \text{for } n \in \mathbb{N}, n > \lambda,$$

where $\lambda > 0$ is a constant. Show that X_n converges in distribution to $Poisson(\lambda)$.

Solution: By Theorem 7.1, it suffices to show that

$$\lim_{n\to\infty} P_{X_n}(k) = P_X(k), \qquad \text{for all } k = 0, 1, 2, \cdots.$$

We have

$$\lim_{n\to\infty} P_{X_n}(k) = \lim_{n\to\infty} \binom{n}{k} \left(\frac{\lambda}{n}\right)^k \left(1 - \frac{\lambda}{n}\right)^{n-k}$$

$$= \lambda^k \lim_{n\to\infty} \frac{n!}{k!(n-k)!} \left(\frac{1}{n^k}\right) \left(1 - \frac{\lambda}{n}\right)^{n-k}$$

$$= \frac{\lambda^k}{k!} \cdot \lim_{n\to\infty} \left(\left[\frac{n(n-1)(n-2)...(n-k+1)}{n^k}\right] \left[\left(1 - \frac{\lambda}{n}\right)^n\right] \left[\left(1 - \frac{\lambda}{n}\right)^{-k}\right]\right).$$

Note that for a fixed k, we have

$$\lim_{n\to\infty} \frac{n(n-1)(n-2)...(n-k+1)}{n^k} = 1,$$

$$\lim_{n\to\infty} \left(1 - \frac{\lambda}{n}\right)^{-k} = 1,$$

$$\lim_{n\to\infty} \left(1 - \frac{\lambda}{n}\right)^n = e^{-\lambda}.$$

Thus, we conclude

$$\lim_{n\to\infty} P_{X_n}(k) = \frac{e^{-\lambda} \lambda^k}{k!}.$$

We end this section by reminding you that the most famous example of convergence in distribution is the central limit theorem (CLT). The CLT states that the normalized average of i.i.d. random variables $X_1, X_2, X_3, \cdots$ converges in distribution to a standard normal random variable.

7.2.5   Convergence in Probability

Convergence in probability is stronger than convergence in distribution. In particular, for a sequence X_1, X_2, X_3, $\cdots$ to converge to a random variable X, we must have that $P(|X_n - X| \geq \epsilon)$ goes to 0 as $n \to \infty$, for any $\epsilon > 0$. To say that X_n converges in probability to X, we write

$$X_n \xrightarrow{p} X.$$

Here is the formal definition of convergence in probability:

Convergence in Probability

A sequence of random variables X_1, X_2, X_3, $\cdots$ converges **in probability** to a random variable X, shown by $X_n \xrightarrow{p} X$, if

$$\lim_{n\to\infty} P\big(|X_n - X| \geq \epsilon\big) = 0, \qquad \text{for all } \epsilon > 0.$$

Example 7.8. Let $X_n \sim Exponential(n)$, show that $X_n \xrightarrow{p} 0$. That is, the sequence X_1, X_2, X_3, $\cdots$ converges in probability to the zero random variable X.

Solution: We have

$$
\begin{aligned}
\lim_{n\to\infty} P\big(|X_n - 0| \geq \epsilon\big) &= \lim_{n\to\infty} P\big(X_n \geq \epsilon\big) && (\text{ since } X_n \geq 0) \\
&= \lim_{n\to\infty} e^{-n\epsilon} && (\text{ since } X_n \sim Exponential(n)) \\
&= 0, && \text{for all } \epsilon > 0.
\end{aligned}
$$

Example 7.9. Let X be a random variable, and $X_n = X + Y_n$, where

$$EY_n = \frac{1}{n}, \qquad \text{Var}(Y_n) = \frac{\sigma^2}{n},$$

where $\sigma > 0$ is a constant. Show that $X_n \xrightarrow{p} X$.

Solution: First note that by the triangle inequality, for all $a, b \in \mathbb{R}$, we have $|a+b| \leq |a|+|b|$. Choosing $a = Y_n - EY_n$ and $b = EY_n$, we obtain

$$|Y_n| \leq |Y_n - EY_n| + \frac{1}{n}.$$

Now, for any $\epsilon > 0$, we have

$$P\big(|X_n - X| \geq \epsilon\big) = P\big(|Y_n| \geq \epsilon\big)$$

$$\leq P\left(|Y_n - EY_n| + \frac{1}{n} \geq \epsilon\right)$$

$$= P\left(|Y_n - EY_n| \geq \epsilon - \frac{1}{n}\right)$$

$$\leq \frac{\text{Var}(Y_n)}{\left(\epsilon - \frac{1}{n}\right)^2} \qquad\qquad \text{(by Chebyshev's inequality)}$$

$$= \frac{\sigma^2}{n\left(\epsilon - \frac{1}{n}\right)^2} \to 0 \quad \text{as } n \to \infty.$$

Therefore, we conclude $X_n \xrightarrow{p} X$.

As we mentioned previously, convergence in probability is stronger than convergence in distribution. That is, if $X_n \xrightarrow{p} X$, then $X_n \xrightarrow{d} X$. The converse is not necessarily true. For example, let $X_1, X_2, X_3, \cdots$ be a sequence of i.i.d. *Bernoulli* $\left(\frac{1}{2}\right)$ random variables. Let also $X \sim Bernoulli\left(\frac{1}{2}\right)$ be independent from the X_i's. Then, $X_n \xrightarrow{d} X$. However, X_n does not converge in probability to X, since $|X_n - X|$ is in fact also a *Bernoulli* $\left(\frac{1}{2}\right)$ random variable and

$$P\big(|X_n - X| \geq \epsilon\big) = \frac{1}{2}, \quad \text{for } 0 < \epsilon < 1.$$

A special case in which the converse is true is when $X_n \xrightarrow{d} c$, where c is a constant. In this case, convergence in distribution implies convergence in probability. We can state the following theorem:

Theorem 7.2. If $X_n \xrightarrow{d} c$, where c is a constant, then $X_n \xrightarrow{p} c$.

Proof. Since $X_n \xrightarrow{d} c$, we conclude that for any $\epsilon > 0$, we have

$$\lim_{n \to \infty} F_{X_n}(c - \epsilon) = 0,$$

$$\lim_{n \to \infty} F_{X_n}\left(c + \frac{\epsilon}{2}\right) = 1.$$

We can write for any $\epsilon > 0$,

$$
\begin{aligned}
\lim_{n \to \infty} P\big(|X_n - c| \geq \epsilon\big) &= \lim_{n \to \infty} \Big[P\big(X_n \leq c - \epsilon\big) + P\big(X_n \geq c + \epsilon\big) \Big] \\
&= \lim_{n \to \infty} P\big(X_n \leq c - \epsilon\big) + \lim_{n \to \infty} P\big(X_n \geq c + \epsilon\big) \\
&= \lim_{n \to \infty} F_{X_n}(c - \epsilon) + \lim_{n \to \infty} P\big(X_n \geq c + \epsilon\big) \\
&= 0 + \lim_{n \to \infty} P\big(X_n \geq c + \epsilon\big) \qquad \big(\text{since } \lim_{n \to \infty} F_{X_n}(c - \epsilon) = 0\big) \\
&\leq \lim_{n \to \infty} P\big(X_n > c + \frac{\epsilon}{2}\big) \\
&= 1 - \lim_{n \to \infty} F_{X_n}(c + \frac{\epsilon}{2}) \\
&= 0 \qquad\qquad \big(\text{since } \lim_{n \to \infty} F_{X_n}(c + \frac{\epsilon}{2}) = 1\big).
\end{aligned}
$$

Since $\lim_{n \to \infty} P\big(|X_n - c| \geq \epsilon\big) \geq 0$, we conclude that

$$
\lim_{n \to \infty} P\big(|X_n - c| \geq \epsilon\big) = 0, \qquad \text{for all } \epsilon > 0,
$$

which means $X_n \xrightarrow{p} c$. ∎

The most famous example of convergence in probability is the weak law of large numbers (WLLN). We proved WLLN in Section 7.1. The WLLN states that if $X_1, X_2, X_3, \cdots$ are i.i.d. random variables with mean $EX_i = \mu < \infty$, then the average sequence defined by

$$
\overline{X}_n = \frac{X_1 + X_2 + ... + X_n}{n}
$$

converges in probability to μ. It is called the "weak" law because it refers to convergence in probability. There is another version of the law of large numbers that is called the strong law of large numbers (SLLN). We will discuss SLLN in Section 7.2.7.

7.2.6 Convergence in Mean

One way of interpreting the convergence of a sequence X_n to X is to say that the "distance" between X and X_n is getting smaller and smaller. For example, if we define the distance between X_n and X as $P\big(|X_n - X| \geq \epsilon\big)$, we have convergence in probability. One way to define the distance between X_n and X is

$$
E\left(|X_n - X|^r\right),
$$

where $r \geq 1$ is a fixed number. This refers to **convergence in mean**.[2] The most common choice is $r = 2$, in which case it is called the **mean-square convergence**. [3]

---

[2] For convergence in mean, it is usually required that $E|X_n^r| < \infty$.

[3] Some authors refer to the case $r = 1$ as convergence in mean.

> ### Convergence in Mean
>
> Let $r \geq 1$ be a fixed number. A sequence of random variables X_1, X_2, X_3, $\cdots$ converges **in the rth mean** or **in the L^r norm** to a random variable X, shown by $X_n \xrightarrow{L^r} X$, if
>
> $$\lim_{n \to \infty} E\left(|X_n - X|^r\right) = 0.$$
>
> If $r = 2$, it is called the **mean-square convergence**, and it is shown by $X_n \xrightarrow{m.s.} X$.

Example 7.10. Let $X_n \sim Uniform\left(0, \frac{1}{n}\right)$. Show that $X_n \xrightarrow{L^r} 0$, for any $r \geq 1$.

Solution: The PDF of X_n is given by

$$f_{X_n}(x) = \begin{cases} n & 0 \leq x \leq \frac{1}{n} \\ \\ 0 & \text{otherwise} \end{cases}$$

We have

$$E\left(|X_n - 0|^r\right) = \int_0^{\frac{1}{n}} x^r n \ dx$$
$$= \frac{1}{(r+1)n^r} \to 0, \qquad \text{for all } r \geq 1.$$

Theorem 7.3. Let $1 \leq r \leq s$. If $X_n \xrightarrow{L^s} X$, then $X_n \xrightarrow{L^r} X$.

Proof. We can use Hölder's inequality, which was proved in Section 6.2.6. Hölder's inequality states that

$$E|XY| \leq \left(E|X|^p\right)^{\frac{1}{p}} \left(E|Y|^q\right)^{\frac{1}{q}},$$

where $1 < p, q < \infty$ and $\frac{1}{p} + \frac{1}{q} = 1$. In Hölder's inequality, choose

$$X = |X_n - X|^r,$$
$$Y = 1,$$
$$p = \frac{s}{r} > 1.$$

We obtain

$$E|X_n - X|^r \leq \left(E|X_n - X|^s\right)^{\frac{1}{p}}.$$

Now, by assumption $X_n \xrightarrow{L^s} X$, which means

$$\lim_{n \to \infty} E\left(|X_n - X|^s\right) = 0.$$

We conclude

$$\lim_{n\to\infty} E\left(|X_n - X|^r\right) \leq \lim_{n\to\infty} \left(E|X_n - X|^s\right)^{\frac{1}{p}}$$
$$= 0.$$

Therefore, $X_n \xrightarrow{L^r} X$. ∎

As we mentioned before, convergence in mean is stronger than convergence in probability. We can prove this using Markov's inequality.

Theorem 7.4. If $X_n \xrightarrow{L^r} X$ for some $r \geq 1$, then $X_n \xrightarrow{p} X$.

Proof. For any $\epsilon > 0$, we have

$$P\left(|X_n - X| \geq \epsilon\right) = P\left(|X_n - X|^r \geq \epsilon^r\right) \qquad \text{(since } r \geq 1)$$
$$\leq \frac{E|X_n - X|^r}{\epsilon^r} \qquad \text{(by Markov's inequality)}$$

Since by assumption $\lim_{n\to\infty} E\left(|X_n - X|^r\right) = 0$, we conclude

$$\lim_{n\to\infty} P\left(|X_n - X| \geq \epsilon\right) = 0, \qquad \text{for all } \epsilon > 0.$$

∎

The converse of Theorem 7.4 is not true in general. That is, there are sequences that converge in probability but not in mean. Let us look at an example.

Example 7.11. Consider a sequence $\{X_n, n = 1, 2, 3, \cdots\}$ such that

$$X_n = \begin{cases} n^2 & \text{with probability } \frac{1}{n} \\ \\ 0 & \text{with probability } 1 - \frac{1}{n} \end{cases}$$

Show that

(a) $X_n \xrightarrow{p} 0$.

(b) X_n does not converge in the rth mean for any $r \geq 1$.

Solution:

(a) To show $X_n \xrightarrow{p} 0$, we can write, for any $\epsilon > 0$

$$\lim_{n\to\infty} P\left(|X_n| \geq \epsilon\right) = \lim_{n\to\infty} P(X_n = n^2)$$
$$= \lim_{n\to\infty} \frac{1}{n}$$
$$= 0.$$

We conclude that $X_n \xrightarrow{p} 0$.

(b) For any $r \geq 1$, we can write

$$\lim_{n \to \infty} E\left(|X_n|^r\right) = \lim_{n \to \infty} \left(n^{2r} \cdot \frac{1}{n} + 0 \cdot \left(1 - \frac{1}{n}\right)\right)$$
$$= \lim_{n \to \infty} n^{2r-1}$$
$$= \infty \qquad (\text{since } r \geq 1).$$

Therefore, X_n does not converge in the rth mean for any $r \geq 1$. In particular, it is interesting to note that, although $X_n \xrightarrow{p} 0$, the expected value of X_n does not converge to 0.

7.2.7 Almost Sure Convergence

Consider a sequence of random variables X_1, X_2, X_3, $\cdots$ that is defined on an underlying sample space S. For simplicity, let us assume that S is a finite set, so we can write

$$S = \{s_1, s_2, \cdots, s_k\}.$$

Remember that each X_n is a function from S to the set of real numbers. Thus, we may write

$$X_n(s_i) = x_{ni}, \qquad \text{for } i = 1, 2, \cdots, k.$$

After this random experiment is performed, one of the s_i's will be the outcome of the experiment, and the values of the X_n's are known. If s_j is the outcome of the experiment, we observe the following sequence:

$$x_{1j}, x_{2j}, x_{3j}, \cdots.$$

Since this is a sequence of real numbers, we can talk about its convergence. Does it converge? If yes, what does it converge to? **Almost sure** convergence is defined based on the convergence of such sequences. Before introducing almost sure convergence let us look at an example.

Example 7.12. Consider the following random experiment: A fair coin is tossed once. Here, the sample space has only two elements $S = \{H, T\}$. We define a sequence of random variables X_1, X_2, X_3, $\cdots$ on this sample space as follows:

$$X_n(s) = \begin{cases} \frac{n}{n+1} & \text{if } s = H \\ \\ (-1)^n & \text{if } s = T \end{cases}$$

(a) For each of the possible outcomes (H or T), determine whether the resulting sequence of real numbers converges or not.

(b) Find

$$P\left(\left\{s_i \in S : \lim_{n \to \infty} X_n(s_i) = 1\right\}\right).$$

Solution:

(a) If the outcome is H, then we have $X_n(H) = \frac{n}{n+1}$, so we obtain the following sequence

$$\frac{1}{2}, \frac{2}{3}, \frac{3}{4}, \frac{4}{5}, \cdots.$$

This sequence converges to 1 as n goes to infinity. If the outcome is T, then we have $X_n(T) = (-1)^n$, so we obtain the following sequence

$$-1, 1, -1, 1, -1, \cdots.$$

This sequence does not converge as it oscillates between -1 and 1 forever.

(b) By part (a), the event $\{s_i \in S : \lim_{n \to \infty} X_n(s_i) = 1\}$ happens if and only if the outcome is H, so

$$P\left(\left\{s_i \in S : \lim_{n \to \infty} X_n(s_i) = 1\right\}\right) = P(H)$$
$$= \frac{1}{2}.$$

In the above example, we saw that the sequence $X_n(s)$ converged when $s = H$ and did not converge when $s = T$. In general, if the probability that the sequence $X_n(s)$ converges to $X(s)$ is equal to 1, we say that X_n converges to X **almost surely** and write [4]

$$X_n \xrightarrow{a.s.} X.$$

Almost Sure Convergence

A sequence of random variables $X_1, X_2, X_3, \cdots$ converges **almost surely** to a random variable X, shown by $X_n \xrightarrow{a.s.} X$, if

$$P\left(\left\{s \in S : \lim_{n \to \infty} X_n(s) = X(s)\right\}\right) = 1.$$

Example 7.13. Consider the sample space $S = [0, 1]$ with a probability measure that is uniform on this space, i.e.,

$$P([a, b]) = b - a, \qquad \text{for all } 0 \leq a \leq b \leq 1.$$

Define the sequence $\{X_n, n = 1, 2, \cdots\}$ as follows:

$$X_n(s) = \begin{cases} 1 & 0 \leq s < \frac{n+1}{2n} \\ 0 & \text{otherwise} \end{cases}$$

---
[4]There are different notations for almost sure convergence in the literature. They have the same meaning as $X_n \xrightarrow{a.s.} X$:

- X_n converges *with probability* 1 to X, written as $X_n \to X$ w.p.1,
- X_n converges *almost everywhere* to X, written as $X_n \xrightarrow{a.e.} X$.

Also, define the random variable X on this sample space as follows:

$$X(s) = \begin{cases} 1 & 0 \le s < \frac{1}{2} \\ 0 & \text{otherwise} \end{cases}$$

Show that $X_n \xrightarrow{a.s.} X$.

Solution: Define the set A as follows:

$$A = \left\{ s \in S : \lim_{n \to \infty} X_n(s) = X(s) \right\}.$$

We need to prove that $P(A) = 1$. Let's first find A. Note that $\frac{n+1}{2n} > \frac{1}{2}$, so for any $s \in [0, \frac{1}{2})$, we have

$$X_n(s) = X(s) = 1.$$

Therefore, we conclude that $[0, 0.5) \subset A$. Now if $s > \frac{1}{2}$, then

$$X(s) = 0.$$

Also, since $2s - 1 > 0$, we can write

$$X_n(s) = 0, \qquad \text{for all } n > \frac{1}{2s-1}.$$

Therefore,

$$\lim_{n \to \infty} X_n(s) = 0 = X(s), \qquad \text{for all } s > \frac{1}{2}.$$

We conclude $(\frac{1}{2}, 1] \subset A$. You can check that $s = \frac{1}{2} \notin A$, since

$$X_n \left(\frac{1}{2} \right) = 1, \qquad \text{for all } n,$$

while $X \left(\frac{1}{2} \right) = 0$. We conclude

$$A = \left[0, \frac{1}{2} \right) \cup \left(\frac{1}{2}, 1 \right] = S - \left\{ \frac{1}{2} \right\}.$$

Since $P(A) = 1$, we conclude $X_n \xrightarrow{a.s.} X$.

In some problems, proving almost sure convergence directly can be difficult. Thus, it is desirable to know some sufficient conditions for almost sure convergence. Here is a result that is sometimes useful when we would like to prove almost sure convergence.

Theorem 7.5. Consider the sequence $X_1, X_2, X_3, \cdots$. If for all $\epsilon > 0$, we have

$$\sum_{n=1}^{\infty} P(|X_n - X| > \epsilon) < \infty,$$

then $X_n \xrightarrow{a.s.} X$.

Example 7.14. Consider a sequence $\{X_n, n = 1, 2, 3, \cdots\}$ such that

$$
X_n = \begin{cases} -\frac{1}{n} & \text{with probability } \frac{1}{2} \\ \frac{1}{n} & \text{with probability } \frac{1}{2} \end{cases}
$$

Show that $X_n \xrightarrow{a.s.} 0$.

Solution: By Theorem 7.5, it suffices to show that

$$
\sum_{n=1}^{\infty} P\big(|X_n| > \epsilon\big) < \infty.
$$

Note that $|X_n| = \frac{1}{n}$. Thus, $|X_n| > \epsilon$ if and only if $n < \frac{1}{\epsilon}$. Thus, we conclude

$$
\sum_{n=1}^{\infty} P\big(|X_n| > \epsilon\big) \leq \sum_{n=1}^{\lfloor \frac{1}{\epsilon} \rfloor} P\big(|X_n| > \epsilon\big)
$$

$$
= \lfloor \frac{1}{\epsilon} \rfloor < \infty.
$$

Theorem 7.5 provides only a sufficient condition for almost sure convergence. In particular, if we obtain

$$
\sum_{n=1}^{\infty} P\big(|X_n - X| > \epsilon\big) = \infty,
$$

then we still don't know whether the X_n's converge to X almost surely or not. Here, we provide a condition that is both necessary and sufficient.

Theorem 7.6. Consider the sequence X_1, X_2, X_3, $\cdots$. For any $\epsilon > 0$, define the set of events

$$
A_m = \{|X_n - X| < \epsilon, \text{ for all } n \geq m\}.
$$

Then $X_n \xrightarrow{a.s.} X$ if and only if for any $\epsilon > 0$, we have

$$
\lim_{m \to \infty} P(A_m) = 1.
$$

Example 7.15. Let X_1, X_2, X_3, $\cdots$ be independent random variables, where

$$
X_n \sim Bernoulli\left(\frac{1}{n}\right)
$$

for $n = 2, 3, \cdots$. The goal here is to check whether $X_n \xrightarrow{a.s.} 0$.

1. Check that $\sum_{n=1}^{\infty} P\big(|X_n| > \epsilon\big) = \infty$.

2. Show that the sequence X_1, X_2, ... does not converge to 0 almost surely using Theorem 7.6.

Solution:

1. We first note that for $0 < \epsilon < 1$, we have

$$\sum_{n=1}^{\infty} P\big(|X_n| > \epsilon\big) = \sum_{n=1}^{\infty} P(X_n = 1)$$
$$= \sum_{n=1}^{\infty} \frac{1}{n} = \infty.$$

2. To use Theorem 7.6, we define

$$A_m = \{|X_n| < \epsilon, \text{ for all } n \geq m\}.$$

Note that for $0 < \epsilon < 1$, we have

$$A_m = \{X_n = 0, \text{ for all } n \geq m\}.$$

According to Theorem 7.6, it suffices to show that

$$\lim_{m \to \infty} P(A_m) < 1.$$

We can in fact show that $\lim_{m \to \infty} P(A_m) = 0$. To show this, we will prove $P(A_m) = 0$, for every $m \geq 2$. For $0 < \epsilon < 1$, we have

$$
\begin{aligned}
P(A_m) &= P\big(\{X_n = 0, \text{ for all } n \geq m\}\big) \\
&\leq P\big(\{X_n = 0, \text{ for } n = m, m+1, \cdots, N\}\big) \quad &&\text{(for every positive integer } N \geq m\text{)} \\
&= P(X_m = 0)P(X_{m+1} = 0)\cdots P(X_N = 0) \quad &&\text{(since } X_i\text{'s are independent)} \\
&= \frac{m-1}{m} \cdot \frac{m}{m+1} \cdots \frac{N-1}{N} \\
&= \frac{m-1}{N}.
\end{aligned}
$$

Thus, by choosing N large enough, we can show that $P(A_m)$ is less than any positive number. Therefore, $P(A_m) = 0$ for all $m \geq 2$. We conclude that $\lim_{m \to \infty} P(A_m) = 0$. Thus, according to Theorem 7.6, the sequence X_1, X_2, ... does not converge to 0 almost surely.

An important example for almost sure convergence is the **strong law of large numbers (SLLN)**. Here, we state the SLLN without proof. The interested reader can find a proof of SLLN in [19]. A simpler proof can be obtained if we assume the finiteness of the fourth moment. (See [20] for example.)

The strong law of large numbers (SLLN)

Let $X_1, X_2, ..., X_n$ be i.i.d. random variables with a finite expected value $EX_i = \mu < \infty$. Let also

$$M_n = \frac{X_1 + X_2 + ... + X_n}{n}.$$

Then $M_n \xrightarrow{a.s.} \mu$.

We end this section by stating a version of the **continuous mapping theorem**. This theorem is sometimes useful when proving the convergence of random variables.

Theorem 7.7. Let $X_1, X_2, X_3, \cdots$ be a sequence of random variables. Let also $h : \mathbb{R} \mapsto \mathbb{R}$ be a <u>continuous</u> function. Then, the following statements are true:

- If $X_n \xrightarrow{d} X$, then $h(X_n) \xrightarrow{d} h(X)$.

- If $X_n \xrightarrow{p} X$, then $h(X_n) \xrightarrow{p} h(X)$.

- If $X_n \xrightarrow{a.s.} X$, then $h(X_n) \xrightarrow{a.s.} h(X)$.

7.2.8 Solved Problems

1. Let $X_1, X_2, X_3, \cdots$ be a sequence of random variables such that

$$X_n \sim Geometric\left(\frac{\lambda}{n}\right), \qquad \text{for } n = 1, 2, 3, \cdots,$$

where $\lambda > 0$ is a constant. Define a new sequence Y_n as

$$Y_n = \frac{1}{n}X_n, \qquad \text{for } n = 1, 2, 3, \cdots.$$

Show that Y_n converges in distribution to $Exponential(\lambda)$.

Solution: Note that if $W \sim Geometric(p)$, then for any positive integer l, we have

$$P(W \leq l) = \sum_{k=1}^{l} (1-p)^{k-1} p$$

$$= p \sum_{k=1}^{l} (1-p)^{k-1}$$

$$= p \cdot \frac{1 - (1-p)^l}{1 - (1-p)}$$

$$= 1 - (1-p)^l.$$

Now, since $Y_n = \frac{1}{n} X_n$, for any positive real number, we can write

$$P(Y_n \leq y) = P(X_n \leq ny)$$

$$= 1 - \left(1 - \frac{\lambda}{n}\right)^{\lfloor ny \rfloor},$$

where $\lfloor ny \rfloor$ is the largest integer less than or equal to ny. We then write

$$\lim_{n \to \infty} F_{Y_n}(y) = \lim_{n \to \infty} 1 - \left(1 - \frac{\lambda}{n}\right)^{\lfloor ny \rfloor}$$

$$= 1 - \lim_{n \to \infty} \left(1 - \frac{\lambda}{n}\right)^{\lfloor ny \rfloor}$$

$$= 1 - e^{-\lambda y}.$$

The last equality holds because $ny - 1 \leq \lfloor ny \rfloor \leq ny$, and

$$\lim_{n \to \infty} \left(1 - \frac{\lambda}{n}\right)^{ny} = e^{-\lambda y}.$$

2. Let X_1, X_2, X_3, $\cdots$ be a sequence of i.i.d. $Uniform(0,1)$ random variables. Define the sequence Y_n as

$$Y_n = \min(X_1, X_2, \cdots, X_n).$$

Prove the following convergence results independently (i.e, do not conclude the weaker convergence modes from the stronger ones).

(a) $Y_n \xrightarrow{d} 0$.

(b) $Y_n \xrightarrow{p} 0$.

(c) $Y_n \xrightarrow{L^r} 0$, for all $r \geq 1$.

(d) $Y_n \xrightarrow{a.s} 0$.

Solution:

(a) $Y_n \xrightarrow{d} 0$: Note that

$$F_{X_n}(x) = \begin{cases} 0 & x < 0 \\ x & 0 \leq x \leq 1 \\ 1 & x > 1 \end{cases}$$

Also, note that $R_{Y_n} = [0, 1]$. For $0 \leq y \leq 1$, we can write

$$\begin{aligned} F_{Y_n}(y) &= P(Y_n \leq y) \\ &= 1 - P(Y_n > y) \\ &= 1 - P(X_1 > y, X_2 > y, \cdots, X_n > y) \\ &= 1 - P(X_1 > y)P(X_2 > y) \cdots P(X_n > y) \qquad \text{(since } X_i\text{'s are independent)} \\ &= 1 - (1 - F_{X_1}(y))(1 - F_{X_2}(y)) \cdots (1 - F_{X_n}(y)) \\ &= 1 - (1 - y)^n. \end{aligned}$$

Therefore, we conclude

$$\lim_{n \to \infty} F_{Y_n}(y) = \begin{cases} 0 & y \leq 0 \\ 1 & y > 0 \end{cases}$$

Therefore, $Y_n \xrightarrow{d} 0$.

(b) $Y_n \xrightarrow{p} 0$: Note that as we found in part (a)

$$F_{Y_n}(y) = \begin{cases} 0 & y < 0 \\ 1 - (1 - y)^n & 0 \leq y \leq 1 \\ 1 & y > 1 \end{cases}$$

In particular, note that Y_n is a continuous random variable. To show $Y_n \xrightarrow{p} 0$, we need to show that

$$\lim_{n \to \infty} P(|Y_n| \geq \epsilon) = 0, \qquad \text{for all } \epsilon > 0.$$

Since $Y_n \geq 0$, it suffices to show that

$$\lim_{n \to \infty} P(Y_n \geq \epsilon) = 0, \qquad \text{for all } \epsilon > 0.$$

For $\epsilon \in (0, 1)$, we have

$$\begin{aligned} P(Y_n \geq \epsilon) &= 1 - P(Y_n < \epsilon) \\ &= 1 - P(Y_n \leq \epsilon) \qquad \text{(since } Y_n \text{ is a continuous random variable)} \\ &= 1 - F_{Y_n}(\epsilon) \\ &= (1 - \epsilon)^n. \end{aligned}$$

Therefore,

$$\lim_{n \to \infty} P(|Y_n| \geq \epsilon) = \lim_{n \to \infty} (1 - \epsilon)^n$$
$$= 0, \qquad \text{for all } \epsilon \in (0, 1].$$

(c) $Y_n \xrightarrow{L^r} 0$, for all $r \geq 1$: By differentiating $F_{Y_n}(y)$, we obtain

$$f_{Y_n}(y) = \begin{cases} n(1-y)^{n-1} & 0 \leq y \leq 1 \\ 0 & \text{otherwise} \end{cases}$$

Thus, for $r \geq 1$, we can write

$$E|Y_n|^r = \int_0^1 ny^r(1-y)^{n-1}dy$$
$$\leq \int_0^1 ny(1-y)^{n-1}dy \qquad (\text{since } r \geq 1)$$
$$= \left[-y(1-y)^n \right]_0^1 + \int_0^1 (1-y)^n dy \qquad (\text{integration by parts})$$
$$= \frac{1}{n+1}.$$

Therefore

$$\lim_{n \to \infty} E\left(|Y_n|^r\right) = 0.$$

(d) $Y_n \xrightarrow{a.s.} 0$: We will prove

$$\sum_{n=1}^{\infty} P(|Y_n| > \epsilon) < \infty,$$

which implies $Y_n \xrightarrow{a.s.} 0$. By our discussion in part (b),

$$\sum_{n=1}^{\infty} P(|Y_n| > \epsilon) = \sum_{n=1}^{\infty} (1 - \epsilon)^n$$
$$= \frac{1 - \epsilon}{\epsilon} < \infty \qquad (\text{geometric series}).$$

3. Let $X_n \sim N(0, \frac{1}{n})$. Show that $X_n \xrightarrow{a.s.} 0$. *Hint:* You may decide to use the inequality given in Equation 4.7, which is

$$1 - \Phi(x) \leq \frac{1}{\sqrt{2\pi}} \frac{1}{x} e^{-\frac{x^2}{2}}.$$

Solution: We will prove

$$\sum_{n=1}^{\infty} P(|X_n| > \epsilon) < \infty,$$

which implies $X_n \xrightarrow{a.s.} 0$. In particular,

$$P(|X_n| > \epsilon) = 2(1 - \Phi(\epsilon n)) \qquad \text{(since } X_n \sim N(0, \frac{1}{n}))$$

$$\leq \frac{1}{\sqrt{2\pi}} \frac{2}{\epsilon n} e^{-\frac{\epsilon^2 n^2}{2}}$$

$$\leq \frac{1}{\sqrt{2\pi}} \frac{2}{\epsilon} e^{-\frac{\epsilon^2 n^2}{2}}$$

$$\leq \frac{1}{\sqrt{2\pi}} \frac{2}{\epsilon} e^{-\frac{\epsilon^2 n}{2}}.$$

Therefore,

$$\sum_{n=1}^{\infty} P(|X_n| > \epsilon) \leq \sum_{n=1}^{\infty} \frac{1}{\sqrt{2\pi}} \frac{2}{\epsilon} e^{-\frac{\epsilon^2 n}{2}}$$

$$= \frac{1}{\sqrt{2\pi}} \frac{2}{\epsilon} \sum_{n=1}^{\infty} e^{-\frac{\epsilon^2 n}{2}}$$

$$= \frac{1}{\sqrt{2\pi}} \frac{2}{\epsilon} \frac{e^{-\frac{\epsilon^2}{2}}}{1 - e^{-\frac{\epsilon^2}{2}}} < \infty \qquad \text{(geometric series)}.$$

4. Consider the sample space $S = [0, 1]$ with uniform probability distribution, i.e.,

$$P([a, b]) = b - a, \qquad \text{for all } 0 \leq a \leq b \leq 1.$$

Define the sequence $\{X_n, n = 1, 2, \cdots\}$ as $X_n(s) = \frac{n}{n+1} s + (1 - s)^n$. Also, define the random variable X on this sample space as $X(s) = s$. Show that $X_n \xrightarrow{a.s.} X$.

Solution: For any $s \in (0, 1]$, we have

$$\lim_{n \to \infty} X_n(s) = \lim_{n \to \infty} \left[\frac{n}{n+1} s + (1 - s)^n \right]$$

$$= s = X(s).$$

However, if $s = 0$, then

$$\lim_{n \to \infty} X_n(0) = \lim_{n \to \infty} \left[\frac{n}{n+1} \cdot 0 + (1-0)^n \right]$$
$$= 1.$$

Thus, we conclude

$$\lim_{n \to \infty} X_n(s) = X(s), \qquad \text{for all } s \in (0,1].$$

Since $P\big((0,1]\big) = 1$, we conclude $X_n \xrightarrow{a.s.} X$.

5. Let $\{X_n, n = 1, 2, \cdots\}$ and $\{Y_n, n = 1, 2, \cdots\}$ be two sequences of random variables, defined on the sample space S. Suppose that we know

$$X_n \xrightarrow{a.s.} X,$$
$$Y_n \xrightarrow{a.s.} Y.$$

Prove that $X_n + Y_n \xrightarrow{a.s.} X + Y$.

Solution: Define the sets A and B as follows:

$$A = \left\{ s \in S : \lim_{n \to \infty} X_n(s) = X(s) \right\},$$
$$B = \left\{ s \in S : \lim_{n \to \infty} Y_n(s) = Y(s) \right\}.$$

By definition of almost sure convergence, we conclude $P(A) = P(B) = 1$. Therefore, $P(A^c) = P(B^c) = 0$. We conclude

$$P(A \cap B) = 1 - P(A^c \cup B^c)$$
$$\geq 1 - P(A^c) - P(B^c)$$
$$= 1.$$

Thus, $P(A \cap B) = 1$. Now, consider the sequence $\{Z_n, n = 1, 2, \cdots\}$, where $Z_n = X_n + Y_n$, and define the set C as

$$C = \left\{ s \in S : \lim_{n \to \infty} Z_n(s) = X(s) + Y(s) \right\}.$$

We claim $A \cap B \subset C$. Specifically, if $s \in A \cap B$, then we have

$$\lim_{n \to \infty} X_n(s) = X(s), \qquad \lim_{n \to \infty} Y_n(s) = Y(s).$$

Therefore,

$$\lim_{n \to \infty} Z_n(s) = \lim_{n \to \infty} \left[X_n(s) + Y_n(s) \right]$$
$$= \lim_{n \to \infty} X_n(s) + \lim_{n \to \infty} Y_n(s)$$
$$= X(s) + Y(s).$$

Thus, $s \in C$. We conclude $A \cap B \subset C$. Thus,

$$P(C) \geq P(A \cap B) = 1,$$

which implies $P(C) = 1$. This means that $Z_n \xrightarrow{a.s.} X + Y$.

6. Let $\{X_n, n = 1, 2, \cdots\}$ and $\{Y_n, n = 1, 2, \cdots\}$ be two sequences of random variables, defined on the sample space S. Suppose that we know

$$X_n \xrightarrow{p} X,$$
$$Y_n \xrightarrow{p} Y.$$

Prove that $X_n + Y_n \xrightarrow{p} X + Y$.

Solution: For $n \in \mathbb{N}$, define the following events

$$A_n = \left\{ |X_n - X| < \frac{\epsilon}{2} \right\},$$
$$B_n = \left\{ |Y_n - Y| < \frac{\epsilon}{2} \right\}.$$

Since $X_n \xrightarrow{p} X$ and $Y_n \xrightarrow{p} Y$, we have for all $\epsilon > 0$

$$\lim_{n \to \infty} P(A_n) = 1,$$
$$\lim_{n \to \infty} P(B_n) = 1.$$

We can also write

$$P(A_n \cap B_n) = P(A_n) + P(B_n) - P(A_n \cup B_n)$$
$$\geq P(A_n) + P(B_n) - 1.$$

Therefore,

$$\lim_{n \to \infty} P(A_n \cap B_n) = 1.$$

Now, let us define the events C_n and D_n as follows:

$$C_n = \left\{ |X_n - X| + |Y_n - Y| < \epsilon \right\},$$
$$D_n = \left\{ |X_n + Y_n - X - Y| < \epsilon \right\}.$$

Now, note that $(A_n \cap B_n) \subset C_n$, thus $P(A_n \cap B_n) \leq P(C_n)$. Also, by the triangle inequality for absolute values, we have

$$|(X_n - X) + (Y_n - Y)| \leq |X_n - X| + |Y_n - Y|.$$

Therefore, $C_n \subset D_n$, which implies

$$P(C_n) \leq P(D_n).$$

We conclude

$$P(A_n \cap B_n) \leq P(C_n) \leq P(D_n).$$

Since $\lim_{n \to \infty} P(A_n \cap B_n) = 1$, we conclude $\lim_{n \to \infty} P(D_n) = 1$. This by definition means that $X_n + Y_n \overset{p}{\to} X + Y$.

7.3   End of Chapter Problems

1. Let X_i be i.i.d. $Uniform(0, 1)$. We define the sample mean as

$$M_n = \frac{X_1 + X_2 + ... + X_n}{n}.$$

(a) Find $E[M_n]$ and $\text{Var}(M_n)$ as a function of n.

(b) Using Chebyshev's inequality, find an upper bound on

$$P\left(\left|M_n - \frac{1}{2}\right| \geq \frac{1}{100}\right).$$

(c) Using your bound, show that

$$\lim_{n \to \infty} P\left(\left|M_n - \frac{1}{2}\right| \geq \frac{1}{100}\right) = 0.$$

2. The number of accidents in a certain city is modeled by a Poisson random variable with an average rate of 10 accidents per day. Suppose that the number of accidents on different days are independent. Use the central limit theorem to find the probability that there will be more than 3800 accidents in a certain year. Assume that there are 365 days in a year.

3. In a communication system, each codeword consists of 1000 bits. Due to the noise, each bit may be received in error with probability 0.1. It is assumed bit errors occur independently. Since error correcting codes are used in this system, each codeword can be decoded reliably if there are less than or equal to 125 errors in the received codeword, otherwise the decoding fails. Using the CLT, find the probability of decoding failure.

4. 50 students live in a dormitory. The parking lot has the capacity for 30 cars. Each student has a car with probability $\frac{1}{2}$, independently from other students. Use the CLT (with continuity correction) to find the probability that there won't be enough parking spaces for all the cars.

5. The amount of time needed for a certain machine to process a job is a random variable with mean $EX_i = 10$ minutes and $\text{Var}(X_i) = 2$ minutes^2. The times needed for different jobs are independent from each other. Find the probability that the machine processes less than or equal to 40 jobs in 7 hours.

6. You have a fair coin. You toss the coin n times. Let X be the portion of times that you observe heads. How large n has to be so that you are 95% sure that $0.45 \leq X \leq 0.55$? In other words, how large n has to be so that

$$P(0.45 \leq X \leq 0.55) \geq .95 \ ?$$

7. An engineer is measuring a quantity q. It is assumed that there is a random error in each measurement, so the engineer will take n measurements and reports the average of the

measurements as the estimated value of q. Specifically, if Y_i is the value that is obtained in the i'th measurement, we assume that

$$Y_i = q + X_i,$$

where X_i is the error in the ith measurement. We assume that X_i's are i.i.d. with $EX_i = 0$ and $\text{Var}(X_i) = 4$ units. The engineer reports the average of measurements

$$M_n = \frac{Y_1 + Y_2 + \dots + Y_n}{n}.$$

How many measurements does the engineer need to make until he is 95% sure that the final error is less than 0.1 units? In other words, what should the value of n be such that

$$P\big(q - 0.1 \leq M_n \leq q + 0.1\big) \geq 0.95 \ ?$$

8. Let $X_2, X_3, X_4, \cdots$ be a sequence of random variables such that

$$F_{X_n}(x) = \begin{cases} \frac{e^{n(x-1)}}{1+e^{n(x-1)}} & x > 0 \\ 0 & \text{otherwise} \end{cases}$$

Show that X_n converges in distribution to $X = 1$.

9. Let $X_2, X_3, X_4, \cdots$ be a sequence of non-negative random variables such that

$$F_{X_n}(x) = \begin{cases} \frac{e^{nx} + xe^n}{e^{nx} + \left(\frac{n+1}{n}\right)e^n} & 0 \leq x \leq 1 \\ \frac{e^{nx} + e^n}{e^{nx} + \left(\frac{n+1}{n}\right)e^n} & x > 1 \end{cases}$$

Show that X_n converges in distribution to $Uniform(0, 1)$.

10. Consider a sequence $\{X_n, n = 1, 2, 3, \cdots\}$ such that

$$X_n = \begin{cases} n & \text{with probability } \frac{1}{n^2} \\ 0 & \text{with probability } 1 - \frac{1}{n^2} \end{cases}$$

Show that

(a) $X_n \xrightarrow{p} 0$;

(b) $X_n \xrightarrow{L^r} 0$, for $r < 2$;

(c) X_n does not converge to 0 in the rth mean for any $r \geq 2$;

(d) $X_n \xrightarrow{a.s.} 0$.

11. We perform the following random experiment. We put $n \geq 10$ blue balls and n red balls in a bag. We pick 10 balls at random (without replacement) from the bag. Let X_n be the number of blue balls. We perform this experiment for $n = 10, 11, 12, \cdots$. Prove that $X_n \xrightarrow{d} Binomial\left(10, \frac{1}{2}\right)$.

12. Find two sequences of random variables $\{X_n, n = 1, 2, \cdots\}$ and $\{Y_n, n = 1, 2, \cdots\}$ such that

$$X_n \overset{d}{\to} X,$$

and

$$Y_n \overset{d}{\to} Y,$$

but $X_n + Y_n$ does not converge in distribution to $X + Y$.

13. Let $X_1, X_2, X_3, \cdots$ be a sequence of continuous random variable such that

$$f_{X_n}(x) = \frac{n}{2}e^{-n|x|}.$$

Show that X_n converges in probability to 0.

14. Let $X_1, X_2, X_3, \cdots$ be a sequence of continuous random variable such that

$$f_{X_n}(x) = \begin{cases} \frac{1}{nx^2} & x > \frac{1}{n} \\ 0 & \text{otherwise} \end{cases}$$

Show that X_n converges in probability to 0.

15. Let $Y_1, Y_2, Y_3, \cdots$ be a sequence of i.i.d. random variables with mean $EY_i = \mu$ and finite variance $\text{Var}(Y_i) = \sigma^2$. Define the sequence $\{X_n, n = 2, 3, ...\}$ as

$$X_n = \frac{Y_1Y_2 + Y_2Y_3 + \cdots Y_{n-1}Y_n + Y_nY_1}{n}, \qquad \text{for } n = 2, 3, \cdots.$$

Show that $X_n \overset{p}{\to} \mu^2$.

16. Let $Y_1, Y_2, Y_3, \cdots$ be a sequence of positive i.i.d. random variables with $0 < E[\ln Y_i] = \gamma < \infty$. Define the sequence $\{X_n, n = 1, 2, 3, ...\}$ as

$$X_n = (Y_1Y_2Y_3 \cdots Y_{n-1}Y_n)^{\frac{1}{n}}, \qquad \text{for } n = 1, 2, 3, \cdots.$$

Show that $X_n \overset{p}{\to} e^{\gamma}$.

17. Let $X_1, X_2, X_3, \cdots$ be a sequence of random variable such that

$$X_n \sim Poisson(n\lambda), \qquad \text{for } n = 1, 2, 3, \cdots,$$

where $\lambda > 0$ is a constant. Define a new sequence Y_n as

$$Y_n = \frac{1}{n}X_n, \qquad \text{for } n = 1, 2, 3, \cdots.$$

Show that Y_n converges in mean square to λ, i.e., $Y_n \overset{m.s.}{\longrightarrow} \lambda$.

18. Let $\{X_n, n = 1, 2, \cdots\}$ and $\{Y_n, n = 1, 2, \cdots\}$ be two sequences of random variables, defined on the sample space S. Suppose that we know

$$X_n \xrightarrow{L^r} X,$$

$$Y_n \xrightarrow{L^r} Y.$$

Prove that $X_n + Y_n \xrightarrow{L^r} X + Y$. *Hint:* You may want to use Minkowski's inequality which states that for two random variables X and Y with finite moments, and $1 \le p < \infty$, we have

$$E\left[|X + Y|^p\right] \le E\left[|X|^p\right]^{\frac{1}{p}} + E\left[|Y|^p\right]^{\frac{1}{p}}.$$

19. Let X_1, X_2, X_3, $\cdots$ be a sequence of random variable such that $X_n \sim Rayleigh(\frac{1}{n})$, i.e.,

$$f_{X_n}(x) = \begin{cases} n^2 x e^{-\frac{n^2 x^2}{2}} & x > 0 \\ 0 & \text{otherwise} \end{cases}$$

Show that $X_n \xrightarrow{a.s.} 0$.

20. Let $Y_1, Y_2, \cdots$ be independent random variables, where $Y_n \sim Bernoulli\left(\frac{n}{n+1}\right)$ for $n = 1, 2, 3, \cdots$. We define the sequence $\{X_n, n = 2, 3, 4, \cdots\}$ as

$$X_{n+1} = Y_1 Y_2 Y_3 \cdots Y_n, \qquad \text{for } n = 1, 2, 3, \cdots.$$

Show that $X_n \xrightarrow{a.s.} 0$.

Chapter 8

Statistical Inference I: Classical Methods

8.1 Introduction

In real life, we work with data that are affected by randomness, and we need to extract information and draw conclusions from the data. The randomness might come from a variety of sources. Here are two examples of such situations:

- Suppose that we would like to predict the outcome of an election. Since we cannot poll the entire population, we will choose a random sample from the population and ask them who they plan to vote for. In this experiment, the randomness comes from the sampling. Note also that if our poll is conducted one month before the election, another source of randomness is that people might change their opinions during the one month period.

- In a wireless communication system, a message is transmitted from a transmitter to a receiver. However, the receiver receives a corrupted version (a noisy version) of the transmitted signal. The receiver needs to extract the original message from the received noisy version. Here, the randomness comes from the noise.

Examples like these are abundant. Dealing with such situations is the subject of the field of statistical inference.

> **Statistical inference** is a collection of methods that deal with drawing conclusions from data that are prone to random variation.

Clearly, we use our knowledge of probability theory when we work on statistical inference problems. However, the big addition here is that we need to work with real **data**. The probability problems that we have seen in this book so far were clearly defined and the probability models were given to us. For example, you might have seen a problem like this:

Let X be a normal random variable with mean $\mu = 100$ and variance $\sigma^2 = 15$. Find the probability that $X > 110$.

In real life, we might not know the distribution of X, so we need to collect data, and from the data we should conclude whether X has a normal distribution or not. Now, suppose that we can use the central limit theorem to argue that X is normally distributed. Even in that case, we need to collect data to be able estimate μ and σ.

Here is a general setup for a statistical inference problem: There is an unknown quantity that we would like to estimate. We get some data. From the data, we estimate the desired quantity. There are two major approaches to this problem:

- **Frequentist (classical) Inference:** In this approach, the unknown quantity θ is assumed to be a fixed quantity. That is, θ is a deterministic (non-random) quantity that is to be estimated by the observed data. For example, in the polling problem stated above we might consider θ as the percentage of people who will vote for a certain candidate, call him/her Candidate A. After asking n randomly chosen voters, we might estimate θ by

$$\hat{\Theta} = \frac{Y}{n},$$

  where Y is the number of people (among the randomly chosen voters) who say they will vote for Candidate A. Although θ is assumed to be a non-random quantity, our estimator of θ, which we show by $\hat{\Theta}$ is a random variable because it depends on our random sample.

- **Bayesian Inference:** In the Bayesian approach, the unknown quantity Θ is assumed to be a random variable, and we assume that we have some initial guess about the distribution of Θ. After observing the data, we update the distribution of Θ using Bayes' Rule.

  As an example, consider the communication system in which the information is transmitted in the form of bits, i.e., 0's and 1's. Let's assume that, in each transmission, the transmitter sends a 1 with probability p, or it sends a 0 with probability $1 - p$. Thus, if Θ is the transmitted bit, then $\Theta \sim Bernoulli(p)$. At the receiver, X, which is a noisy version of Θ, is received. The receiver has to recover Θ from X. Here, to estimate Θ, we use our prior knowledge that $\Theta \sim Bernoulli(p)$.

In summary, you may say that frequentist (classical) inference deals with estimating non-random quantities, while Bayesian inference deals with estimating random variables. We will discuss frequentist and Bayesian approaches more in detail in this and the next chapter. Nevertheless, it is important to note that both approaches are very useful and widely used in practice. In this chapter, we will focus on frequentist methods, while in the next chapter we will discuss Bayesian methods.

8.1.1   Random Sampling

When collecting data, we often make several observations on a random variable. For example, suppose that our goal is to investigate the height distribution of people in a well defined population (i.e., adults between 25 and 50 in a certain country). To do this, we define random variables $X_1, X_2, X_3, ..., X_n$ as follows: We choose a random sample of size n with replacement from the population and let X_i be the height of the ith chosen person. More specifically,

1. We chose a person uniformly at random from the population and let X_1 be the height of that person. Here, every person in the population has the same chance of being chosen.

2. To determine the value of X_2, again we choose a person uniformly (and independently from the first person) at random and let X_2 be the height of that person. Again, every person in the population has the same chance of being chosen.

3. In general, X_i is the height of the ith person that is chosen uniformly and independently from the population.

You might ask why we do the sampling with replacement. In practice, we often do the sampling without replacement. That is, we do not allow one person to be chosen twice. However, if the population is large, then the probability of choosing one person twice is extremely low, and it can be shown that the results obtained from sampling with replacement are very close to the results obtained using sampling without replacement. The big advantage of sampling with replacement (the above procedure) is that X_i's will be independent and this makes the analysis much simpler.

Now, for example, if we would like to estimate the average height in the population, we may define an estimator as

$$\hat{\Theta} = \frac{X_1 + X_2 + \cdots + X_n}{n}.$$

The random variables X_1, X_2, X_3, ..., X_n defined above are independent and identically distributed (i.i.d.) and we refer to them collectively as a (simple) random sample.

The collection of random variables X_1, X_2, X_3, ..., X_n is said to be a **random sample** of size n if they are independent and identically distributed (i.i.d.), i.e.,

1. X_1, X_2, X_3, ..., X_n are independent random variables, and

2. they have the same distribution, i.e, $F_{X_1}(x) = F_{X_2}(x) = \ldots = F_{X_n}(x)$, for all $x \in \mathbb{R}$.

In the above example, the random variable $\hat{\Theta} = \frac{X_1 + X_2 + \cdots + X_n}{n}$ is called a **point estimator** for the average height in the population. After performing the above experiment, we will obtain $\hat{\Theta} = \hat{\theta}$. Here, $\hat{\theta}$ is called an **estimate** of the average height in the population. In general, a point estimator is a function of the random sample $\hat{\Theta} = h(X_1, X_2, \cdots, X_n)$ that is used to estimate an unknown quantity.

It is worth noting that there are different methods for sampling from a population. We refer to the above sampling method as *simple random sampling*. In general, "sampling is concerned with the selection of a subset of individuals from within a statistical population to estimate characteristics of the whole population" [18]. Nevertheless, for the material that we cover in this book simple random sampling is sufficient. Unless otherwise stated, when we refer to random samples, we assume they are simple random samples.

Some Properties of Random Samples:

Since we will be working with random samples, we would like to review some properties of random samples in this section. Here, we assume that X_1, X_2, X_3, ..., X_n are a random sample. Specifically, we assume

1. the X_i's are independent;

2. $F_{X_1}(x) = F_{X_2}(x) = ... = F_{X_n}(x) = F_X(x)$;

3. $EX_i = EX = \mu < \infty$;

4. $0 < \text{Var}(X_i) = \text{Var}(X) = \sigma^2 < \infty$.

Sample Mean:

The sample mean is defined as

$$\overline{X} = \frac{X_1 + X_2 + ... + X_n}{n}.$$

Another common notation for the sample mean is M_n. Since X_i are assumed to have the CDF $F_X(x)$, the sample mean is sometimes denoted by $M_n(X)$ to indicate the distribution of X_i's.

<u>Properties of the sample mean</u>

1. $E\overline{X} = \mu$.

2. $\text{Var}(\overline{X}) = \frac{\sigma^2}{n}$.

3. Weak Law of Large Numbers (WLLN):

$$\lim_{n \to \infty} P(|\overline{X} - \mu| \geq \epsilon) = 0.$$

4. Central Limit Theorem: The random variable

$$Z_n = \frac{\overline{X} - \mu}{\sigma/\sqrt{n}} = \frac{X_1 + X_2 + ... + X_n - n\mu}{\sqrt{n}\sigma}$$

converges in distribution to the standard normal random variable as n goes to infinity, that is

$$\lim_{n \to \infty} P(Z_n \leq x) = \Phi(x), \qquad \text{for all } x \in \mathbb{R}$$

where $\Phi(x)$ is the standard normal CDF.

Order Statistics:

Given a random sample, we might be interested in quantities such as the largest, the smallest, or the middle value in the sample. Thus, we often order the observed data from the smallest to the largest. We call the resulting ordered random variables *order statistics*.

More specifically, let X_1, X_2, X_3, ..., X_n be a random sample from a continuous distribution with CDF $F_X(x)$. Let us order X_i's from the smallest to the largest and denote the resulting sequence of random variables as

$$X_{(1)}, X_{(2)}, \cdots, X_{(n)}.$$

Thus, we have

$$X_{(1)} = \min\left(X_1, X_2, \cdots, X_n\right);$$

and

$$X_{(n)} = \max\left(X_1, X_2, \cdots, X_n\right).$$

We call $X_{(1)}, X_{(2)}, \cdots, X_{(n)}$ the **order statistics** of the random sample X_1, X_2, X_3, ..., X_n. We are often interested in the PDFs or CDFs of the $X_{(i)}$'s. The following theorem provides these functions.

Theorem 8.1. Let X_1, X_2, ..., X_n be a random sample from a continuous distribution with CDF $F_X(x)$ and PDF $f_X(x)$. Let $X_{(1)}, X_{(2)}, \cdots, X_{(n)}$ be the order statistics of X_1, X_2, X_3, ..., X_n. Then the CDF and PDF of $X_{(i)}$ are given by

$$f_{X_{(i)}}(x) = \frac{n!}{(i-1)!(n-i)!} f_X(x) \left[F_X(x)\right]^{i-1} \left[1 - F_X(x)\right]^{n-i},$$

$$F_{X_{(i)}}(x) = \sum_{k=i}^{n} \binom{n}{k} \left[F_X(x)\right]^k \left[1 - F_X(x)\right]^{n-k}.$$

Also, the joint PDF of $X_{(1)}, X_{(2)}, \cdots, X_{(n)}$ is given by

$$f_{X_{(1)}, \cdots, X_{(n)}}(x_1, x_2, \cdots, x_n) = \begin{cases} n! f_X(x_1) f_X(x_2) \cdots f_X(x_n) & \text{for } x_1 \leq x_2 \leq x_2 \cdots \leq x_n \\ \\ 0 & \text{otherwise} \end{cases}$$

A method to prove the above theorem is outlined in the End of Chapter Problems section. Let's look at an example.

Example 8.1. Let X_1, X_2, X_3, X_4 be a random sample from the $Uniform(0, 1)$ distribution, and let $X_{(1)}$, $X_{(2)}$, $X_{(3)}$, $X_{(4)}$. Find the PDFs of $X_{(1)}$, $X_{(2)}$, and $X_{(4)}$.

Solution: Here, the ranges of the random variables are $[0, 1]$, so the PDFs and CDFs are zero outside of $[0, 1]$. We have

$$f_X(x) = 1, \qquad \text{for } x \in [0, 1],$$

and

$$F_X(x) = x, \qquad \text{for } x \in [0, 1].$$

By Theorem 8.1, we obtain

$$\begin{aligned} f_{X_{(1)}}(x) &= \frac{4!}{(1-1)!(4-1)!} f_X(x) \big[F_X(x)\big]^{1-1} \big[1 - F_X(x)\big]^{4-1} \\ &= 4 f_X(x) \big[1 - F_X(x)\big]^3 \\ &= 4(1-x)^3, \qquad \text{for } x \in [0, 1]. \end{aligned}$$

$$\begin{aligned} f_{X_{(2)}}(x) &= \frac{4!}{(2-1)!(4-2)!} f_X(x) \big[F_X(x)\big]^{2-1} \big[1 - F_X(x)\big]^{4-2} \\ &= 12 f_X(x) F_X(x) \big[1 - F_X(x)\big]^2 \\ &= 12x(1-x)^2, \qquad \text{for } x \in [0, 1]. \end{aligned}$$

$$\begin{aligned} f_{X_{(4)}}(x) &= \frac{4!}{(4-1)!(4-4)!} f_X(x) \big[F_X(x)\big]^{4-1} \big[1 - F_X(x)\big]^{4-4} \\ &= 4 f_X(x) \big[F_X(x)\big]^3 \\ &= 4x^3, \qquad \text{for } x \in [0, 1]. \end{aligned}$$

8.2   Point Estimation

Here, we assume that θ is an unknown parameter to be estimated. For example, θ might be the expected value of a random variable, $\theta = EX$. The important assumption here is that θ is a fixed (non-random) quantity. To estimate θ, we need to collect some data. Specifically, we get a random sample X_1, X_2, X_3, ..., X_n such that X_i's have the same distribution as X. To estimate θ, we define a point estimator $\hat{\Theta}$ that is a function of the random sample, i.e.,

$$\hat{\Theta} = h(X_1, X_2, \cdots, X_n).$$

For example, if $\theta = EX$, we may choose $\hat{\Theta}$ to be the sample mean

$$\hat{\Theta} = \overline{X} = \frac{X_1 + X_2 + \dots + X_n}{n}.$$

There are infinitely many possible estimators for θ, so how can we make sure that we have chosen a good estimator? How do we compare different possible estimators? To do this, we provide a list of some desirable properties that we would like our estimators to have. Intuitively, we know that a good estimator should be able to give us values that are "close" to the real value of θ. To make this notion more precise we provide some definitions.

8.2.1   Evaluating Estimators

We define three main desirable properties for point estimators. The first one is related to the estimator's *bias*. The bias of an estimator $\hat{\Theta}$ tells us on average how far $\hat{\Theta}$ is from the real value of θ.

Let $\hat{\Theta} = h(X_1, X_2, \cdots, X_n)$ be a point estimator for θ. The **bias** of point estimator $\hat{\Theta}$ is defined by

$$B(\hat{\Theta}) = E[\hat{\Theta}] - \theta.$$

In general, we would like to have a bias that is close to 0, indicating that on average, $\hat{\Theta}$ is close to θ. It is worth noting that $B(\hat{\Theta})$ might depend on the actual value of θ. In other words, you might have an estimator for which $B(\hat{\Theta})$ is small for some values of θ and large for some other values of θ. A desirable scenario is when $B(\hat{\Theta}) = 0$, i.e, $E[\hat{\Theta}] = \theta$, for all values of θ. In this case, we say that $\hat{\Theta}$ is an *unbiased* estimator of θ.

Let $\hat{\Theta} = h(X_1, X_2, \cdots, X_n)$ be a point estimator for a parameter θ. We say that $\hat{\Theta}$ is an **unbiased** estimator of θ if

$$B(\hat{\Theta}) = 0, \qquad \text{for all possible values of } \theta.$$

Example 8.2. Let X_1, X_2, X_3, ..., X_n be a random sample. Show that the sample mean

$$\hat{\Theta} = \overline{X} = \frac{X_1 + X_2 + ... + X_n}{n}$$

is an unbiased estimator of $\theta = EX_i$.

Solution: We have

$$\begin{aligned}
B(\hat{\Theta}) &= E[\hat{\Theta}] - \theta \\
&= E\left[\overline{X}\right] - \theta \\
&= EX_i - \theta \\
&= 0.
\end{aligned}$$

Note that if an estimator is unbiased, it is not necessarily a good estimator. In the above example, if we choose $\hat{\Theta}_1 = X_1$, then $\hat{\Theta}_1$ is also an unbiased estimator of θ:

$$\begin{aligned}
B(\hat{\Theta}_1) &= E[\hat{\Theta}_1] - \theta \\
&= EX_1 - \theta \\
&= 0.
\end{aligned}$$

Nevertheless, we suspect that $\hat{\Theta}_1$ is probably not as good as the sample mean $\overline{X}$. Therefore, we need other measures to ensure that an estimator is a "good" estimator. A very common measure is the *mean squared error* defined by $E\left[(\hat{\Theta} - \theta)^2\right]$.

> The **mean squared error** (MSE) of a point estimator $\hat{\Theta}$, shown by $MSE(\hat{\Theta})$, is defined as
>
> $$MSE(\hat{\Theta}) = E\big[(\hat{\Theta} - \theta)^2\big].$$

Note that $\hat{\Theta} - \theta$ is the error that we make when we estimate θ by $\hat{\Theta}$. Thus, the MSE is a measure of the distance between $\hat{\Theta}$ and θ, and a smaller MSE is generally indicative of a better estimator.

Example 8.3. Let X_1, X_2, X_3, ..., X_n be a random sample from a distribution with mean $EX_i = \theta$, and variance $\text{Var}(X_i) = \sigma^2$. Consider the following two estimators for θ:

1. $\hat{\Theta}_1 = X_1$.

2. $\hat{\Theta}_2 = \overline{X} = \frac{X_1 + X_2 + ... + X_n}{n}$.

Find $MSE(\hat{\Theta}_1)$ and $MSE(\hat{\Theta}_2)$ and show that for $n > 1$, we have

$$MSE(\hat{\Theta}_1) > MSE(\hat{\Theta}_2).$$

Solution: We have

$$
\begin{aligned}
MSE(\hat{\Theta}_1) &= E\big[(\hat{\Theta}_1 - \theta)^2\big] \\
&= E[(X_1 - EX_1)^2] \\
&= \text{Var}(X_1) \\
&= \sigma^2.
\end{aligned}
$$

To find $MSE(\hat{\Theta}_2)$, we can write

$$
\begin{aligned}
MSE(\hat{\Theta}_2) &= E\big[(\hat{\Theta}_2 - \theta)^2\big] \\
&= E[(\overline{X} - \theta)^2] \\
&= \text{Var}(\overline{X} - \theta) + \big(E[\overline{X} - \theta]\big)^2.
\end{aligned}
$$

The last equality results from $EY^2 = \text{Var}(Y) + (EY)^2$, where $Y = \overline{X} - \theta$. Now, note that

$$\text{Var}(\overline{X} - \theta) = \text{Var}(\overline{X})$$

since θ is a constant. Also, $E[\overline{X} - \theta] = 0$. Thus, we conclude

$$
\begin{aligned}
MSE(\hat{\Theta}_2) &= \text{Var}(\overline{X}) \\
&= \frac{\sigma^2}{n}.
\end{aligned}
$$

Thus, we conclude for $n > 1$,

$$MSE(\hat{\Theta}_1) > MSE(\hat{\Theta}_2).$$

From the above example, we conclude that although both $\hat{\Theta}_1$ and $\hat{\Theta}_2$ are unbiased estimators of the mean, $\hat{\Theta}_2 = \overline{X}$ is probably a better estimator since it has a smaller MSE. In general, if $\hat{\Theta}$ is a point estimator for θ, we can write

$$MSE(\hat{\Theta}) = E\big[(\hat{\Theta} - \theta)^2\big]$$
$$= \mathrm{Var}(\hat{\Theta} - \theta) + \big(E[\hat{\Theta} - \theta]\big)^2$$
$$= \mathrm{Var}(\hat{\Theta}) + B(\hat{\Theta})^2.$$

If $\hat{\Theta}$ is a point estimator for θ,

$$MSE(\hat{\Theta}) = \mathrm{Var}(\hat{\Theta}) + B(\hat{\Theta})^2,$$

where $B(\hat{\Theta}) = E[\hat{\Theta}] - \theta$ is the bias of $\hat{\Theta}$.

The last property that we discuss for point estimators is *consistency*. Loosely speaking, we say that an estimator is consistent, if as the sample size n gets larger, $\hat{\Theta}$ converges to the real value of θ. More precisely, we have the following definition:

Let $\hat{\Theta}_1$, $\hat{\Theta}_2$, $\cdots$, $\hat{\Theta}_n$, $\cdots$, be a sequence of point estimators of θ. We say that $\hat{\Theta}_n$ is a **consistent** estimator of θ, if

$$\lim_{n \to \infty} P\big(|\hat{\Theta}_n - \theta| \geq \epsilon\big) = 0, \qquad \text{for all } \epsilon > 0.$$

Example 8.4. Let X_1, X_2, X_3, ..., X_n be a random sample with mean $EX_i = \theta$, and variance $\mathrm{Var}(X_i) = \sigma^2$. Show that $\hat{\Theta}_n = \overline{X}$ is a consistent estimator of θ.

Solution: We need to show that

$$\lim_{n \to \infty} P\big(|\overline{X} - \theta| \geq \epsilon\big) = 0, \qquad \text{for all } \epsilon > 0.$$

But this is true because of the weak law of large numbers. In particular, we can use Chebyshev's inequality to write

$$P(|\overline{X} - \theta| \geq \epsilon) \leq \frac{\mathrm{Var}(\overline{X})}{\epsilon^2}$$
$$= \frac{\sigma^2}{n\epsilon^2},$$

which goes to 0 as $n \to \infty$.

We could also show the consistency of $\hat{\Theta}_n = \overline{X}$ by looking at the MSE. As we found previously, the MSE of $\hat{\Theta}_n = \overline{X}$ is given by

$$MSE(\hat{\Theta}_n) = \frac{\sigma^2}{n}.$$

Thus, $MSE(\hat{\Theta}_n)$ goes to 0 as $n \to \infty$. From this, we can conclude that $\hat{\Theta}_n = \overline{X}$ is a consistent estimator for θ. In fact, we can state the following theorem:

Theorem 8.2. Let $\hat{\Theta}_1$, $\hat{\Theta}_2$, $\cdots$ be a sequence of point estimators of θ. If

$$\lim_{n \to \infty} MSE(\hat{\Theta}_n) = 0,$$

then $\hat{\Theta}_n$ is a consistent estimator of θ.

Proof. We can write

$$P(|\hat{\Theta}_n - \theta| \geq \epsilon) = P(|\hat{\Theta}_n - \theta|^2 \geq \epsilon^2)$$

$$\leq \frac{E[\hat{\Theta}_n - \theta]^2}{\epsilon^2} \quad \text{(by Markov's inequality)}$$

$$= \frac{MSE(\hat{\Theta}_n)}{\epsilon^2},$$

which goes to 0 as $n \to \infty$ by the assumption. ■

8.2.2 Point Estimators for Mean and Variance

The above discussion suggests that the sample mean, $\overline{X}$, is often a reasonable point estimator for the mean. Now, suppose that we would like to estimate the variance of a distribution σ^2. Assuming $0 < \sigma^2 < \infty$, by definition

$$\sigma^2 = E[(X - \mu)^2].$$

Thus, the variance itself is the mean of the random variable $Y = (X - \mu)^2$. This suggests the following estimator for the variance

$$\hat{\sigma}^2 = \frac{1}{n} \sum_{k=1}^{n} (X_k - \mu)^2.$$

By linearity of expectation, $\hat{\sigma}^2$ is an unbiased estimator of σ^2. Also, by the weak law of large numbers, $\hat{\sigma}^2$ is also a consistent estimator of σ^2. However, in practice we often do not know the value of μ. Thus, we may replace μ by our estimate of the μ, the sample mean, to obtain the following estimator for σ^2:

$$\overline{S}^2 = \frac{1}{n} \sum_{k=1}^{n} (X_k - \overline{X})^2.$$

Using a little algebra, you can show that

$$\overline{S}^2 = \frac{1}{n} \left(\sum_{k=1}^{n} X_k^2 - n\overline{X}^2 \right).$$

Example 8.5. Let X_1, X_2, X_3, ..., X_n be a random sample with mean $EX_i = \mu$, and variance $\text{Var}(X_i) = \sigma^2$. Suppose that we use

$$\overline{S}^2 = \frac{1}{n} \sum_{k=1}^{n} (X_k - \overline{X})^2 = \frac{1}{n} \left(\sum_{k=1}^{n} X_k^2 - n\overline{X}^2 \right)$$

to estimate σ^2. Find the bias of this estimator

$$B(\overline{S}^2) = E[\overline{S}^2] - \sigma^2.$$

Solution: First note that

$$EX^2 = (E\overline{X})^2 + \text{Var}(\overline{X})$$
$$= \mu^2 + \frac{\sigma^2}{n}.$$

Thus,

$$E[\overline{S}^2] = \frac{1}{n}\left(\sum_{k=1}^{n} EX_k^2 - nE\overline{X}^2\right)$$
$$= \frac{1}{n}\left(n(\mu^2 + \sigma^2) - n\left(\mu^2 + \frac{\sigma^2}{n}\right)\right)$$
$$= \frac{n-1}{n}\sigma^2.$$

Therefore,

$$B(\overline{S}^2) = E[\overline{S}^2] - \sigma^2$$
$$= -\frac{\sigma^2}{n}.$$

We conclude that $\overline{S}^2$ is a biased estimator of the variance. Nevertheless, note that if n is relatively large, the bias is very small. Since $E[\overline{S}^2] = \frac{n-1}{n}\sigma^2$, we can obtain an unbiased estimator of σ^2 by multiplying $\overline{S}^2$ by $\frac{n}{n-1}$. Thus, we define

$$S^2 = \frac{1}{n-1}\sum_{k=1}^{n}(X_k - \overline{X})^2 = \frac{1}{n-1}\left(\sum_{k=1}^{n} X_k^2 - n\overline{X}^2\right).$$

By the above discussion, S^2 is an unbiased estimator of the variance. We call it the **sample variance**. We should note that if n is large, the difference between S^2 and $\overline{S}^2$ is very small. We also define the **sample standard deviation** as

$$S = \sqrt{S^2}.$$

Although the sample standard deviation is usually used as an estimator for the standard deviation, it is a biased estimator. To see this, note that S is random, so $\text{Var}(S) > 0$. Thus,

$$0 < \text{Var}(S) = ES^2 - (ES)^2$$
$$= \sigma^2 - (ES)^2.$$

Therefore, $ES < \sigma$, which means that S is a biased estimator of σ.

Let X_1, X_2, X_3, ..., X_n be a random sample with mean $EX_i = \mu < \infty$, and variance $0 < \text{Var}(X_i) = \sigma^2 < \infty$. The **sample variance** of this random sample is defined as

$$S^2 = \frac{1}{n-1}\sum_{k=1}^{n}(X_k - \overline{X})^2 = \frac{1}{n-1}\left(\sum_{k=1}^{n}X_k^2 - n\overline{X}^2\right).$$

The sample variance is an unbiased estimator of σ^2. The **sample standard deviation** is defined as

$$S = \sqrt{S^2},$$

and is commonly used as an estimator for σ. Nevertheless, S is a biased estimator of σ.

You can use the `mean` command in MATLAB to compute the sample mean for a given sample. More specifically, for a given vector $x = [x_1, x_2, \cdots, x_n]$, `mean(x)` returns the sample average

$$\frac{x_1 + x_2 + \cdots + x_n}{n}.$$

Also, the functions `var` and `std` can be used to compute the sample variance and the sample standard deviation respectively.

Example 8.6. Let T be the time that is needed for a specific task in a factory to be completed. In order to estimate the mean and variance of T, we observe a random sample $T_1, T_2, \cdots, T_6$. Thus, T_i's are i.i.d. and have the same distribution as T. We obtain the following values (in minutes):

$$18, 21, 17, 16, 24, 20.$$

Find the values of the sample mean, the sample variance, and the sample standard deviation for the observed sample.

Solution: The sample mean is

$$\overline{T} = \frac{T_1 + T_2 + T_3 + T_4 + T_5 + T_6}{6}$$
$$= \frac{18 + 21 + 17 + 16 + 24 + 20}{6}$$
$$= 19.33$$

The sample variance is given by

$$S^2 = \frac{1}{6-1}\sum_{k=1}^{6}(T_k - 19.333)^2 = 8.67$$

Finally, the sample standard deviation is given by

$$S = \sqrt{S^2} = 2.94$$

You can use the following MATLAB code to compute the above values:

```
t=[18, 21, 17, 16, 24, 20];
m=mean(t);
v=var(t);
s=std(t);
```

8.2.3   Maximum Likelihood Estimation (MLE)

So far, we have discussed estimating the mean and variance of a distribution. Our methods have been somewhat ad hoc. More specifically, it is not clear how we can estimate other parameters. Now we would like to talk about a systematic way of parameter estimation. Specifically, we would like to introduce an estimation method called *maximum likelihood estimation* (MLE). To give you the idea behind MLE let us look at an example.

Example 8.7. I have a bag that contains 3 balls. Each ball is either red or blue, but I have no information in addition to this. Thus, the number of blue balls, call it θ, might be 0, 1, 2, or 3. I am allowed to choose 4 balls at random from the bag <u>with</u> replacement. We define the random variables X_1, X_2, X_3, and X_4 as follows

$$X_i = \begin{cases} 1 & \text{if the } i\text{th chosen ball is blue} \\ 0 & \text{if the } i\text{th chosen ball is red} \end{cases}$$

Note that X_i's are i.i.d. and $X_i \sim Bernoulli(\frac{\theta}{3})$. After doing my experiment, I observe the following values for X_i's.

$$x_1 = 1, x_2 = 0, x_3 = 1, x_4 = 1.$$

Thus, I observe 3 blue balls and 1 red balls.

1. For each possible value of θ, find the probability of the observed sample, $(x_1, x_2, x_3, x_4) = (1, 0, 1, 1)$.

2. For which value of θ is the probability of the observed sample the largest?

Solution: Since $X_i \sim Bernoulli(\frac{\theta}{3})$, we have

$$P_{X_i}(x) = \begin{cases} \frac{\theta}{3} & \text{for } x = 1 \\ 1 - \frac{\theta}{3} & \text{for } x = 0 \end{cases}$$

Since X_i's are independent, the joint PMF of X_1, X_2, X_3, and X_4 can be written as

$$P_{X_1 X_2 X_3 X_4}(x_1, x_2, x_3, x_4) = P_{X_1}(x_1) P_{X_2}(x_2) P_{X_3}(x_3) P_{X_4}(x_4)$$

Therefore,

$$P_{X_1 X_2 X_3 X_4}(1, 0, 1, 1) = \frac{\theta}{3} \cdot \left(1 - \frac{\theta}{3}\right) \cdot \frac{\theta}{3} \cdot \frac{\theta}{3}$$

$$= \left(\frac{\theta}{3}\right)^3 \left(1 - \frac{\theta}{3}\right).$$

Note that the joint PMF depends on θ, so we write it as $P_{X_1 X_2 X_3 X_4}(x_1, x_2, x_3, x_4; \theta)$. We obtain the values given in Table 8.1 for the probability of $(1, 0, 1, 1)$.

Table 8.1: Values of $P_{X_1 X_2 X_3 X_4}(1, 0, 1, 1; \theta)$ for Example 8.7

θ	$P_{X_1 X_2 X_3 X_4}(1, 0, 1, 1; \theta)$
0	0
1	0.0247
2	0.0988
3	0

The probability of observed sample for $\theta = 0$ and $\theta = 3$ is zero. This makes sense because our sample included both red and blue balls. From the table we see that the probability of the observed data is maximized for $\theta = 2$. This means that the observed data is most likely to occur for $\theta = 2$. For this reason, we may choose $\hat{\theta} = 2$ as our estimate of θ. This is called the maximum likelihood estimate (MLE) of θ.

The above example gives us the idea behind the maximum likelihood estimation. Here, we introduce this method formally. To do so, we first define the **likelihood** function. Let X_1, X_2, X_3, ..., X_n be a random sample from a distribution with a parameter θ (In general, θ might be a vector, $\theta = (\theta_1, \theta_2, \cdots, \theta_k)$.) Suppose that x_1, x_2, x_3, ..., x_n are the observed values of X_1, X_2, X_3, ..., X_n. If X_i's are discrete random variables, we define the *likelihood* function as the probability of the observed sample as a function of θ:

$$L(x_1, x_2, \cdots, x_n; \theta) = P(X_1 = x_1, X_2 = x_2, \cdots, X_n = x_n; \theta)$$

$$= P_{X_1 X_2 \cdots X_n}(x_1, x_2, \cdots, x_n; \theta).$$

To get a more compact formula, we may use the vector notation, $\mathbf{X} = (X_1, X_2, \cdots, X_n)$. Thus, we may write

$$L(\mathbf{x}; \theta) = P_{\mathbf{X}}(\mathbf{x}; \theta).$$

If X_1, X_2, X_3, ..., X_n are jointly continuous, we use the joint PDF instead of the joint PMF. Thus, the likelihood is defined by

$$L(x_1, x_2, \cdots, x_n; \theta) = f_{X_1 X_2 \cdots X_n}(x_1, x_2, \cdots, x_n; \theta).$$

Let X_1, X_2, X_3, ..., X_n be a random sample from a distribution with a parameter θ. Suppose that we have observed $X_1 = x_1$, $X_2 = x_2$, $\cdots$, $X_n = x_n$.

- If X_i's are discrete, then the **likelihood function** is defined as

$$L(x_1, x_2, \cdots, x_n; \theta) = P_{X_1 X_2 \cdots X_n}(x_1, x_2, \cdots, x_n; \theta).$$

- If X_i's are jointly continuous, then the likelihood function is defined as

$$L(x_1, x_2, \cdots, x_n; \theta) = f_{X_1 X_2 \cdots X_n}(x_1, x_2, \cdots, x_n; \theta).$$

In some problems, it is easier to work with the **log likelihood function** given by

$$\ln L(x_1, x_2, \cdots, x_n; \theta).$$

Example 8.8. For the following random samples, find the likelihood function:

1. $X_i \sim Binomial(3, \theta)$ and we have observed $(x_1, x_2, x_3, x_4) = (1, 3, 2, 2)$.

2. $X_i \sim Exponential(\theta)$ and we have observed $(x_1, x_2, x_3, x_4) = (1.23, 3.32, 1.98, 2.12)$.

Solution: Remember that when we have a random sample, X_i's are i.i.d., so we can obtain the joint PMF and PDF by multiplying the marginal (individual) PMFs and PDFs.

1. If $X_i \sim Binomial(3, \theta)$, then

$$P_{X_i}(x; \theta) = \binom{3}{x} \theta^x (1 - \theta)^{3-x}$$

Thus,

$$\begin{aligned} L(x_1, x_2, x_3, x_4; \theta) &= P_{X_1 X_2 X_3 X_4}(x_1, x_2, x_3, x_4; \theta) \\ &= P_{X_1}(x_1; \theta) P_{X_2}(x_2; \theta) P_{X_3}(x_3; \theta) P_{X_4}(x_4; \theta) \\ &= \binom{3}{x_1}\binom{3}{x_2}\binom{3}{x_3}\binom{3}{x_4} \theta^{x_1+x_2+x_3+x_4}(1-\theta)^{12-(x_1+x_2+x_3+x_4)}. \end{aligned}$$

Since we have observed $(x_1, x_2, x_3, x_4) = (1, 3, 2, 2)$, we have

$$\begin{aligned} L(1, 3, 2, 2; \theta) &= \binom{3}{1}\binom{3}{3}\binom{3}{2}\binom{3}{2} \theta^8 (1-\theta)^4 \\ &= 27\, \theta^8 (1-\theta)^4. \end{aligned}$$

2. If $X_i \sim Exponential(\theta)$, then

$$f_{X_i}(x; \theta) = \theta e^{-\theta x} u(x),$$

where $u(x)$ is the unit step function, i.e., $u(x) = 1$ for $x \geq 0$ and $u(x) = 0$ for $x < 0$. Thus, for $x_i \geq 0$, we can write

$$L(x_1, x_2, x_3, x_4; \theta) = f_{X_1 X_2 X_3 X_4}(x_1, x_2, x_3, x_4; \theta)$$
$$= f_{X_1}(x_1; \theta) f_{X_2}(x_2; \theta) f_{X_3}(x_3; \theta) f_{X_4}(x_4; \theta)$$
$$= \theta^4 e^{-(x_1 + x_2 + x_3 + x_4)\theta}.$$

Since we have observed $(x_1, x_2, x_3, x_4) = (1.23, 3.32, 1.98, 2.12)$, we have

$$L(1.23, 3.32, 1.98, 2.12; \theta) = \theta^4 e^{-8.65\theta}.$$

Now that we have defined the likelihood function, we are ready to define maximum likelihood estimation. Let X_1, X_2, X_3, ..., X_n be a random sample from a distribution with a parameter θ. Suppose that we have observed $X_1 = x_1$, $X_2 = x_2$, $\cdots$, $X_n = x_n$. The maximum likelihood estimate of θ, shown by $\hat{\theta}_{ML}$ is the value that maximizes the likelihood function

$$L(x_1, x_2, \cdots, x_n; \theta).$$

Figure 8.1 illustrates finding the maximum likelihood estimate as the maximizing value of θ for the likelihood function. There are two cases shown in the figure: In the first graph, θ is a discrete-valued parameter, such as the one in Example 8.7. In the second one, θ is a continuous-valued parameter, such as the ones in Example 8.8. In both cases, the maximum likelihood estimate of θ is the value that maximizes the likelihood function.

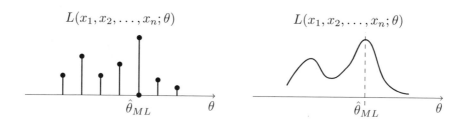

Figure 8.1: The maximum likelihood estimate for θ.

Let us find the maximum likelihood estimates for the observations of Example 8.8.

Example 8.9. For the following random samples, find the maximum likelihood estimate of θ:

1. $X_i \sim Binomial(3, \theta)$, and we have observed $(x_1, x_2, x_3, x_4) = (1, 3, 2, 2)$.

2. $X_i \sim Exponential(\theta)$ and we have observed $(x_1, x_2, x_3, x_4) = (1.23, 3.32, 1.98, 2.12)$.

Solution:

1. In Example 8.8, we found the likelihood function as

$$L(1, 3, 2, 2; \theta) = 27 \, \theta^8 (1 - \theta)^4.$$

To find the value of θ that maximizes the likelihood function, we can take the derivative and set it to zero. We have

$$\frac{dL(1, 3, 2, 2; \theta)}{d\theta} = 27 \big[8\theta^7 (1 - \theta)^4 - 4\theta^8 (1 - \theta)^3 \big].$$

Thus, we obtain

$$\hat{\theta}_{ML} = \frac{2}{3}.$$

2. In Example 8.8, we found the likelihood function as

$$L(1.23, 3.32, 1.98, 2.12; \theta) = \theta^4 e^{-8.65\theta}.$$

Here, it is easier to work with the log likelihood function, $\ln L(1.23, 3.32, 1.98, 2.12; \theta)$. Specifically,

$$\ln L(1.23, 3.32, 1.98, 2.12; \theta) = 4 \ln \theta - 8.65\theta.$$

By differentiating, we obtain

$$\frac{4}{\theta} - 8.65 = 0,$$

which results in

$$\hat{\theta}_{ML} = 0.46$$

It is worth noting that, technically, we need to look at the second derivatives and endpoints to make sure that the values that we obtained above are the maximizing values. For this example, it turns out that the obtained values are indeed the maximizing values.

Note that the value of the maximum likelihood estimate is a function of the observed data. Thus, as any other estimator, the maximum likelihood estimator (MLE), shown by $\hat{\Theta}_{ML}$, is indeed a random variable. The MLE estimates $\hat{\theta}_{ML}$ that we found above were the values of the random variable $\hat{\Theta}_{ML}$ for the specified observed data.

The Maximum Likelihood Estimator (MLE)

Let X_1, X_2, X_3, ..., X_n be a random sample from a distribution with a parameter θ. Given that we have observed $X_1 = x_1$, $X_2 = x_2$, $\cdots$, $X_n = x_n$, a maximum likelihood estimate of θ, shown by $\hat{\theta}_{ML}$, is a value of θ that maximizes the likelihood function

$$L(x_1, x_2, \cdots, x_n; \theta).$$

A maximum likelihood estimator (MLE) of the parameter θ, shown by $\hat{\Theta}_{ML}$ is a random variable $\hat{\Theta}_{ML} = \hat{\Theta}_{ML}(X_1, X_2, \cdots, X_n)$ whose value when $X_1 = x_1$, $X_2 = x_2$, $\cdots$, $X_n = x_n$ is given by $\hat{\theta}_{ML}$.

Example 8.10. For the following examples, find the maximum likelihood estimator (MLE) of θ:

1. $X_i \sim Binomial(m, \theta)$ and we have observed X_1, X_2, X_3, ..., X_n.

2. $X_i \sim Exponential(\theta)$ and we have observed X_1, X_2, X_3, ..., X_n.

Solution:

1. Similar to our calculation in Example 8.8, for the observed values of $X_1 = x_1$, $X_2 = x_2$, $\cdots$, $X_n = x_n$, the likelihood function is given by

$$L(x_1, x_2, \cdots, x_n; \theta) = P_{X_1 X_2 \cdots X_n}(x_1, x_2, \cdots, x_n; \theta)$$

$$= \prod_{i=1}^{n} P_{X_i}(x_i; \theta)$$

$$= \prod_{i=1}^{n} \binom{m}{x_i} \theta^{x_i} (1 - \theta)^{m - x_i}$$

$$= \left[\prod_{i=1}^{n} \binom{m}{x_i} \right] \theta^{\sum_{i=1}^{n} x_i} (1 - \theta)^{mn - \sum_{i=1}^{n} x_i}.$$

Note that the first term does not depend on θ, so we may write $L(x_1, x_2, \cdots, x_n; \theta)$ as

$$L(x_1, x_2, \cdots, x_n; \theta) = c\, \theta^s (1 - \theta)^{mn - s},$$

where c does not depend on θ, and $s = \sum_{k=1}^{n} x_i$. By differentiating and setting the derivative to 0 we obtain

$$\hat{\theta}_{ML} = \frac{1}{mn} \sum_{k=1}^{n} x_i.$$

This suggests that the MLE can be written as

$$\hat{\Theta}_{ML} = \frac{1}{mn} \sum_{k=1}^{n} X_i.$$

2. Similar to our calculation in Example 8.8, for the observed values of $X_1 = x_1$, $X_2 = x_2$, $\cdots$, $X_n = x_n$, the likelihood function is given by

$$L(x_1, x_2, \cdots, x_n; \theta) = \prod_{i=1}^{n} f_{X_i}(x_i; \theta)$$

$$= \prod_{i=1}^{n} \theta e^{-\theta x_i}$$

$$= \theta^n e^{-\theta \sum_{k=1}^{n} x_i}.$$

Therefore,

$$\ln L(x_1, x_2, \cdots, x_n; \theta) = n \ln \theta - \sum_{k=1}^{n} x_i \theta.$$

By differentiating and setting the derivative to 0 we obtain

$$\hat{\theta}_{ML} = \frac{n}{\sum_{k=1}^{n} x_i}.$$

This suggests that the MLE can be written as

$$\hat{\Theta}_{ML} = \frac{n}{\sum_{k=1}^{n} X_i}.$$

The examples that we have discussed had only one unknown parameter θ. In general, θ could be a vector of parameters, and we can apply the same methodology to obtain the MLE. More specifically, if we have k unknown parameters θ_1, θ_2, $\cdots$, θ_k, then we need to maximize the likelihood function

$$L(x_1, x_2, \cdots, x_n; \theta_1, \theta_2, \cdots, \theta_k) \tag{8.1}$$

to obtain the maximum likelihood estimators $\hat{\Theta}_1$, $\hat{\Theta}_2$, $\cdots$, $\hat{\Theta}_k$. Let's look at an example.

Example 8.11. Suppose that we have observed the random sample X_1, X_2, X_3, ..., X_n, where $X_i \sim N(\theta_1, \theta_2)$, so

$$f_{X_i}(x_i; \theta_1, \theta_2) = \frac{1}{\sqrt{2\pi\theta_2}} e^{-\frac{(x_i - \theta_1)^2}{2\theta_2}}.$$

Find the maximum likelihood estimators for θ_1 and θ_2.

Solution: The likelihood function is given by

$$L(x_1, x_2, \cdots, x_n; \theta_1, \theta_2) = \frac{1}{(2\pi)^{\frac{n}{2}} \theta_2^{\frac{n}{2}}} \exp\left(-\frac{1}{2\theta_2} \sum_{i=1}^{n} (x_i - \theta_1)^2\right).$$

Here again, it is easier to work with the log likelihood function

$$\ln L(x_1, x_2, \cdots, x_n; \theta_1, \theta_2) = -\frac{n}{2} \ln(2\pi) - \frac{n}{2} \ln \theta_2 - \frac{1}{2\theta_2} \sum_{i=1}^{n} (x_i - \theta_1)^2.$$

We take the derivatives with respect to θ_1 and θ_2 and set them to zero:

$$\frac{\partial}{\partial \theta_1} \ln L(x_1, x_2, \cdots, x_n; \theta_1, \theta_2) = \frac{1}{\theta_2} \sum_{i=1}^{n} (x_i - \theta_1) = 0$$

$$\frac{\partial}{\partial \theta_2} \ln L(x_1, x_2, \cdots, x_n; \theta_1, \theta_2) = -\frac{n}{2\theta_2} + \frac{1}{2\theta_2^2} \sum_{i=1}^{n} (x_i - \theta_1)^2 = 0.$$

By solving the above equations, we obtain the following maximum likelihood estimates for θ_1 and θ_2:

$$\hat{\theta}_1 = \frac{1}{n} \sum_{i=1}^{n} x_i,$$

$$\hat{\theta}_2 = \frac{1}{n} \sum_{i=1}^{n} (x_i - \theta_1)^2.$$

We can write the MLE of θ_1 and θ_2 as random variables $\hat{\Theta}_1$ and $\hat{\Theta}_2$:

$$\hat{\Theta}_1 = \frac{1}{n} \sum_{i=1}^{n} X_i,$$

$$\hat{\Theta}_2 = \frac{1}{n} \sum_{i=1}^{n} (X_i - \Theta_1)^2.$$

Note that $\hat{\Theta}_1$ is the sample mean, $\overline{X}$, and therefore it is an unbiased estimator of the mean. Here, $\hat{\Theta}_2$ is very close to the sample variance which we defined as

$$S^2 = \frac{1}{n-1} \sum_{i=1}^{n} (X_i - \overline{X})^2.$$

In fact,

$$\hat{\Theta}_2 = \frac{n-1}{n} S^2.$$

Since we already know that the sample variance of unbiased estimator of the variance, we conclude that $\hat{\Theta}_2$ is a biased estimator of the variance:

$$E\hat{\Theta}_2 = \frac{n-1}{n} \theta_2.$$

Nevertheless, the bias is very small here and it goes to zero as n gets large.

Note: Here, we caution that we cannot always find the maximum likelihood estimator by setting the derivative to zero. For example, if θ is an integer-valued parameter (such as the number of blue balls in Example 8.7), then we cannot use differentiation and we need to find the maximizing value in another way. Even if θ is a real-valued parameter, we cannot always find the MLE by setting the derivative to zero. For example, the maximum might be obtained at the endpoints of the acceptable ranges. We will see an example of such scenarios in the Solved Problems section (Section 8.2.5).

8.2.4 Asymptotic Properties of MLEs

We end this section by mentioning that MLEs have some nice asymptotic properties. By asymptotic properties we mean properties that are true when the sample size becomes large. Here, we state these properties without proofs.

<u>Asymptotic Properties of MLEs</u>

Let X_1, X_2, X_3, ..., X_n be a random sample from a distribution with a parameter θ. Let $\hat{\Theta}_{ML}$ denote the maximum likelihood estimator (MLE) of θ. Then, under some mild regularity conditions,

1. $\hat{\Theta}_{ML}$ is asymptotically consistent, i.e.,

$$\lim_{n \to \infty} P(|\hat{\Theta}_{ML} - \theta| > \epsilon) = 0.$$

2. $\hat{\Theta}_{ML}$ is asymptotically unbiased, i.e.,

$$\lim_{n \to \infty} E[\hat{\Theta}_{ML}] = \theta.$$

3. As n becomes large, $\hat{\Theta}_{ML}$ is approximately a normal random variable. More precisely, the random variable

$$\frac{\hat{\Theta}_{ML} - \theta}{\sqrt{\text{Var}(\hat{\Theta}_{ML})}}$$

converges in distribution to $N(0, 1)$.

8.2.5 Solved Problems

1. Let X be the height of a randomly chosen individual from a population. In order to estimate the mean and variance of X, we observe a random sample $X_1, X_2, \cdots, X_7$. Thus, X_i's are i.i.d. and have the same distribution as X. We obtain the following values (in centimeters):

$$166.8, 171.4, 169.1, 178.5, 168.0, 157.9, 170.1$$

Find the values of the sample mean, the sample variance, and the sample standard deviation for the observed sample.

Solution: The sample mean is

$$\overline{X} = \frac{X_1 + X_2 + X_3 + X_4 + X_5 + X_6 + X_7}{7}$$

$$= \frac{166.8 + 171.4 + 169.1 + 178.5 + 168.0 + 157.9 + 170.1}{7}$$

$$= 168.8$$

The sample variance is given by

$$S^2 = \frac{1}{7-1} \sum_{k=1}^{7} (X_k - 168.8)^2 = 37.7$$

Finally, the sample standard deviation is given by

$$S = \sqrt{S^2} = 6.1$$

The following MATLAB code can be used to obtain these values:

```
x=[166.8, 171.4, 169.1, 178.5, 168.0, 157.9, 170.1];
m=mean(x);
v=var(x);
s=std(x);
```

2. Prove the following:

(a) If $\hat{\Theta}_1$ is an unbiased estimator for θ, and W is a zero mean random variable, then

$$\hat{\Theta}_2 = \hat{\Theta}_1 + W$$

is also an unbiased estimator for θ.

(b) If $\hat{\Theta}_1$ is an estimator for θ such that $E[\hat{\Theta}_1] = a\theta + b$, where $a \neq 0$, show that

$$\hat{\Theta}_2 = \frac{\hat{\Theta}_1 - b}{a}$$

is an unbiased estimator for θ.

Solution:

(a) We have

$$E[\hat{\Theta}_2] = E[\hat{\Theta}_1] + E[W] \qquad \text{(by linearity of expectation)}$$
$$= \theta + 0 \qquad \text{(since } \hat{\Theta}_1 \text{ is unbiased and } EW = 0)$$
$$= \theta.$$

Thus, $\hat{\Theta}_2$ is an unbiased estimator for θ.

(b) We have

$$E[\hat{\Theta}_2] = \frac{E[\hat{\Theta}_1] - b}{a} \qquad \text{(by linearity of expectation)}$$
$$= \frac{a\theta + b - b}{a}$$
$$= \theta.$$

Thus, $\hat{\Theta}_2$ is an unbiased estimator for θ.

3. Let X_1, X_2, X_3, ..., X_n be a random sample from a $Uniform(0, \theta)$ distribution, where θ is unknown. Define the estimator

$$\hat{\Theta}_n = \max\{X_1, X_2, \cdots, X_n\}.$$

(a) Find the bias of $\hat{\Theta}_n$, $B(\hat{\Theta}_n)$.

(b) Find the MSE of $\hat{\Theta}_n$, $MSE(\hat{\Theta}_n)$.

(c) Is $\hat{\Theta}_n$ a consistent estimator of θ?

Solution: If $X \sim Uniform(0, \theta)$, then the PDF and CDF of X are given by

$$f_X(x) = \begin{cases} \frac{1}{\theta} & 0 \le x \le \theta \\ 0 & \text{otherwise} \end{cases}$$

and

$$F_X(x) = \begin{cases} 0 & x < 0 \\ \frac{x}{\theta} & 0 \le x \le \theta \\ 1 & x > 1 \end{cases}$$

By Theorem 8.1, the PDF of $\hat{\Theta}_n$ is given by

$$f_{\hat{\Theta}_n}(y) = n f_X(x) \big[F_X(x) \big]^{n-1}$$
$$= \begin{cases} \frac{n y^{n-1}}{\theta^n} & 0 \le y \le \theta \\ 0 & \text{otherwise} \end{cases}$$

(a) To find the bias of $\hat{\Theta}_n$, we have

$$E[\hat{\Theta}_n] = \int_0^\theta y \cdot \frac{ny^{n-1}}{\theta^n} dy$$

$$= \frac{n}{n+1}\theta.$$

Thus, the bias is given by

$$B(\hat{\Theta}_n) = E[\hat{\Theta}_n] - \theta$$

$$= \frac{n}{n+1}\theta - \theta$$

$$= -\frac{\theta}{n+1}.$$

(b) To find $MSE(\hat{\Theta}_n)$, we can write

$$MSE(\hat{\Theta}_n) = \text{Var}(\hat{\Theta}_n) + B(\hat{\Theta}_n)^2$$

$$= \text{Var}(\hat{\Theta}_n) + \frac{\theta^2}{(n+1)^2}.$$

Thus, we need to find $\text{Var}(\hat{\Theta})$. We have

$$E\left[\hat{\Theta}_n^2\right] = \int_0^\theta y^2 \cdot \frac{ny^{n-1}}{\theta^n} dy$$

$$= \frac{n}{n+2}\theta^2.$$

Thus,

$$\text{Var}(\hat{\Theta}_n) = E\left[\hat{\Theta}_n^2\right] - \left(E[\hat{\Theta}_n]\right)^2$$

$$= \frac{n}{(n+2)(n+1)^2}\theta^2.$$

Therefore,

$$MSE(\hat{\Theta}_n) = \frac{n}{(n+2)(n+1)^2}\theta^2 + \frac{\theta^2}{(n+1)^2}$$

$$= \frac{2\theta^2}{(n+2)(n+1)}.$$

(c) Note that

$$\lim_{n\to\infty} MSE(\hat{\Theta}_n) = \lim_{n\to\infty} \frac{2\theta^2}{(n+2)(n+1)} = 0.$$

Thus, by Theorem 8.2, $\hat{\Theta}_n$ is a consistent estimator of θ.

4. Let X_1, X_2, X_3, ..., X_n be a random sample from a *Geometric(θ)* distribution, where θ is unknown. Find the maximum likelihood estimator (MLE) of θ based on this random sample.

Solution: If $X_i \sim Geometric(\theta)$, then

$$P_{X_i}(x; \theta) = (1 - \theta)^{x-1}\theta.$$

Thus, the likelihood function is given by

$$
\begin{aligned}
L(x_1, x_2, \cdots, x_n; \theta) &= P_{X_1 X_2 \cdots X_n}(x_1, x_2, \cdots, x_n; \theta) \\
&= P_{X_1}(x_1; \theta) P_{X_2}(x_2; \theta) \cdots P_{X_n}(x_n; \theta) \\
&= (1 - \theta)^{\left[\sum_{i=1}^n x_i - n\right]} \theta^n.
\end{aligned}
$$

Then, the log likelihood function is given by

$$\ln L(x_1, x_2, \cdots, x_n; \theta) = \left(\sum_{i=1}^n x_i - n\right) \ln(1 - \theta) + n \ln \theta.$$

Thus,

$$\frac{d \ln L(x_1, x_2, \cdots, x_n; \theta)}{d\theta} = \left(\sum_{i=1}^n x_i - n\right) \cdot \frac{-1}{1 - \theta} + \frac{n}{\theta}.$$

By setting the derivative to zero, we can check that the maximizing value of θ is given by

$$\hat{\theta}_{ML} = \frac{n}{\sum_{i=1}^n x_i}.$$

Thus, the MLE can be written as

$$\hat{\Theta}_{ML} = \frac{n}{\sum_{i=1}^n X_i}.$$

5. Let X_1, X_2, X_3, ..., X_n be a random sample from a *Uniform($0, \theta$)* distribution, where θ is unknown. Find the maximum likelihood estimator (MLE) of θ based on this random sample.

Solution: If $X_i \sim Uniform(0, \theta)$, then

$$
f_X(x) = \begin{cases} \frac{1}{\theta} & 0 \le x \le \theta \\ 0 & \text{otherwise} \end{cases}
$$

The likelihood function is given by

$$L(x_1, x_2, \cdots, x_n; \theta) = f_{X_1 X_2 \cdots X_n}(x_1, x_2, \cdots, x_n; \theta)$$

$$= f_{X_1}(x_1; \theta) f_{X_2}(x_2; \theta) \cdots f_{X_n}(x_n; \theta)$$

$$= \begin{cases} \frac{1}{\theta^n} & 0 \leq x_1, x_2, \cdots, x_n \leq \theta \\ \\ 0 & \text{otherwise} \end{cases}$$

Note that $\frac{1}{\theta^n}$ is a decreasing function of θ. Thus, to maximize it, we need to choose the smallest possible value for θ. For $i = 1, 2, ..., n$, we need to have $\theta \geq x_i$. Thus, the smallest possible value for θ is

$$\hat{\theta}_{ML} = \max(x_1, x_2, \cdots, x_n).$$

Therefore, the MLE can be written as

$$\hat{\Theta}_{ML} = \max(X_1, X_2, \cdots, X_n).$$

Note that this is one of those cases wherein $\hat{\theta}_{ML}$ cannot be obtained by setting the derivative of the likelihood function to zero. Here, the maximum is achieved at an endpoint of the acceptable interval.

8.3 Interval Estimation (Confidence Intervals)

Let X_1, X_2, X_3, ..., X_n be a random sample from a distribution with a parameter θ that is to be estimated. Suppose that we have observed $X_1 = x_1$, $X_2 = x_2$, $\cdots$, $X_n = x_n$. So far, we have discussed point estimation for θ. The point estimate $\hat{\theta}$ alone does not give much information about θ. In particular, without additional information, we do not know how close $\hat{\theta}$ is to the real θ.

Here, we will introduce the concept of **interval estimation**. In this approach, instead of giving just one value $\hat{\theta}$ as the estimate for θ, we will produce an interval that is likely to include the true value of θ. Thus, instead of saying

$$\hat{\theta} = 34.25,$$

we might report the interval

$$[\hat{\theta}_l, \hat{\theta}_h] = [30.69, 37.81],$$

which we hope includes the real value of θ. That is, we produce two estimates for θ, a *high estimate* $\hat{\theta}_h$ and a low estimate $\hat{\theta}_l$.

In interval estimation, there are two important concepts. One is the **length** of the reported interval, $\hat{\theta}_h - \hat{\theta}_l$. The length of the interval shows the precision with which we can estimate θ. The smaller the interval, the higher the precision with which we can estimate θ. The second important factor is the **confidence level** that shows how confident we are about the interval. The confidence level is the probability that the interval that we construct includes the real value of θ. Therefore, high confidence levels are desirable. We will discuss these concepts in this section.

8.3.1 The general framework of Interval Estimation

Let X_1, X_2, X_3, ..., X_n be a random sample from a distribution with a parameter θ that is to be estimated. Our goal is to find two estimators for θ:

1. the low estimator, $\hat{\Theta}_l = \hat{\Theta}_l(X_1, X_2, \cdots, X_n)$, and

2. the high estimator, $\hat{\Theta}_h = \hat{\Theta}_h(X_1, X_2, \cdots, X_n)$.

The interval estimator is given by the interval $[\hat{\Theta}_l, \hat{\Theta}_h]$. The estimators $\hat{\Theta}_l$ and $\hat{\Theta}_h$ are chosen such that the probability that the interval $[\hat{\Theta}_l, \hat{\Theta}_h]$ includes θ is larger than $1 - \alpha$. Here, $1 - \alpha$ is said to be **confidence level**. We would like α to be small. Common values for α are 0.1, .05, and .01 which correspond to confidence levels 90%, 95%, and 99% respectively. Thus, when we are asked to find a 95% confidence interval for a parameter θ, we need to find $\hat{\Theta}_l$ and $\hat{\Theta}_h$ such that

$$P\left(\hat{\Theta}_l < \theta \text{ and } \hat{\Theta}_h > \theta\right) \geq 0.95$$

The above discussion will become clearer as we go through examples. Before doing that let's formally define interval estimation.

<u>Interval Estimation</u>

Let X_1, X_2, X_3, ..., X_n be a random sample from a distribution with a parameter θ that is to be estimated. An **interval estimator** with **confidence level** $1 - \alpha$ consists of two estimators $\hat{\Theta}_l(X_1, X_2, \cdots, X_n)$ and $\hat{\Theta}_h(X_1, X_2, \cdots, X_n)$ such that

$$P\left(\hat{\Theta}_l \leq \theta \text{ and } \hat{\Theta}_h \geq \theta\right) \geq 1 - \alpha,$$

for every possible value of θ. Equivalently, we say that $[\hat{\Theta}_l, \hat{\Theta}_h]$ is a $(1 - \alpha)100\%$ **confidence interval** for θ.

Note that the condition

$$P\left(\hat{\Theta}_l \leq \theta \text{ and } \hat{\Theta}_h \geq \theta\right) \geq 1 - \alpha$$

can be equivalently written as

$$P\left(\hat{\Theta}_l \leq \theta \leq \hat{\Theta}_h\right) \geq 1 - \alpha, \quad \text{or} \quad P\left(\theta \in [\hat{\Theta}_l, \hat{\Theta}_h]\right) \geq 1 - \alpha.$$

The randomness in these terms is due to $\hat{\Theta}_l$ and $\hat{\Theta}_h$, not θ. Here, θ is the unknown quantity which is assumed to be non-random (frequentist inference). On the other hand, $\hat{\Theta}_l$ and $\hat{\Theta}_h$ are random variables because they are functions of the observed random variables X_1, X_2, X_3, ..., X_n.

8.3.2   Finding Interval Estimators

Here, we would like to discuss how we find interval estimators. Before doing so, let's review a simple fact from random variables and their distributions. Let X be a continuous random variable with CDF $F_X(x) = P(X \leq x)$. Suppose that we are interested in finding two values x_h and x_l such that

$$P\left(x_l \leq X \leq x_h\right) = 1 - \alpha.$$

One way to do this is to chose x_l and x_h such that

$$P(X \leq x_l) = \frac{\alpha}{2}, \qquad \text{and} \qquad P(X \geq x_h) = \frac{\alpha}{2}.$$

Equivalently,

$$F_X(x_l) = \frac{\alpha}{2}, \qquad \text{and} \qquad F_X(x_h) = 1 - \frac{\alpha}{2}.$$

We can rewrite these equations by using the inverse function F_X^{-1} as

$$x_l = F_X^{-1}\left(\frac{\alpha}{2}\right), \qquad \text{and} \qquad x_h = F_X^{-1}\left(1 - \frac{\alpha}{2}\right).$$

We call the interval $[x_l, x_h]$ a $(1 - \alpha)$ interval for X. Figure 8.2 shows the values of x_l and x_h using the CDF of X, and also using the PDF of X.

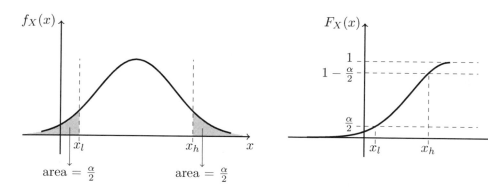

Figure 8.2: $[x_l, x_h]$ is a $(1 - \alpha)$ interval for X, that is, $P\left(x_l \leq X \leq x_h\right) = 1 - \alpha$.

Example 8.12. Let $Z \sim N(0, 1)$, find x_l and x_h such that

$$P\left(x_l \leq Z \leq x_h\right) = 0.95$$

Solution: Here, $\alpha = 0.05$ and the CDF of Z is given by the Φ function. Thus, we can choose

$$x_l = \Phi^{-1}(0.025) = -1.96, \qquad \text{and} \qquad x_h = \Phi^{-1}(1 - 0.025) = 1.96$$

Thus, for a standard normal random variable Z, we have

$$P\left(-1.96 \leq Z \leq 1.96\right) = 0.95$$

More generally, we can find a $(1 - \alpha)$ interval for the standard normal random variable. Assume $Z \sim N(0,1)$. Let us define a notation that is commonly used. For any $p \in [0,1]$, we define z_p as the real value for which

$$P(Z > z_p) = p.$$

Therefore,

$$\Phi(z_p) = 1 - p, \qquad z_p = \Phi^{-1}(1 - p).$$

By symmetry of the normal distribution, we also conclude

$$z_{1-p} = -z_p.$$

Figure 8.3 shows z_p and $z_{1-p} = -z_p$ on the real line. In MATLAB, to compute z_p you can use the following command: norminv(1-p).

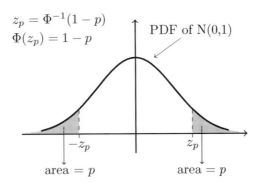

Figure 8.3: By definition, z_p is the real number, for which we have $\Phi(z_p) = 1 - p$.

Now, using the z_p notation, we can state a $(1 - \alpha)$ interval for the standard normal random variable Z as

$$P\left(-z_{\frac{\alpha}{2}} \leq Z \leq z_{\frac{\alpha}{2}}\right) = 1 - \alpha.$$

Figure 8.4 shows the $(1 - \alpha)$ interval for the standard normal random variable Z.

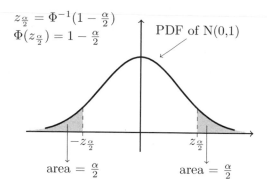

Figure 8.4: A $(1 - \alpha)$ interval for $N(0,1)$ distribution. In particular, in this figure, we have $P\left(Z \in \left[-z_{\frac{\alpha}{2}}, z_{\frac{\alpha}{2}}\right]\right) = 1 - \alpha$.

Now, let's talk about how we can find interval estimators. A general approach is to start with a point estimator $\hat{\Theta}$, such as the MLE, and create the interval $[\hat{\Theta}_l, \hat{\Theta}_h]$ around it such that $P\left(\theta \in [\hat{\Theta}_l, \hat{\Theta}_h]\right) \geq 1 - \alpha$. How do we do this? Let's look at an example.

Example 8.13. Let X_1, X_2, X_3, ..., X_n be a random sample from a normal distribution $N(\theta, 1)$. Find a 95% confidence interval for θ.

Solution: Let's start with a point estimator $\hat{\Theta}$ for θ. Since θ is the mean of the distribution, we can use the sample mean

$$\hat{\Theta} = \overline{X} = \frac{X_1 + X_2 + ... + X_n}{n}.$$

Since $X_i \sim N(\theta, 1)$ and the X_i's are independent, we conclude that

$$\overline{X} \sim N\left(\theta, \frac{1}{n}\right).$$

By normalizing $\overline{X}$, we conclude that the random variable

$$\frac{\overline{X} - \theta}{\frac{1}{\sqrt{n}}} = \sqrt{n}(\overline{X} - \theta)$$

has a $N(0,1)$ distribution. Therefore, by Example 8.12, we conclude

$$P\left(-1.96 \leq \sqrt{n}(\overline{X} - \theta) \leq 1.96\right) = 0.95$$

which is equivalent to (by rearranging the terms)

$$P\left(\overline{X} - \frac{1.96}{\sqrt{n}} \leq \theta \leq \overline{X} + \frac{1.96}{\sqrt{n}}\right) = 0.95$$

Therefore, we can report the interval

$$[\hat{\Theta}_l, \hat{\Theta}_h] = \left[\overline{X} - \frac{1.96}{\sqrt{n}}, \overline{X} + \frac{1.96}{\sqrt{n}}\right]$$

as our 95% confidence interval for θ.

At first, it might seem that our solution to Example 8.13 is not based on a systematic method. You might have asked: "How should I know that I need to work with the normalized $\overline{X}$?" However, by thinking more deeply about the way we solved this example, we can suggest a general method to solve confidence interval problems.

The crucial fact about the random variable

$$\overline{X} - \theta$$

is that its distribution does not depend on the unknown parameter θ. Thus, we could easily find a 95% interval for the random variable $\sqrt{n}(\overline{X} - \theta)$ that did not depend on θ. Such a random variable is called a **pivot** or a **pivotal quantity**. Let us define this more precisely.

Pivotal Quantity

Let X_1, X_2, X_3, ..., X_n be a random sample from a distribution with a parameter θ that is to be estimated. The random variable Q is said to be a *pivot* or a *pivotal quantity*, if it has the following properties:

1. It is a function of the observed data X_1, X_2, X_3, ..., X_n and the unknown parameter θ, but it does not depend on any other unknown parameters:

$$Q = Q(X_1, X_2, \cdots, X_n, \theta).$$

2. The probability distribution of Q does not depend on θ or any other unknown parameters.

Example 8.14. Check that the random variables $Q_1 = \overline{X} - \theta$ and $Q_2 = \sqrt{n}(\overline{X} - \theta)$ are both valid pivots in Example 8.13.

Solution: We note that Q_1 and Q_2 by definitions are functions of $\overline{X}$ and θ. Since

$$\overline{X} = \frac{X_1 + X_2 + ... + X_n}{n},$$

we conclude Q_1 and Q_2 are both functions of the observed data X_1, X_2, X_3, ..., X_n and the unknown parameter θ, and they do not depend on any other unknown parameters. Also,

$$Q_1 \sim N(0, \frac{1}{n}), \qquad Q_2 \sim N(0, 1).$$

Thus, their distributions do not depend on θ or any other unknown parameters. We conclude that Q_1 and Q_2 are both valid pivots.

To summarize, here are the steps in the pivotal method for finding confidence intervals:

1. First, find a pivotal quantity $Q(X_1, X_2, \cdots, X_n, \theta)$.

2. Find an interval for Q such that

$$P\big(q_l \leq Q \leq q_h\big) = 1 - \alpha.$$

3. Using algebraic manipulations, convert the above equation to an equation of the form

$$P\big(\hat{\Theta}_l \leq \theta \leq \hat{\Theta}_h\big) = 1 - \alpha.$$

You are probably still not sure how exactly you can perform these steps. The most crucial one is the first step. How do we find a pivotal quantity? Luckily, for many important cases that appear frequently in practice, statisticians have already found the pivotal quantities, so we can use their results directly. In practice, many of the interval estimation problems you encounter are of the forms for which general confidence intervals have been found previously. Therefore, to solve many confidence interval problems, it suffices to write the problem in a format similar to a previously solved problem. As you see more examples, you will feel more confident about solving confidence interval problems.

Example 8.15. Let X_1, X_2, X_3, ..., X_n be a random sample from a distribution with known variance $\text{Var}(X_i) = \sigma^2$, and unknown mean $EX_i = \theta$. Find a $(1 - \alpha)$ confidence interval for θ. Assume that n is large.

Solution: As usual, to find a confidence interval, we start with a point estimate. Since $\theta = EX_i$, a natural choice is the sample mean

$$\overline{X} = \frac{X_1 + X_2 + \dots + X_n}{n}.$$

Since n is large, by the Central Limit Theorem (CLT), we conclude that

$$Q = \frac{\overline{X} - \theta}{\frac{\sigma}{\sqrt{n}}}$$

has approximately $N(0, 1)$ distribution. In particular, Q is a function of the X_i's and θ, and its distribution does not depend on θ, or any other unknown parameters. Thus, Q is a pivotal quantity. The next step is to find a $(1 - \alpha)$ interval for Q. As we saw before, a $(1 - \alpha)$ interval for the standard normal random variable Q can be stated as

$$P\left(-z_{\frac{\alpha}{2}} \leq Q \leq z_{\frac{\alpha}{2}}\right) = 1 - \alpha.$$

Therefore,

$$P\left(-z_{\frac{\alpha}{2}} \leq \frac{\overline{X} - \theta}{\frac{\sigma}{\sqrt{n}}} \leq z_{\frac{\alpha}{2}}\right) = 1 - \alpha.$$

which is equivalent to

$$P\left(\overline{X} - z_{\frac{\alpha}{2}}\frac{\sigma}{\sqrt{n}} \leq \theta \leq \overline{X} + z_{\frac{\alpha}{2}}\frac{\sigma}{\sqrt{n}}\right) = 1 - \alpha.$$

We conclude that $\left[\overline{X} - z_{\frac{\alpha}{2}}\frac{\sigma}{\sqrt{n}}, \overline{X} + z_{\frac{\alpha}{2}}\frac{\sigma}{\sqrt{n}}\right]$ is a $(1-\alpha)100\%$ confidence interval for θ.

The above example is our first important case of known interval estimators, so let's summarize what we have shown:

Assumptions: A random sample X_1, X_2, X_3, ..., X_n is given from a distribution with known variance $\text{Var}(X_i) = \sigma^2 < \infty$; n is large.

Parameter to be Estimated: $\theta = EX_i$.

Confidence Interval: $\left[\overline{X} - z_{\frac{\alpha}{2}}\frac{\sigma}{\sqrt{n}}, \overline{X} + z_{\frac{\alpha}{2}}\frac{\sigma}{\sqrt{n}}\right]$ is approximately a $(1-\alpha)100\%$ confidence interval for θ.

Note that to obtain the above interval, we used the CLT. Thus, what we found is an approximate confidence interval. Nevertheless, for large n, the approximation is very good.

Example 8.16. An engineer is measuring a quantity θ. It is assumed that there is a random error in each measurement, so the engineer will take n measurements and report the average of the measurements as the estimated value of θ. Here, n is assumed to be large enough so that the central limit theorem applies. If X_i is the value that is obtained in the ith measurement, we assume that

$$X_i = \theta + W_i,$$

where W_i is the error in the ith measurement. We assume that the W_i's are i.i.d. with $EW_i = 0$ and $\text{Var}(W_i) = 4$ square units. The engineer reports the average of the measurements

$$\overline{X} = \frac{X_1 + X_2 + ... + X_n}{n}.$$

How many measurements does the engineer need to make until he is 90% sure that the final error is less than 0.25 units? In other words, what should the value of n be such that

$$P\left(\theta - 0.25 \leq \overline{X} \leq \theta + 0.25\right) \geq .90 \ ?$$

Solution: Note that, here, the X_i's are i.i.d. with mean

$$EX_i = \theta + EW_i$$
$$= \theta,$$

and variance

$$\mathrm{Var}(X_i) = \mathrm{Var}(W_i)$$
$$= 4.$$

Thus, we can restate the problem using our confidence interval terminology: "Let X_1, X_2, X_3, ..., X_n be a random sample from a distribution with known variance $\mathrm{Var}(X_i) = \sigma^2 = 4$. How large n should be so that the interval

$$\left[\overline{X} - 0.25, \overline{X} + 0.25 \right]$$

is a 90% confidence interval for $\theta = EX_i$?"

By our discussion above, the 90% confidence interval for $\theta = EX_i$ is given by

$$\left[\overline{X} - z_{\frac{\alpha}{2}} \frac{\sigma}{\sqrt{n}}, \overline{X} + z_{\frac{\alpha}{2}} \frac{\sigma}{\sqrt{n}} \right]$$

Thus, we need

$$z_{\frac{\alpha}{2}} \frac{\sigma}{\sqrt{n}} = 0.25,$$

where $\sigma = 2$, $\alpha = 1 - 0.90 = 0.1$. In particular,

$$z_{\frac{\alpha}{2}} = z_{0.05} = \Phi^{-1}(1 - 0.05) = 1.645$$

Thus, we need to have

$$1.645 \frac{2}{\sqrt{n}} = 0.25$$

We conclude that $n \geq 174$ is sufficient.

Now suppose that X_1, X_2, X_3, ..., X_n is a random sample from a distribution with *unknown* variance $\mathrm{Var}(X_i) = \sigma^2$. Our goal is to find a $1 - \alpha$ confidence interval for $\theta = EX_i$. We also assume that n is large. By the above discussion, we can say

$$P\left(\overline{X} - z_{\frac{\alpha}{2}} \frac{\sigma}{\sqrt{n}} \leq \theta \leq \overline{X} + z_{\frac{\alpha}{2}} \frac{\sigma}{\sqrt{n}} \right) = 1 - \alpha.$$

However, there is a problem here. We do not know the value of σ. How do we deal with this issue? There are two general approaches: we can either find an upper bound for σ, or we can estimate σ.

1. An upper bound for σ^2: Suppose that we can somehow show that

$$\sigma \leq \sigma_{max},$$

where $\sigma_{max} < \infty$ is a real number. Then, if we replace σ in $\left[\overline{X} - z_{\frac{\alpha}{2}} \frac{\sigma}{\sqrt{n}}, \overline{X} + z_{\frac{\alpha}{2}} \frac{\sigma}{\sqrt{n}} \right]$ by σ_{max}, the interval gets bigger. In other words, the interval

$$\left[\overline{X} - z_{\frac{\alpha}{2}} \frac{\sigma_{max}}{\sqrt{n}}, \overline{X} + z_{\frac{\alpha}{2}} \frac{\sigma_{max}}{\sqrt{n}} \right]$$

is still a valid $(1 - \alpha)100\%$ confidence interval for θ.

2. Estimate σ^2: Note that here, since n is large, we should be able to find a relatively good estimate for σ^2. After estimating σ^2, we can use that estimate and $\left[\overline{X} - z_{\frac{\alpha}{2}}\frac{\sigma}{\sqrt{n}}, \overline{X} + z_{\frac{\alpha}{2}}\frac{\sigma}{\sqrt{n}}\right]$ to find an approximate $(1 - \alpha)100\%$ confidence interval for θ.

We now provide examples of each approach.

Example 8.17. (Public Opinion Polling) We would like to estimate the portion of people who plan to vote for Candidate A in an upcoming election. It is assumed that the number of voters is large, and θ is the portion of voters who plan to vote for Candidate A. We define the random variable X as follows. A voter is chosen uniformly at random among all voters and we ask her/him: "Do you plan to vote for Candidate A?" If she/he says "yes," then $X = 1$, otherwise $X = 0$. Then,

$$X \sim Bernoulli(\theta).$$

Let X_1, X_2, X_3, ..., X_n be a random sample from this distribution, which means that the X_i's are i.i.d. and $X_i \sim Bernoulli(\theta)$. In other words, we randomly select n voters (with replacement) and we ask each of them if they plan to vote for Candidate A. Find a $(1-\alpha)100\%$ confidence interval for θ based on X_1, X_2, X_3, ..., X_n.

Solution: Note that, here,

$$EX_i = \theta.$$

Thus, we want to estimate the mean of the distribution. Note also that

$$Var(X_i) = \sigma^2 = \theta(1 - \theta).$$

Thus, to find σ, we need to know θ. But θ is the parameter that we would like to estimate in the first place. By the above discussion, we know that if we can find an upper bound for σ, we can use it to build a confidence interval for θ. Luckily, it is easy to find an upper bound for σ in this problem. More specifically, if you define

$$f(\theta) = \theta(1 - \theta), \qquad \text{for } \theta \in [0, 1].$$

By taking derivatives, you can show that the maximum value for $f(\theta)$ is obtained at $\theta = \frac{1}{2}$ and that

$$f(\theta) \leq f\left(\frac{1}{2}\right) = \frac{1}{4}, \qquad \text{for } \theta \in [0, 1].$$

We conclude that

$$\sigma_{max} = \frac{1}{2}$$

is an upper bound for σ. We conclude that the interval

$$\left[\overline{X} - z_{\frac{\alpha}{2}}\frac{\sigma_{max}}{\sqrt{n}}, \overline{X} + z_{\frac{\alpha}{2}}\frac{\sigma_{max}}{\sqrt{n}}\right]$$

is a $(1-\alpha)100\%$ confidence interval for θ, where $\sigma_{max} = \frac{1}{2}$. Thus,

$$\left[\overline{X} - \frac{z_{\frac{\alpha}{2}}}{2\sqrt{n}}, \overline{X} + \frac{z_{\frac{\alpha}{2}}}{2\sqrt{n}}\right]$$

is a $(1-\alpha)100\%$ confidence interval for θ. Note that we obtained the interval by using the CLT, so it is an approximate interval. Nevertheless, for large n, the approximation is very good. Also, since we have used an upper bound for σ, this confidence interval might be too conservative, specifically if θ is far from $\frac{1}{2}$.

The above setting is another important case of known interval estimators, so let's summarize it:

Assumptions: A random sample X_1, X_2, X_3, ..., X_n is given from a $Bernoulli(\theta)$; n is large.

Parameter to be Estimated: θ

Confidence Interval: $\left[\overline{X} - \frac{z_{\frac{\alpha}{2}}}{2\sqrt{n}}, \overline{X} + \frac{z_{\frac{\alpha}{2}}}{2\sqrt{n}}\right]$ is approximately a $(1-\alpha)100\%$ confidence interval for θ. This is a conservative confidence interval as it is obtained using an upper bound for σ.

Example 8.18. There are two candidates in a presidential election: Candidate A and Candidate B. Let θ be the portion of people who plan to vote for Candidate A. Our goal is to find a confidence interval for θ. Specifically, we choose a random sample (with replacement) of n voters and ask them if they plan to vote for Candidate A. Our goal is to estimate the θ such that the margin of error is 3 percentage points. Assume a 95% confidence level. That is, we would like to choose n such that

$$P\left(\overline{X} - 0.03 \le \theta \le \overline{X} + 0.03\right) \ge 0.95,$$

where $\overline{X}$ is the portion of people in our random sample that say they plan to vote for Candidate A. How large does n need to be?

Solution: Based on the above discussion,

$$\left[\overline{X} - \frac{z_{\frac{\alpha}{2}}}{2\sqrt{n}}, \overline{X} + \frac{z_{\frac{\alpha}{2}}}{2\sqrt{n}}\right]$$

is a valid $(1-\alpha)100\%$ confidence interval for θ. Therefore, we need to have

$$\frac{z_{\frac{\alpha}{2}}}{2\sqrt{n}} = 0.03$$

Here, $\alpha = 0.05$, so $z_{\frac{\alpha}{2}} = z_{0.025} = 1.96$. Therefore, we obtain

$$n = \left(\frac{1.96}{2 \times 0.03} \right)^2.$$

We conclude $n \geq 1068$ is enough. The above calculation provides a reason why most polls before elections are conducted with a sample size of around one thousand.

As we mentioned, the above calculation might be a little conservative. Another approach would be to estimate σ^2 instead of using an upper bound. In this example, the structure of the problem suggests a way to estimate σ^2. Specifically, since

$$\sigma^2 = \theta(1 - \theta),$$

we may use

$$\hat{\sigma}^2 = \hat{\theta}(1 - \hat{\theta})$$
$$= \overline{X}(1 - \overline{X})$$

as an estimate for θ, where $\hat{\theta} = \overline{X}$. The rationale behind this approximation is that since n is large, $\overline{X}$ is likely a good estimate of θ, thus $\hat{\sigma}^2 = \hat{\theta}(1 - \hat{\theta})$ is a good estimate of σ^2. After estimating σ^2, we can use $\left[\overline{X} - z_{\frac{\alpha}{2}} \frac{\hat{\sigma}}{\sqrt{n}}, \overline{X} + z_{\frac{\alpha}{2}} \frac{\hat{\sigma}}{\sqrt{n}} \right]$ as an approximate $(1 - \alpha)100\%$ confidence interval for θ. To summarize, we have the following confidence interval rule:

Assumptions: A random sample X_1, X_2, X_3, ..., X_n is given from a *Bernoulli*(θ); n is large.

Parameter to be Estimated: θ

Confidence Interval: $\left[\overline{X} - z_{\frac{\alpha}{2}} \sqrt{\frac{\overline{X}(1 - \overline{X})}{n}}, \overline{X} + z_{\frac{\alpha}{2}} \sqrt{\frac{\overline{X}(1 - \overline{X})}{n}} \right]$ is approximately a $(1 - \alpha)100\%$ confidence interval for θ.

Again, the above confidence interval is an approximate confidence interval because we used two approximations: the CLT and an approximation for σ^2.

The above scenario is a special case $(Bernoulli(\theta))$ for which we could come up with a point estimator for σ^2. Can we have a more general estimator for σ^2 that we can use for any distribution? We have already discussed such a point estimator and we called it the sample variance:

$$S^2 = \frac{1}{n-1} \sum_{k=1}^{n} (X_k - \overline{X})^2 = \frac{1}{n-1} \left(\sum_{k=1}^{n} X_k^2 - n\overline{X}^2 \right).$$

Thus, using the sample variance, S^2, we can have an estimate for σ^2. If n is large, this estimate is likely to be close to the real value of σ^2. So let us summarize this discussion as follows:

Assumptions: A random sample X_1, X_2, X_3, ..., X_n is given from a distribution with unknown variance $\text{Var}(X_i) = \sigma^2 < \infty$; n is large.

Parameter to be Estimated: $\theta = EX_i$.

Confidence Interval: If S is the sample standard deviation

$$S = \sqrt{\frac{1}{n-1}\sum_{k=1}^{n}(X_k - \overline{X})^2} = \sqrt{\frac{1}{n-1}\left(\sum_{k=1}^{n}X_k^2 - n\overline{X}^2\right)},$$

then the interval

$$\left[\overline{X} - z_{\frac{\alpha}{2}}\frac{S}{\sqrt{n}}, \overline{X} + z_{\frac{\alpha}{2}}\frac{S}{\sqrt{n}}\right]$$

is approximately a $(1-\alpha)100\%$ confidence interval for θ.

Example 8.19. We have collected a random sample X_1, X_2, X_3, ..., X_{100} from an unknown distribution. The sample mean and the sample variance for this random sample are given by

$$\overline{X} = 15.6, S^2 = 8.4$$

Construct an approximate 99% confidence interval for $\theta = EX_i$.

  Solution: Here, the interval

$$\left[\overline{X} - z_{\frac{\alpha}{2}}\frac{S}{\sqrt{n}}, \overline{X} + z_{\frac{\alpha}{2}}\frac{S}{\sqrt{n}}\right]$$

is approximately a $(1-\alpha)100\%$ confidence interval for θ. Since $\alpha = 0.01$, we have

$$z_{\frac{\alpha}{2}} = z_{0.005} = 2.576$$

Using $n = 100$, $\overline{X} = 15.6$, $S^2 = 8.4$, we obtain the following interval

$$\left[15.6 - 2.576\frac{\sqrt{8.4}}{\sqrt{100}}, 15.6 + 2.576\frac{\sqrt{8.4}}{\sqrt{100}}\right] = [14.85, 16.34].$$

8.3.3 Confidence Intervals for Normal Samples

In the above discussion, we assumed n to be large so that we could use the CLT. An interesting aspect of the confidence intervals that we obtained was that they often did not depend on the details of the distribution from which we obtained the random sample. That is, the confidence intervals only depended on statistics such as $\overline{X}$ and S^2.

What if n is not large? In this case, we cannot use the CLT, so we need to use the probability distribution from which the random sample is obtained. A very important case is when we have a sample X_1, X_2, X_3, ..., X_n from a normal distribution. Here, we would like to discuss how to find interval estimators for the mean and the variance of a normal distribution. Before doing so, we need to introduce two probability distributions that are related to the normal distribution. These distributions are useful when finding interval estimators for the mean and the variance of a normal distribution.

Chi-Squared Distribution:

Let us remember the gamma distribution. A continuous random variable X is said to have a *gamma* distribution with parameters $\alpha > 0$ and $\lambda > 0$, shown as $X \sim Gamma(\alpha, \lambda)$, if its PDF is given by

$$f_X(x) = \begin{cases} \frac{\lambda^\alpha x^{\alpha-1} e^{-\lambda x}}{\Gamma(\alpha)} & x > 0 \\ 0 & \text{otherwise} \end{cases}$$

Now, we would like to define a closely related distribution, called the chi-squared distribution. We know that if Z_1, Z_2, $\cdots$, Z_n are independent standard normal random variables, then the random variable

$$X = Z_1 + Z_2 + \cdots + Z_n$$

is also normal. More specifically, $X \sim N(0, n)$. Now, if we define a random variable Y as

$$Y = Z_1^2 + Z_2^2 + \cdots + Z_n^2,$$

then Y is said to have a **chi-squared** distribution with n **degrees of freedom** shown by

$$Y \sim \chi^2(n).$$

It can be shown that the random variable Y has, in fact, a gamma distribution with parameters $\alpha = \frac{n}{2}$ and $\lambda = \frac{1}{2}$,

$$Y \sim Gamma\left(\frac{n}{2}, \frac{1}{2}\right).$$

Figure 8.5 shows the PDF of $\chi^2(n)$ distribution for some values of n.

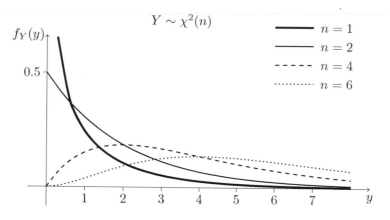

Figure 8.5: The PDF of $\chi^2(n)$ distribution for some values of n.

So, let us summarize the definition and some properties of the chi-squared distribution.

The Chi-Squared Distribution

Definition 8.1. If Z_1, Z_2, $\cdots$, Z_n are independent standard normal random variables, the random variable Y defined as

$$Y = Z_1^2 + Z_2^2 + \cdots + Z_n^2$$

is said to have a *chi-squared* distribution with n *degrees of freedom* shown by

$$Y \sim \chi^2(n).$$

Properties:

1. The chi-squared distribution is a special case of the gamma distribution. More specifically,

$$Y \sim Gamma\left(\frac{n}{2}, \frac{1}{2}\right).$$

Thus,

$$f_Y(y) = \frac{1}{2^{\frac{n}{2}}\Gamma\left(\frac{n}{2}\right)} y^{\frac{n}{2}-1} e^{-\frac{y}{2}}, \qquad \text{for } y > 0.$$

2. $EY = n$, $\text{Var}(Y) = 2n$.

3. For any $p \in [0,1]$ and $n \in \mathbb{N}$, we define $\chi_{p,n}^2$ as the real value for which

$$P(Y > \chi_{p,n}^2) = p,$$

where $Y \sim \chi^2(n)$. Figure 8.6 shows $\chi_{p,n}^2$. In MATLAB, to compute $\chi_{p,n}^2$ you can use the following command: chi2inv(1-p,n).

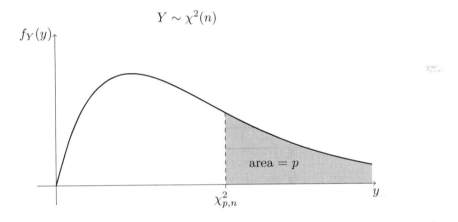

Figure 8.6: The definition of $\chi_{p,n}^2$.

Now, why do we need the chi-squared distribution? One reason is the following theorem, which we will use in estimating the variance of normal random variables.

Theorem 8.3. Let X_1, X_2, $\cdots$, X_n be i.i.d. $N(\mu, \sigma^2)$ random variables. Also, let S^2 be the sample variance for this random sample. Then, the random variable Y defined as

$$Y = \frac{(n-1)S^2}{\sigma^2} = \frac{1}{\sigma^2} \sum_{i=1}^{n} (X_i - \overline{X})^2$$

has a chi-squared distribution with $n-1$ degrees of freedom, i.e., $Y \sim \chi^2(n-1)$. Moreover, $\overline{X}$ and S^2 are independent random variables.

The t-Distribution:

The next distribution we need is the **Student's t-distribution** (or simply the **t-distribution**). Here, we provide the definition and some properties of the t-distribution.

The t-Distribution

Definition 8.2. Let $Z \sim N(0,1)$, and $Y \sim \chi^2(n)$, where $n \in \mathbb{N}$. Also assume that Z and Y are independent. The random variable T defined as

$$T = \frac{Z}{\sqrt{Y/n}}$$

is said to have a t-distribution with n *degrees of freedom* shown by

$$T \sim T(n).$$

Properties:

1. The t-distribution has a bell-shaped PDF centered at 0, but its PDF is more spread out than the normal PDF (Figure 8.7).

2. $ET = 0$, for $n > 0$. But ET, is undefined for $n = 1$.

3. $\text{Var}(T) = \frac{n}{n-2}$, for $n > 2$. But, $\text{Var}(T)$ is undefined for $n = 1, 2$.

4. As n becomes large, the t density approaches the standard normal PDF. More formally, we can write

$$T(n) \xrightarrow{d} N(0,1).$$

5. For any $p \in [0,1]$ and $n \in \mathbb{N}$, we define $t_{p,n}$ as the real value for which

$$P(T > t_{p,n}) = p.$$

Since the t-distribution has a symmetric PDF, we have

$$t_{1-p,n} = -t_{p,n}.$$

In MATLAB, to compute $t_{p,n}$ you can use the following command: tinv(1-p,n).

Figure 8.7 shows the PDF of t-distribution for some values of n and compares them with the PDF of the standard normal distribution. As we see, the t density is more spread out than the standard normal PDF. Figure 8.8 shows $t_{p,n}$.

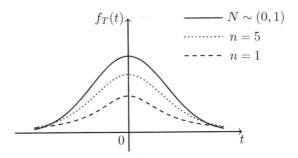

Figure 8.7: The PDF of t-distribution for some values of n compared with the standard normal PDF.

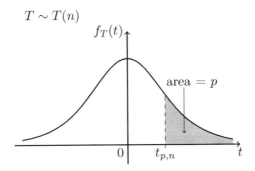

Figure 8.8: The definition of $t_{p,n}$.

Why do we need the t-distribution? One reason is the following theorem which we will use in estimating the mean of normal random variables.

Theorem 8.4. Let X_1, X_2, $\cdots$, X_n be i.i.d. $N(\mu, \sigma^2)$ random variables. Also, let S^2 be the sample variance for this random sample. Then, the random variable T defined as

$$T = \frac{\overline{X} - \mu}{S/\sqrt{n}}$$

has a t-distribution with $n - 1$ degrees of freedom, i.e., $T \sim T(n - 1)$.

Proof. Define the random variable Z as

$$Z = \frac{\overline{X} - \mu}{\sigma/\sqrt{n}}.$$

Then, $Z \sim N(0, 1)$. Also, define the random variable Y as

$$Y = \frac{(n-1)S^2}{\sigma^2}.$$

Then by Theorem 8.3, $Y \sim \chi^2(n-1)$. We conclude that the random variable

$$T = \frac{Z}{\sqrt{\frac{Y}{n-1}}} = \frac{\overline{X} - \mu}{S/\sqrt{n}}$$

has a t-distribution with $n - 1$ degrees of freedom.   ■

Confidence Intervals for the Mean of Normal Random Variables:

Here, we assume that X_1, X_2, X_3, ..., X_n is a random sample from a normal distribution $N(\mu, \sigma^2)$, and our goal is to find an interval estimator for μ. We no longer require n to be large. Thus, n could be any positive integer. There are two possible scenarios depending on whether σ^2 is known or not.

If the value of σ^2 is known, we can easily find a confidence interval for μ. This can be done using exactly the same method that we used to estimate μ for a general distribution for the case of large n. More specifically, we know that the random variable

$$Q = \frac{\overline{X} - \mu}{\sigma/\sqrt{n}}$$

has $N(0, 1)$ distribution. In particular, Q is a function of the X_i's and μ, and its distribution does not depend on μ. Thus, Q is a pivotal quantity, and we conclude that $\left[\overline{X} - z_{\frac{\alpha}{2}} \frac{\sigma}{\sqrt{n}}, \overline{X} + z_{\frac{\alpha}{2}} \frac{\sigma}{\sqrt{n}} \right]$ is $(1 - \alpha)100\%$ confidence interval for μ.

Assumptions: A random sample X_1, X_2, X_3, ..., X_n is given from a $N(\mu, \sigma^2)$ distribution, where $\text{Var}(X_i) = \sigma^2$ <u>is known</u>.

Parameter to be Estimated: $\mu = EX_i$.

Confidence Interval: $\left[\overline{X} - z_{\frac{\alpha}{2}} \frac{\sigma}{\sqrt{n}}, \overline{X} + z_{\frac{\alpha}{2}} \frac{\sigma}{\sqrt{n}} \right]$ is a $(1 - \alpha)100\%$ confidence interval for μ.

The more interesting case is when we do not know the variance σ^2. More specifically, we are given X_1, X_2, X_3, ..., X_n, which is a random sample from a normal distribution $N(\mu, \sigma^2)$, and our goal is to find an interval estimator for μ. However, σ^2 is also unknown. In this case, using Theorem 8.4, we conclude that the random variable T defined as

$$T = \frac{\overline{X} - \mu}{S/\sqrt{n}}$$

has a t-distribution with $n-1$ degrees of freedom, i.e., $T \sim T(n-1)$. Here, the random variable T is a pivotal quantity, since it is a function of the X_i's and μ, and its distribution does not depend on μ or any other unknown parameters. Now that we have a pivot, the next step is to find a $(1-\alpha)$ interval for T. Using the definition of $t_{p,n}$, a $(1-\alpha)$ interval for T can be stated as

$$P\left(-t_{\frac{\alpha}{2},n-1} \leq T \leq t_{\frac{\alpha}{2},n-1}\right) = 1-\alpha.$$

Therefore,

$$P\left(-t_{\frac{\alpha}{2},n-1} \leq \frac{\overline{X} - \mu}{S/\sqrt{n}} \leq t_{\frac{\alpha}{2},n-1}\right) = 1-\alpha,$$

which is equivalent to

$$P\left(\overline{X} - t_{\frac{\alpha}{2},n-1}\frac{S}{\sqrt{n}} \leq \mu \leq \overline{X} + t_{\frac{\alpha}{2},n-1}\frac{S}{\sqrt{n}}\right) = 1-\alpha.$$

We conclude that $\left[\overline{X} - t_{\frac{\alpha}{2},n-1}\frac{S}{\sqrt{n}}, \overline{X} + t_{\frac{\alpha}{2},n-1}\frac{S}{\sqrt{n}}\right]$ is $(1-\alpha)100\%$ confidence interval for μ.

Assumptions: A random sample X_1, X_2, X_3, ..., X_n is given from a $N(\mu, \sigma^2)$ distribution, where $\mu = EX_i$ and $\text{Var}(X_i) = \sigma^2$ <u>are unknown</u>.

Parameter to be Estimated: $\mu = EX_i$.

Confidence Interval: $\left[\overline{X} - t_{\frac{\alpha}{2},n-1}\frac{S}{\sqrt{n}}, \overline{X} + t_{\frac{\alpha}{2},n-1}\frac{S}{\sqrt{n}}\right]$ is a $(1-\alpha)$ confidence interval for μ.

Example 8.20. A farmer weighs 10 randomly chosen watermelons from his farm and he obtains the following values (in lbs):

| 7.72 | 9.58 | 12.38 | 7.77 | 11.27 | 8.80 | 11.10 | 7.80 | 10.17 | 6.00 |

Assuming that the weight is normally distributed with mean μ and variance σ^2, find a 95% confidence interval for μ.

Solution: Using the data we obtain

$$\overline{X} = 9.26,$$
$$S^2 = 3.96$$

Here, $n = 10$, $\alpha = 0.05$, so we need

$$t_{0.025,9} \approx 2.262$$

The above value can be obtained in MATLAB using the command tinv(0.975,9). Thus, we can obtain a 95% confidence interval for μ as

$$\left[\overline{X} - t_{\frac{\alpha}{2},n-1} \frac{S}{\sqrt{n}}, \overline{X} + t_{\frac{\alpha}{2},n-1} \frac{S}{\sqrt{n}} \right] = \left[9.26 - 2.26 \cdot \frac{\sqrt{3.96}}{\sqrt{10}}, 9.26 + 2.26 \cdot \frac{\sqrt{3.96}}{\sqrt{10}} \right]$$
$$= [7.84, 10.68].$$

Therefore, $[7.84, 10.68]$ is a 95% confidence interval for μ.

Confidence Intervals for the Variance of Normal Random Variables:

Now, suppose that we would like to estimate the variance of a normal distribution. More specifically, assume that X_1, X_2, X_3, ..., X_n is a random sample from a normal distribution $N(\mu, \sigma^2)$, and our goal is to find an interval estimator for σ^2. We assume that μ is also unknown. Again, n could be any positive integer.

By Theorem 8.3, the random variable Q defined as

$$Q = \frac{(n-1)S^2}{\sigma^2} = \frac{1}{\sigma^2} \sum_{i=1}^{n} (X_i - \overline{X})^2$$

has a chi-squared distribution with $n-1$ degrees of freedom, i.e., $Q \sim \chi^2(n-1)$. In particular, Q is a pivotal quantity since it is a function of the X_i's and σ^2, and its distribution does not depend on σ^2 or any other unknown parameters. Using the definition of $\chi^2_{p,n}$, a $(1-\alpha)$ interval for Q can be stated as

$$P\left(\chi^2_{1-\frac{\alpha}{2},n-1} \leq Q \leq \chi^2_{\frac{\alpha}{2},n-1} \right) = 1 - \alpha.$$

Therefore,

$$P\left(\chi^2_{1-\frac{\alpha}{2},n-1} \leq \frac{(n-1)S^2}{\sigma^2} \leq \chi^2_{\frac{\alpha}{2},n-1} \right) = 1 - \alpha.$$

which is equivalent to

$$P\left(\frac{(n-1)S^2}{\chi^2_{\frac{\alpha}{2},n-1}} \leq \sigma^2 \leq \frac{(n-1)S^2}{\chi^2_{1-\frac{\alpha}{2},n-1}} \right) = 1 - \alpha.$$

We conclude that $\left[\frac{(n-1)S^2}{\chi^2_{\frac{\alpha}{2},n-1}}, \frac{(n-1)S^2}{\chi^2_{1-\frac{\alpha}{2},n-1}} \right]$ is a $(1-\alpha)100\%$ confidence interval for σ^2.

Assumptions: A random sample X_1, X_2, X_3, ..., X_n is given from a $N(\mu, \sigma^2)$ distribution, where $\mu = EX_i$ and $\text{Var}(X_i) = \sigma^2$ are unknown.

Parameter to be Estimated: $\text{Var}(X_i) = \sigma^2$.

Confidence Interval: $\left[\frac{(n-1)S^2}{\chi^2_{\frac{\alpha}{2}, n-1}}, \frac{(n-1)S^2}{\chi^2_{1-\frac{\alpha}{2}, n-1}} \right]$ is a $(1-\alpha)100\%$ confidence interval for σ^2.

Example 8.21. For the data given in Example 8.20, find a 95% confidence interval for σ^2. Again, assume that the weight is normally distributed with mean μ and variance σ^2, where μ and σ are unknown.

Solution: As before, using the data we obtain

$$\overline{X} = 9.26,$$
$$S^2 = 3.96$$

Here, $n = 10$, $\alpha = 0.05$, so we need

$$\chi^2_{0.025,9} = 19.02, \qquad \chi^2_{0.975,9} = 2.70$$

The above values can obtained in MATLAB using the commands chi2inv(0.975,9) and chi2inv(0.025,9), respectively. Thus, we can obtain a 95% confidence interval for σ^2 as

$$\left[\frac{(n-1)S^2}{\chi^2_{\frac{\alpha}{2}, n-1}}, \frac{(n-1)S^2}{\chi^2_{1-\frac{\alpha}{2}, n-1}} \right] = \left[\frac{9 \times 3.96}{19.02}, \frac{9 \times 3.96}{2.70} \right]$$
$$= [1.87, 13.20].$$

Therefore, $[1.87, 13.20]$ is a 95% confidence interval for σ^2.

8.3.4 Solved Problems

1. Let X_1, X_2, X_3, ..., X_n be a random sample from an exponential distribution with parameter θ, i.e.,

$$f_{X_i}(x; \theta) = \theta e^{-\theta x} u(x).$$

Our goal is to find a $(1-\alpha)100\%$ confidence interval for θ. To do this, we need to remember a few facts about the gamma distribution. More specifically, if $Y = X_1 + X_2 + \cdots + X_n$, where the X_i's are independent *Exponential*(θ) random variables, then $Y \sim Gamma(n, \theta)$. Thus, the random variable Q defined as

$$Q = \theta(X_1 + X_2 + \cdots + X_n)$$

has a *Gamma*$(n, 1)$ distribution. Let us define $\gamma_{p,n}$ as follows. For any $p \in [0, 1]$ and $n \in \mathbb{N}$, we define $\gamma_{p,n}$ as the real value for which

$$P(Q > \gamma_{p,n}) = p,$$

where $Q \sim Gamma(n, 1)$.

(a) Explain why $Q = \theta(X_1 + X_2 + \cdots + X_n)$ is a pivotal quantity.

(b) Using Q and the definition of $\gamma_{p,n}$, construct a $(1-\alpha)100\%$ confidence interval for θ.

Solution:

(a) Q is a function of the X_i's and θ, and its distribution does not depend on θ or any other unknown parameters. Thus, Q is a pivotal quantity.

(b) Using the definition of $\gamma_{p,n}$, a $(1-\alpha)$ interval for Q can be stated as

$$P\left(\gamma_{1-\frac{\alpha}{2},n-1} \leq Q \leq \gamma_{\frac{\alpha}{2},n-1}\right) = 1 - \alpha.$$

Therefore,

$$P\left(\gamma_{1-\frac{\alpha}{2},n-1} \leq \theta(X_1 + X_2 + \cdots + X_n) \leq \gamma_{\frac{\alpha}{2},n-1}\right) = 1 - \alpha.$$

Since $X_1 + X_2 + \cdots + X_n$ is always a positive quantity, the above equation is equivalent to

$$P\left(\frac{\gamma_{1-\frac{\alpha}{2},n-1}}{X_1 + X_2 + \cdots + X_n} \leq \theta \leq \frac{\gamma_{\frac{\alpha}{2},n-1}}{X_1 + X_2 + \cdots + X_n}\right) = 1 - \alpha.$$

We conclude that $\left[\frac{\gamma_{1-\frac{\alpha}{2},n-1}}{X_1+X_2+\cdots+X_n}, \frac{\gamma_{\frac{\alpha}{2},n-1}}{X_1+X_2+\cdots+X_n}\right]$ is a $(1-\alpha)100\%$ confidence interval for θ.

2. A random sample X_1, X_2, X_3, ..., X_{100} is given from a distribution with known variance $\text{Var}(X_i) = 16$. For the observed sample, the sample mean is $\overline{X} = 23.5$. Find an approximate 95% confidence interval for $\theta = EX_i$.

Solution: Here, $\left[\overline{X} - z_{\frac{\alpha}{2}}\frac{\sigma}{\sqrt{n}}, \overline{X} + z_{\frac{\alpha}{2}}\frac{\sigma}{\sqrt{n}}\right]$ is an approximate $(1-\alpha)100\%$ confidence interval. Since $\alpha = 0.05$, we have

$$z_{\frac{\alpha}{2}} = z_{0.025} = \Phi^{-1}(1 - 0.025) = 1.96$$

Also, $\sigma = 4$. Therefore, the approximate confidence interval is

$$\left[23.5 - 1.96\frac{4}{\sqrt{100}}, 23.5 - 1.96\frac{4}{\sqrt{100}}\right] \approx [22.7, 24.3].$$

3. To estimate the portion of voters who plan to vote for Candidate A in an election, a random sample of size n from the voters is chosen. The sampling is done with replacement. Let θ be the portion of voters who plan to vote for Candidate A among all voters. How large does n need to be so that we can obtain a 90% confidence interval with 3% margin of error? That is, how large n needs to be such that

$$P\left(\overline{X} - 0.03 \leq \theta \leq \overline{X} + 0.03\right) \geq 0.90,$$

where $\overline{X}$ is the portion of people in our random sample that say they plan to vote for Candidate A.

Solution: Here,

$$\left[\overline{X} - \frac{z_{\frac{\alpha}{2}}}{2\sqrt{n}}, \overline{X} + \frac{z_{\frac{\alpha}{2}}}{2\sqrt{n}}\right]$$

is an approximate $(1 - \alpha)100\%$ confidence interval for θ. Since $\alpha = 0.1$, we have

$$z_{\frac{\alpha}{2}} = z_{0.05} = \Phi^{-1}(1 - 0.05) = 1.645$$

Therefore, we need to have

$$\frac{1.645}{2\sqrt{n}} = 0.03$$

Therefore, we obtain

$$n = \left(\frac{1.645}{2 \times 0.03}\right)^2.$$

We conclude $n \geq 752$ is enough.

4. (a) Let X be a random variable such that $R_X \subset [a, b]$, i.e., we always have $a \leq X \leq b$. Show that
$$\text{Var}(X) \leq \frac{(b - a)^2}{4}.$$

   (b) Let X_1, X_2, X_3, ..., X_n be a random sample from an unknown distribution with CDF $F_X(x)$ such that $R_X \subset [a, b]$. Specifically, EX and $\text{Var}(X)$ are unknown. Find a $(1 - \alpha)100\%$ confidence interval for $\theta = EX$. Assume that n is large.

Solution:

(a) Define $Y = X - \frac{a+b}{2}$. Thus, $R_Y \subset [-\frac{b-a}{2}, \frac{b-a}{2}]$. Then,

$$
\begin{aligned}
\text{Var}(X) &= \text{Var}(Y) \\
&= E[Y^2] - \mu_Y^2 \\
&\leq E[Y^2] \\
&\leq \left(\frac{b-a}{2}\right)^2 \qquad \left(\text{since } Y^2 \leq \left(\frac{b-a}{2}\right)^2\right) \\
&= \frac{(b-a)^2}{4}.
\end{aligned}
$$

(b) Here, we have an upper bound on σ, which is $\sigma_{max} = \frac{(b-a)}{2}$. Thus, the interval

$$
\left[\overline{X} - z_{\frac{\alpha}{2}} \frac{\sigma_{max}}{\sqrt{n}}, \overline{X} + z_{\frac{\alpha}{2}} \frac{\sigma_{max}}{\sqrt{n}}\right]
$$

is a $(1-\alpha)100\%$ confidence interval for θ. More specifically,

$$
\left[\overline{X} - z_{\frac{\alpha}{2}} \frac{b-a}{2\sqrt{n}}, \overline{X} + z_{\frac{\alpha}{2}} \frac{b-a}{2\sqrt{n}}\right]
$$

is a $(1-\alpha)100\%$ confidence interval for θ.

5. A random sample $X_1, X_2, X_3, ..., X_{144}$ is given from a distribution with unknown variance $\text{Var}(X_i) = \sigma^2$. For the observed sample, the sample mean is $\overline{X} = 55.2$, and the sample variance is $S^2 = 34.5$. Find a 99% confidence interval for $\theta = EX_i$.

Solution: The interval

$$
\left[\overline{X} - z_{\frac{\alpha}{2}} \frac{S}{\sqrt{n}}, \overline{X} + z_{\frac{\alpha}{2}} \frac{S}{\sqrt{n}}\right]
$$

is approximately a $(1-\alpha)100\%$ confidence interval for θ. Here, $n = 144$, $\alpha = 0.01$, so we need

$$
z_{\frac{\alpha}{2}} = z_{0.005} = \Phi^{-1}(1 - 0.005) \approx 2.58
$$

Thus, we can obtain a 99% confidence interval for θ as

$$
\left[\overline{X} - z_{\frac{\alpha}{2}} \frac{S}{\sqrt{n}}, \overline{X} + z_{\frac{\alpha}{2}} \frac{S}{\sqrt{n}}\right] = \left[55.2 - 2.58 \cdot \frac{\sqrt{34.5}}{12}, 55.2 + 2.58 \cdot \frac{\sqrt{34.5}}{12}\right]
$$

$$
\approx [53.94, 56.46].
$$

Therefore, $[53.94, 56.46]$ is an approximate 99% confidence interval for θ.

6. A random sample X_1, X_2, X_3, ..., X_{16} is given from a normal distribution with unknown mean $\mu = EX_i$ and unknown variance $\text{Var}(X_i) = \sigma^2$. For the observed sample, the sample mean is $\overline{X} = 16.7$, and the sample variance is $S^2 = 7.5$.

   (a) Find a 95% confidence interval for μ.

   (b) Find a 95% confidence interval for σ^2.

Solution:

   (a) Here, the interval

$$\left[\overline{X} - t_{\frac{\alpha}{2},n-1}\frac{S}{\sqrt{n}}, \overline{X} + t_{\frac{\alpha}{2},n-1}\frac{S}{\sqrt{n}} \right]$$

is a $(1-\alpha)100\%$ confidence interval for μ. Let $n = 16$, $\alpha = 0.05$, then

$$t_{0.025,15} \approx 2.13$$

The above value can obtained in MATLAB using the command tinv(0.975,15). Thus, we can obtain a 95% confidence interval for μ as

$$\left[16.7 - 2.13\frac{\sqrt{7.5}}{4}, 16.7 + 2.13\frac{\sqrt{7.5}}{4} \right] \approx [15.24, 18.16].$$

Therefore, $[15.24, 18.16]$ is a 95% confidence interval for μ.

   (b) Here, $\left[\frac{(n-1)S^2}{\chi^2_{\frac{\alpha}{2},n-1}}, \frac{(n-1)S^2}{\chi^2_{1-\frac{\alpha}{2},n-1}} \right]$ is a $(1-\alpha)100\%$ confidence interval for σ^2. In this problem, $n = 16$, $\alpha = .05$, so we need

$$\chi^2_{0.025,15} \approx 27.49, \qquad \chi^2_{0.975,15} \approx 6.26$$

The above values can obtained in MATLAB using the commands chi2inv(0.975,15) and chi2inv(0.025,15), respectively. Thus, we can obtain a 95% confidence interval for σ^2 as

$$\left[\frac{(n-1)S^2}{\chi^2_{\frac{\alpha}{2},n-1}}, \frac{(n-1)S^2}{\chi^2_{1-\frac{\alpha}{2},n-1}} \right] = \left[\frac{15 \times 7.5}{27.49}, \frac{15 \times 7.5}{6.26} \right]$$
$$\approx [4.09, 17.97].$$

Therefore, $[4.09, 17.97]$ is a 95% confidence interval for σ^2.

8.4   Hypothesis Testing

8.4.1   Introduction

Often, we need to test whether a hypothesis is true or false. For example, a pharmaceutical company might be interested in knowing if a new drug is effective in treating a disease. Here, there are two hypotheses. The first one is that the drug is not effective, while the second hypothesis is that the drug is effective. We call these hypotheses H_0 and H_1 respectively. As another example, consider a radar system that uses radio waves to detect aircraft. The system receives a signal and, based on the received signal, it needs to decide whether an aircraft is present or not. Here, there are again two opposing hypotheses:

H_0: No aircraft is present.

H_1: An aircraft is present.

The hypothesis H_0 is called the *null hypothesis* and the hypothesis H_1 is called the *alternative hypothesis*. The null hypothesis, H_0, is usually referred to as the default hypothesis, i.e., the hypothesis that is initially assumed to be true. The alternative hypothesis, H_1, is the statement contradictory to H_0. Based on the observed data, we need to decide either to accept H_0, or to reject it, in which case we say we accept H_1. These are problems of *hypothesis testing*. In this section, we will discuss how to approach such problems from a classical (frequentist) point of view. We will start with an example, and then provide a general framework to approach hypothesis testing problems. When looking at the example, we will introduce some terminology that is commonly used in hypothesis testing. Do not worry much about the terminology when reading this example as we will provide more precise definitions later on.

Example 8.22. You have a coin and you would like to check whether it is fair or not. More specifically, let θ be the probability of heads, $\theta = P(H)$. You have two hypotheses:

H_0 (the null hypothesis): The coin is fair, i.e. $\theta = \theta_0 = \frac{1}{2}$.

H_1 (the alternative hypothesis): The coin is not fair, i.e., $\theta \neq \frac{1}{2}$.

We need to design a test to either accept H_0 or H_1. To check whether the coin is fair or not, we perform the following experiment. We toss the coin 100 times and record the number of heads. Let X be the number of heads that we observe, so

$$X \sim Binomial(100, \theta).$$

Now, if H_0 is true, then $\theta = \theta_0 = \frac{1}{2}$, so we expect the number of heads to be close to 50. Thus, intuitively we can say that if we observe close to 50 heads we should accept H_0, otherwise we should reject it. More specifically, we suggest the following criteria: If $|X - 50|$ is less than or equal to some threshold, we accept H_0. On the other hand, if $|X - 50|$ is larger than the threshold we reject H_0 and accept H_1. Let's call that threshold t.

If $|X - 50| \leq t$, accept H_0.

If $|X - 50| > t$, accept H_1.

But how do we choose the threshold t? To choose t properly, we need to state some requirements for our test. An important factor here is probability of error. One way to make an error is when we reject H_0 while in fact it is true. We call this *type I error*. More specifically, this is the event that $|X - 50| > t$ when H_0 is true. Thus,

$$P(\text{type I error}) = P(|X - 50| > t \mid H_0).$$

We read this as the probability that $|X - 50| > t$ <u>when</u> H_0 is true.[1] To be able to decide what t needs to be, we can choose a desired value for $P(\text{type I error})$. For example, we might want to have a test for which

$$P(\text{type I error}) \leq \alpha = 0.05$$

Here, α is called *the level of significance*. We can choose

$$P(|X - 50| > t \mid H_0) = \alpha = 0.05 \tag{8.2}$$

to satisfy the desired level of significance. Since we know the distribution of X under H_0, i.e., $X|H_0 \sim Binomial(100, \theta = \frac{1}{2})$, we should be able to choose t such that Equation 8.2 holds. Note that by the central limit theorem (CLT), for large values of n, we can approximate a $Binomial(n, \theta)$ distribution by a normal distribution. More specifically, we can say that for large values of n, if $X \sim Binomial(n, \theta_0 = \frac{1}{2})$, then

$$Y = \frac{X - n\theta_0}{\sqrt{n\theta_0(1 - \theta_0)}} = \frac{X - 50}{5} \tag{8.3}$$

is (approximately) a standard normal random variable, $N(0, 1)$. Thus, to be able to use the CLT, instead of looking at X directly, we can look at Y. Note that

$$P(\text{type I error}) = P(|X - 50| > t|H_0) = P\left(\left|\frac{X - 50}{5}\right| > \frac{t}{5} \mid H_0\right)$$
$$= P\left(|Y| > \frac{t}{5} \mid H_0\right).$$

For simplicity, let's put $c = \frac{t}{5}$, so we can summarize our test as follows:

If $|Y| \leq c$, accept H_0.

If $|Y| > c$, accept H_1.

where $Y = \frac{X-50}{5}$. Now, we need to decide what c should be. We need to have

$$\alpha = P(|Y| > c)$$
$$= 1 - P(-c \leq Y \leq c)$$
$$\approx 2 - 2\Phi(c) \qquad \left(\text{using } \Phi(x) = 1 - \Phi(-x)\right).$$

Thus, we need to have

$$2 - 2\Phi(c) = 0.05$$

---

[1] Note that, here, $P(|X - 50| > t \mid H_0)$ is not a conditional probability, since in classical statistics we do not treat H_0 and H_1 as random events. Another common notation is $P(|X - 50| > t$ when H_0 is true).

So we obtain

$$c = \Phi^{-1}(0.975) = 1.96$$

Thus, we conclude the following test

If $|Y| \leq 1.96$, accept H_0.

If $|Y| > 1.96$, accept H_1.

The set $A = [-1.96, 1.96]$ is called the *acceptance region*, because it includes the points that result in accepting H_0. The set $R = (-\infty, -1.96) \cup (1.96, \infty)$ is called the *rejection region* because it includes the points that correspond to rejecting H_0. Figure 8.9 summarizes these concepts.

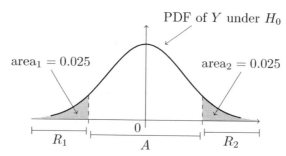

A = Acceptance Region

$R = R_1 \cup R_2$ = Rejection Region

$\alpha = P(\text{type I error}) = \text{area}_1 + \text{area}_2 = 0.05$

Figure 8.9: Acceptance rejection, rejection region, and type I error for Example 8.22.

Note that since $Y = \frac{X-50}{5}$, we can equivalently state the test as

If $|X - 50| \leq 9.8$, accept H_0.

If $|X - 50| > 9.8$, accept H_1.

Or equivalently,

If the observed number of heads is in $\{41, 42, \cdots, 59\}$, accept H_0.

If the observed number of heads is in $\{0, 1, \cdots, 40\} \cup \{60, 61, \cdots, 100\}$, reject H_0 (accept H_1).

In summary, if the observed number of heads is more than 9 counts away from 50, we reject H_0.

Before ending our discussion on this example, we would like to mention another point. Suppose that we toss the coin 100 times and observe 55 heads. Based on the above discussion we should accept H_0. However, it is often recommended to say "we failed to reject H_0" instead of saying "we are accepting H_0." The reason is that we have not really proved that H_0 is true. In fact, all we know is that the result of our experiment was not statistically contradictory to H_0. Nevertheless, we will not worry about this terminology in this book.

8.4.2 General Setting and Definitions

Example 8.22 provided a basic introduction to hypothesis testing. Here, we would like to provide a general setting for problems of hypothesis testing and formally define the terminology that is used in hypothesis testing. Although there are several new phrases such as null hypothesis, type I error, significance level, etc., there are not many new concepts or tools here. Thus, after going through a few examples, the concepts should become clear.

Suppose that θ is an unknown parameter. A hypothesis is a statement such as $\theta = 1$, $\theta > 1.3$, $\theta \neq 0.5$, etc. In hypothesis testing problems, we need to decide between two contradictory hypotheses. More precisely, let S be the set of possible values for θ. Suppose that we can partition S into two disjoint sets S_0 and S_1. Let H_0 be the hypothesis that $\theta \in S_0$, and let H_1 be the hypothesis that $\theta \in S_1$.

H_0 (the **null** hypothesis): $\theta \in S_0$.

H_1 (the **alternative** hypothesis): $\theta \in S_1$.

In Example 8.22, $S = [0,1]$, $S_0 = \{\frac{1}{2}\}$, and $S_1 = [0,1] - \{\frac{1}{2}\}$. Here, H_0 is an example of a **simple** hypothesis because S_0 contains only one value of θ. On the other hand, H_1 is an example of **composite** hypothesis since S_1 contains more than one element. It is often the case that the null hypothesis is chosen to be a simple hypothesis.

Often, to decide between H_0 and H_1, we look at a function of the observed data. For instance, in Example 8.22, we looked at the random variable Y, defined as

$$Y = \frac{X - n\theta_0}{\sqrt{n\theta_0(1 - \theta_0)}},$$

where X was the total number of heads. Here, X is a function of the observed data (sequence of heads and tails), and thus Y is a function of the observed data. We call Y a *statistic*.

Definition 8.3. Let $X_1, X_2, \cdots, X_n$ be a random sample of interest. A **statistic** is a real-valued function of the data. For example, the sample mean, defined as

$$W(X_1, X_2, \cdots, X_n) = \frac{X_1 + X_2 + ... + X_n}{n},$$

is a statistic. A **test statistic** is a statistic based on which we build our test.

To decide whether to choose H_0 or H_1, we choose a test statistic, $W = W(X_1, X_2, \cdots, X_n)$. Now, assuming H_0, we can define the set $A \subset \mathbb{R}$ as the set of possible values of W for which we would accept H_0. The set A is called the **acceptance region**, while the set $R = \mathbb{R} - A$ is said to be the **rejection region**. In Example 8.22, the acceptance region was found to be the set $A = [-1.96, 1.96]$, and the set $R = (-\infty, -1.96) \cup (1.96, \infty)$ was the rejection region.

There are two possible errors that we can make. We define **type I error** as the event that we reject H_0 when H_0 is true. Note that the probability of type I error in general depends on the real value of θ. More specifically,

$$P(\text{type I error} \mid \theta) = P(\text{Reject } H_0 \mid \theta)$$
$$= P(W \in R \mid \theta), \qquad \text{for } \theta \in S_0.$$

If the probability of type I error satisfies

$$P(\text{type I error}) \leq \alpha, \qquad \text{for all } \theta \in S_0,$$

then we say the test has **significance level** α or simply the test is a **level** α test. Note that it is often the case that the null hypothesis is a simple hypothesis, so S_0 has only one element (as in Example 8.22).

The second possible error that we can make is to accept H_0 when H_0 is false. This is called the **type II error**. Since the alternative hypothesis, H_1, is usually a composite hypothesis (so it includes more than one value of θ), the probability of type II error is usually a function of θ. The probability of type II error is usually shown by β:

$$\beta(\theta) = P(\text{Accept } H_0 \mid \theta), \qquad \text{for } \theta \in S_1.$$

We now go through an example to practice the above concepts.

Example 8.23. Consider a radar system that uses radio waves to detect aircraft. The system receives a signal and, based on the received signal, it needs to decide whether an aircraft is present or not. Let X be the received signal. Suppose that we know

$$X = W, \qquad \text{if no aircraft is present.}$$
$$X = 1 + W, \qquad \text{if an aircraft is present.}$$

where $W \sim N(0, \sigma^2 = \frac{1}{9})$. Thus, we can write $X = \theta + W$, where $\theta = 0$ if there is no aircraft, and $\theta = 1$ if there is an aircraft. Suppose that we define H_0 and H_1 as follows:

H_0 (null hypothesis): No aircraft is present.

H_1 (alternative hypothesis): An aircraft is present.

(a) Write the null hypothesis, H_0, and the alternative hypothesis, H_1, in terms of possible values of θ.

(b) Design a level 0.05 test ($\alpha = 0.05$) to decide between H_0 and H_1.

(c) Find the probability of type II error, β, for the above test. Note that this is the probability of missing a present aircraft.

(d) If we observe $X = 0.6$, is there enough evidence to reject H_0 at significance level $\alpha = 0.01$?

(e) If we would like the probability of missing a present aircraft to be less than 5%, what is the smallest significance level that we can achieve?

Solution:

(a) The null hypothesis corresponds to $\theta = 0$ and the alternative hypothesis corresponds to $\theta = 1$. Thus, we can write

H_0 (null hypothesis): No aircraft is present: $\theta = 0$.

H_1 (alternative hypothesis): An aircraft is present: $\theta = 1$.

Note that here both hypotheses are simple.

(b) To decide between H_0 and H_1, we look at the observed data. Here, the situation is relatively simple. The observed data is just the random variable X. Under H_0, $X \sim N(0, \frac{1}{9})$, and under H_1, $X \sim N(1, \frac{1}{9})$. Thus, we can suggest the following test: We choose a threshold c. If the observed value of X is less than c, we choose H_0 (i.e., $\theta = EX = 0$). If the observed value of X is larger than c, we choose H_1 (i.e., $\theta = EX = 1$). To choose c, we use the required α:

$$
\begin{aligned}
P(\text{type I error}) &= P(\text{Reject } H_0 \mid H_0) \\
&= P(X > c \mid H_0) \\
&= P(W > c) \\
&= 1 - \Phi(3c) \qquad \left(\text{since assuming } H_0, X \sim N\left(0, \frac{1}{9}\right)\right).
\end{aligned}
$$

Letting $P(\text{type I error}) = \alpha$, we obtain

$$
c = \frac{1}{3}\Phi^{-1}(1 - \alpha).
$$

Letting $\alpha = 0.05$, we obtain

$$
c = \frac{1}{3}\Phi^{-1}(0.95) = 0.548
$$

(c) Note that, here, the alternative hypothesis is a simple hypothesis. That is, it includes only one value of θ (i.e., $\theta = 1$). Thus, we can write

$$
\begin{aligned}
\beta = P(\text{type II error}) &= P(\text{accept } H_0 \mid H_1) \\
&= P(X < c \mid H_1) \\
&= P(1 + W < c) \\
&= P(W < c - 1) \\
&= \Phi(3(c - 1)).
\end{aligned}
$$

Since $c = 0.548$, we obtain $\beta = 0.088$.

(d) In part (b), we obtained

$$
c = \frac{1}{3}\Phi^{-1}(1 - \alpha).
$$

For $\alpha = 0.01$, we have $c = \frac{1}{3}\Phi^{-1}(0.99) = 0.775$ which is larger than 0.6. Thus, we cannot reject H_0 at significance level $\alpha = 0.01$.

(e) In part (c), we obtained

$$
\beta = \Phi(3(c - 1)).
$$

To have $\beta = 0.05$, we obtain

$$c = 1 + \frac{1}{3}\Phi^{-1}(\beta)$$
$$= 1 + \frac{1}{3}\Phi^{-1}(0.05)$$
$$= 0.452$$

Thus, we need to have $c \leq 0.452$ to obtain $\beta \leq 0.05$. Therefore,

$$P(\text{type I error}) = 1 - \Phi(3c)$$
$$= 1 - \Phi(3 \times 0.452)$$
$$= 0.0875,$$

which means that the smallest significance level that we can achieve is $\alpha = 0.0875$.

Trade-off Between α and β: Since α and β indicate error probabilities, we would ideally like both of them to be small. However, there is in fact a trade-off between α and β. That is, if we want to decrease the probability of type I error (α), then the probability of type II error (β) increases, and vise versa. To see this, we can look at our analysis in Example 8.23. In that example, we found

$$\alpha = 1 - \Phi(3c),$$
$$\beta = \Phi(3(c - 1)).$$

Note that $\Phi(x)$ is an increasing function. If we make c larger, α becomes smaller, and β becomes larger. On the other hand, if we make c smaller, α becomes larger, and β becomes smaller. Figure 8.10 shows type I and type II error probabilities for Example 8.23.

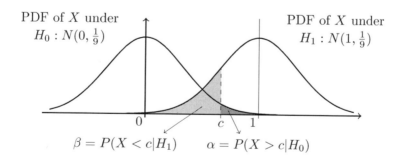

Figure 8.10: Type I and type II errors in Example 8.23.

8.4.3 Hypothesis Testing for the Mean

Here, we would like to discuss some common hypothesis testing problems. We assume that we have a random sample $X_1, X_2, ..., X_n$ from a distribution and our goal is to make inference about the mean of the distribution μ. We consider three hypothesis testing problems. The first one is a test to decide between the following hypotheses:

H_0: $\mu = \mu_0$,

H_1: $\mu \neq \mu_0$.

In this case, the null hypothesis is a simple hypothesis and the alternative hypothesis is a *two-sided* hypothesis (i.e., it includes both $\mu < \mu_0$ and $\mu > \mu_0$). We call this hypothesis test a *two-sided* test. The second and the third cases are *one-sided* tests. More specifically, the second case is

H_0: $\mu \leq \mu_0$,

H_1: $\mu > \mu_0$.

Here, both H_0 and H_1 are one-sided, so we call this test a *one-sided* test. The third case is very similar to the second case. More specifically, the third scenario is

H_0: $\mu \geq \mu_0$,

H_1: $\mu < \mu_0$.

In all of the three cases, we use the sample mean

$$\overline{X} = \frac{X_1 + X_2 + ... + X_n}{n}$$

to define our statistic. In particular, if we know the variance of the X_i's, $\mathrm{Var}(X_i) = \sigma^2$, then we define our test statistic as the normalized sample mean (assuming H_0):

$$W(X_1, X_2, \cdots, X_n) = \frac{\overline{X} - \mu_0}{\sigma/\sqrt{n}}.$$

If we do not know the variance of the X_i's, we use

$$W(X_1, X_2, \cdots, X_n) = \frac{\overline{X} - \mu_0}{S/\sqrt{n}},$$

where S is the sample standard deviation,

$$S = \sqrt{\frac{1}{n-1}\sum_{k=1}^{n}(X_k - \overline{X})^2} = \sqrt{\frac{1}{n-1}\left(\sum_{k=1}^{n}X_k^2 - n\overline{X}^2\right)}.$$

In any case, we will be able to find the distribution of W, and thus we can design our tests by calculating error probabilities. Let us start with the first case.

Two-sided Tests for the Mean:

Here, we are given a random sample $X_1, X_2, ..., X_n$ from a distribution. Let $\mu = EX_i$. Our goal is to decide between

H_0: $\mu = \mu_0$,

H_1: $\mu \neq \mu_0$.

Example 8.22, which we saw previously, is an instance of this case. If H_0 is true, we expect $\overline{X}$ to be close to μ_0, and so we expect $W(X_1, X_2, \cdots, X_n)$ to be close to 0 (see the definition of W above).

Therefore, we can suggest the following test. Choose a threshold, and call it c. If $|W| \leq c$, accept H_0, and if $|W| > c$, accept H_1. How do we choose c? If α is the required significance level, we must have

$$P(\text{type I error}) = P(\text{Reject } H_0 \mid H_0)$$
$$= P(|W| > c \mid H_0) \leq \alpha.$$

Thus, we can choose c such that $P(|W| > c \mid H_0) = \alpha$. Let us look at an example.

Example 8.24. Let $X_1, X_2, ..., X_n$ be a random sample from a $N(\mu, \sigma^2)$ distribution, where μ is unknown but σ is known. Design a level α test to choose between

H_0: $\mu = \mu_0$,

H_1: $\mu \neq \mu_0$.

Solution: As discussed above, we let

$$W(X_1, X_2, \cdots, X_n) = \frac{\overline{X} - \mu_0}{\sigma/\sqrt{n}}.$$

Note that, assuming H_0, $W \sim N(0, 1)$. We will choose a threshold, c. If $|W| \leq c$, we accept H_0, and if $|W| > c$, accept H_1. To choose c, we let

$$P(|W| > c \mid H_0) = \alpha.$$

Since the standard normal PDF is symmetric around 0, we have

$$P(|W| > c \mid H_0) = 2P(W > c \mid H_0).$$

Thus, we conclude $P(W > c \mid H_0) = \frac{\alpha}{2}$. Therefore,

$$c = z_{\frac{\alpha}{2}}.$$

Therefore, we accept H_0 if

$$\left| \frac{\overline{X} - \mu_0}{\sigma/\sqrt{n}} \right| \leq z_{\frac{\alpha}{2}},$$

and reject it otherwise.

Relation to Confidence Intervals: It is interesting to examine the above acceptance region. Here, we accept H_0 if

$$\left| \frac{\overline{X} - \mu_0}{\sigma/\sqrt{n}} \right| \leq z_{\frac{\alpha}{2}}.$$

We can rewrite the above condition as

$$\mu_0 \in \left[\overline{X} - z_{\frac{\alpha}{2}} \frac{\sigma}{\sqrt{n}}, \overline{X} + z_{\frac{\alpha}{2}} \frac{\sigma}{\sqrt{n}} \right].$$

The above interval should look familiar to you. It is the $(1 - \alpha)100\%$ confidence interval for μ_0. This is not a coincidence as there is a general relationship between confidence interval problems and hypothesis testing problems.

Example 8.25. For the above example (Example 8.24), find β, the probability of type II error, as a function of μ.

Solution: We have

$$\beta(\mu) = P(\text{type II error}) = P(\text{accept } H_0 \mid \mu)$$

$$= P \left(\left| \frac{\overline{X} - \mu_0}{\sigma/\sqrt{n}} \right| < z_{\frac{\alpha}{2}} \mid \mu \right).$$

If $X_i \sim N(\mu, \sigma^2)$, then $\overline{X} \sim N(\mu, \frac{\sigma^2}{n})$. Thus,

$$\beta(\mu) = P \left(\left| \frac{\overline{X} - \mu_0}{\sigma/\sqrt{n}} \right| < z_{\frac{\alpha}{2}} \mid \mu \right)$$

$$= P \left(\mu_0 - z_{\frac{\alpha}{2}} \frac{\sigma}{\sqrt{n}} \leq \overline{X} \leq \mu_0 + z_{\frac{\alpha}{2}} \frac{\sigma}{\sqrt{n}} \right)$$

$$= \Phi \left(z_{\frac{\alpha}{2}} + \frac{\mu_0 - \mu}{\sigma/\sqrt{n}} \right) - \Phi \left(-z_{\frac{\alpha}{2}} + \frac{\mu_0 - \mu}{\sigma/\sqrt{n}} \right).$$

Unknown variance: The above results (Example 8.25) can be extended to the case when we do not know the variance using the t-distribution. More specifically, consider the following example.

Example 8.26. Let $X_1, X_2, ..., X_n$ be a random sample from a $N(\mu, \sigma^2)$ distribution, where μ and σ are unknown. Design a level α test to choose between

H_0: $\mu = \mu_0$,

H_1: $\mu \neq \mu_0$.

Solution: Let S^2 be the sample variance for this random sample. Then, the random variable W defined as

$$W(X_1, X_2, \cdots, X_n) = \frac{\overline{X} - \mu_0}{S/\sqrt{n}}$$

has a t-distribution with $n - 1$ degrees of freedom, i.e., $W \sim T(n-1)$. Thus, we can repeat the analysis of Example 8.24 here. The only difference is that we need to replace σ by S and $z_{\frac{\alpha}{2}}$ by $t_{\frac{\alpha}{2}, n-1}$. Therefore, we accept H_0 if

$$|W| \leq t_{\frac{\alpha}{2}, n-1},$$

and reject it otherwise. Let us look at a numerical example of this case.

Example 8.27. The average adult male height in a certain country is 170 cm. We suspect that the men in a certain city in that country might have a different average height due to some environmental factors. We pick a random sample of size 9 from the adult males in the city and obtain the following values for their heights (in cm):

176.2   157.9   160.1   180.9   165.1   167.2   162.9   155.7   166.2

Assume that the height distribution in this population is normally distributed. Here, we need to decide between

H_0: $\mu = 170$,

H_1: $\mu \neq 170$.

Based on the observed data, is there enough evidence to reject H_0 at significance level $\alpha = 0.05$?

Solution: Let's first compute the sample mean and the sample standard deviation. The sample mean is

$$\overline{X} = \frac{X_1 + X_2 + X_3 + X_4 + X_5 + X_6 + X_7 + X_8 + X_9}{9}$$
$$= 165.8$$

The sample variance is given by

$$S^2 = \frac{1}{9-1} \sum_{k=1}^{9} (X_k - \overline{X})^2 = 68.01$$

The sample standard deviation is given by

$$S = \sqrt{S^2} = 8.25$$

The following MATLAB code can be used to obtain these values:

```
x=[176.2,157.9,160.1,180.9,165.1,167.2,162.9,155.7,166.2];
m=mean(x);
v=var(x);
s=std(x);
```

Now, our test statistic is

$$W(X_1, X_2, \cdots, X_9) = \frac{\overline{X} - \mu_0}{S/\sqrt{n}}$$
$$= \frac{165.8 - 170}{8.25/3} = -1.52$$

Thus, $|W| = 1.52$. Also, we have

$$t_{\frac{\alpha}{2},n-1} = t_{0.025,8} \approx 2.31$$

The above value can be obtained in MATLAB using the command tinv(0.975,8). Thus, we conclude

$$|W| \leq t_{\frac{\alpha}{2},n-1}.$$

Therefore, we accept H_0. In other words, we do not have enough evidence to conclude that the average height in the city is different from the average height in the country.

What if the sample is not from a normal distribution? In the case that n is large, we can say that

$$W(X_1, X_2, \cdots, X_n) = \frac{\overline{X} - \mu_0}{S/\sqrt{n}}$$

is approximately standard normal. Therefore, we accept $H_0 : \mu = \mu_0$ if

$$\left| \frac{\overline{X} - \mu_0}{S/\sqrt{n}} \right| \leq z_{\frac{\alpha}{2}},$$

and reject it otherwise (i.e., accept $H_1 : \mu \neq \mu_0$).

Let us summarize what we have obtained for the two-sided test for the mean.

Table 8.2: Two-sided hypothesis testing for the mean: $H_0 : \mu = \mu_0$, H_1: $\mu \neq \mu_0$.

Case	Test Statistic	Acceptance Region
$X_i \sim N(\mu, \sigma^2)$, σ known	$W = \frac{\overline{X} - \mu_0}{\sigma/\sqrt{n}}$	$|W| \leq z_{\frac{\alpha}{2}}$
n large, X_i non-normal	$W = \frac{\overline{X} - \mu_0}{S/\sqrt{n}}$	$|W| \leq z_{\frac{\alpha}{2}}$
$X_i \sim N(\mu, \sigma^2)$, σ unknown	$W = \frac{\overline{X} - \mu_0}{S/\sqrt{n}}$	$|W| \leq t_{\frac{\alpha}{2},n-1}$

One-sided Tests for the Mean:

We can provide a similar analysis when we have a one-sided test. Let's show this by an example.

Example 8.28. Let $X_1, X_2, ..., X_n$ be a random sample from a $N(\mu, \sigma^2)$ distribution, where μ is unknown and σ is known. Design a level α test to choose between

H_0: $\mu \leq \mu_0$,

H_1: $\mu > \mu_0$.

Solution: As before, we define the test statistic as

$$W(X_1, X_2, \cdots, X_n) = \frac{\overline{X} - \mu_0}{\sigma/\sqrt{n}}.$$

If H_0 is true (i.e., $\mu \leq \mu_0$), we expect $\overline{X}$ (and thus W) to be relatively small, while if H_1 is true, we expect $\overline{X}$ (and thus W) to be larger. This suggests the following test: Choose a threshold, and call it c. If $W \leq c$, accept H_0, and if $W > c$, accept H_1. How do we choose c? If α is the required significance level, we must have

$$\begin{aligned} P(\text{type I error}) &= P(\text{Reject } H_0 \mid H_0) \\ &= P(W > c \mid \mu \leq \mu_0) \leq \alpha. \end{aligned}$$

Here, the probability of type I error depends on μ. More specifically, for any $\mu \leq \mu_0$, we can write

$$\begin{aligned} P(\text{type I error} \mid \mu) &= P(\text{Reject } H_0 \mid \mu) \\ &= P(W > c \mid \mu) \\ &= P\left(\frac{\overline{X} - \mu_0}{\sigma/\sqrt{n}} > c \mid \mu\right) \\ &= P\left(\frac{\overline{X} - \mu}{\sigma/\sqrt{n}} + \frac{\mu - \mu_0}{\sigma/\sqrt{n}} > c \mid \mu\right) \\ &= P\left(\frac{\overline{X} - \mu}{\sigma/\sqrt{n}} > c + \frac{\mu_0 - \mu}{\sigma/\sqrt{n}} \mid \mu\right) \\ &\leq P\left(\frac{\overline{X} - \mu}{\sigma/\sqrt{n}} > c \mid \mu\right) \qquad (\text{ since } \mu \leq \mu_0) \\ &= 1 - \Phi(c) \qquad \left(\text{ since given } \mu, \frac{\overline{X} - \mu}{\sigma/\sqrt{n}} \sim N(0,1)\right). \end{aligned}$$

Thus, we can choose $\alpha = 1 - \Phi(c)$, which results in

$$c = z_\alpha.$$

Therefore, we accept H_0 if

$$\frac{\overline{X} - \mu_0}{\sigma/\sqrt{n}} \leq z_\alpha,$$

and reject it otherwise.

The above analysis can be repeated for other cases. More generally, suppose that we are given a random sample $X_1, X_2, ..., X_n$ from a distribution. Let $\mu = EX_i$. Our goal is to decide between

H_0: $\mu \leq \mu_0$,

H_1: $\mu > \mu_0$.

We define the test statistic as before, i.e., we define W as

$$W(X_1, X_2, \cdots, X_n) = \frac{\overline{X} - \mu_0}{\sigma/\sqrt{n}},$$

if $\sigma = \sqrt{\mathrm{Var}(X_i)}$ is known, and as

$$W(X_1, X_2, \cdots, X_n) = \frac{\overline{X} - \mu_0}{S/\sqrt{n}},$$

if σ is unknown. If H_0 is true (i.e., $\mu \leq \mu_0$), we expect that $\overline{X}$ (and thus W) to be relatively small, while if H_1 is true, we expect $\overline{X}$ (and thus W) to be larger. This suggests the following test: Choose a threshold c. If $W \leq c$, accept H_0, and if $W > c$, accept H_1. To choose c, note that

$$P(\text{type I error}) = P(\text{Reject } H_0 \mid H_0)$$
$$= P(W > c \mid \mu \leq \mu_0)$$
$$\leq P(W > c \mid \mu = \mu_0).$$

Note that the last inequality resulted because if we make μ larger, the probability of $W > c$ can only increase. In other words, we assumed the worst case scenario, i.e, $\mu = \mu_0$ for the probability of error. Thus, we can choose c such that $P(W > c \mid \mu = \mu_0) = \alpha$. By doing this procedure, we obtain the acceptance regions reflected in Table 8.3.

Note that the tests mentioned in Table 8.3 remain valid if we replace the null hypothesis by $\mu = \mu_0$. The reason for this is that in choosing the threshold c, we assumed the worst case scenario, i.e., $\mu = \mu_0$.

Finally, if we need to decide between

H_0: $\mu \geq \mu_0$,

H_1: $\mu < \mu_0$,

we can again repeat the above analysis and we obtain the acceptance regions reflected in Table 8.4.

8.4.4 P-Values

In the above discussions, we only reported an "accept" or a "reject" decision as the conclusion of a hypothesis test. However, we can provide more information using what we call *P-values*. In other words, we could indicate how close the decision was. More specifically, suppose we end up rejecting H_0 at at significance level $\alpha = 0.05$. Then we could ask: "How about if we require significance level $\alpha = 0.01$?" Can we still reject H_0? More specifically, we can ask the following question:

Table 8.3: One-sided hypothesis testing for the mean: $H_0 : \mu \leq \mu_0$, $H_1: \mu > \mu_0$.

Case	Test Statistic	Acceptance Region
$X_i \sim N(\mu, \sigma^2)$, σ known	$W = \dfrac{\overline{X} - \mu_0}{\sigma/\sqrt{n}}$	$W \leq z_\alpha$
n large, X_i non-normal	$W = \dfrac{\overline{X} - \mu_0}{S/\sqrt{n}}$	$W \leq z_\alpha$
$X_i \sim N(\mu, \sigma^2)$, σ unknown	$W = \dfrac{\overline{X} - \mu_0}{S/\sqrt{n}}$	$W \leq t_{\alpha,n-1}$

Table 8.4: One-sided hypothesis testing for the mean: $H_0 : \mu \geq \mu_0$, $H_1: \mu < \mu_0$.

Case	Test Statistic	Acceptance Region
$X_i \sim N(\mu, \sigma^2)$, σ known	$W = \dfrac{\overline{X} - \mu_0}{\sigma/\sqrt{n}}$	$W \geq -z_\alpha$
n large, X_i non-normal	$W = \dfrac{\overline{X} - \mu_0}{S/\sqrt{n}}$	$W \geq -z_\alpha$
$X_i \sim N(\mu, \sigma^2)$, σ unknown	$W = \dfrac{\overline{X} - \mu_0}{S/\sqrt{n}}$	$W \geq -t_{\alpha,n-1}$

What is the lowest significance level α that results in rejecting the null hypothesis?

The answer to the above question is called the *P-value*.

P-value is the lowest significance level α that results in rejecting the null hypothesis.

Intuitively, if the *P*-value is small, it means that the observed data is very unlikely to have occurred under H_0, so we are more confident in rejecting the null hypothesis. How do we find *P*-values? Let's look at an example.

Example 8.29. You have a coin and you would like to check whether it is fair or biased. More specifically, let θ be the probability of heads, $\theta = P(H)$. Suppose that you need to choose between the following hypotheses:

H_0 (the null hypothesis): The coin is fair, i.e., $\theta = \theta_0 = \frac{1}{2}$.

H_1 (the alternative hypothesis): The coin is not fair, i.e., $\theta > \frac{1}{2}$.

We toss the coin 100 times and observe 60 heads.

1. Can we reject H_0 at significance level $\alpha = 0.05$?

2. Can we reject H_0 at significance level $\alpha = 0.01$?

3. What is the P-value?

Solution: Let X be the random variable showing the number of observed heads. In our experiment, we observed $X = 60$. Since $n = 100$ is relatively large, assuming H_0 is true, the random variable

$$W = \frac{X - n\theta_0}{\sqrt{n\theta_0(1 - \theta_0)}} = \frac{X - 50}{5} \tag{8.4}$$

is (approximately) a standard normal random variable, $N(0, 1)$. If H_0 is true, we expect X to be close to 50, while if H_1 is true, we expect X to be larger. Thus, we can suggest the following test: We choose a threshold c. If $W \leq c$, we accept H_0; otherwise, we accept H_1. To calculate $P(\text{type I error})$, we can write

$$P(\text{type I error}) = P(\text{Reject } H_0 \mid H_0)$$
$$= P(W > c \mid H_0).$$

Since $W \sim N(0, 1)$ under H_0, we need to choose

$$c = z_\alpha,$$

to ensure significance level α. In this example, we obtain

$$W = \frac{X - 50}{5} = \frac{60 - 50}{5} = 2. \tag{8.5}$$

1. If we require significance level $\alpha = 0.05$, then

$$c = z_{0.05} = 1.645$$

The above value can be obtained in MATLAB using the following command: norminv(1-0.05). Since we have $W = 2 > 1.645$, we reject H_0, and accept H_1.

2. If we require significance level $\alpha = 0.01$, then

$$c = z_{0.01} = 2.33$$

The above value can be obtained in MATLAB using the following command: norminv(1-0.01). Since we have $W = 2 \leq 2.33$, we fail to reject H_0, so we accept H_0.

3. P-value is the lowest significance level α that results in rejecting H_0. Here, since $W = 2$, we will reject H_0 if and only if $c < 2$. Note that $z_\alpha = c$, thus

$$\alpha = 1 - \Phi(c).$$

If $c = 2$, we obtain

$$\alpha = 1 - \Phi(2) = 0.023$$

Therefore, we reject H_0 for $\alpha > 0.023$. Thus, the P-value is equal to 0.023.

The above example suggests the following way to compute P-values:

Computing P-Values

Consider a hypothesis test for choosing between H_0 and H_1. Let W be the test statistic, and w_1 be the observed value of W.

1. Assume H_0 is true.

2. The P-value is $P(\text{type I error})$ when the test threshold c is chosen to be $c = w_1$.

To see how we can use the above method, again consider Example 8.29. Here,

$$W = \frac{X - 50}{5},$$

which is approximately $N(0, 1)$ under H_0. The observed value of W is

$$w_1 = \frac{60 - 50}{5} = 2.$$

Thus,

$$
\begin{aligned}
P - \text{value} &= P(\text{type I error when } c = 2) \\
&= P(W > 2) \\
&= 1 - \Phi(2) = 0.023
\end{aligned}
$$

8.4.5   Likelihood Ratio Tests

So far we have focused on specific examples of hypothesis testing problems. Here, we would like to introduce a relatively general hypothesis testing procedure called the *likelihood ratio test*. Before doing so, let us quickly review the definition of the likelihood function, which was previously discussed in Section 8.2.3.

Review of the Likelihood Function:

Let X_1, X_2, X_3, ..., X_n be a random sample from a distribution with a parameter θ. Suppose that we have observed $X_1 = x_1$, $X_2 = x_2$, $\cdots$, $X_n = x_n$.

- If the X_i's are discrete, then the **likelihood function** is defined as

$$L(x_1, x_2, \cdots, x_n; \theta) = P_{X_1 X_2 \cdots X_n}(x_1, x_2, \cdots, x_n; \theta).$$

- If the X_i's are jointly continuous, then the likelihood function is defined as

$$L(x_1, x_2, \cdots, x_n; \theta) = f_{X_1 X_2 \cdots X_n}(x_1, x_2, \cdots, x_n; \theta).$$

Likelihood Ratio Tests:

Consider a hypothesis testing problem in which both the null and the alternative hypotheses are simple. That is

H_0: $\theta = \theta_0$,

H_1: $\theta = \theta_1$.

Now, let X_1, X_2, X_3, ..., X_n be a random sample from a distribution with a parameter θ. Suppose that we have observed $X_1 = x_1$, $X_2 = x_2$, $\cdots$, $X_n = x_n$. One way to decide between H_0 and H_1 is to compare the corresponding likelihood functions:

$$l_0 = L(x_1, x_2, \cdots, x_n; \theta_0), \qquad l_1 = L(x_1, x_2, \cdots, x_n; \theta_1).$$

More specifically, if l_0 is much larger than l_1, we should accept H_0. On the other hand if l_1 is much larger, we tend to reject H_0. Therefore, we can look at the ratio $\frac{l_0}{l_1}$ to decide between H_0 and H_1. This is the idea behind *likelihood ratio tests*.

Likelihood Ratio Test for Simple Hypotheses

Let X_1, X_2, X_3, ..., X_n be a random sample from a distribution with a parameter θ. Suppose that we have observed $X_1 = x_1$, $X_2 = x_2$, $\cdots$, $X_n = x_n$. To decide between two simple hypotheses

H_0: $\theta = \theta_0$,

H_1: $\theta = \theta_1$,

we define

$$\lambda(x_1, x_2, \cdots, x_n) = \frac{L(x_1, x_2, \cdots, x_n; \theta_0)}{L(x_1, x_2, \cdots, x_n; \theta_1)}.$$

To perform a **likelihood ratio test (LRT)**, we choose a constant c. We reject H_0 if $\lambda < c$ and accept it if $\lambda \geq c$. The value of c can be chosen based on the desired α.

Let's look at an example to see how we can perform a likelihood ratio test.

Example 8.30. Here, we look again at the radar problem (Example 8.23). More specifically, we observe the random variable X:

$$X = \theta + W,$$

where $W \sim N(0, \sigma^2 = \frac{1}{9})$. We need to decide between

$\quad H_0$: $\theta = \theta_0 = 0$,

$\quad H_1$: $\theta = \theta_1 = 1$.

Let $X = x$. Design a level 0.05 test ($\alpha = 0.05$) to decide between H_0 and H_1.

Solution: If $\theta = \theta_0 = 0$, then $X \sim N(0, \sigma^2 = \frac{1}{9})$. Therefore,

$$L(x; \theta_0) = f_X(x; \theta_0) = \frac{3}{\sqrt{2\pi}} e^{-\frac{9x^2}{2}}.$$

On the other hand, if $\theta = \theta_1 = 1$, then $X \sim N(1, \sigma^2 = \frac{1}{9})$. Therefore,

$$L(x; \theta_1) = f_X(x; \theta_1) = \frac{3}{\sqrt{2\pi}} e^{-\frac{9(x-1)^2}{2}}.$$

Therefore,

$$\lambda(x) = \frac{L(x; \theta_0)}{L(x; \theta_1)} = \exp\left\{ -\frac{9x^2}{2} + \frac{9(x-1)^2}{2} \right\}$$

$$= \exp\left\{ \frac{9(1 - 2x)}{2} \right\}.$$

Thus, we accept H_0 if

$$\exp\left\{ \frac{9(1 - 2x)}{2} \right\} \geq c,$$

where c is the threshold. Equivalently, we accept H_0 if

$$x \leq \frac{1}{2}\left(1 - \frac{2}{9} \ln c \right).$$

Let us define $c' = \frac{1}{2}\left(1 - \frac{2}{9} \ln c \right)$, where c' is a new threshold. Remember that x is the observed value of the random variable X. Thus, we can summarize the decision rule as follows. We accept H_0 if

$$X \leq c'.$$

How to do we choose c'? We use the required α.

$$P(\text{type I error}) = P(\text{Reject } H_0 \mid H_0)$$
$$= P(X > c' \mid H_0)$$
$$= P(X > c') \qquad \left(\text{where } X \sim N\left(0, \frac{1}{9}\right)\right)$$
$$= 1 - \Phi(3c').$$

Letting $P(\text{type I error}) = \alpha$, we obtain

$$c' = \frac{1}{3}\Phi^{-1}(1 - \alpha).$$

Letting $\alpha = 0.05$, we obtain

$$c' = \frac{1}{3}\Phi^{-1}(.95) = 0.548$$

As we see, in this case, the likelihood ratio test is exactly the same test that we obtained in Example 8.23.

How do we perform the likelihood ratio test if the hypotheses are not simple? Suppose that θ is an unknown parameter. Let S be the set of possible values for θ and suppose that we can partition S into two disjoint sets S_0 and S_1. Consider the following hypotheses:

H_0: $\theta \in S_0$,

H_1: $\theta \in S_1$.

The idea behind the general likelihood ratio test can be explained as follows: We first find the likelihoods corresponding to the most likely values of θ in S_0 and S_1 respectively. That is, we find

$$l_0 = \max\{L(x_1, x_2, \cdots, x_n; \theta) : \theta \in S_0\},$$
$$l = \max\{L(x_1, x_2, \cdots, x_n; \theta) : \theta \in S\}.$$

(To be more accurate, we need to replace max by sup.) Let us consider two extreme cases. First, if $l_0 = l$, then we can say that the most likely value of θ belongs to S_0. This indicates that we should not reject H_0. On the other hand, if $\frac{l_0}{l_1}$ is much smaller than 1, we should probably reject H_0 in favor of H_1. To conduct a likelihood ratio test, we choose a threshold $0 \le c \le 1$ and compare $\frac{l_0}{l}$ to c. If $\frac{l_0}{l} \ge c$, we accept H_0. If $\frac{l_0}{l} < c$, we reject H_0. The value of c can be chosen based on the desired α.

Likelihood Ratio Tests

Let X_1, X_2, X_3, ..., X_n be a random sample from a distribution with a parameter θ. Suppose that we have observed $X_1 = x_1$, $X_2 = x_2$, $\cdots$, $X_n = x_n$. Define

$$\lambda(x_1, x_2, \cdots, x_n) = \frac{\sup\{L(x_1, x_2, \cdots, x_n; \theta) : \theta \in S_0\}}{\sup\{L(x_1, x_2, \cdots, x_n; \theta) : \theta \in S\}}.$$

To perform a **likelihood ratio test (LRT)**, we choose a constant c in $[0, 1]$. We reject H_0 if $\lambda < c$ and accept it if $\lambda \geq c$. The value of c can be chosen based on the desired α.

8.4.6   Solved Problems

1. Let $X \sim Geometric(\theta)$. We observe X and we need to decide between

   H_0: $\theta = \theta_0 = 0.5$,

   H_1: $\theta = \theta_1 = 0.1$

   (a) Design a level 0.05 test ($\alpha = 0.05$) to decide between H_0 and H_1.

   (b) Find the probability of type-II error β.

Solution:

   (a) We choose a threshold $c \in \mathbb{N}$ and compare the observed value of $X = x$ to c. We accept H_0 if $x \leq c$ and reject it if $x > c$. The probability of type I error is given by

   $$\begin{aligned}
   P(\text{type I error}) &= P(\text{Reject } H_0 \mid H_0) \\
   &= P(\text{Reject } H_0 \mid \theta = 0.5) \\
   &= P(X > c \mid \theta = 0.5) \\
   &= \sum_{k=c+1}^{\infty} P(X = k) \qquad \left(\text{where } X \sim Geometric(\theta_0 = 0.5)\right) \\
   &= \sum_{k=c+1}^{\infty} (1 - \theta_0)^{k-1} \theta_0 \\
   &= (1 - \theta_0)^c \theta_0 \sum_{l=0}^{\infty} (1 - \theta_0)^l \\
   &= (1 - \theta_0)^c.
   \end{aligned}$$

To have $\alpha = 0.05$, we need to choose c such that $(1 - \theta_0)^c \leq \alpha = 0.05$, so we obtain

$$
\begin{aligned}
c &\geq \frac{\ln \alpha}{\ln(1 - \theta_0)} \\
&= \frac{\ln(0.05)}{\ln(.5)} \\
&= 4.32
\end{aligned}
$$

Since we would like $c \in \mathbb{N}$, we can let $c = 5$. To summarize, we have the following decision rule: Accept H_0 if the observed value of X is in the set $A = \{1, 2, 3, 4, 5\}$, and reject H_0 otherwise.

(b) Since the alternative hypothesis H_1 is a simple hypothesis ($\theta = \theta_1$), there is only one value for β,

$$
\begin{aligned}
\beta &= P(\text{type II error}) = P(\text{accept } H_0 \mid H_1) \\
&= P(X \leq c \mid H_1) \\
&= 1 - (1 - \theta_1)^c \\
&= 1 - (0.9)^5 \\
&= 0.41
\end{aligned}
$$

2. Let X_1, X_2, X_3, X_4 be a random sample from a $N(\mu, 1)$ distribution, where μ is unknown. Suppose that we have observed the following values

$$2.82 \quad 2.71 \quad 3.22 \quad 2.67$$

We would like to decide between

H_0: $\mu = \mu_0 = 2$,

H_1: $\mu \neq 2$.

(a) Assuming $\alpha = 0.1$, Do you accept H_0 or H_1?

(b) If we require significance level α, find β as a function of μ and α.

Solution:

(a) We have a sample from a normal distribution with known variance, so using the first row in Table 8.2, we define the test statistic as

$$
W = \frac{\overline{X} - \mu_0}{\sigma/\sqrt{n}}.
$$

We have $\overline{X} = 2.85$, $\mu_0 = 2$, $\sigma = 1$, and $n = 4$. So, we obtain

$$W = \frac{2.85 - 2}{1/2}$$

$$= 1.7$$

Here, $\alpha = 0.1$, so $z_{\frac{\alpha}{2}} = z_{0.05} = 1.645$. Since

$$|W| > z_{\frac{\alpha}{2}},$$

we reject H_0 and accept H_1.

(b) Here, the test statistic W is

$$W \sim = 2(\overline{X} - 2).$$

If $X \sim N(\mu, 1)$, then

$$\overline{X} \sim N\left(\mu, \frac{1}{4}\right),$$

and

$$W \sim N(2(\mu - 2), 1).$$

Thus, we have

$$\begin{aligned}
\beta &= P(\text{type II error}) = P(\text{accept } H_0 \mid \mu) \\
&= P(|W| < z_{\frac{\alpha}{2}} \mid \mu) \\
&= P(|W| < z_{\frac{\alpha}{2}}) \quad \left(\text{when } W \sim N(2(\mu - 2), 1)\right) \\
&= \Phi\left(z_{\frac{\alpha}{2}} - 2\mu + 4\right) - \Phi\left(-z_{\frac{\alpha}{2}} - 2\mu + 4\right).
\end{aligned}$$

3. Let $X_1, X_2, ..., X_{100}$ be a random sample from an unknown distribution. After observing this sample, the sample mean and the sample variance are calculated to be

$$\overline{X} = 21.32, \qquad S^2 = 27.6$$

Design a level 0.05 test to choose between

H_0: $\mu = 20$,

H_1: $\mu > 20$.

Do you accept or reject H_0?

Solution: Here, we have a non-normal sample, where $n = 100$ is large. As we have discussed previously, to test for the above hypotheses, we can use the results of Table 8.3. More specifically, using the second row of Table 8.3, we define the test statistic as

$$W = \frac{\overline{X} - \mu_0}{S/\sqrt{n}}$$
$$= \frac{21.32 - 20}{\sqrt{27.6}/\sqrt{100}}$$
$$= 2.51$$

Here, $\alpha = 0.05$, so $z_\alpha = z_{0.05} = 1.645$. Since

$$W > z_\alpha,$$

we reject H_0 and accept H_1.

4. Let X_1, X_2, X_3, X_4 be a random sample from a $N(\mu, \sigma^2)$ distribution, where μ and σ are unknown. Suppose that we have observed the following values

$$3.58 \qquad 10.03 \qquad 4.77 \qquad 14.66$$

We would like to decide between

H_0: $\mu \geq 10$,

H_1: $\mu < 10$.

Assuming $\alpha = 0.05$, Do you accept H_0 or H_1?

Solution: Here, we have a sample from a normal distribution with unknown mean and unknown variance. Thus, using the third row in Table 8.4, we define the test statistic as

$$W = \frac{\overline{X} - \mu_0}{S/\sqrt{n}}.$$

Using the data we obtain

$$\overline{X} = 8.26, \qquad S = 5.10$$

Therefore, we obtain

$$W = \frac{8.26 - 10}{5.10/2}$$
$$= -0.68$$

Here, $\alpha = 0.05$, so $n = 4$, $t_{\alpha,n-1} = t_{0.05,3} = 2.35$. Since

$$W > -t_{\alpha,n-1},$$

we fail to reject H_0, so we accept H_0.

5. Let $X_1, X_2, ..., X_{81}$ be a random sample from an unknown distribution. After observing this sample, the sample mean and the sample variance are calculated to be

$$\overline{X} = 8.25, \qquad S^2 = 14.6$$

Design a test to decide between

$$H_0: \mu = 9,$$
$$H_1: \mu < 9,$$

and calculate the P-value for the observed data.

Solution: Here, we have a non-normal sample, where $n = 81$ is large. As we have discussed previously, to test for the above hypotheses, we can use the results of Table 8.4. More specifically, using the second row of Table 8.4, we define the test statistic as

$$W = \frac{\overline{X} - \mu_0}{S/\sqrt{n}}$$
$$= \frac{8.25 - 9}{\sqrt{14.6}/\sqrt{81}}$$
$$= -1.767$$

The P-value is $P(\text{type I error})$ when the test threshold c is chosen to be $c = -1.767$. Since the threshold for this test (as indicated by Table 8.4) is $-z_\alpha$, we obtain

$$-z_\alpha = -1.767$$

Noting that by definition $z_\alpha = \Phi^{-1}(1 - \alpha)$, we obtain $P(\text{type I error})$ as

$$\alpha = 1 - \Phi(1.767) \approx 0.0386$$

Therefore,

$$P - \text{value} \approx 0.0386$$

8.5   Linear Regression

Sometimes we are interested in obtaining a simple model that explains the relationship between two or more variables. For example, suppose that we are interested in studying the relationship between the income of parents and the income of their children in a certain country. In general, many factors can impact the income of a person. Nevertheless, we suspect that children from wealthier families generally tend to become wealthier when they grow up. Here, we can consider two variables:

1. The family income can be defined as the average income of parents at a certain period.

2. The child income can be defined as his/her average income at a certain period (e.g, age).

To examine the relationship between the two variables, we collect some data

$$(x_i, y_i), \qquad \text{for } i = 1, 2, \cdots, n,$$

where y_i is the average income of the ith child, and x_i is the average income of his/her parents. We are often interested in finding a simple model. A **linear** model is probably the simplest model that we can define, where we write

$$y_i \approx \beta_0 + \beta_1 x_i.$$

Of course, there are other factors that impact each child's future income, so we might write

$$y_i = \beta_0 + \beta_1 x_i + \epsilon_i,$$

where ϵ_i is modeled as a random variable. More specifically, if we approximate a child's future income by $\hat{y}_i = \beta_0 + \beta_1 x_i$, then ϵ_i indicates the error in our approximation. The goal here is to obtain the best values of β_0 and β_1 that result in the smallest errors. In other words, we would like to draw a "line" in the $x - y$ plane that best fits our data points. The line

$$\hat{y} = \beta_0 + \beta_1 x$$

is called the **regression line**. Figure 8.11 shows the regression line.

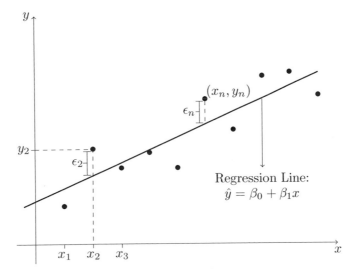

Figure 8.11: Regression line is the line that best represents the data points (x_i, y_i).

We may summarize our model as

$$Y = \beta_0 + \beta_1 x + \epsilon.$$

Note that since ϵ is a random variable, Y is also a random variable. The variable x is called the **predictor** or the **explanatory variable**, and the random variable Y is called the **response** variable. That is, here, we use x to predict/estimate Y.

8.5.1   Simple Linear Regression Model

Here, we provide a model that is called the **simple linear regression** model. Our model is

$$Y_i = \beta_0 + \beta_1 x_i + \epsilon_i,$$

where we model ϵ_i's as independent and zero-mean normal random variables,

$$\epsilon_i \sim N(0, \sigma^2).$$

The parameters β_0, β_1, and σ^2 are considered fixed but unknown. The assumption is that we have data points (x_1, y_1), (x_2, y_2), $\cdots$, (x_n, y_n) and our goal is to find the "best" values for β_0 and β_1 resulting in the line that provides the "best" fit for the data points. Here, y_i's are the observed values of the random variables Y_i's. To have a well-defined problem we add the following assumptions. We assume $n \geq 3$. We also assume that not all x_i's are identical.

There are several common methods for finding good values for β_0 and β_1. These methods will result in the same answers; however, they are philosophically based on different ideas. Here, we will provide two methods for estimating β_0 and β_1. A third method will be discussed in the Solved Problems section.

8.5.2   The First Method for Finding β_0 and β_1

Here, we assume that x_i's are observed values of a random variable X. Therefore, we can summarize our model as

$$Y = \beta_0 + \beta_1 X + \epsilon,$$

where ϵ is a $N(0, \sigma^2)$ random variable independent of X. First, we take expectation from both sides to obtain

$$EY = \beta_0 + \beta_1 EX + E[\epsilon]$$
$$= \beta_0 + \beta_1 EX$$

Thus,

$$\beta_0 = EY - \beta_1 EX.$$

Next, we look at $\text{Cov}(X, Y)$,

$$\begin{aligned}
\text{Cov}(X, Y) &= \text{Cov}(X, \beta_0 + \beta_1 X + \epsilon) \\
&= \beta_0 \text{Cov}(X, 1) + \beta_1 \text{Cov}(X, X) + \text{Cov}(X, \epsilon) \\
&= 0 + \beta_1 \text{Cov}(X, X) + 0 \qquad \text{(since } X \text{ and } \epsilon \text{ are independent)} \\
&= \beta_1 \text{Var}(X).
\end{aligned}$$

Therefore, we obtain

$$\beta_1 = \frac{\text{Cov}(X,Y)}{\text{Var}(X)}, \qquad \beta_0 = EY - \beta_1 EX.$$

Now, we can find β_0 and β_1 if we know EX, EY, $\frac{\text{Cov}(X,Y)}{\text{Var}(X)}$. Here, we have the observed pairs (x_1, y_1), (x_2, y_2), $\cdots$, (x_n, y_n), so we may estimate these quantities. More specifically, we define

$$\bar{x} = \frac{x_1 + x_2 + ... + x_n}{n},$$
$$\bar{y} = \frac{y_1 + y_2 + ... + y_n}{n},$$
$$s_{xx} = \sum_{i=1}^{n}(x_i - \bar{x})^2,$$
$$s_{xy} = \sum_{i=1}^{n}(x_i - \bar{x})(y_i - \bar{y}).$$

We can then estimate β_0 and β_1 as

$$\hat{\beta}_1 = \frac{s_{xy}}{s_{xx}},$$
$$\hat{\beta}_0 = \bar{y} - \hat{\beta}_1\bar{x}.$$

The above formulas give us the regression line

$$\hat{y} = \hat{\beta}_0 + \hat{\beta}_1 x.$$

For each x_i, the **fitted value** $\hat{y}_i$ is obtained by

$$\hat{y}_i = \hat{\beta}_0 + \hat{\beta}_1 x_i.$$

Here, $\hat{y}_i$ is the predicted value of y_i using the regression formula. The errors in this prediction are given by

$$e_i = y_i - \hat{y}_i,$$

which are called the **residuals**.

Simple Linear Regression

Given the observations (x_1, y_1), (x_2, y_2), $\cdots$, (x_n, y_n), we can write the regression line as

$$\hat{y} = \beta_0 + \beta_1 x.$$

We can estimate β_0 and β_1 as

$$\hat{\beta}_1 = \frac{s_{xy}}{s_{xx}},$$

$$\hat{\beta}_0 = \bar{y} - \hat{\beta}_1 \bar{x},$$

where

$$s_{xx} = \sum_{i=1}^{n}(x_i - \bar{x})^2,$$

$$s_{xy} = \sum_{i=1}^{n}(x_i - \bar{x})(y_i - \bar{y}).$$

For each x_i, the **fitted value** $\hat{y}_i$ is obtained by

$$\hat{y}_i = \hat{\beta}_0 + \hat{\beta}_1 x_i.$$

The quantities

$$e_i = y_i - \hat{y}_i$$

are called the **residuals**.

Example 8.31. Consider the following observed values of (x_i, y_i):

$$(1, 3) \qquad (2, 4) \qquad (3, 8) \qquad (4, 9)$$

1. Find the estimated regression line

$$\hat{y} = \hat{\beta}_0 + \hat{\beta}_1 x,$$

based on the observed data.

2. For each x_i, compute the fitted value of y_i using

$$\hat{y}_i = \hat{\beta}_0 + \hat{\beta}_1 x_i.$$

3. Compute the residuals, $e_i = y_i - \hat{y}_i$ and note that

$$\sum_{i=1}^{4} e_i = 0.$$

Solution:

1. We have

$$\bar{x} = \frac{1+2+3+4}{4} = 2.5,$$

$$\bar{y} = \frac{3+4+8+9}{4} = 6,$$

$$s_{xx} = (1-2.5)^2 + (2-2.5)^2 + (3-2.5)^2 + (4-2.5)^2 = 5,$$

$$s_{xy} = (1-2.5)(3-6) + (2-2.5)(4-6) + (3-2.5)(8-6) + (4-2.5)(9-6) = 11.$$

Therefore, we obtain

$$\hat{\beta}_1 = \frac{s_{xy}}{s_{xx}} = \frac{11}{5} = 2.2$$

$$\hat{\beta}_0 = 6 - (2.2)(2.5) = 0.5$$

2. The fitted values are given by

$$\hat{y}_i = 0.5 + 2.2x_i,$$

so we obtain

$$\hat{y}_1 = 2.7, \qquad \hat{y}_2 = 4.9, \qquad \hat{y}_3 = 7.1, \qquad \hat{y}_4 = 9.3$$

3. We have

$$e_1 = y_1 - \hat{y}_1 = 3 - 2.7 = 0.3,$$
$$e_2 = y_2 - \hat{y}_2 = 4 - 4.9 = -0.9,$$
$$e_3 = y_3 - \hat{y}_3 = 8 - 7.1 = 0.9,$$
$$e_4 = y_4 - \hat{y}_4 = 9 - 9.3 = -0.3$$

So, $e_1 + e_2 + e_3 + e_4 = 0$.

We can use MATLAB or other software packages to do regression analysis. For example, the following MATLAB code can be used to obtain the estimated regression line in Example 8.31.

```
x=[1;2;3;4];
x0=ones(size(x));
y=[3;4;8;9];
beta = regress(y,[x0,x]);
```

Coefficient of Determination (R-Squared):

Let's look again at the above model for regression. We wrote

$$Y = \beta_0 + \beta_1 X + \epsilon,$$

where ϵ is a $N(0, \sigma^2)$ random variable independent of X. Note that, here, X is the only variable that we observe, so we estimate Y using X. That is, we can write

$$\hat{Y} = \beta_0 + \beta_1 X.$$

The error in our estimate is

$$Y - \hat{Y} = \epsilon.$$

Note that the randomness in Y comes from two sources: X and ϵ. More specifically, if we look at $\text{Var}(Y)$, we can write

$$\text{Var}(Y) = \beta_1^2 \text{Var}(X) + \text{Var}(\epsilon) \qquad \text{(since } X \text{ and } \epsilon \text{ are assumed to be independent)}.$$

The above equation can be interpreted as follows. The total variation in Y can be divided into two parts. The first part, $\beta_1^2 \text{Var}(X)$, is due to variation in X. The second part, $\text{Var}(\epsilon)$, is the variance of error. In other words, $\text{Var}(\epsilon)$ is the variance left in Y after we know X. If the variance of error, $\text{Var}(\epsilon)$, is small, then Y is close to $\hat{Y}$, so our regression model will be successful in estimating Y.

From the above discussion, we can define

$$\rho^2 = \frac{\beta_1^2 \text{Var}(X)}{\text{Var}(Y)}$$

as the *portion* of variance of Y that is explained by variation in X. From the above discussion, we can also conclude that $0 \leq \rho^2 \leq 1$. More specifically, if ρ^2 is close to 1, Y can be estimated very well as a linear function of X. On the other hand if ρ^2 is small, then the variance of error is large and Y cannot be accurately estimated as a linear function of X.

Since $\beta_1 = \frac{\text{Cov}(X,Y)}{\text{Var}(X)}$, we can write

$$\rho^2 = \frac{\beta_1^2 \text{Var}(X)}{\text{Var}(Y)} = \frac{[\text{Cov}(X,Y)]^2}{\text{Var}(X)\text{Var}(Y)} \tag{8.6}$$

The above equation should look familiar to you. Here, ρ is the correlation coefficient that we have seen before. Here, we are basically saying that if X and Y are highly correlated (i.e., $\rho(X,Y)$ is large), then Y can be well approximated by a linear function of X, i.e., $Y \approx \hat{Y} = \beta_0 + \beta_1 X$.

We conclude that ρ^2 is an indicator showing the strength of our regression model in estimating (predicting) Y from X. In practice, we often do not have ρ but we have the observed pairs (x_1, y_1), (x_2, y_2), $\cdots$, (x_n, y_n). We can estimate ρ^2 from the observed data. We show it by r^2 and call it *R-squared* or *coefficient of determination*.

<u>Coefficient of Determination</u>

For the observed data pairs, $(x_1, y_1), (x_2, y_2), \cdots, (x_n, y_n)$, we define **coefficient of determination**, r^2 as

$$r^2 = \frac{s_{xy}^2}{s_{xx} s_{yy}},$$

where

$$s_{xx} = \sum_{i=1}^{n} (x_i - \overline{x})^2, \qquad s_{yy} = \sum_{i=1}^{n} (y_i - \overline{y})^2, \qquad s_{xy} = \sum_{i=1}^{n} (x_i - \overline{x})(y_i - \overline{y}).$$

We have $0 \leq r^2 \leq 1$. Larger values of r^2 generally suggest that our linear model

$$\hat{y}_i = \hat{\beta}_0 + \hat{\beta}_1 x_i$$

is a good fit for the data.

Two sets of data pairs are shown in Figure 8.12. In both data sets, the values of the y_i's (the heights of the data points) have considerable variation. The data points shown in (a) are very close to the regression line. Therefore, most of the variation in y is explained by the regression formula. That is, here, the $\hat{y}_i$'s are relatively close to the y_i's, so r^2 is close to 1. On the other hand, for the data shown in (b), a lot of variation in y is left unexplained by the regression model. Therefore, r^2 for this data set is much smaller than r^2 for the data set in (a).

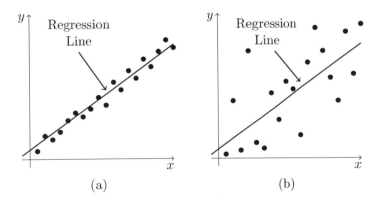

Figure 8.12: The data in (a) results in a high value of r^2, while the data shown in (b) results in a low value of r^2.

Example 8.32. For the data in Example 8.31, find the coefficient of determination.

Solution: In Example 8.31, we found

$$s_{xx} = 5, \qquad s_{xy} = 11.$$

We also have

$$s_{yy} = (3-6)^2 + (4-6)^2 + (8-6)^2 + (9-6)^2 = 26.$$

We conclude

$$r^2 = \frac{11^2}{5 \times 26} \approx 0.93$$

8.5.3 The Method of Least Squares

Here, we use a different method to estimate β_0 and β_1. This method will result in the same estimates as before; however, it is based on a different idea. Suppose that we have data points (x_1, y_1), (x_2, y_2), $\cdots$, (x_n, y_n). Consider the model

$$\hat{y} = \beta_0 + \beta_1 x.$$

The errors (residuals) are given by

$$e_i = y_i - \hat{y}_i = y_i - \beta_0 - \beta_1 x_i.$$

The *sum of the squared errors* is given by

$$g(\beta_0, \beta_1) = \sum_{i=1}^{n} e_i^2 = \sum_{i=1}^{n} (y_i - \beta_0 - \beta_1 x_i)^2. \tag{8.7}$$

To find the best fit for the data, we find the values of $\hat{\beta}_0$ and $\hat{\beta}_1$ such that $g(\beta_0, \beta_1)$ is minimized. This can be done by taking partial derivatives with respect to β_0 and β_1, and setting them to zero. We obtain

$$\frac{\partial g}{\partial \beta_0} = \sum_{i=1}^{n} 2(-1)(y_i - \beta_0 - \beta_1 x_i) = 0, \tag{8.8}$$

$$\frac{\partial g}{\partial \beta_1} = \sum_{i=1}^{n} 2(-x_i)(y_i - \beta_0 - \beta_1 x_i) = 0. \tag{8.9}$$

By solving the above equations, we obtain the same values of $\hat{\beta}_0$ and $\hat{\beta}_1$ as before

$$\hat{\beta}_1 = \frac{s_{xy}}{s_{xx}},$$

$$\hat{\beta}_0 = \overline{y} - \hat{\beta}_1 \overline{x},$$

where

$$s_{xx} = \sum_{i=1}^{n} (x_i - \overline{x})^2,$$

$$s_{xy} = \sum_{i=1}^{n} (x_i - \overline{x})(y_i - \overline{y}).$$

This method is called the method of **least squares**, and for this reason, we call the above values of $\hat{\beta}_0$ and $\hat{\beta}_1$ the **least squares estimates** of β_0 and β_1.

8.5.4 Extensions and Issues

In practice, the models that we use might be more complicated than the simple linear regression model. We almost always use computers to do regression analysis (MATLAB, EXCEL, etc). Therefore, when we understand the basics of the simple linear regression, we can simply use computers to do more complicated analyses. A natural extension to the simple linear regression is *multiple linear regression*.

Multiple Linear Regression:

In the above discussion, our model had only one predictor (explanatory variable), x. We can consider models with more than one explanatory variable. For example, suppose that you would like to have a model to predict house prices based on square footage, age, number of bedrooms, etc. Here, your response variable y is the house price. Your goal is to have a linear model

$$y = \beta_0 + \beta_1 x + \beta_2 z + \cdots + \beta_k w + \epsilon,$$

where $x, z, \cdots, w$ are the explanatory variables (square footage, age, number of bedrooms, etc). Such a model is an example of a **multiple linear regression** models. It is possible to extend the method of least squares to this case to compute estimates of $\beta_0, \beta_1, \cdots, \beta_k$. In MATLAB, the command regress can be used for multiple linear regression.

It is worth noting that when we say <u>linear</u> regression, we mean linear in the unknown parameters β_i. For example, the model

$$y = \beta_0 + \beta_1 x + \beta_2 x^2 + \epsilon$$

is a linear regression model since it is linear in β_0, β_1, and β_2.

We would like to end this section by mentioning that when running regression algorithms, one needs to be careful about some practical considerations. Issues such as overfitting [21], heteroscedasticity [22], and multicollinearity [23] might cause problems in regression analysis.

8.5.5 Solved Problems

1. Consider the following observed values of (x_i, y_i):

$$(-1, 6), \qquad (0, 3), \qquad (1, 2), \qquad (2, -1)$$

(a) Find the estimated regression line

$$\hat{y} = \hat{\beta}_0 + \hat{\beta}_1 x,$$

based on the observed data.

(b) For each x_i, compute the fitted value of y_i using

$$\hat{y}_i = \hat{\beta}_0 + \hat{\beta}_1 x_i.$$

(c) Compute the residuals, $e_i = y_i - \hat{y}_i$.

(d) Find R-squared (the coefficient of determination).

Solution:

(a) We have

$$\overline{x} = \frac{-1 + 0 + 1 + 2}{4} = 0.5,$$

$$\overline{y} = \frac{6 + 3 + 2 + (-1)}{4} = 2.5,$$

$$s_{xx} = (-1 - 0.5)^2 + (0 - 0.5)^2 + (1 - 0.5)^2 + (2 - 0.5)^2 = 5,$$

$$s_{xy} = (-1 - 0.5)(6 - 2.5) + (0 - 0.5)(3 - 2.5)$$
$$+ (1 - 0.5)(2 - 2.5) + (2 - 0.5)(-1 - 2.5) = -11.$$

Therefore, we obtain

$$\hat{\beta}_1 = \frac{s_{xy}}{s_{xx}} = \frac{-11}{5} = -2.2,$$

$$\hat{\beta}_0 = 2.5 - (-2.2)(0.5) = 3.6$$

The following MATLAB code can be used to obtain the estimated regression line

```
x=[-1;0;1;2];
x0=ones(size(x));
y=[6;3;2;-1];
beta = regress(y,[x0,x]);
```

(b) The fitted values are given by

$$\hat{y}_i = 3.6 - 2.2x_i,$$

so we obtain

$$\hat{y}_1 = 5.8, \qquad \hat{y}_2 = 3.6, \qquad \hat{y}_3 = 1.4, \qquad \hat{y}_4 = -0.8$$

(c) We have

$$e_1 = y_1 - \hat{y}_1 = 6 - 5.8 = 0.2,$$
$$e_2 = y_2 - \hat{y}_2 = 3 - 3.6 = -0.6,$$
$$e_3 = y_3 - \hat{y}_3 = 2 - 1.4 = 0.6,$$
$$e_4 = y_4 - \hat{y}_4 = -1 - (-0.8) = -0.2$$

(d) We have

$$s_{yy} = (6 - 2.5)^2 + (3 - 2.5)^2 + (2 - 2.5)^2 + (-1 - 2.5)^2 = 25.$$

We conclude

$$r^2 = \frac{(-11)^2}{5 \times 25} \approx 0.968$$

2. Consider the model

$$Y = \beta_0 + \beta_1 X + \epsilon,$$

where ϵ is a $N(0, \sigma^2)$ random variable independent of X. Let also

$$\hat{Y} = \beta_0 + \beta_1 X.$$

Show that

$$E[(Y - EY)^2] = E[(\hat{Y} - EY)^2] + E[(Y - \hat{Y})^2].$$

Solution: Since X and ϵ are independent, we can write

$$\text{Var}(Y) = \beta_1^2 \text{Var}(X) + \text{Var}(\epsilon) \qquad (8.10)$$

Note that,

$$\hat{Y} - EY = (\beta_0 + \beta_1 X) - (\beta_0 + \beta_1 EX)$$
$$= \beta_1 (X - EX).$$

Therefore,

$$E[(\hat{Y} - EY)^2] = \beta_1^2 \text{Var}(X).$$

Also,

$$E[(Y - EY)^2] = \text{Var}(Y), \qquad E[(Y - \hat{Y})^2] = \text{Var}(\epsilon).$$

Combining with Equation 8.10, we conclude

$$E[(Y - EY)^2] = E[(\hat{Y} - EY)^2] + E[(Y - \hat{Y})^2].$$

3. Show that, in a simple linear regression, the estimated coefficients $\hat{\beta}_0$ and $\hat{\beta}_1$ (least squares estimates of β_0 and β_1) satisfy the following equations

$$\sum_{i=1}^{n} e_i = 0, \qquad \sum_{i=1}^{n} e_i x_i = 0, \qquad \sum_{i=1}^{n} e_i \hat{y}_i = 0,$$

where $e_i = y_i - \hat{y}_i = y_i - \hat{\beta}_0 - \hat{\beta}_1 x$.

Hint: $\hat{\beta}_0$ and $\hat{\beta}_1$ satisfy Equations 8.8 and 8.9. By cancelling the (-2) factor, you can write

$$\sum_{i=1}^{n}(y_i - \hat{\beta}_0 - \hat{\beta}_1 x_i) = 0,$$

$$\sum_{i=1}^{n}(y_i - \hat{\beta}_0 - \hat{\beta}_1 x_i)x_i = 0.$$

Use the above equations to show the desired equations.

Solution: We have

$$\sum_{i=1}^{n}(y_i - \hat{\beta}_0 - \hat{\beta}_1 x_i) = 0,$$

$$\sum_{i=1}^{n}(y_i - \hat{\beta}_0 - \hat{\beta}_1 x_i)x_i = 0.$$

Since $e_i = y_i - \hat{\beta}_0 - \hat{\beta}_1 x$, we conclude

$$\sum_{i=1}^{n} e_i = 0,$$

$$\sum_{i=1}^{n} e_i x_i = 0.$$

Moreover,

$$\sum_{i=1}^{n} e_i \hat{y}_i = \sum_{i=1}^{n} e_i(\hat{\beta}_0 + \hat{\beta}_1 x_i)$$

$$= \hat{\beta}_0 \sum_{i=1}^{n} e_i + \hat{\beta}_1 \sum_{i=1}^{n} e_i x_i$$

$$= 0 + 0 = 0.$$

4. Show that the coefficient of determination can also be obtained as

$$r^2 = \frac{\sum_{i=1}^{n}(\hat{y}_i - \overline{y})^2}{\sum_{i=1}^{n}(y_i - \overline{y})^2}.$$

Solution: We know

$$\hat{y}_i = \beta_0 + \beta_1 x_i,$$
$$\bar{y} = \beta_0 + \beta_1 \bar{x}.$$

Therefore,

$$\sum_{i=1}^{n} (\hat{y}_i - \bar{y})^2 = \sum_{i=1}^{n} (\beta_1 x_i - \beta_1 \bar{x})^2$$
$$= \beta_1^2 \sum_{i=1}^{n} (x_i - \bar{x})^2$$
$$= \beta_1^2 s_{xx}.$$

Therefore,

$$\frac{\sum_{i=1}^{n} (\hat{y}_i - \bar{y})^2}{\sum_{i=1}^{n} (y_i - \bar{y})^2} = \frac{\beta_1^2 s_{xx}}{s_{yy}}$$
$$= \frac{s_{xy}^2}{s_{xx} s_{yy}} \qquad \left(\text{since } \beta_1 = \frac{s_{xy}}{s_{xx}}\right)$$
$$= r^2.$$

5. (**The Method of Maximum Likelihood**) This problem assumes that you are familiar with the maximum likelihood method discussed in Section 8.2.3. Consider the model

$$Y_i = \beta_0 + \beta_1 x_i + \epsilon_i,$$

where ϵ_i's are independent $N(0, \sigma^2)$ random variables. Our goal is to estimate β_0 and β_1. We have the observed data pairs (x_1, y_1), (x_2, y_2), $\cdots$, (x_n, y_n).

(a) Argue that, for given values of β_0, β_1, and x_i, Y_i is a normal random variable with mean $\beta_0 + \beta_1 x_i$ and variance σ^2. Moreover, show that the Y_i's are independent.

(b) Find the likelihood function

$$L(y_1, y_2, \cdots, y_n; \beta_0, \beta_1) = f_{Y_1 Y_2 \cdots Y_n}(y_1, y_2, \cdots, y_n; \beta_0, \beta_1).$$

(c) Show that the maximum likelihood estimates of β_0 and β_1 are the same as the ones we obtained using the least squares method.

Solution:

(a) Given values of β_0, β_1, and x_i, $c = \beta_0 + \beta_1 x_i$ is a constant. Therefore, $Y_i = c + \epsilon_i$ is a normal random variable with mean c and variance σ^2. Also, since the ϵ_i's are independent, we conclude that Y_i's are also independent random variables.

(b) By the previous part, for given values of β_0, β_1, and x_i,

$$f_{Y_i}(y; \beta_0, \beta_1) = \frac{1}{\sqrt{2\pi\sigma^2}} \exp\left\{-\frac{1}{2\sigma^2}(y - \beta_0 - \beta_1 x_i)^2\right\}.$$

Therefore, the likelihood function is given by

$$\begin{aligned}
L(y_1, y_2, \cdots, y_n; \beta_0, \beta_1) &= f_{Y_1 Y_2 \cdots Y_n}(y_1, y_2, \cdots, y_n; \beta_0, \beta_1) \\
&= f_{Y_1}(y_1; \beta_0, \beta_1) f_{Y_2}(y_2; \beta_0, \beta_1) \cdots f_{Y_n}(y_n; \beta_0, \beta_1) \\
&= \frac{1}{(2\pi\sigma^2)^{\frac{n}{2}}} \exp\left\{-\frac{1}{2\sigma^2} \sum_{i=1}^{n}(y - \beta_0 - \beta_1 x_i)^2\right\}.
\end{aligned}$$

(c) To find the maximum likelihood estimates (MLE) of β_0 and β_1, we need to find $\hat{\beta}_0$ and $\hat{\beta}_1$ such that the likelihood function

$$L(y_1, y_2, \cdots, y_n; \beta_0, \beta_1) = \frac{1}{(2\pi\sigma^2)^{\frac{n}{2}}} \exp\left\{-\frac{1}{2\sigma^2} \sum_{i=1}^{n}(y - \beta_0 - \beta_1 x_i)^2\right\}$$

is maximized. This is equivalent to minimizing

$$\sum_{i=1}^{n}(y - \beta_0 - \beta_1 x_i)^2.$$

The above expression is the sum of the squared errors, $g(\beta_0, \beta_1)$ (Equation 8.7). Therefore, the maximum likelihood estimation for this model is the same as the least squares method.

8.6 End of Chapter Problems

1. Let X be the weight of a randomly chosen individual from a population of adult men. In order to estimate the mean and variance of X, we observe a random sample $X_1, X_2, \cdots, X_{10}$. Thus, the X_i's are i.i.d. and have the same distribution as X. We obtain the following values (in pounds):

$$165.5, \ 175.4, \ 144.1, \ 178.5, \ 168.0, \ 157.9, \ 170.1, \ 202.5, \ 145.5, \ 135.7$$

Find the values of the sample mean, the sample variance, and the sample standard deviation for the observed sample.

2. Let X_1, X_2, X_3, ..., X_n be a random sample with unknown mean $EX_i = \mu$, and unknown variance $\text{Var}(X_i) = \sigma^2$. Suppose that we would like to estimate $\theta = \mu^2$. We define the estimator $\hat{\Theta}$ as

$$\hat{\Theta} = \left(\overline{X}\right)^2 = \left[\frac{1}{n}\sum_{k=1}^{n} X_k\right]^2$$

to estimate θ. Is $\hat{\Theta}$ an unbiased estimator of θ? Why?

3. Let X_1, X_2, X_3, ..., X_n be a random sample from the following distribution

$$f_X(x) = \begin{cases} \theta\left(x - \frac{1}{2}\right) + 1 & \text{for } 0 \le x \le 1 \\ 0 & \text{otherwise} \end{cases}$$

where $\theta \in [-2, 2]$ is an unknown parameter. We define the estimator $\hat{\Theta}_n$ as

$$\hat{\Theta}_n = 12\overline{X} - 6$$

to estimate θ.

(a) Is $\hat{\Theta}_n$ an unbiased estimator of θ?

(b) Is $\hat{\Theta}_n$ a consistent estimator of θ?

(c) Find the mean squared error (MSE) of $\hat{\Theta}_n$.

4. Let $X_1, \ldots, X_4$ be a random sample from a *Geometric(p)* distribution. Suppose we observed $(x_1, x_2, x_3, x_4) = (2, 3, 3, 5)$. Find the likelihood function using $P_{X_i}(x_i; p) = p(1-p)^{x_i - 1}$ as the PMF.

5. Let $X_1, \ldots, X_4$ be a random sample from an *Exponential(θ)* distribution. Suppose we observed $(x_1, x_2, x_3, x_4) = (2.35, 1.55, 3.25, 2.65)$. Find the likelihood function using

$$f_{X_i}(x_i; \theta) = \theta e^{-\theta x_i}, \qquad \text{for } x_i \ge 0$$

as the PDF.

6. Often when working with maximum likelihood functions, out of ease we maximize the log-likelihood rather than the likelihood to find the maximum likelihood estimator. Why is maximizing $L(\mathbf{x}; \theta)$ as a function of θ equivalent to maximizing $\log L(\mathbf{x}; \theta)$?

7. Let X be one observation from a $N(0, \sigma^2)$ distribution.

   (a) Find an unbiased estimator of σ^2.

   (b) Find the log likelihood, $\log(L(x; \sigma^2))$, using

   $$f_X(x; \sigma^2) = \frac{1}{\sqrt{2\pi}\sigma} exp\left\{-\frac{x^2}{2\sigma^2}\right\}$$

   as the PDF.

   (c) Find the Maximum Likelihood Estimate (MLE) for the standard deviation σ, $\hat{\sigma}_{ML}$.

8. Let $X_1, \ldots, X_n$ be a random sample from a $Poisson(\lambda)$ distribution.

   (a) Find the likelihood equation, $L(x_1, \ldots, x_n; \lambda)$, using

   $$P_{X_i}(x_1, \ldots, x_n; \lambda) = \frac{e^{-\lambda}\lambda^{x_i}}{x_i!}$$

   as the PMF.

   (b) Find the log likelihood function and use that to obtain the MLE for λ, $\hat{\lambda}_{ML}$.

9. In this problem, we would like to find the CDFs of the order statistics. Let $X_1, \ldots, X_n$ be a random sample from a continuous distribution with CDF $F_X(x)$ and PDF $f_X(x)$. Define $X_{(1)}, \ldots, X_{(n)}$ as the order statistics and show that

   $$F_{X_{(i)}}(x) = \sum_{k=i}^{n} \binom{n}{k} \left[F_X(x)\right]^k \left[1 - F_X(x)\right]^{n-k}.$$

   Hint: Fix $x \in \mathbb{R}$. Let Y be a random variable that counts the number of X_j's $\leq x$. Define $\{X_j \leq x\}$ as a "success" and $\{X_j > x\}$ as a "failure," and show that $Y \sim Binomial(n, p = F_X(x))$.

10. In this problem, we would like to find the PDFs of order statistics. Let $X_1, \ldots, X_n$ be a random sample from a continuous distribution with CDF $F_X(x)$ and PDF $f_X(x)$. Define $X_{(1)}, \ldots, X_{(n)}$ as the order statistics. Our goal here is to show that

    $$f_{X_{(i)}}(x) = \frac{n!}{(i-1)!(n-i)!} f_X(x) \left[F_X(x)\right]^{i-1} \left[1 - F_X(x)\right]^{n-i}.$$

    One way to do this is to differentiate the CDF (found in Problem 9). However, here, we would like to derive the PDF directly. Let $f_{X_{(i)}}(x)$ be the PDF of $X_{(i)}$. By definition of the PDF, for small δ, we can write

    $$f_{X_{(i)}}(x)\delta \approx P(x \leq X_{(i)} \leq x + \delta).$$

    Note that the event $\{x \leq X_{(i)} \leq x + \delta\}$ occurs if $i - 1$ of the X_j's are less than x, one of them is in $[x, x + \delta]$, and $n - i$ of them are larger than $x + \delta$. Using this, find $f_{X_{(i)}}(x)$.

    Hint: Remember the multinomial distribution. More specifically, suppose that an experiment has 3 possible outcomes, so the sample space is given by

    $$S = \{s_1, s_2, s_3\}.$$

Also, suppose that $P(s_i) = p_i$ for $i = 1, 2, 3$. Then for $n = n_1 + n_2 + n_3$ independent trials of this experiment, the probability that each s_i appears n_i times is given by

$$\binom{n}{n_1, n_2, n_3} p_1^{n_1} p_2^{n_2} p_3^{n_3} = \frac{n!}{n_1! n_2! n_3!} p_1^{n_1} p_2^{n_2} p_3^{n_3}.$$

11. A random sample X_1, X_2, X_3, ..., X_{100} is given from a distribution with known variance $\mathrm{Var}(X_i) = 81$. For the observed sample, the sample mean is $\overline{X} = 50.1$. Find an approximate 95% confidence interval for $\theta = EX_i$.

12. To estimate the portion of voters who plan to vote for Candidate A in an election, a random sample of size n from the voters is chosen. The sampling is done with replacement. Let θ be the portion of voters who plan to vote for Candidate A among all voters.

   (a) How large does n need to be so that we can obtain a 90% confidence interval with 3% margin of error?

   (b) How large does n need to be so that we can obtain a 99% confidence interval with 3% margin of error?

13. Let X_1, X_2, X_3, ..., X_{100} be a random sample from a distribution with unknown variance $\mathrm{Var}(X_i) = \sigma^2 < \infty$. For the observed sample, the sample mean is $\overline{X} = 110.5$, and the sample variance is $S^2 = 45.6$. Find a 95% confidence interval for $\theta = EX_i$.

14. A random sample X_1, X_2, X_3, ..., X_{36} is given from a normal distribution with unknown mean $\mu = EX_i$ and unknown variance $\mathrm{Var}(X_i) = \sigma^2$. For the observed sample, the sample mean is $\overline{X} = 35.8$, and the sample variance is $S^2 = 12.5$.

   (a) Find and compare 90%, 95%, and 99% confidence interval for μ.

   (b) Find and compare 90%, 95%, and 99% confidence interval for σ^2.

15. Let X_1, X_2, X_3, X_4, X_5 be a random sample from a $N(\mu, 1)$ distribution, where μ is unknown. Suppose that we have observed the following values

$$5.45, \quad 4.23, \quad 7.22, \quad 6.94, \quad 5.98$$

   We would like to decide between

   H_0: $\mu = \mu_0 = 5$,
   H_1: $\mu \neq 5$.

   (a) Define a test statistic to test the hypotheses and draw a conclusion assuming $\alpha = 0.05$.

   (b) Find a 95% confidence interval around $\overline{X}$. Is μ_0 included in the interval? How does the exclusion of μ_0 in the interval relate to the hypotheses we are testing?

16. Let $X_1, \ldots, X_9$ be a random sample from a $N(\mu, 1)$ distribution, where μ is unknown. Suppose that we have observed the following values

$$16.34, \quad 18.57, \quad 18.22, \quad 16.94, \quad 15.98, \quad 15.23, \quad 17.22, \quad 16.54, \quad 17.54$$

We would like to decide between

H_0: $\mu = \mu_0 = 16$,

H_1: $\mu \neq 16$.

(a) Find a 90% confidence interval around $\overline{X}$. Is μ_0 included in the interval? How does this relate to our hypothesis test?

(b) Define a test statistic to test the hypotheses and draw a conclusion assuming $\alpha = 0.1$.

17. Let $X_1, X_2, ..., X_{150}$ be a random sample from an unknown distribution. After observing this sample, the sample mean and the sample variance are calculated to be

$$\overline{X} = 52.28, \qquad S^2 = 30.9$$

Design a level 0.05 test to choose between

H_0: $\mu = 50$,

H_1: $\mu > 50$.

Do you accept or reject H_0?

18. Let X_1, X_2, X_3, X_4, X_5 be a random sample from a $N(\mu, \sigma^2)$ distribution, where μ and σ are both unknown. Suppose that we have observed the following values

$$27.72, \qquad 22.24, \qquad 32.86, \qquad 19.66, \qquad 35.34$$

We would like to decide between

H_0: $\mu \geq 30$,

H_1: $\mu < 30$.

Assuming $\alpha = 0.05$, what do you conclude?

19. Let $X_1, X_2, ..., X_{121}$ be a random sample from an unknown distribution. After observing this sample, the sample mean and the sample variance are calculated to be

$$\overline{X} = 29.25, \qquad S^2 = 20.7$$

Design a test to decide between

H_0: $\mu = 30$,

H_1: $\mu < 30$,

and calculate the P-value for the observed data.

20. Suppose we would like to test the hypothesis that at least 10% of students suffer from allergies. We collect a random sample of 225 students and 21 of them suffer from allergies.

(a) State the null and alternative hypotheses.

(b) Obtain a test statistic and a P-value.

(c) State the conclusion at the $\alpha = 0.05$ level.

21. Consider the following observed values of (x_i, y_i):

$$(-5, -2), \quad (-3, 1), \quad (0, 4), \quad (2, 6), \quad (1, 3).$$

   (a) Find the estimated regression line

$$\hat{y} = \hat{\beta}_0 + \hat{\beta}_1 x$$

   based on the observed data.

   (b) For each x_i, compute the fitted value of y_i using

$$\hat{y}_i = \hat{\beta}_0 + \hat{\beta}_1 x_i.$$

   (c) Compute the residuals, $e_i = y_i - \hat{y}_i$.

   (d) Calculate R-squared.

22. Consider the following observed values of (x_i, y_i):

$$(1, 3), \quad (3, 7).$$

   (a) Find the estimated regression line

$$\hat{y} = \hat{\beta}_0 + \hat{\beta}_1 x$$

   based on the observed data.

   (b) For each x_i, compute the fitted value of y_i using

$$\hat{y}_i = \hat{\beta}_0 + \hat{\beta}_1 x_i.$$

   (c) Compute the residuals, $e_i = y_i - \hat{y}_i$.

   (d) Calculate R-squared.

   (e) Explain the above results. In particular, can you conclude that the obtained regression line is a good model here?

23. Consider the simple linear regression model

$$Y_i = \beta_0 + \beta_1 x_i + \epsilon_i,$$

where ϵ_i's are independent $N(0, \sigma^2)$ random variables. Therefore, Y_i is a normal random variable with mean $\beta_0 + \beta_1 x_i$ and variance σ^2. Moreover, Y_i's are independent. As usual, we have the observed data pairs (x_1, y_1), (x_2, y_2), $\cdots$, (x_n, y_n) from which we would like to estimate β_0 and β_1. In this chapter, we found the following estimators

$$\hat{\beta}_1 = \frac{s_{xy}}{s_{xx}},$$
$$\hat{\beta}_0 = \overline{Y} - \hat{\beta}_1 \overline{x}.$$

where

$$s_{xx} = \sum_{i=1}^{n} (x_i - \overline{x})^2,$$
$$s_{xy} = \sum_{i=1}^{n} (x_i - \overline{x})(Y_i - \overline{Y}).$$

(a) Show that $\hat{\beta}_1$ is a normal random variable.

(b) Show that $\hat{\beta}_1$ is an unbiased estimator of β_1, i.e.,

$$E[\hat{\beta}_1] = \beta_1.$$

(c) Show that

$$\text{Var}(\hat{\beta}_1) = \frac{\sigma^2}{s_{xx}}.$$

24. Again consider the simple linear regression model

$$Y_i = \beta_0 + \beta_1 x_i + \epsilon_i,$$

where ϵ_i's are independent $N(0, \sigma^2)$ random variables, and

$$\hat{\beta}_1 = \frac{s_{xy}}{s_{xx}},$$
$$\hat{\beta}_0 = \overline{Y} - \hat{\beta}_1 \overline{x}.$$

(a) Show that $\hat{\beta}_0$ is a normal random variable.

(b) Show that $\hat{\beta}_0$ is an unbiased estimator of β_0, i.e.,

$$E[\hat{\beta}_0] = \beta_0.$$

(c) For any $i = 1, 2, 3, ..., n$, show that

$$\text{Cov}(\hat{\beta}_1, Y_i) = \frac{x_i - \overline{x}}{s_{xx}} \sigma^2.$$

(d) Show that

$$\text{Cov}(\hat{\beta}_1, \overline{Y}) = 0.$$

(e) Show that

$$\text{Var}(\hat{\beta}_0) = \frac{\sum_{i=1}^{n} x_i^2}{n s_{xx}} \sigma^2.$$

Chapter 9

Statistical Inference II: Bayesian Inference

9.1 Bayesian Inference

The following is a general setup for a statistical inference problem: There is an unknown quantity that we would like to estimate. We get some data. From the data, we estimate the desired quantity. In the previous chapter, we discussed the **frequentist** approach to this problem. In that approach, the unknown quantity θ is assumed to be a fixed (non-random) quantity that is to be estimated by the observed data.

In this chapter, we would like to discuss a different framework for inference, namely the **Bayesian** approach. In the Bayesian framework, we treat the unknown quantity, Θ, as a random variable. More specifically, we assume that we have some initial guess about the distribution of Θ. This distribution is called the *prior distribution*. After observing some data, we update the distribution of Θ (based on the observed data). This step is usually done using *Bayes' Rule*. That is why this approach is called the Bayesian approach. The details of this approach will be clearer as you go through the chapter. Here, to motivate the Bayesian approach, we will provide two examples of statistical problems that might be solved using the Bayesian approach.

Example 9.1. Suppose that you would like to estimate the portion of voters in your town that plan to vote for Party A in an upcoming election. To do so, you take a random sample of size n from the likely voters in the town. Since you have a limited amount of time and resources, your sample is relatively small. Specifically, suppose that $n = 20$. After doing your sampling, you find out that 6 people in your sample say they will vote for Party A.

Let θ be the true portion of voters in your town who plan to vote for Party A. You might want to estimate θ as

$$\hat{\theta} = \frac{6}{20} = 0.3$$

In fact, in absence of any other data, that seems to be a reasonable estimate. However, you might feel that $n = 20$ is too small. Thus, your guess is that the error in your estimation might be too high. While thinking about this problem, you remember that the data from the previous election is available to you. You look at that data and find out that, in the previous election,

40% of the people in your town voted for Party A. How can you use this data to possibly improve your estimate of θ? You might argue as follows:

Although the portion of votes for Party A changes from one election to another, the change is not usually very drastic. Therefore, given that in the previous election 40% of the voters voted for Party A, you might want to *model* the portion of votes for Party A in the next election as a random variable Θ with a probability density function, $f_\Theta(\theta)$, that is mostly concentrated around $\theta = 0.4$. For example, you might want to choose the density such that

$$E[\Theta] = 0.4$$

Figure 9.1 shows an example of such density functions. Such a distribution shows your *prior* belief about Θ in the absence of any additional data. That is, before taking your random sample of size $n = 20$, this is your guess about the distribution of Θ.

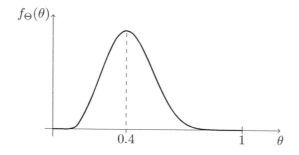

Figure 9.1: An example of a prior distribution for Θ in Example 9.1.

Therefore, you initially have the prior distribution $f_\Theta(\theta)$. Then you collect some data, shown by D. More specifically, here your data is a random sample of size $n = 20$ voters, 6 of whom are voting for Party A. As we will discuss in more detail, you can then proceed to find an updated distribution for Θ, called the *posterior* distribution, using Bayes' rule:

$$f_{\Theta|D}(\theta|D) = \frac{P(D|\theta)f_\Theta(\theta)}{P(D)}.$$

We can now use the posterior density, $f_{\Theta|D}(\theta|D)$, to further draw inferences about Θ. More specifically, we might use it to find point or interval estimates of Θ.

Example 9.2. Consider a communication channel as shown in Figure 9.2. We can model the communication over this channel as follows. At time n, a random variable X_n is generated and is transmitted over the channel. However, the channel is noisy. Thus, at the receiver, a noisy version of X_n is received. More specifically, the received signal is

$$Y_n = X_n + W_n,$$

where $W_n \sim N(0, \sigma^2)$ is the noise added to X_n. We assume that the receiver knows the distribution of X_n. The goal here is to recover (estimate) the value of X_n based on the observed value of Y_n.

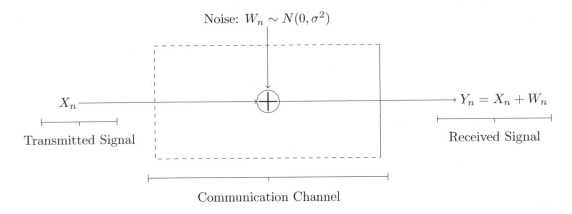

Figure 9.2: Noisy communication channel in Example 9.2.

Again, we are dealing with estimating a random variable (X_n). In this case, the *prior* distribution is $f_X(x)$. After observing Y_n, the *posterior* distribution can be written as

$$f_{X_n|Y_n}(x|y) = \frac{f_{Y_n|X_n}(y|x) f_X(x)}{f_Y(y)}.$$

Here, we have assumed both X and Y are continuous random variables. The above formula is a version of Bayes' rule. We will discuss the details of this approach shortly; however, as you'll notice, we are using the same framework as Example 9.1. After finding the posterior distribution, $f_{X_n|Y_n}(x|y)$, we can then use it to estimate the value of X_n.

If you think about Examples 9.1 and 9.2 carefully, you will notice that they have similar structures. Basically, in both problems, our goal is to draw an inference about the value of an unobserved random variable (Θ or X_n). We observe some data (D or Y_n). We then use Bayes' rule to make inference about the unobserved random variable. This is generally how we approach inference problems in *Bayesian statistics*.

It is worth noting that Examples 9.1 and 9.2 are conceptually different in the following sense: In Example 9.1, the choice of prior distribution $f_\Theta(\theta)$ is somewhat unclear. That is, different people might use different prior distributions. In other words, the choice of prior distribution is *subjective* here. On the other hand, in Example 9.2, the prior distribution $f_{X_n}(x)$ might be determined as a part of the communication system design. In other words, for this example, the prior distribution might be known without any ambiguity. Nevertheless, once the prior distribution is determined, then one uses similar methods to attack both problems. For this reason, we study both problems under the umbrella of Bayesian statistics.

Bayesian Statistical Inference

The goal is to draw inferences about an unknown variable X by observing a related random variable Y. The unknown variable is modeled as a random variable X, with **prior distribution**

$$f_X(x), \qquad \text{if } X \text{ is continuous,}$$
$$P_X(x), \qquad \text{if } X \text{ is discrete.}$$

After observing the value of the random variable Y, we find the **posterior** distribution of X. This is the conditional PDF (or PMF) of X given $Y = y$,

$$f_{X|Y}(x|y) \qquad \text{or} \qquad P_{X|Y}(x|y).$$

The posterior distribution is usually found using Bayes' formula. Using the posterior distribution, we can then find point or interval estimates of X.

Note that in the above setting, X or Y (or possibly both) could be random vectors. For example, $X = (X_1, X_2, \cdots, X_n)$ might consist of several random variables. However, the general idea of Bayesian statistics stays the same. We will specifically talk about estimating random vectors in Section 9.1.7.

9.1.1 Prior and Posterior

Let X be the random variable whose value we try to estimate. Let Y be the observed random variable. That is, we have observed $Y = y$, and we would like to estimate X. Assuming both X and Y are discrete, we can write

$$P(X = x | Y = y) = \frac{P(X = x, Y = y)}{P(Y = y)}$$
$$= \frac{P(Y = y | X = x) P(X = x)}{P(Y = y)}.$$

Using our notation for PMF and conditional PMF, the above equation can be rewritten as

$$P_{X|Y}(x|y) = \frac{P_{Y|X}(y|x) P_X(x)}{P_Y(y)}.$$

The above equation, as we have seen before, is just one way of writing Bayes' rule. If either X or Y are continuous random variables, we can replace the corresponding PMF with PDF in the above formula. For example, if X is a continuous random variable, while Y is discrete we can write

$$f_{X|Y}(x|y) = \frac{P_{Y|X}(y|x) f_X(x)}{P_Y(y)}.$$

To find the denominator ($P_Y(y)$ or $f_Y(y)$), we often use the law of total probability. Let's look at an example.

Example 9.3. Let $X \sim Uniform(0,1)$. Suppose that we know

$$Y \mid X = x \quad \sim \quad Geometric(x).$$

Find the posterior density of X given $Y = 2$, $f_{X|Y}(x|2)$.

Solution: Using Bayes' rule we have

$$f_{X|Y}(x|2) = \frac{P_{Y|X}(2|x)f_X(x)}{P_Y(2)}.$$

We know $Y \mid X = x \quad \sim \quad Geometric(x)$, so

$$P_{Y|X}(y|x) = x(1-x)^{y-1}, \qquad \text{for } y = 1, 2, \cdots.$$

Therefore,

$$P_{Y|X}(2|x) = x(1-x).$$

To find $P_Y(2)$, we can use the law of total probability

$$
\begin{aligned}
P_Y(2) &= \int_{-\infty}^{\infty} P_{Y|X}(2|x)f_X(x) \quad dx \\
&= \int_0^1 x(1-x) \cdot 1 \quad dx \\
&= \frac{1}{6}.
\end{aligned}
$$

Therefore, we obtain

$$
\begin{aligned}
f_{X|Y}(x|2) &= \frac{x(1-x) \cdot 1}{\frac{1}{6}} \\
&= 6x(1-x), \qquad \text{for } 0 \le x \le 1.
\end{aligned}
$$

For the remainder of this chapter, for simplicity, we often write the posterior PDF as

$$f_{X|Y}(x|y) = \frac{f_{Y|X}(y|x)f_X(x)}{f_Y(y)},$$

which implies that both X and Y are continuous. Nevertheless, we understand that if either X or Y is discrete, we need to replace the PDF by the corresponding PMF.

9.1.2 Maximum A Posteriori (MAP) Estimation

The posterior distribution, $f_{X|Y}(x|y)$ (or $P_{X|Y}(x|y)$), contains all the knowledge about the unknown quantity X. Therefore, we can use the posterior distribution to find point or interval estimates of X. One way to obtain a point estimate is to choose the value of x that maximizes the posterior PDF (or PMF). This is called the *maximum a posteriori (MAP) estimation*.

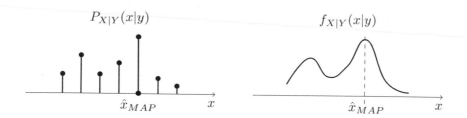

Figure 9.3: The maximum a posteriori (MAP) estimate of X given $Y = y$ is the value of x that maximizes the posterior PDF or PMF. The MAP estimate of X is usually shown by $\hat{x}_{MAP}$.

Maximum A Posteriori (MAP) Estimation

The MAP estimate of the random variable X, given that we have observed $Y = y$, is given by the value of x that maximizes

$$f_{X|Y}(x|y) \text{ if } X \text{ is a continuous random variable,}$$
$$P_{X|Y}(x|y) \text{ if } X \text{ is a discrete random variable.}$$

The MAP estimate is shown by $\hat{x}_{MAP}$.

To find the MAP estimate, we need to find the value of x that maximizes

$$f_{X|Y}(x|y) = \frac{f_{Y|X}(y|x)f_X(x)}{f_Y(y)}.$$

Note that $f_Y(y)$ does not depend on the value of x. Therefore, we can equivalently find the value of x that maximizes

$$f_{Y|X}(y|x)f_X(x).$$

This can simplify finding the MAP estimate significantly, because finding $f_Y(y)$ might be complicated. More specifically, finding $f_Y(y)$ usually is done using the law of total probability, which involves integration or summation, such as the one in Example 9.3.

To find the MAP estimate of X given that we have observed $Y = y$, we find the value of x that maximizes

$$f_{Y|X}(y|x)f_X(x).$$

If either X or Y is discrete, we replace its PDF in the above expression by the corresponding PMF.

Example 9.4. Let X be a continuous random variable with the following PDF:

$$f_X(x) = \begin{cases} 2x & \text{if } 0 \le x \le 1 \\ 0 & \text{otherwise} \end{cases}$$

Also, suppose that

$$Y \mid X = x \quad \sim \quad Geometric(x).$$

Find the MAP estimate of X given $Y = 3$.

Solution: We know that $Y \mid X = x \quad \sim \quad Geometric(x)$, so

$$P_{Y|X}(y|x) = x(1-x)^{y-1}, \qquad \text{for } y = 1, 2, \cdots .$$

Therefore,

$$P_{Y|X}(3|x) = x(1-x)^2.$$

We need to find the value of $x \in [0,1]$ that maximizes

$$P_{Y|X}(y|x)f_X(x) = x(1-x)^2 \cdot 2x$$
$$= 2x^2(1-x)^2.$$

We can find the maximizing value by differentiation. We obtain

$$\frac{\mathrm{d}}{\mathrm{d}x}\left[x^2(1-x)^2\right] = 2x(1-x)^2 - 2(1-x)x^2 = 0.$$

Solving for x (and checking for maximization criteria), we obtain the MAP estimate as

$$\hat{x}_{MAP} = \frac{1}{2}.$$

9.1.3   Comparison to ML Estimation

We discussed maximum likelihood estimation in the previous chapter. Assuming that we have observed $Y = y$, the maximum likelihood (ML) estimate of X is the value of x that maximizes

$$f_{Y|X}(y|x) \tag{9.1}$$

We show the ML estimate of X by $\hat{x}_{ML}$. On the other hand, the MAP estimate of X is the value of x that maximizes

$$f_{Y|X}(y|x)f_X(x) \tag{9.2}$$

The two expressions in Equations 9.1 and 9.2 are somewhat similar. The difference is that Equation 9.2 has an extra term, $f_X(x)$. For example, if X is uniformly distributed over a finite interval, then the ML and the MAP estimate will be the same.

Example 9.5. Suppose that the signal $X \sim N(0, \sigma_X^2)$ is transmitted over a communication channel. Assume that the received signal is given by

$$Y = X + W,$$

where $W \sim N(0, \sigma_W^2)$ is independent of X.

1. Find the ML estimate of X, given $Y = y$ is observed.

2. Find the MAP estimate of X, given $Y = y$ is observed.

Solution: Here, we have

$$f_X(x) = \frac{1}{\sqrt{2\pi}\sigma_X} e^{-\frac{x^2}{2\sigma_X^2}}.$$

We also have, $Y|X = x \quad \sim \quad N(x, \sigma_W^2)$, so

$$f_{Y|X}(y|x) = \frac{1}{\sqrt{2\pi}\sigma_W} e^{-\frac{(y-x)^2}{2\sigma_W^2}}.$$

1. The ML estimate of X, given $Y = y$, is the value of x that maximizes

$$f_{Y|X}(y|x) = \frac{1}{\sqrt{2\pi}\sigma_W} e^{-\frac{(y-x)^2}{2\sigma_W^2}}.$$

To maximize the above function, we should minimize $(y - x)^2$. Therefore, we conclude

$$\hat{x}_{ML} = y.$$

2. The MAP estimate of X, given $Y = y$, is the value of x that maximizes

$$f_{Y|X}(y|x)f_X(x) = c \exp\left\{-\left[\frac{(y-x)^2}{2\sigma_W^2} + \frac{x^2}{2\sigma_X^2}\right]\right\},$$

where c is a constant. To maximize the above function, we should minimize

$$\frac{(y-x)^2}{2\sigma_W^2} + \frac{x^2}{2\sigma_X^2}.$$

By differentiation, we obtain the MAP estimate of x as

$$\hat{x}_{MAP} = \frac{\sigma_X^2}{\sigma_X^2 + \sigma_W^2} y.$$

9.1.4   Conditional Expectation (MMSE)

Remember that the posterior distribution, $f_{X|Y}(x|y)$, contains all the knowledge that we have about the unknown quantity X. Therefore, to find a point estimate of X, we can just choose a summary statistic of the posterior such as its mean, median, or mode. If we choose the mode (the value of x that maximizes $f_{X|Y}(x|y)$), we obtain the MAP estimate of X. Another option would be to choose the posterior mean, i.e.,

$$\hat{x} = E[X|Y = y].$$

We will show that $E[X|Y = y]$ will give us the best estimate of X in terms of the *mean squared error*. For this reason, the conditional expectation is called the *minimum mean squared error (MMSE) estimate* of X. It is also called the *least mean squares (LMS) estimate* or simply the *Bayes' estimate* of X.

Minimum Mean Squared Error (MMSE) Estimation

The **minimum mean squared error (MMSE)** estimate of the random variable X, given that we have observed $Y = y$, is given by

$$\hat{x}_M = E[X|Y = y].$$

Example 9.6. Let X be a continuous random variable with the following PDF

$$f_X(x) = \begin{cases} 2x & \text{if } 0 \le x \le 1 \\ \\ 0 & \text{otherwise} \end{cases}$$

We also know that

$$f_{Y|X}(y|x) = \begin{cases} 2xy - x + 1 & \text{if } 0 \le y \le 1 \\ \\ 0 & \text{otherwise} \end{cases}$$

Find the MMSE estimate of X, given $Y = y$ is observed.

Solution: First we need to find the posterior density, $f_{X|Y}(x|y)$. We have

$$f_{X|Y}(x|y) = \frac{f_{Y|X}(y|x)f_X(x)}{f_Y(y)}.$$

We can find $f_Y(y)$ as

$$\begin{aligned} f_Y(y) &= \int_0^1 f_{Y|X}(y|x)f_X(x)dx \\ &= \int_0^1 (2xy - x + 1)2x\,dx \\ &= \frac{4}{3}y + \frac{1}{3}, \qquad \text{for } 0 \le y \le 1. \end{aligned}$$

We conclude

$$f_{X|Y}(x|y) = \frac{6x(2xy - x + 1)}{4y + 1}, \qquad \text{for } 0 \le x \le 1.$$

The MMSE estimate of X given $Y = y$ is then given by

$$\hat{x}_M = E[X|Y = y]$$
$$= \int_0^1 x f_{X|Y}(x|y) dx$$
$$= \frac{1}{4y + 1} \int_0^1 6x^2(2xy - x + 1) dx$$
$$= \frac{3y + \frac{1}{2}}{4y + 1}.$$

9.1.5   Mean Squared Error (MSE)

Suppose that we would like to estimate the value of an unobserved random variable X given that we have observed $Y = y$. In general, our estimate $\hat{x}$ is a function of y:

$$\hat{x} = g(y).$$

The error in our estimate is given by

$$\tilde{X} = X - \hat{x}$$
$$= X - g(y).$$

Often, we are interested in the *mean squared error* (MSE) given by

$$E[(X - \hat{x})^2|Y = y] = E[(X - g(y))^2|Y = y].$$

One way of finding a point estimate $\hat{x} = g(y)$ is to find a function $g(Y)$ that minimizes the *mean squared error* (MSE). Here, we show that $g(y) = E[X|Y = y]$ has the lowest MSE among all possible estimators. That is why it is called the *minimum mean squared error (MMSE) estimate*.

For simplicity, let us first consider the case that we would like to estimate X without observing anything. What would be our best estimate of X in that case? Let a be our estimate of X. Then, the MSE is given by

$$h(a) = E[(X - a)^2]$$
$$= EX^2 - 2aEX + a^2.$$

This is a quadratic function of a, and we can find the minimizing value of a by differentiation:

$$h'(a) = -2EX + 2a.$$

Therefore, we conclude the minimizing value of a is

$$a = EX.$$

Now, if we have observed $Y = y$, we can repeat the above argument. The only difference is that everything is conditioned on $Y = y$. More specifically, the MSE is given by

$$h(a) = E[(X - a)^2 | Y = y]$$
$$= E[X^2 | Y = y] - 2aE[X | Y = y] + a^2.$$

Again, we obtain a quadratic function of a, and by differentiation we obtain the MMSE estimate of X given $Y = y$ as

$$\hat{x}_M = E[X | Y = y].$$

Suppose that we would like to estimate the value of an unobserved random variable X, by observing the value of a random variable $Y = y$. In general, our estimate $\hat{x}$ is a function of y, so we can write

$$\hat{X} = g(Y).$$

Note that, since Y is a random variable, the estimator $\hat{X} = g(Y)$ is also a random variable. The error in our estimate is given by

$$\tilde{X} = X - \hat{X}$$
$$= X - g(Y),$$

which is also a random variable. We can then define the *mean squared error* (MSE) of this estimator by

$$E[(X - \hat{X})^2] = E[(X - g(Y))^2].$$

From our discussion above we can conclude that the conditional expectation $\hat{X}_M = E[X|Y]$ has the lowest MSE among all other estimators $g(Y)$.

Mean Squared Error (MSE) of an Estimator

Let $\hat{X} = g(Y)$ be an estimator of the random variable X, given that we have observed the random variable Y. The **mean squared error (MSE)** of this estimator is defined as

$$E[(X - \hat{X})^2] = E[(X - g(Y))^2].$$

The MMSE estimator of X,

$$\hat{X}_M = E[X|Y],$$

has the lowest MSE among all possible estimators.

Properties of the Estimation Error:

Here, we would like to study the MSE of the conditional expectation. First, note that

$$E[\hat{X}_M] = E[E[X|Y]]$$
$$= E[X] \quad \text{(by the law of iterated expectations)}.$$

Therefore, $\hat{X}_M = E[X|Y]$ is an *unbiased* estimator of X. In other words, for $\hat{X}_M = E[X|Y]$, the estimation error, $\tilde{X}$, is a zero-mean random variable

$$E[\tilde{X}] = EX - E[\hat{X}_M] = 0.$$

Before going any further, let us state and prove a useful lemma.

Lemma 9.1. Define the random variable $W = E[\tilde{X}|Y]$. Let $\hat{X}_M = E[X|Y]$ be the MMSE estimator of X given Y, and let $\tilde{X} = X - \hat{X}_M$ be the estimation error. Then, we have

(a) $W = 0$.

(b) For any function $g(Y)$, we have $E[\tilde{X} \cdot g(Y)] = 0$.

Proof. (a) We can write

$$W = E[\tilde{X}|Y]$$
$$= E[X - \hat{X}_M|Y]$$
$$= E[X|Y] - E[\hat{X}_M|Y]$$
$$= \hat{X}_M - E[\hat{X}_M|Y]$$
$$= \hat{X}_M - \hat{X}_M = 0.$$

The last line resulted because $\hat{X}_M$ is a function of Y, so $E[\hat{X}_M|Y] = \hat{X}_M$.

(b) First, note that

$$E[\tilde{X} \cdot g(Y)|Y] = g(Y)E[\tilde{X}|Y]$$
$$= g(Y) \cdot W = 0.$$

Next, by the law of iterated expectations, we have

$$E[\tilde{X} \cdot g(Y)] = E\big[E[\tilde{X} \cdot g(Y)|Y]\big] = 0.$$

$$\blacksquare$$

We are now ready to state a very interesting property of the estimation error for the MMSE estimator. Namely, we show that the estimation error, $\tilde{X}$, and $\hat{X}_M$ are uncorrelated. To see this, note that

$$\text{Cov}(\tilde{X}, \hat{X}_M) = E[\tilde{X} \cdot \hat{X}_M] - E[\tilde{X}]E[\hat{X}_M]$$
$$= E[\tilde{X} \cdot \hat{X}_M] \quad \text{(since } E[\tilde{X}] = 0\text{)}$$
$$= E[\tilde{X} \cdot g(Y)] \quad \text{(since } \hat{X}_M \text{ is a function of } Y\text{)}$$
$$= 0 \quad \text{(by Lemma 9.1)}.$$

Now, let us look at $\text{Var}(X)$. The estimation error is $\tilde{X} = X - \hat{X}_M$, so

$$X = \tilde{X} + \hat{X}_M.$$

Since $\text{Cov}(\tilde{X}, \hat{X}_M) = 0$, we conclude

$$\text{Var}(X) = \text{Var}(\hat{X}_M) + \text{Var}(\tilde{X}). \tag{9.3}$$

The above formula can be interpreted as follows. Part of the variance of X is explained by the variance in $\hat{X}_M$. The remaining part is the variance in estimation error. In other words, if $\hat{X}_M$ captures most of the variation in X, then the error will be small. Note also that we can rewrite Equation 9.3 as

$$E[X^2] - E[X]^2 = E[\hat{X}_M^2] - E[\hat{X}_M]^2 + E[\tilde{X}^2] - E[\tilde{X}]^2.$$

Note that

$$E[\hat{X}_M] = E[X], \qquad E[\tilde{X}] = 0.$$

We conclude

$$E[X^2] = E[\hat{X}_M^2] + E[\tilde{X}^2].$$

Some Additional Properties of the MMSE Estimator

- The MMSE estimator, $\hat{X}_M = E[X|Y]$, is an unbiased estimator of X, i.e.,

$$E[\hat{X}_M] = EX, \quad E[\tilde{X}] = 0.$$

- The estimation error, $\tilde{X}$, and $\hat{X}_M$ are uncorrelated

$$\text{Cov}(\tilde{X}, \hat{X}_M) = 0.$$

- We have

$$\text{Var}(X) = \text{Var}(\hat{X}_M) + \text{Var}(\tilde{X}),$$
$$E[X^2] = E[\hat{X}_M^2] + E[\tilde{X}^2].$$

Let us look at an example to practice the above concepts. This is an example involving jointly normal random variables. Thus, before solving the example, it is useful to remember the properties of jointly normal random variables. Remember that two random variables X and Y are jointly normal if $aX + bY$ has a normal distribution for all $a, b \in \mathbb{R}$. As we have seen before, if X and Y are jointly normal random variables with parameters μ_X, σ_X^2, μ_Y, σ_Y^2, and ρ, then, given $Y = y$, X is normally distributed with

$$E[X|Y = y] = \mu_X + \rho\sigma_X\frac{y - \mu_Y}{\sigma_Y},$$
$$\mathrm{Var}(X|Y = y) = (1 - \rho^2)\sigma_X^2.$$

Example 9.7. Let $X \sim N(0, 1)$ and

$$Y = X + W,$$

where $W \sim N(0, 1)$ is independent of X.

(a) Find the MMSE estimator of X given Y, $(\hat{X}_M)$.

(b) Find the MSE of this estimator, using $MSE = E[(X - \hat{X}_M)^2]$.

(c) Check that $E[X^2] = E[\hat{X}_M^2] + E[\tilde{X}^2]$.

Solution: Since X and W are independent and normal, Y is also normal. Moreover, X and Y are also jointly normal, since for all $a, b \in \mathbb{R}$, we have

$$aX + bY = (a + b)X + bW,$$

which is also a normal random variable. Note also,

$$\begin{aligned}
\mathrm{Cov}(X, Y) &= \mathrm{Cov}(X, X + W) \\
&= \mathrm{Cov}(X, X) + \mathrm{Cov}(X, W) \\
&= \mathrm{Var}(X) = 1.
\end{aligned}$$

Therefore,

$$\begin{aligned}
\rho(X, Y) &= \frac{\mathrm{Cov}(X, Y)}{\sigma_X\sigma_Y} \\
&= \frac{1}{1 \cdot \sqrt{2}} = \frac{1}{\sqrt{2}}.
\end{aligned}$$

(a) The MMSE estimator of X given Y is

$$\begin{aligned}
\hat{X}_M &= E[X|Y] \\
&= \mu_X + \rho\sigma_X\frac{Y - \mu_Y}{\sigma_Y} \\
&= \frac{Y}{2}.
\end{aligned}$$

(b) The MSE of this estimator is given by

$$E[(X - \hat{X}_M)^2] = E\left[\left(X - \frac{Y}{2}\right)^2\right]$$

$$= E\left[X^2 - XY + \frac{Y^2}{4}\right]$$

$$= EX^2 - E[X(X + W)] + \frac{EY^2}{4}$$

$$= EX^2 - EX^2 - EXEW + \frac{EY^2}{4}$$

$$= \frac{\text{Var}(Y) + (EY)^2}{4}$$

$$= \frac{2 + 0}{4} = \frac{1}{2}.$$

(c) Note that $E[X^2] = 1$. Also,

$$E[\hat{X}_M^2] = \frac{EY^2}{4} = \frac{1}{2}.$$

In the above, we also found $MSE = E[\tilde{X}^2] = \frac{1}{2}$. Therefore, we have

$$E[X^2] = E[\hat{X}_M^2] + E[\tilde{X}^2].$$

9.1.6 Linear MMSE Estimation of Random Variables

Suppose that we would like to estimate the value of an unobserved random variable X, given that we have observed $Y = y$. In general, our estimate $\hat{x}$ is a function of y

$$\hat{x} = g(y).$$

For example, the MMSE estimate of X given $Y = y$ is

$$g(y) = E[X|Y = y].$$

We might face some difficulties if we want to use the MMSE in practice. First, the function $g(y) = E[X|Y = y]$ might have a complicated form. Specifically, if X and Y are random vectors, computing $E[X|Y = y]$ might not be easy. Moreover, to find $E[X|Y = y]$ we need to know $f_{X|Y}(y)$, which might not be easy to find in some problems. To address these issues, we might want to use a simpler function $g(y)$ to estimate X. In particular, we might want $g(y)$ to be a linear function of y.

Suppose that we would like to have an estimator for X of the form

$$\hat{X}_L = g(Y) = aY + b,$$

where a and b are some real numbers to be determined. More specifically, our goal is to choose a and b such that the MSE of the above estimator

$$MSE = E[(X - \hat{X}_L)^2]$$

is minimized. We call the resulting estimator the **linear MMSE** estimator. The following theorem gives us the optimal values for a and b.

Theorem 9.1. Let X and Y be two random variables with finite means and variances. Also, let ρ be the correlation coefficient of X and Y. Consider the function

$$h(a, b) = E[(X - aY - b)^2].$$

Then,

1. The function $h(a, b)$ is minimized if

$$a = a^* = \frac{\text{Cov}(X, Y)}{\text{Var}(Y)}, \qquad b = b^* = EX - aEY.$$

2. We have $h(a^*, b^*) = (1 - \rho^2)\text{Var}(X)$.

3. $E[(X - a^*Y - b^*)Y] = 0$ (orthogonality principle).

Proof. We have

$$\begin{aligned}
h(a, b) &= E[(X - aY - b)^2] \\
&= E[X^2 + a^2 Y^2 + b^2 - 2aXY - 2bX + 2abY] \\
&= EX^2 + a^2 EY^2 + b^2 - 2aEXY - 2bEX + 2abEY.
\end{aligned}$$

Thus, $h(a, b)$ is a quadratic function of a and b. We take the derivatives with respect to a and b and set them to zero, so we obtain

$$EY^2 \cdot a + EY \cdot b = EXY \tag{9.4}$$

$$EY \cdot a + b = EX \tag{9.5}$$

Solving for a and b, we obtain

$$a^* = \frac{\text{Cov}(X, Y)}{\text{Var}(Y)}, \qquad b^* = EX - aEY.$$

It can be verified that the above values do in fact minimize $h(a, b)$. Note that Equation 9.5 implies that $E[X - a^*Y - b^*] = 0$. Therefore,

$$\begin{aligned}
h(a^*, b^*) &= E[(X - a^*Y - b^*)^2] \\
&= \text{Var}(X - a^*Y - b^*) \\
&= \text{Var}(X - a^*Y) \\
&= \text{Var}(X) + a^{*2}\text{Var}(Y) - 2a^*\text{Cov}(X, Y) \\
&= \text{Var}(X) + \frac{\text{Cov}(X, Y)^2}{\text{Var}(Y)^2}\text{Var}(Y) - 2\frac{\text{Cov}(X, Y)}{\text{Var}(Y)}\text{Cov}(X, Y) \\
&= \text{Var}(X) - \frac{\text{Cov}(X, Y)^2}{\text{Var}(Y)} \\
&= (1 - \rho^2)\text{Var}(X).
\end{aligned}$$

Finally, note that

$$E[(X - a^*Y - b^*)Y] = EXY - a^*EY^2 - b^*EY$$
$$= 0 \quad \text{(by Equation 9.4)}.$$

$\blacksquare$

Note that $\tilde{X} = X - a^*Y - b^*$ is the error in the linear MMSE estimation of X given Y. From the above theorem, we conclude that

$$E[\tilde{X}] = 0,$$
$$E[\tilde{X}Y] = 0.$$

In sum, we can write the linear MMSE estimator of X given Y as

$$\hat{X}_L = \frac{\text{Cov}(X,Y)}{\text{Var}(Y)}(Y - EY) + EX.$$

If $\rho = \rho(X,Y)$ is the correlation coefficient of X and Y, then $\text{Cov}(X,Y) = \rho\sigma_X\sigma_Y$, so the above formula can be written as

$$\hat{X}_L = \frac{\rho\sigma_X}{\sigma_Y}(Y - EY) + EX.$$

Linear MMSE Estimator

The **linear MMSE** estimator of the random variable X, given that we have observed Y, is given by

$$\hat{X}_L = \frac{\text{Cov}(X,Y)}{\text{Var}(Y)}(Y - EY) + EX$$
$$= \frac{\rho\sigma_X}{\sigma_Y}(Y - EY) + EX.$$

The estimation error, defined as $\tilde{X} = X - \hat{X}_L$, satisfies the **orthogonality principle**:

$$E[\tilde{X}] = 0,$$
$$\text{Cov}(\tilde{X}, Y) = E[\tilde{X}Y] = 0.$$

The MSE of the linear MMSE is given by

$$E[(X - X_L)^2] = E[\tilde{X}^2] = (1 - \rho^2)\text{Var}(X).$$

Note that to compute the linear MMSE estimates, we only need to know expected values, variances, and the covariance. Let us look at an example.

Example 9.8. Suppose $X \sim Uniform(1,2)$, and given $X = x$, Y is exponential with parameter $\lambda = \frac{1}{x}$.

(a) Find the linear MMSE estimate of X given Y.

(b) Find the MSE of this estimator.

(c) Check that $E[\tilde{X}Y] = 0$.

Solution: We have

$$\hat{X}_L = \frac{\text{Cov}(X,Y)}{\text{Var}(Y)}(Y - EY) + EX.$$

Therefore, we need to find EX, EY, $\text{Var}(Y)$, and $\text{Cov}(X,Y)$. First, note that we have $EX = \frac{3}{2}$, and

$$
\begin{aligned}
EY &= E[E[Y|X]] && \text{(law of iterated expectations)} \\
&= E[X] && \left(\text{since } Y|X \sim Exponential(\frac{1}{X})\right) \\
&= \frac{3}{2}.
\end{aligned}
$$

$$
\begin{aligned}
EY^2 &= E[E[Y^2|X]] && \text{(law of iterated expectations)} \\
&= E[2X^2] && \left(\text{since } Y|X \sim Exponential(\frac{1}{X})\right) \\
&= \int_1^2 2x^2 dx \\
&= \frac{14}{3}.
\end{aligned}
$$

Therefore,

$$
\begin{aligned}
\text{Var}(Y) &= EY^2 - (EY)^2 \\
&= \frac{14}{3} - \frac{9}{4} \\
&= \frac{29}{12}.
\end{aligned}
$$

We also have

$$
\begin{aligned}
EXY &= E[E[XY|X]] && \text{(law of iterated expectations)} \\
EXY &= E[XE[Y|X]] && \text{(given } X, X \text{ is a constant)} \\
&= E[X \cdot X] && \left(\text{since } Y|X \sim Exponential(\frac{1}{X})\right) \\
&= \int_1^2 x^2 dx \\
&= \frac{7}{3}.
\end{aligned}
$$

Thus,

$$\text{Cov}(X,Y) = E[XY] - (EX)(EY)$$
$$= \frac{7}{3} - \frac{3}{2} \cdot \frac{3}{2}$$
$$= \frac{1}{12}.$$

(a) The linear MMSE estimate of X given Y is

$$\hat{X}_L = \frac{\text{Cov}(X,Y)}{\text{Var}(Y)}(Y - EY) + EX$$
$$= \frac{1}{29}\left(Y - \frac{3}{2}\right) + \frac{3}{2}$$
$$= \frac{Y}{29} + \frac{42}{29}.$$

(b) The MSE of $\hat{X}_L$ is

$$MSE = (1 - \rho^2)\text{Var}(X).$$

Since $X \sim Uniform(1,2)$, $\text{Var}(X) = \frac{1}{12}$. Also,

$$\rho^2 = \frac{\text{Cov}^2(X,Y)}{\text{Var}(X)\text{Var}(Y)}$$
$$= \frac{1}{29}.$$

Thus,

$$MSE = \left(1 - \frac{1}{29}\right)\frac{1}{12} = \frac{7}{87}.$$

(c) We have

$$\tilde{X} = X - \hat{X}_L$$
$$= X - \frac{Y}{29} - \frac{42}{29}.$$

Therefore,

$$E[\tilde{X}Y] = E\left[\left(X - \frac{Y}{29} - \frac{42}{29}\right)Y\right]$$
$$= E[XY] - \frac{EY^2}{29} - \frac{42}{29}EY$$
$$= \frac{7}{3} - \frac{14}{3 \cdot 29} - \frac{42}{29} \cdot \frac{3}{2}$$
$$= 0.$$

9.1.7   Estimation for Random Vectors

The examples that we have seen so far involved only two random variables X and Y. In practice, we often need to estimate several random variables and we might observe several random variables. In other words, we might want to estimate the value of an unobserved random vector $\mathbf{X}$:

$$\mathbf{X} = \begin{bmatrix} X_1 \\ X_2 \\ \vdots \\ X_m \end{bmatrix},$$

given that we have observed the random vector $\mathbf{Y}$,

$$\mathbf{Y} = \begin{bmatrix} Y_1 \\ Y_2 \\ \vdots \\ Y_n \end{bmatrix}.$$

Almost everything that we have discussed can be extended to the case of random vectors. For example, to find the MMSE estimate of $\mathbf{X}$ given $\mathbf{Y} = \mathbf{y}$, we can write

$$\hat{\mathbf{X}}_M = E[\mathbf{X}|\mathbf{Y}] = \begin{bmatrix} E[X_1|Y_1, Y_2, \cdots, Y_n] \\ E[X_2|Y_1, Y_2, \cdots, Y_n] \\ \vdots \\ E[X_m|Y_1, Y_2, \cdots, Y_n] \end{bmatrix}.$$

However, the above conditional expectations might be too complicated computationally. Therefore, for random vectors, it is very common to consider simpler estimators such as the linear MMSE. Let's now discuss linear MMSE for random vectors.

Linear MMSE for Random Vectors:

Suppose that we would like to have an estimator for the random vector $\mathbf{X}$ in the form of

$$\hat{\mathbf{X}}_L = \mathbf{AY} + \mathbf{b},$$

where $\mathbf{A}$ and $\mathbf{b}$ are fixed matrices to be determined. Remember that for two random variables X and Y, the linear MMSE estimator of X given Y is

$$\hat{X}_L = \frac{\text{Cov}(X,Y)}{\text{Var}(Y)}(Y - EY) + EX$$
$$= \frac{\text{Cov}(X,Y)}{\text{Cov}(Y,Y)}(Y - EY) + EX.$$

We can extend this result to the case of random vectors. More specifically, we can show that the linear MMSE estimator of the random vector $\mathbf{X}$ given the random vector $\mathbf{Y}$ is given by

$$\hat{\mathbf{X}}_L = \mathbf{C_{XY}C_Y}^{-1}(\mathbf{Y} - E[\mathbf{Y}]) + E[\mathbf{X}].$$

In the above equation, $\mathbf{C_Y}$ is the covariance matrix of $\mathbf{Y}$, defined as

$$\mathbf{C_Y} = E[(\mathbf{Y} - E\mathbf{Y})(\mathbf{Y} - E\mathbf{Y})^T],$$

and $\mathbf{C_{XY}}$ is the cross covariance matrix of $\mathbf{X}$ and $\mathbf{Y}$, defined as

$$\mathbf{C_{XY}} = E[(\mathbf{X} - E\mathbf{X})(\mathbf{Y} - E\mathbf{Y})^T].$$

The above calculations can easily be done using MATLAB or other packages. However, it is sometimes easier to use the orthogonality principle to find $\hat{\mathbf{X}}_L$. We now explain how to use the orthogonality principle to find linear MMSE estimators.

Using the Orthogonality Principle to Find Linear MMSE Estimators for Random Vectors:

Suppose that we are estimating a vector $\mathbf{X}$:

$$\mathbf{X} = \begin{bmatrix} X_1 \\ X_2 \\ \vdots \\ X_m \end{bmatrix}$$

given that we have observed the random vector $\mathbf{Y}$. Let

$$\hat{\mathbf{X}}_L = \begin{bmatrix} \hat{X}_1 \\ \hat{X}_2 \\ \vdots \\ \hat{X}_m \end{bmatrix}$$

be the vector estimate. We define the MSE as

$$MSE = \sum_{k=1}^{m} E[(X_k - \hat{X}_k)^2].$$

Therefore, to minimize the MSE, it suffices to minimize each $E[(X_k - \hat{X}_k)^2]$ individually. This means that we only need to discuss estimating a random variable X given that we have observed the random vector $\mathbf{Y}$. Since we would like our estimator to be linear, we can write

$$\hat{X}_L = \sum_{k=1}^{n} a_k Y_k + b.$$

The error in our estimate $\tilde{X}$ is then given by

$$\tilde{X} = X - \hat{X}_L$$
$$= X - \sum_{k=1}^{n} a_k Y_k - b.$$

Similar to the proof of Theorem 9.1, we can show that the linear MMSE should satisfy

$$E[\tilde{X}] = 0,$$

$$\text{Cov}(\tilde{X}, Y_j) = E[\tilde{X}Y_j] = 0, \qquad \text{for all } j = 1, 2, \cdots, n.$$

The above equations are called the **orthogonality principle**. The orthogonality principle is often stated as follows: The error $(\tilde{X})$ must be orthogonal to the observations $(Y_1, Y_2, \cdots, Y_n)$. Note that there are $n+1$ unknowns $(a_1, a_2, \cdots, a_n$ and $b)$ and $n+1$ equations. Let us look at an example to see how we can apply the orthogonality principle.

Example 9.9. Let X be an unobserved random variable with $EX = 0$, $\text{Var}(X) = 4$. Assume that we have observed Y_1 and Y_2 given by

$$Y_1 = X + W_1,$$
$$Y_2 = X + W_2,$$

where $EW_1 = EW_2 = 0$, $\text{Var}(W_1) = 1$, and $\text{Var}(W_2) = 4$. Assume that W_1, W_2, and X are independent random variables. Find the linear MMSE estimator of X, given Y_1 and Y_2.

Solution: The linear MMSE of X given Y has the form

$$\hat{X}_L = aY_1 + bY_2 + c.$$

We use the orthogonality principle. We have

$$E[\tilde{X}] = aEY_1 + bEY_2 + c$$
$$= a \cdot 0 + b \cdot 0 + c = c.$$

Using $E[\tilde{X}] = 0$, we conclude $c = 0$. Next, we note

$$\begin{aligned}
\text{Cov}(\hat{X}_L, Y_1) &= \text{Cov}(aY_1 + bY_2, Y_1) \\
&= a\text{Cov}(Y_1, Y_1) + b\text{Cov}(Y_1, Y_2) \\
&= a\text{Cov}(X + W_1, X + W_1) + b\text{Cov}(X + W_1, X + W_2) \\
&= a(\text{Var}(X) + \text{Var}(W_1)) + b\text{Var}(X) \\
&= 5a + 4b.
\end{aligned}$$

Similarly, we find

$$\begin{aligned}
\text{Cov}(\hat{X}_L, Y_2) &= \text{Cov}(aY_1 + bY_2, Y_2) \\
&= a\text{Var}(X) + b(\text{Var}(X) + \text{Var}(W_2)) \\
&= 4a + 8b.
\end{aligned}$$

We need to have

$$\text{Cov}(\tilde{X}, Y_j) = 0, \qquad \text{for } j = 1, 2,$$

which is equivalent to

$$\text{Cov}(\hat{X}_L, Y_j) = \text{Cov}(X, Y_j), \qquad \text{for } j = 1, 2.$$

Since $\text{Cov}(X, Y_1) = \text{Cov}(X, Y_2) = \text{Var}(X) = 4$, we conclude

$$5a + 4b = 4,$$
$$4a + 8b = 4.$$

Solving for a and b, we obtain $a = \frac{2}{3}$, and $b = \frac{1}{6}$. Therefore, the linear MMSE estimator of X, given Y_1 and Y_2, is

$$\hat{X}_L = \frac{2}{3}Y_1 + \frac{1}{6}Y_2.$$

9.1.8 Bayesian Hypothesis Testing

Suppose that we need to decide between two hypotheses H_0 and H_1. In the Bayesian setting, we assume that we know prior probabilities of H_0 and H_1. That is, we know $P(H_0) = p_0$ and $P(H_1) = p_1$, where $p_0 + p_1 = 1$. We observe the random variable (or the random vector) Y. We know the distribution of Y under the two hypotheses, i.e, we know

$$f_Y(y|H_0), \quad \text{and} \quad f_Y(y|H_1).$$

Using Bayes' rule, we can obtain the posterior probabilities of H_0 and H_1:

$$P(H_0|Y = y) = \frac{f_Y(y|H_0)P(H_0)}{f_Y(y)},$$
$$P(H_1|Y = y) = \frac{f_Y(y|H_1)P(H_1)}{f_Y(y)}.$$

One way to decide between H_0 and H_1 is to compare $P(H_0|Y = y)$ and $P(H_1|Y = y)$, and accept the hypothesis with the higher posterior probability. This is the idea behind the *maximum a posteriori (MAP) test*. Here, since we are choosing the hypothesis with the highest probability, it is relatively easy to show that the error probability is minimized.

To be more specific, according to the MAP test, we choose H_0 if and only if

$$P(H_0|Y = y) \geq P(H_1|Y = y).$$

In other words, we choose H_0 if and only if

$$f_Y(y|H_0)P(H_0) \geq f_Y(y|H_1)P(H_1).$$

Note that as always, we use the PMF instead of the PDF if Y is a discrete random variable. We can generalize the MAP test to the case where you have more than two hypotheses. In that case, again we choose the hypothesis with the highest posterior probability.

MAP Hypothesis Test

Choose the hypothesis with the highest posterior probability, $P(H_i|Y = y)$. Equivalently, choose hypothesis H_i with the highest $f_Y(y|H_i)P(H_i)$.

Example 9.10. Suppose that the random variable X is transmitted over a communication channel. Assume that the received signal is given by

$$Y = X + W,$$

where $W \sim N(0, \sigma^2)$ is independent of X. Suppose that $X = 1$ with probability p, and $X = -1$ with probability $1 - p$. The goal is to decide between $X = 1$ and $X = -1$ by observing the random variable Y. Find the MAP test for this problem.

Solution: Here, we have two hypotheses:

H_0: $X = 1$,

H_1: $X = -1$.

Under H_0, $Y = 1 + W$, so $Y|H_0 \sim N(1, \sigma^2)$. Therefore,

$$f_Y(y|H_0) = \frac{1}{\sigma\sqrt{2\pi}} e^{-\frac{(y-1)^2}{2\sigma^2}}.$$

Under H_1, $Y = -1 + W$, so $Y|H_1 \sim N(-1, \sigma^2)$. Therefore,

$$f_Y(y|H_1) = \frac{1}{\sigma\sqrt{2\pi}} e^{-\frac{(y+1)^2}{2\sigma^2}}.$$

Thus, we choose H_0 if and only if

$$\frac{1}{\sigma\sqrt{2\pi}} e^{-\frac{(y-1)^2}{2\sigma^2}} P(H_0) \geq \frac{1}{\sigma\sqrt{2\pi}} e^{-\frac{(y+1)^2}{2\sigma^2}} P(H_1).$$

We have $P(H_0) = p$, and $P(H_1) = 1 - p$. Therefore, we choose H_0 if and only if

$$\exp\left(\frac{2y}{\sigma^2}\right) \geq \frac{1-p}{p}.$$

Equivalently, we choose H_0 if and only if

$$y \geq \frac{\sigma^2}{2} \ln\left(\frac{1-p}{p}\right).$$

Note that the average error probability for a hypothesis test can be written as

$$P_e = P(\text{choose } H_1|H_0)P(H_0) + P(\text{choose } H_0|H_1)P(H_1). \tag{9.6}$$

As we mentioned earlier, the MAP test achieves the minimum possible average error probability.

Example 9.11. Find the average error probability in Example 9.10

Solution: in Example 9.10, we arrived at the following decision rule: We choose H_0 if and only if

$$y \geq c,$$

where

$$c = \frac{\sigma^2}{2} \ln\left(\frac{1-p}{p}\right).$$

Since $Y|H_0 \sim N(1, \sigma^2)$,

$$P(\text{choose } H_1|H_0) = P(Y < c|H_0)$$
$$= \Phi\left(\frac{c-1}{\sigma}\right)$$
$$= \Phi\left(\frac{\sigma}{2} \ln\left(\frac{1-p}{p}\right) - \frac{1}{\sigma}\right).$$

Since $Y|H_1 \sim N(-1, \sigma^2)$,

$$P(\text{choose } H_0|H_1) = P(Y \geq c|H_1)$$
$$= 1 - \Phi\left(\frac{c+1}{\sigma}\right)$$
$$= 1 - \Phi\left(\frac{\sigma}{2} \ln\left(\frac{1-p}{p}\right) + \frac{1}{\sigma}\right).$$

Figure 9.4 shows the two error probabilities for this example. Therefore, the average error probability is given by

$$P_e = P(\text{choose } H_1|H_0)P(H_0) + P(\text{choose } H_0|H_1)P(H_1)$$
$$= p \cdot \Phi\left(\frac{\sigma}{2} \ln\left(\frac{1-p}{p}\right) - \frac{1}{\sigma}\right) + (1-p) \cdot \left[1 - \Phi\left(\frac{\sigma}{2} \ln\left(\frac{1-p}{p}\right) + \frac{1}{\sigma}\right)\right].$$

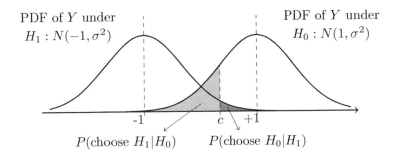

Figure 9.4: Error probabilities for Examples 9.10 and 9.11.

Minimum Cost Hypothesis Test:

Suppose that you are building a sensor network to detect fires in a forest. Based on the information collected by the sensors, the system needs to decide between two opposing hypotheses:

H_0: There is no fire,

H_1: There is a fire.

There are two possible types of errors that we can make: We might accept H_0 while H_1 is true, or we might accept H_1 while H_0 is true. Note that the cost associated with these two errors are not the same. In other words, if there is a fire and we miss it, we will be making a costlier error. To address situations like this, we associate a cost to each error type:

C_{10}: The cost of choosing H_1, given that H_0 is true.

C_{01}: The cost of choosing H_0, given that H_1 is true.

Then, the average cost can be written as

$$C = C_{10}P(\text{choose } H_1|H_0)P(H_0) + C_{01}P(\text{choose } H_0|H_1)P(H_1).$$

The goal of *minimum cost hypothesis testing* is to minimise the above expression. Luckily, this can be done easily. Note that we can rewrite the average cost as

$$C = P(\text{choose } H_1|H_0) \cdot [P(H_0)C_{10}] + P(\text{choose } H_0|H_1) \cdot [P(H_1)C_{01}].$$

The above expression is very similar to the average error probability of the MAP test (Equation 9.6). The only difference is that we have $P(H_0)C_{10}$ instead of $P(H_0)$, and we have $P(H_1)C_{01}$ instead of $P(H_1)$. Therefore, we can use a decision rule similar to the MAP decision rule. More specifically, we choose H_0 if and only if

$$f_Y(y|H_0)P(H_0)C_{10} \geq f_Y(y|H_1)P(H_1)C_{01} \tag{9.7}$$

Here is another way to interpret the above decision rule. If we divide both sides of Equation 9.7 by $f_Y(y)$ and apply Bayes' rule, we conclude the following: We choose H_0 if and only if

$$P(H_0|y)C_{10} \geq P(H_1|y)C_{01}.$$

Note that $P(H_0|y)C_{10}$ is the expected cost of accepting H_1. We call this the **posterior risk** of accepting H_1. Similarly, $P(H_1|y)C_{01}$ is the posterior risk (expected cost) of accepting H_0. Therefore, we can summarize the minimum cost test as follows: We accept the hypothesis with the lowest posterior risk.

$$\text{Minimum Cost Hypothesis Test}$$

Assuming the following costs

C_{10}: The cost of choosing H_1, given that H_0 is true.

C_{01}: The cost of choosing H_0, given that H_1 is true.

We choose H_0 if and only if

$$\frac{f_Y(y|H_0)}{f_Y(y|H_1)} \geq \frac{P(H_1)C_{01}}{P(H_0)C_{10}}.$$

Equivalently, we choose H_0 if and only if

$$P(H_0|y)C_{10} \geq P(H_1|y)C_{01}.$$

Example 9.12. A surveillance system is in charge of detecting intruders to a facility. There are two hypotheses to choose from:

H_0: No intruder is present.

H_1: There is an intruder.

The system sends an alarm message if it accepts H_1. Suppose that after processing the data, we obtain $P(H_1|y) = 0.05$. Also, assume that the cost of missing an intruder is 10 times the cost of a false alarm. Should the system send an alarm message (accept H_1)?

Solution: First note that

$$P(H_0|y) = 1 - P(H_1|y) = 0.95$$

The posterior risk of accepting H_1 is

$$P(H_0|y)C_{10} = 0.95C_{10}.$$

We have $C_{01} = 10C_{10}$, so the posterior risk of accepting H_0 is

$$P(H_1|y)C_{01} = (0.05)(10C_{10})$$
$$= 0.5C_{10}.$$

Since $P(H_0|y)C_{10} \geq P(H_1|y)C_{01}$, we accept H_0, so no alarm message needs to be sent.

9.1.9   Bayesian Interval Estimation

Interval estimation has a very natural interpretation in Bayesian inference. Suppose that we would like to estimate the value of an unobserved random variable X, given that we have observed $Y = y$. After calculating the posterior density $f_{X|Y}(x|y)$, we can simply find an interval $[a, b]$ for which we have

$$P(a \leq X \leq b|Y = y) = 1 - \alpha.$$

Such an interval is said to be a $(1 - \alpha)100\%$ *credible* interval for X.

Bayesian Credible Intervals

Given the observation $Y = y$, the interval $[a, b]$ is said to be a $(1 - \alpha)100\%$ **credible interval** for X, if the posterior probability of X being in $[a, b]$ is equal to $1 - \alpha$. In other words,

$$P(a \leq X \leq b|Y = y) = 1 - \alpha.$$

Example 9.13. Let X and Y be jointly normal and $X \sim N(0, 1)$, $Y \sim N(1, 4)$, and $\rho(X, Y) = \frac{1}{2}$. Find a 95% credible interval for X, given $Y = 2$ is observed.

Solution: As we have seen before, if X and Y are jointly normal random variables with parameters μ_X, σ_X^2, μ_Y, σ_Y^2, and ρ, then, given $Y = y$, X is normally distributed with

$$E[X|Y = y] = \mu_X + \rho\sigma_X \frac{y - \mu_Y}{\sigma_Y},$$
$$\text{Var}(X|Y = y) = (1 - \rho^2)\sigma_X^2.$$

Therefore, $X|Y = 2$ is normal with

$$E[X|Y = y] = 0 + \frac{1}{2} \cdot \frac{2 - 1}{2} = \frac{1}{4},$$

$$\text{Var}(X|Y = y) = \left(1 - \frac{1}{4}\right) \cdot 1 = \frac{3}{4}.$$

Here $\alpha = 0.05$, so we need an interval $[a, b]$ for which

$$P(a \leq X \leq b|Y = 2) = 0.95$$

We usually choose a symmetric interval around the expected value $E[X|Y = y] = \frac{1}{4}$. That is, we choose the interval in the form of

$$\left[\frac{1}{4} - c, \frac{1}{4} + c\right].$$

Thus, we need to have

$$P\left(\frac{1}{4} - c \leq X \leq \frac{1}{4} + c \mid Y = 2\right) = \Phi\left(\frac{c}{\sqrt{3/4}}\right) - \Phi\left(\frac{-c}{\sqrt{3/4}}\right)$$

$$= 2\Phi\left(\frac{c}{\sqrt{3/4}}\right) - 1 = 0.95$$

Solving for c, we obtain

$$c = \sqrt{3/4}\,\Phi^{-1}(0.975) \approx 1.70$$

Therefore, the 95% credible interval for X is

$$\left[\frac{1}{4} - c, \frac{1}{4} + c\right] \approx [-1.45, 1.95].$$

9.1.10   Solved Problems

1. Let $X \sim N(0,1)$. Suppose that we know

$$Y \mid X = x \quad \sim \quad N(x,1).$$

Show that the posterior density of X given $Y = y$, $f_{X|Y}(x|y)$, is given by

$$X \mid Y = y \quad \sim \quad N\left(\frac{y}{2}, \frac{1}{2}\right).$$

Solution: Our goal is to show that $f_{X|Y}(x|y)$ is normal with mean $\frac{y}{2}$ and variance $\frac{1}{2}$. Therefore, it suffices to show that

$$f_{X|Y}(x|y) = c(y) \exp\left\{-\left(x - \frac{y}{2}\right)^2\right\},$$

where $c(y)$ is just a function of y. That is, for a given y, $c(y)$ is just the normalizing constant ensuring that $f_{X|Y}(x|y)$ integrates to one. By the assumptions,

$$f_{Y|X}(y|x) = \frac{1}{\sqrt{2\pi}} \exp\left\{-\frac{(y-x)^2}{2}\right\},$$

$$f_X(x) = \frac{1}{\sqrt{2\pi}} \exp\left\{-\frac{x^2}{2}\right\}.$$

Therefore,

$$f_{X|Y}(x|y) = \frac{f_{Y|X}(y|x)f_X(x)}{f_Y(y)}$$

$$= (\text{a function of } y) \cdot f_{Y|X}(y|x)f_X(x)$$

$$= (\text{a function of } y) \cdot \exp\left\{-\frac{(y-x)^2 + x^2}{2}\right\}$$

$$= (\text{a function of } y) \cdot \exp\left\{-\left(x - \frac{y}{2}\right)^2 + \frac{y^2}{4}\right\}$$

$$= (\text{a function of } y) \cdot \exp\left\{-\left(x - \frac{y}{2}\right)^2\right\}.$$

2. We can generalize the result of Problem 1 using the same method. In particular, assuming

$$X \sim N(\mu, \tau^2) \qquad \text{and} \qquad Y \mid X = x \quad \sim \quad N(x, \sigma^2),$$

it can be shown that the posterior density of X given $Y = y$ is given by

$$X \mid Y = y \quad \sim \quad N\left(\frac{y/\sigma^2 + \mu/\tau^2}{1/\sigma^2 + 1/\tau^2}, \frac{1}{1/\sigma^2 + 1/\tau^2}\right).$$

In this problem, you can use the above result. Let $X \sim N(\mu, \tau^2)$ and

$$Y \mid X = x \quad \sim \quad N(x, \sigma^2).$$

Suppose that we have observed the random sample $Y_1, Y_2, \cdots, Y_n$ such that, given $X = x$, the Y_i's are i.i.d. and have the same distribution as $Y \mid X = x$.

(a) Show that the posterior density of X given $\overline{Y}$ (the sample mean) is

$$X \mid \overline{Y} \quad \sim \quad N\left(\frac{n\overline{Y}/\sigma^2 + \mu/\tau^2}{n/\sigma^2 + 1/\tau^2}, \frac{1}{n/\sigma^2 + 1/\tau^2}\right).$$

(b) Find the MAP and the MMSE estimates of X given $\overline{Y}$.

Solution:

(a) Since $Y \mid X = x \quad \sim \quad N(x, \sigma^2)$, we conclude

$$\overline{Y} \mid X = x \quad \sim \quad N\left(x, \frac{\sigma^2}{n}\right).$$

Therefore, we can use the posterior density given in the problem statement (we need to replace σ^2 by $\frac{\sigma^2}{n}$). Thus, the posterior density of X given $\overline{Y}$ is

$$X \mid \overline{Y} \quad \sim \quad N\left(\frac{n\overline{Y}/\sigma^2 + \mu/\tau^2}{n/\sigma^2 + 1/\tau^2}, \frac{1}{n/\sigma^2 + 1/\tau^2}\right).$$

(b) To find the MAP estimate of X given $\overline{Y}$, we need to find the value that maximizes the posterior density. Since the posterior density is normal, the maximum value is obtained at the mean which is

$$\hat{X}_{MAP} = \frac{n\overline{Y}/\sigma^2 + \mu/\tau^2}{n/\sigma^2 + 1/\tau^2}.$$

Also, the MMSE estimate of X given $\overline{Y}$ is

$$\hat{X}_M = E[X|\overline{Y}] = \frac{n\overline{Y}/\sigma^2 + \mu/\tau^2}{n/\sigma^2 + 1/\tau^2}.$$

3. Let $\hat{X}_M$ be the MMSE estimate of X given Y. Show that the MSE of this estimator is

$$MSE = E\big[\text{Var}(X|Y)\big].$$

Solution: We have

$$
\begin{aligned}
\text{Var}(X|Y) &= E[(X - E[X|Y])^2|Y] \qquad &&\text{(by definition of Var}(X|Y)) \\
&= E[(X - \hat{X}_M)^2|Y].
\end{aligned}
$$

Therefore,

$$
\begin{aligned}
E[\text{Var}(X|Y)] &= E\big[E[(X - \hat{X}_M)^2|Y]\big] \\
&= E[(X - \hat{X}_M)^2] \qquad &&\text{(by the law of iterated expectations)} \\
&= MSE \qquad &&\text{(by definition of MSE).}
\end{aligned}
$$

4. Consider two random variables X and Y with the joint PMF given in Table 9.1.

(a) Find the linear MMSE estimator of X given Y, $(\hat{X}_L)$.

(b) Find the MMSE estimator of X given Y, $(\hat{X}_M)$.

(c) Find the MSE of $\hat{X}_M$.

Table 9.1: Joint PMF of X and Y for Problem 4

	$Y = 0$	$Y = 1$
$X = 0$	$\frac{1}{5}$	$\frac{2}{5}$
$X = 1$	$\frac{2}{5}$	0

Solution: Using the table we find out

$$P_X(0) = \frac{1}{5} + \frac{2}{5} = \frac{3}{5},$$
$$P_X(1) = \frac{2}{5} + 0 = \frac{2}{5},$$
$$P_Y(0) = \frac{1}{5} + \frac{2}{5} = \frac{3}{5},$$
$$P_Y(1) = \frac{2}{5} + 0 = \frac{2}{5}.$$

Thus, the marginal distributions of X and Y are both *Bernoulli*$(\frac{2}{5})$. Therefore, we have

$$EX = EY = \frac{2}{5},$$
$$\text{Var}(X) = \text{Var}(Y) = \frac{2}{5} \cdot \frac{3}{5} = \frac{6}{25}.$$

(a) To find the linear MMSE estimator of X given Y, we also need $\text{Cov}(X, Y)$. We have

$$EXY = \sum x_i y_j P_{XY}(x, y) = 0.$$

Therefore,

$$\text{Cov}(X, Y) = EXY - EXEY$$
$$= -\frac{4}{25}.$$

The linear MMSE estimator of X given Y is

$$\hat{X}_L = \frac{\text{Cov}(X, Y)}{\text{Var}(Y)}(Y - EY) + EX$$
$$= \frac{-4/25}{6/25}\left(Y - \frac{2}{5}\right) + \frac{2}{5}$$
$$= -\frac{2}{3}Y + \frac{2}{3}.$$

Since Y can only take two values, we can summarize $\hat{X}_L$ in the following table.

Table 9.2: The linear MMSE estimator of X given Y for Problem 4

	$Y = 0$	$Y = 1$
$\hat{X}_L$	$\frac{2}{3}$	0

(b) To find the MMSE estimator of X given Y, we need the conditional PMFs. We have

$$P_{X|Y}(0|0) = \frac{P_{XY}(0,0)}{P_Y(0)}$$

$$= \frac{\frac{1}{5}}{\frac{3}{5}} = \frac{1}{3}.$$

Thus,

$$P_{X|Y}(1|0) = 1 - \frac{1}{3} = \frac{2}{3}.$$

We conclude

$$X|Y = 0 \sim Bernoulli\left(\frac{2}{3}\right).$$

Similarly, we find

$$P_{X|Y}(0|1) = 1,$$
$$P_{X|Y}(1|1) = 0.$$

Thus, given $Y = 1$, we have always $X = 0$. The MMSE estimator of X given Y is

$$\hat{X}_M = E[X|Y].$$

We have

$$E[X|Y = 0] = \frac{2}{3},$$
$$E[X|Y = 1] = 0.$$

Thus, we can summarize $\hat{X}_M$ in the following table.

Table 9.3: The MMSE estimator of X given Y for Problem 4

	$Y = 0$	$Y = 1$
$\hat{X}_M$	$\frac{2}{3}$	0

We notice that, for this problem, the MMSE and the linear MMSE estimators are the same. In fact, this is not surprising since here, Y can only take two possible values, and for each value we have a corresponding MMSE estimator. The linear MMSE estimator is just the line passing through the two resulting points.

(c) The MSE of $\hat{X}_M$ can be obtained as

$$MSE = E[\tilde{X}^2]$$
$$= EX^2 - E[\hat{X}_M^2]$$
$$= \frac{2}{5} - E[\hat{X}_M^2].$$

From the table for $\hat{X}_M$, we obtain $E[\hat{X}_M^2] = \frac{4}{15}$. Therefore,

$$MSE = \frac{2}{15}.$$

Note that here the MMSE and the linear MMSE estimators are equal, so they have the same MSE. Thus, we can use the formula for the MSE of $\hat{X}_L$ as well:

$$MSE = (1 - \rho(X,Y)^2)\mathrm{Var}(X)$$
$$= \left(1 - \frac{\mathrm{Cov}(X,Y)^2}{\mathrm{Var}(X)\mathrm{Var}(Y)}\right)\mathrm{Var}(X)$$
$$= \left(1 - \frac{(-4/25)^2}{6/25 \cdot 6/25}\right)\frac{6}{25}$$
$$= \frac{2}{15}.$$

5. Consider Example 9.9 in which X is an unobserved random variable with $EX = 0$, $\mathrm{Var}(X) = 4$. Assume that we have observed Y_1 and Y_2 given by

$$Y_1 = X + W_1,$$
$$Y_2 = X + W_2,$$

where $EW_1 = EW_2 = 0$, $\mathrm{Var}(W_1) = 1$, and $\mathrm{Var}(W_2) = 4$. Assume that W_1, W_2 , and X are independent random variables. Find the linear MMSE estimator of X given Y_1 and Y_2 using the vector formula

$$\hat{\mathbf{X}}_L = \mathbf{C_{XY}}\mathbf{C_Y}^{-1}(\mathbf{Y} - E[\mathbf{Y}]) + E[\mathbf{X}].$$

Solution: Note that, here, X is a one dimensional vector, and $\mathbf{Y}$ is a two dimensional vector

$$\mathbf{Y} = \begin{bmatrix} Y_1 \\ Y_2 \end{bmatrix} = \begin{bmatrix} X + W_1 \\ X + W_2 \end{bmatrix}.$$

We have

$$\mathbf{C_Y} = \begin{bmatrix} \text{Var}(Y_1) & \text{Cov}(Y_1, Y_2) \\ \text{Cov}(Y_2, Y_1) & \text{Var}(Y_2) \end{bmatrix} = \begin{bmatrix} 5 & 4 \\ 4 & 8 \end{bmatrix},$$

$$\mathbf{C_{XY}} = \begin{bmatrix} \text{Cov}(X, Y_1) & \text{Cov}(X, Y_2) \end{bmatrix} = \begin{bmatrix} 4 & 4 \end{bmatrix}.$$

Therefore,

$$\begin{aligned} \hat{\mathbf{X}}_L &= \begin{bmatrix} 4 & 4 \end{bmatrix} \begin{bmatrix} 5 & 4 \\ 4 & 8 \end{bmatrix}^{-1} \left(\begin{bmatrix} Y_1 \\ Y_2 \end{bmatrix} - \begin{bmatrix} 0 \\ 0 \end{bmatrix} \right) + 0 \\ &= \begin{bmatrix} \frac{2}{3} & \frac{1}{6} \end{bmatrix} \begin{bmatrix} Y_1 \\ Y_2 \end{bmatrix} \\ &= \frac{2}{3} Y_1 + \frac{1}{6} Y_2, \end{aligned}$$

which is the same as the result that we obtained using the orthogonality principle in Example 9.9.

6. Suppose that we need to decide between two opposing hypotheses H_0 and H_1. Let C_{ij} be the cost of accepting H_i given that H_j is true. That is

   C_{00}: The cost of choosing H_0, given that H_0 is true.

   C_{10}: The cost of choosing H_1, given that H_0 is true.

   C_{01}: The cost of choosing H_0, given that H_1 is true.

   C_{11}: The cost of choosing H_1, given that H_1 is true.

   It is reasonable to assume that the associated cost to a correct decision is less than the cost of an incorrect decision. That is, $c_{00} < c_{10}$ and $c_{11} < c_{01}$. The average cost can be written as

   $$C = \sum_{i,j} C_{ij} P(\text{choose } H_i | H_j) P(H_j)$$

   $$\begin{aligned} = &C_{00} P(\text{choose } H_0 | H_0) P(H_0) + C_{01} P(\text{choose } H_0 | H_1) P(H_1) \\ &+ C_{10} P(\text{choose } H_1 | H_0) P(H_0) + C_{11} P(\text{choose } H_1 | H_1) P(H_1). \end{aligned}$$

   Our goal is to find the decision rule such that the average cost is minimized. Show that the decision rule can be stated as follows: Choose H_0 if and only if

   $$f_Y(y|H_0) P(H_0)(C_{10} - C_{00}) \geq f_Y(y|H_1) P(H_1)(C_{01} - C_{11}) \tag{9.8}$$

   Solution: First, note that

   $$P(\text{choose } H_0 | H_0) = 1 - P(\text{choose } H_1 | H_0),$$
   $$P(\text{choose } H_1 | H_1) = 1 - P(\text{choose } H_0 | H_1).$$

Therefore,

$$C = C_{00}\big[1 - P(\text{choose } H_1|H_0)\big]P(H_0) + C_{01}P(\text{choose } H_0|H_1)P(H_1)$$
$$+ C_{10}P(\text{choose } H_1|H_0)P(H_0) + C_{11}\big[1 - P(\text{choose } H_0|H_1)\big]P(H_1)$$
$$= (C_{10} - C_{00})P(\text{choose } H_1|H_0)P(H_0) + (C_{01} - C_{11})P(\text{choose } H_0|H_1)P(H_1)$$
$$+ C_{00}P(H_0) + C_{11}P(H_1).$$

The term $C_{00}P(H_0) + C_{11}P(H_1)$ is constant (i.e., it does not depend on the decision rule). Therefore, to minimize the cost, we need to minimize

$$D = P(\text{choose } H_1|H_0)P(H_0)(C_{10} - C_{00}) + P(\text{choose } H_0|H_1)P(H_1)(C_{01} - C_{11}).$$

The above expression is very similar to the average error probability of the MAP test (Equation 9.6). The only difference is that we have $P(H_0)(C_{10} - C_{00})$ instead of $P(H_0)$, and we have $P(H_1)(C_{01} - C_{11})$ instead of $P(H_1)$. Therefore, we can use a decision rule similar to the MAP decision rule. More specifically, we choose H_0 if and only if

$$f_Y(y|H_0)P(H_0)(C_{10} - C_{00}) \geq f_Y(y|H_1)P(H_1)(C_{01} - C_{11}).$$

7. Let

$$X \sim N(0, 4) \qquad \text{and} \qquad Y \mid X = x \quad \sim \quad N(x, 1).$$

Suppose that we have observed the random sample $Y_1, Y_2, \cdots, Y_{25}$ such that given $X = x$, the Y_i's are i.i.d. and have the same distribution as $Y \mid X = x$. Find a 95% credible interval for X, given that we have observed

$$\overline{Y} = \frac{Y_1 + Y_2 + \ldots + Y_n}{n} = 0.56$$

Hint: Use the result of Problem 2.

Solution: By part (a) of Problem 2, we have

$$X \mid \overline{Y} \quad \sim \quad N\left(\frac{25(0.56)/1 + 0/4}{25/1 + 1/4}, \frac{1}{25/1 + 1/4}\right)$$
$$= N\big(0.5545, 0.0396\big).$$

Therefore, we choose the interval in the form of

$$\big[0.5545 - c, 0.5545 + c\big].$$

We need to have

$$P\left(0.5545 - c \leq X \leq 0.5545 + c \big| \overline{Y} = 0.56\right) = \Phi\left(\frac{c}{\sqrt{0.0396}}\right) - \Phi\left(\frac{-c}{\sqrt{0.0396}}\right)$$
$$= 2\Phi\left(\frac{c}{\sqrt{0.0396}}\right) - 1 = 0.95$$

Solving for c, we obtain

$$c = \sqrt{0.0396}\,\Phi^{-1}(0.975) \approx 0.39$$

Therefore, the 95% credible interval for X is

$$[0.5545 - 0.39, 0.5545 + 0.39] \approx [0.1645, 0.9445].$$

9.2   End of Chapter Problems

1. Let X be a continuous random variable with the following PDF

$$f_X(x) = \begin{cases} 6x(1-x) & \text{if } 0 \le x \le 1 \\ 0 & \text{otherwise} \end{cases}$$

Suppose that we know

$$Y \mid X = x \quad \sim \quad Geometric(x).$$

Find the posterior density of X given $Y = 2$, $f_{X|Y}(x|2)$.

2. Let X be a continuous random variable with the following PDF

$$f_X(x) = \begin{cases} 3x^2 & \text{if } 0 \le x \le 1 \\ 0 & \text{otherwise} \end{cases}$$

Also, suppose that

$$Y \mid X = x \quad \sim \quad Geometric(x).$$

Find the MAP estimate of X given $Y = 5$.

3. Let X and Y be two jointly continuous random variables with joint PDF

$$f_{XY}(x, y) = \begin{cases} x + \frac{3}{2}y^2 & 0 \le x, y \le 1 \\ 0 & \text{otherwise.} \end{cases}$$

Find the MAP and the ML estimates of X given $Y = y$.

4. Let X be a continuous random variable with the following PDF

$$f_X(x) = \begin{cases} 2x^2 + \frac{1}{3} & \text{if } 0 \le x \le 1 \\ 0 & \text{otherwise} \end{cases}$$

We also know that

$$f_{Y|X}(y|x) = \begin{cases} xy - \frac{x}{2} + 1 & \text{if } 0 \le y \le 1 \\ 0 & \text{otherwise} \end{cases}$$

Find the MMSE estimate of X, given $Y = y$ is observed.

5. Let $X \sim N(0, 1)$ and

$$Y = 2X + W,$$

where $W \sim N(0, 1)$ is independent of X.

(a) Find the MMSE estimator of X given Y, $(\hat{X}_M)$.

(b) Find the MSE of this estimator, using $MSE = E[(X - \hat{X}_M)^2]$.

(c) Check that $E[X^2] = E[\hat{X}_M^2] + E[\tilde{X}^2]$.

6. Suppose $X \sim Uniform(0, 1)$, and given $X = x$, $Y \sim Exponential(\lambda = \frac{1}{2x})$.

(a) Find the linear MMSE estimate of X given Y.

(b) Find the MSE of this estimator.

(c) Check that $E[\tilde{X}Y] = 0$.

7. Suppose that the signal $X \sim N(0, \sigma_X^2)$ is transmitted over a communication channel. Assume that the received signal is given by

$$Y = X + W,$$

where $W \sim N(0, \sigma_W^2)$ is independent of X.

(a) Find the MMSE estimator of X given Y, $(\hat{X}_M)$.

(b) Find the MSE of this estimator.

8. Let X be an unobserved random variable with $EX = 0$, $\text{Var}(X) = 5$. Assume that we have observed Y_1 and Y_2 given by

$$Y_1 = 2X + W_1,$$
$$Y_2 = X + W_2,$$

where $EW_1 = EW_2 = 0$, $\text{Var}(W_1) = 2$, and $\text{Var}(W_2) = 5$. Assume that W_1, W_2, and X are independent random variables. Find the linear MMSE estimator of X, given Y_1 and Y_2.

9. Consider again Problem 8, in which X is an unobserved random variable with $EX = 0$, $\text{Var}(X) = 5$. Assume that we have observed Y_1 and Y_2 given by

$$Y_1 = 2X + W_1,$$
$$Y_2 = X + W_2,$$

where $EW_1 = EW_2 = 0$, $\text{Var}(W_1) = 2$, and $\text{Var}(W_2) = 5$. Assume that W_1, W_2, and X are independent random variables. Find the linear MMSE estimator of X, given Y_1 and Y_2, using the vector formula

$$\hat{\mathbf{X}}_L = \mathbf{C_{XY}}\mathbf{C_Y}^{-1}(\mathbf{Y} - E[\mathbf{Y}]) + E[\mathbf{X}].$$

10. Let X be an unobserved random variable with $EX = 0$, $\text{Var}(X) = 5$. Assume that we have observed Y_1, Y_2, and Y_3 given by

$$Y_1 = 2X + W_1,$$
$$Y_2 = X + W_2,$$
$$Y_3 = X + 2W_3,$$

where $EW_1 = EW_2 = EW_3 = 0$, $\text{Var}(W_1) = 2$, $\text{Var}(W_2) = 5$, and $\text{Var}(W_3) = 3$. Assume that W_1, W_2, W_3, and X are independent random variables. Find the linear MMSE estimator of X, given Y_1, Y_2, and Y_3.

11. Consider two random variables X and Y with the joint PMF given by the table below.

	$Y = 0$	$Y = 1$
$X = 0$	$\frac{1}{7}$	$\frac{3}{7}$
$X = 1$	$\frac{3}{7}$	0

(a) Find the linear MMSE estimator of X given Y, $(\hat{X}_L)$.

(b) Find the MMSE estimator of X given Y, $(\hat{X}_M)$.

(c) Find the MSE of $\hat{X}_M$.

12. Consider two random variables X and Y with the joint PMF given by the table below.

	$Y = 0$	$Y = 1$	$Y = 2$
$X = 0$	$\frac{1}{6}$	$\frac{1}{3}$	0
$X = 1$	$\frac{1}{3}$	0	$\frac{1}{6}$

(a) Find the linear MMSE estimator of X given Y, $(\hat{X}_L)$.

(b) Find the MSE of $\hat{X}_L$.

(c) Find the MMSE estimator of X given Y, $(\hat{X}_M)$.

(d) Find the MSE of $\hat{X}_M$.

13. Suppose that the random variable X is transmitted over a communication channel. Assume that the received signal is given by

$$Y = 2X + W,$$

where $W \sim N(0, \sigma^2)$ is independent of X. Suppose that $X = 1$ with probability p, and $X = -1$ with probability $1 - p$. The goal is to decide between $X = -1$ and $X = 1$ by observing the random variable Y. Find the MAP test for this problem.

14. Find the average error probability in Problem 13.

15. A monitoring system is in charge of detecting malfunctioning machinery in a facility. There are two hypotheses to choose from:

H_0: There is not a malfunction,

H_1: There is a malfunction.

The system notifies a maintenance team if it accepts H_1. Suppose that, after processing the data, we obtain $P(H_1|y) = 0.10$. Also, assume that the cost of missing a malfunction is 30 times the cost of a false alarm. Should the system alert a maintenance team (accept H_1)?

16. Let X and Y be jointly normal and $X \sim N(2, 1)$, $Y \sim N(1, 5)$, and $\rho(X, Y) = \frac{1}{4}$. Find a 90% credible interval for X, given $Y = 1$ is observed.

17. When the choice of a prior distribution is subjective, it is often advantageous to choose a prior distribution that will result in a posterior distribution of the same distributional family. When the prior and posterior distributions share the same distributional family, they are called *conjugate distributions*, and the prior is called a *conjugate prior*. Conjugate priors are used out of ease because they always result in a closed form posterior distribution. One example of this is to use a gamma prior for Poisson distributed data.

Assume our data Y given X is distributed $Y \mid X = x \sim Poisson(\lambda = x)$ and we chose the prior to be $X \sim Gamma(\alpha, \beta)$. Then the PMF for our data is

$$P_{Y|X}(y|x) = \frac{e^{-x} x^y}{y!}, \quad \text{for } x > 0, y \in \{0, 1, 2, \dots\},$$

and the PDF of the prior is given by

$$f_X(x) = \frac{\beta^\alpha x^{\alpha-1} e^{-\beta x}}{\Gamma(\alpha)}, \quad \text{for } x > 0, \ \alpha, \beta > 0.$$

(a) Show that the posterior distribution is $Gamma(\alpha + y, \beta + 1)$.
(*Hint: Remove all the terms not containing x by putting them into some normalizing constant, c, and noting that $f_{X|Y}(x|y) \propto P_{Y|X}(y|x) f_X(x)$.*)

(b) Write out the PDF for the posterior distribution, $f_{X|Y}(x|y)$.

(c) Find mean and variance of the posterior distribution, $E[X|Y]$ and $Var(X|Y)$.

18. Assume our data Y given X is distributed $Y \mid X = x \sim Binomial(n, p = x)$ and we chose the prior to be $X \sim Beta(\alpha, \beta)$. Then the PMF for our data is

$$P_{Y|X}(y|x) = \binom{n}{y} x^y (1 - x)^{n-y}, \quad \text{for } x \in [0, 1], y \in \{0, 1, \dots, n\},$$

and the PDF of the prior is given by

$$f_X(x) = \frac{\Gamma(\alpha + \beta)}{\Gamma(\alpha)\Gamma(\beta)} x^{\alpha-1}(1 - x)^{\beta-1}, \quad \text{for } 0 \le x \le 1, \alpha > 0, \beta > 0.$$

Note that, $EX = \frac{\alpha}{\alpha+\beta}$ and $\text{Var}(X) = \frac{\alpha\beta}{(\alpha+\beta)^2(\alpha+\beta+1)}$.

(a) Show that the posterior distribution is $Beta(\alpha + y, \beta + n - y)$.

(b) Write out the PDF for the posterior distribution, $f_{X|Y}(x|y)$.

(c) Find mean and variance of the posterior distribution, $E[X|Y]$ and $\text{Var}(X|Y)$.

19. Assume our data Y given X is distributed $Y \mid X = x \sim Geometric(p = x)$ and we chose the prior to be $X \sim Beta(\alpha, \beta)$. Refer to Problem 18 for the PDF and moments of the *Beta* distribution.

(a) Show that the posterior distribution is $Beta(\alpha + 1, \beta + y - 1)$.

(b) Write out the PDF for the posterior distribution, $f_{X|Y}(x|y)$.

(c) Find mean and variance of the posterior distribution, $E[X|Y]$ and $\text{Var}(X|Y)$.

20. Assume our data $\mathbf{Y} = (y_1, y_2, \dots, y_n)^T$ given X is independently identically distributed, $\mathbf{Y} \mid X = x \overset{i.i.d.}{\sim} Exponential(\lambda = x)$, and we chose the prior to be $X \sim Gamma(\alpha, \beta)$.

(a) Find the likelihood of the function, $L(\mathbf{Y}; X) = f_{Y_1, Y_2, \dots, Y_n | X}(y_1, y_2, \dots, y_n | x)$.

(b) Using the likelihood function of the data, show that the posterior distribution is $Gamma(\alpha + n, \beta + \sum_{i=1}^{n} y_i)$.

(c) Write out the PDF for the posterior distribution, $f_{X|Y}(x|\mathbf{y})$.

(d) Find mean and variance of the posterior distribution, $E[X|\mathbf{Y}]$ and $\text{Var}(X|\mathbf{Y})$.

Chapter 10

Introduction to Random Processes

10.1 Basic Concepts

In real-life applications, we are often interested in multiple observations of random values over a period of time. For example, suppose that you are observing the stock price of a company over the next few months. In particular, let $S(t)$ be the stock price at time $t \in [0, \infty)$. Here, we assume $t = 0$ refers to current time. Figure 10.1 shows a possible outcome of this random experiment from time $t = 0$ to time $t = 1$.

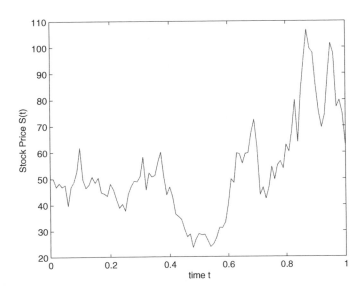

Figure 10.1: A possible realization of values of a stock observed as a function of time. Here, $S(t)$ is an example of a random process.

Note that at any fixed time $t_1 \in [0, \infty)$, $S(t_1)$ is a random variable. Based on your knowledge of finance and the historical data, you might be able to provide a PDF for $S(t_1)$. If you choose another time $t_2 \in [0, \infty)$, you obtain another random variable $S(t_2)$ that could potentially have a different PDF. When we consider the values of $S(t)$ for $t \in [0, \infty)$ collectively, we say $S(t)$ is a **random process** or a **stochastic process**. We may show this process by

$$\{S(t), t \in [0, \infty)\}.$$

Therefore, a random process is a collection of random variables usually indexed by time (or sometimes by space).

A random process is a collection of random variables usually indexed by time.

The process $S(t)$ mentioned here is an example of a **continuous-time** random process. In general, when we have a random process $X(t)$ where t can take real values in an interval on the real line, then $X(t)$ is a continuous-time random process. Here are a few more examples of continuous-time random processes:

- Let $N(t)$ be the number of customers who have visited a bank from $t = 9$ (when the bank opens at 9:00 am) until time t, on a given day, for $t \in [9, 16]$. Here, we measure t in hours, but t can take any real value between 9 and 16. We assume that $N(9) = 0$, and $N(t) \in \{0, 1, 2, ...\}$ for all $t \in [9, 16]$. Note that for any time t_1, the random variable $N(t_1)$ is a discrete random variable. Thus, $N(t)$ is a *discrete-valued* random process. However, since t can take any real value between 9 and 16, $N(t)$ is a continuous-time random process.

- Let $W(t)$ be the thermal noise voltage generated across a resistor in an electric circuit at time t, for $t \in [0, \infty)$. Here, $W(t)$ can take real values.

- Let $T(t)$ be the temperature in New York City at time $t \in [0, \infty)$. We can assume here that t is measured in hours and $t = 0$ refers to the time we start measuring the temperature.

In all of these examples, we are dealing with an uncountable number of random variables. For example, for any given $t_1 \in [9, 16]$, $N(t_1)$ is a random variable. Thus, the random process $N(t)$ consists of an uncountable number of random variables. A random process can be defined on the entire real line, i.e., $t \in (-\infty, \infty)$. In fact, it is sometimes convenient to assume that the process starts at $t = -\infty$ even if we are interested in $X(t)$ only on a finite interval. For example, we can assume that the $T(t)$ defined above is a random process defined for all $t \in \mathbb{R}$ although we get to observe only a finite portion of it.

On the other hand, you can have a **discrete-time** random process. A discrete-time random process is a process

$$\{X(t), t \in J\},$$

where J is a countable set. Since J is countable, we can write $J = \{t_1, t_2, \cdots\}$. We usually define $X(t_n) = X(n)$ or $X(t_n) = X_n$, for $n = 1, 2, \cdots$, (the index values n could be from any countable set such as $\mathbb{N}$ or $\mathbb{Z}$). Therefore, a discrete-time random process is just a sequence of random variables. For this reason, discrete-time random processes are sometimes referred to as **random sequences**. We can denote such a discrete-time process as

$$\big\{X(n), n = 0, 1, 2, \dots\big\} \qquad \text{or} \qquad \big\{X_n, n = 0, 1, 2, \dots\big\}.$$

Or, if the process is defined for all integers, then we may show the process by

$$\big\{X(n), n \in \mathbb{Z}\big\} \qquad \text{or} \qquad \big\{X_n, n \in \mathbb{Z}\big\}.$$

Here is an example of a discrete-time random process. Suppose that we are observing customers who visit a bank starting at a given time. Let X_n for $n \in \mathbb{N}$ be the amount of time the ith customer spends at the bank. This process consists of a countable number of random variables

$$X_1, X_2, X_3, \dots$$

Thus, we say that the process $\big\{X_n, n = 1, 2, 3..\big\}$ is a discrete-time random process. Discrete-time processes are sometimes obtained from continuous-time processes by discretizing time (sampling at specific times). For example, if you only record the temperature in New York City once a day (let's say at noon), then you can define a process

$$X_1 = T(12) \qquad\qquad \text{(temperature at noon on day 1, } t = 12)$$
$$X_2 = T(36) \qquad\qquad \text{(temperature at noon on day 2, } t = 12 + 24)$$
$$X_3 = T(60) \qquad\qquad \text{(temperature at noon on day 3, } t = 12 + 24 + 24)$$
$$\dots$$

And, in general, $X_n = T(t_n)$ where $t_n = 24(n - 1) + 12$ for $n \in \mathbb{N}$. Here, X_n is a discrete-time random process. Figure 10.2 shows a possible realization of this random process.

A **continuous-time** random process is a random process $\big\{X(t), t \in J\big\}$, where J is an interval on the real line such as $[-1, 1]$, $[0, \infty)$, $(-\infty, \infty)$, etc.

A **discrete-time** random process (or a **random sequence**) is a random process $\big\{X(n) = X_n, n \in J\big\}$, where J is a countable set such as $\mathbb{N}$ or $\mathbb{Z}$.

Random Processes as Random Functions:

Consider a random process $\big\{X(t), t \in J\big\}$. This random process is resulted from a random experiment, e.g., observing the stock prices of a company over a period of time. Remember that any random experiment is defined on a sample space S. After observing the values of $X(t)$, we obtain a function of time such as the one showed in Figure 10.1. The function shown in this figure is just one of the many possible outcomes of this random experiment. We call each of these possible functions of $X(t)$ a **sample function** or **sample path**. It is also called a **realization** of $X(t)$.

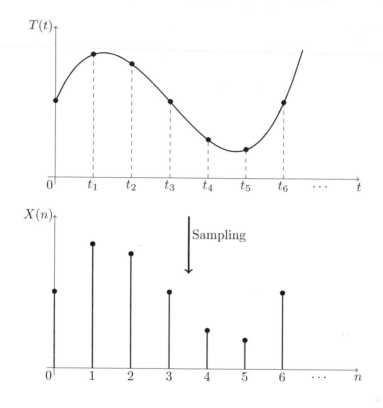

Figure 10.2: Possible realization of the random process $\{X_n, n = 1, 2, 3, \cdots\}$ where X_n shows the temperature in New York City at noon on day n.

From this point of view, a random process can be thought of as a random function of time. You are familiar with the concept of functions. The difference here is that $\{X(t), t \in J\}$ will be equal to one of many possible sample functions after we are done with our random experiment. In engineering applications, random processes are often referred to as *random signals*.

A random process is a random function of time.

Example 10.1. You have 1000 dollars to put in an account with interest rate R, compounded annually. That is, if X_n is the value of the account at year n, then

$$X_n = 1000(1 + R)^n, \qquad \text{for } n = 0, 1, 2, \cdots.$$

The value of R is a random variable that is determined when you put the money in the bank, but it does not not change after that. In particular, assume that $R \sim Uniform(0.04, 0.05)$.

(a) Find all possible sample functions for the random process $\{X_n, n = 0, 1, 2, ...\}$.

(b) Find the expected value of your account at year three. That is, find $E[X_3]$.

Solution:

(a) Here, the randomness in X_n comes from the random variable R. As soon as you know R, you know the entire sequence X_n for $n = 0, 1, 2, \cdots$. In particular, if $R = r$, then

$$X_n = 1000(1 + r)^n, \qquad \text{for all } n \in \{0, 1, 2, \cdots\}.$$

Thus, here sample functions are of the form $f(n) = 1000(1 + r)^n$, $n = 0, 1, 2, \cdots$, where $r \in [0.04, 0.05]$. For any $r \in [0.04, 0.05]$, you obtain a sample function for the random process X_n.

(b) The random variable X_3 is given by

$$X_3 = 1000(1 + R)^3.$$

If you let $Y = 1 + R$, then $Y \sim Uniform(1.04, 1.05)$, so

$$f_Y(y) = \begin{cases} 100 & 1.04 \leq y \leq 1.05 \\ 0 & \text{otherwise} \end{cases}$$

To obtain $E[X_3]$, we can write

$$
\begin{aligned}
E[X_3] &= 1000 E[Y^3] \\
&= 1000 \int_{1.04}^{1.05} 100 y^3 \ dy \qquad \text{(by LOTUS)} \\
&= \frac{10^5}{4} \left[y^4 \right]_{1.04}^{1.05} \\
&= \frac{10^5}{4} \left[(1.05)^4 - (1.04)^4 \right] \\
&\approx 1,141.2
\end{aligned}
$$

Example 10.2. Let $\{X(t), t \in [0, \infty)\}$ be defined as

$$X(t) = A + Bt, \qquad \text{for all } t \in [0, \infty),$$

where A and B are independent normal $N(1, 1)$ random variables.

(a) Find all possible sample functions for this random process.

(b) Define the random variable $Y = X(1)$. Find the PDF of Y.

(c) Let also $Z = X(2)$. Find $E[YZ]$.

Solution:

(a) Here, we note that the randomness in $X(t)$ comes from the two random variables A and B. The random variable A can take any real value $a \in \mathbb{R}$. The random variable B can also take any real value $b \in \mathbb{R}$. As soon as we know the values of A and B, the entire process $X(t)$ is known. In particular, if $A = a$ and $B = b$, then

$$X(t) = a + bt, \qquad \text{for all } t \in [0, \infty).$$

Thus, here, sample functions are of the form $f(t) = a + bt$, $t \geq 0$, where $a, b \in \mathbb{R}$. For any $a, b \in \mathbb{R}$ you obtain a sample function for the random process $X(t)$.

(b) We have

$$Y = X(1) = A + B.$$

Since A and B are independent $N(1,1)$ random variables, $Y = A + B$ is also normal with

$$\begin{aligned}
EY &= E[A + B] \\
&= E[A] + E[B] \\
&= 1 + 1 \\
&= 2,
\end{aligned}$$

$$\begin{aligned}
\text{Var}(Y) &= \text{Var}(A + B) \\
&= \text{Var}(A) + \text{Var}(B) \qquad \text{(since } A \text{ and } B \text{ are independent)} \\
&= 1 + 1 \\
&= 2.
\end{aligned}$$

Thus, we conclude that $Y \sim N(2, 2)$:

$$f_Y(y) = \frac{1}{\sqrt{4\pi}} e^{-\frac{(y-2)^2}{4}}.$$

(c) We have

$$\begin{aligned}
E[YZ] &= E[(A + B)(A + 2B)] \\
&= E[A^2 + 3AB + 2B^2] \\
&= E[A^2] + 3E[AB] + 2E[B^2] \\
&= 2 + 3E[A]E[B] + 2 \cdot 2 \qquad \text{(since } A \text{ and } B \text{ are independent)} \\
&= 9.
\end{aligned}$$

The random processes in the above examples were relatively simple in the sense that the randomness in the process originated from one or two random variables. We will see more complicated examples later on.

10.1.1 PDFs and CDFs

Consider the random process $\{X(t), t \in J\}$. For any $t_0 \in J$, $X(t_0)$ is a random variable, so we can write its CDF

$$F_{X(t_0)}(x) = P\big(X(t_0) \leq x\big).$$

If $t_1, t_2 \in J$, then we can find the joint CDF of $X(t_1)$ and $X(t_2)$ by

$$F_{X(t_1)X(t_2)}(x_1, x_2) = P\big(X(t_1) \leq x_1, X(t_2) \leq x_2\big).$$

More generally for $t_1, t_2, \cdots, t_n \in J$, we can write

$$F_{X(t_1)X(t_2)\cdots X(t_n)}(x_1, x_2, \cdots, x_n) = P\big(X(t_1) \leq x_1, X(t_2) \leq x_2, \cdots, X(t_n) \leq x_n\big).$$

Similarly, we can write joint PDFs or PMFs depending on whether $X(t)$ is continuous-valued (the $X(t_i)$'s are continuous random variables) or discrete-valued (the $X(t_i)$'s are discrete random variables).

Example 10.3. Consider the random process $\{X_n, n = 0, 1, 2, \cdots\}$, in which X_i's are i.i.d. standard normal random variables.

1. Write down $f_{X_n}(x)$ for $n = 0, 1, 2, \cdots$.

2. Write down $f_{X_m X_n}(x_1, x_2)$ for $m \neq n$.

Solution:

1. Since $X_n \sim N(0, 1)$, we have

$$f_{X_n}(x) = \frac{1}{\sqrt{2\pi}} e^{-\frac{x^2}{2}}, \qquad \text{for all } x \in \mathbb{R}.$$

2. If $m \neq n$, then X_m and X_n are independent (because of the i.i.d. assumption), so

$$
\begin{aligned}
f_{X_m X_n}(x_1, x_2) &= f_{X_m}(x_1) f_{X_n}(x_2) \\
&= \frac{1}{\sqrt{2\pi}} e^{-\frac{x_1^2}{2}} \cdot \frac{1}{\sqrt{2\pi}} e^{-\frac{x_2^2}{2}} \\
&= \frac{1}{2\pi} \exp\left\{ -\frac{x_1^2 + x_2^2}{2} \right\}, \qquad \text{for all } x_1, x_2 \in \mathbb{R}.
\end{aligned}
$$

10.1.2 Mean and Correlation Functions

Since random processes are collections of random variables, you already possess the theoretical knowledge necessary to analyze random processes. From now on, we would like to discuss methods and tools that are useful in studying random processes. Remember that expectation and variance were among the important statistics that we considered for random variables. Here, we would like to extend those concepts to random processes.

Mean Function of a Random Process:

Mean Function of a Random Process

For a random process $\{X(t), t \in J\}$, the **mean function** $\mu_X(t) : J \to \mathbb{R}$, is defined as

$$\mu_X(t) = E[X(t)]$$

The above definition is valid for both continuous-time and discrete-time random processes. In particular, if $\{X_n, n \in J\}$ is a discrete-time random process, then

$$\mu_X(n) = E[X_n], \qquad \text{for all } n \in J.$$

The mean function[1] gives us an idea about how the random process behaves on average as time evolves. For example, if $X(t)$ is the temperature in a certain city, the mean function $\mu_X(t)$ might look like the function shown in Figure 10.3. As we see, the expected value of $X(t)$ is lowest in the winter and highest in summer.

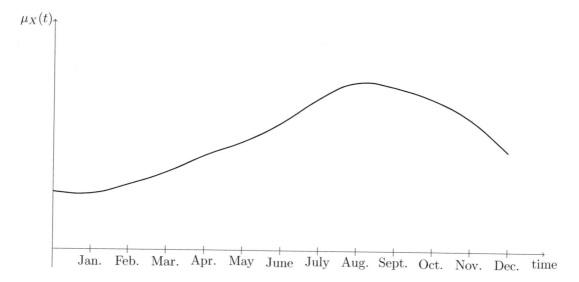

Figure 10.3: The mean function, $\mu_X(t)$, for the temperature in a certain city.

Example 10.4. Find the mean functions for the random processes given in Examples 10.1 and 10.2.

---

[1] Some books show the mean function by $m_X(t)$ or $M_X(t)$. Here, we chose $\mu_X(t)$ to avoid confusion with moment generating functions.

Solution: For $\{X_n, n = 0, 1, 2, \cdots\}$ given in Example 10.1, we have

$$
\begin{aligned}
\mu_X(n) &= E[X_n] \\
&= 1000 E[Y^n] \qquad \left(\text{where } Y = 1 + R \quad \sim \quad Uniform(1.04, 1.05)\right) \\
&= 1000 \int_{1.04}^{1.05} 100 y^n \; dy \qquad \text{(by LOTUS)} \\
&= \frac{10^5}{n+1} \left[y^{n+1} \right]_{1.04}^{1.05} \\
&= \frac{10^5}{n+1} \left[(1.05)^{n+1} - (1.04)^{n+1} \right], \qquad \text{for all } n \in \{0, 1, 2, \cdots\}.
\end{aligned}
$$

For $\{X(t), t \in [0, \infty)\}$ given in Example 10.2, we have

$$
\begin{aligned}
\mu_X(t) &= E[X(t)] \\
&= E[A + Bt] \\
&= E[A] + E[B]t \\
&= 1 + t, \qquad \text{for all } t \in [0, \infty).
\end{aligned}
$$

Autocorrelation and Autocovariance:

The mean function $\mu_X(t)$ gives us the expected value of $X(t)$ at time t, but it does not give us any information about how $X(t_1)$ and $X(t_2)$ are related. To get some insight on the relation between $X(t_1)$ and $X(t_2)$, we define correlation and covariance functions.

For a random process $\{X(t), t \in J\}$, the **autocorrelation function** or, simply, the **correlation function**, $R_X(t_1, t_2)$, is defined by

$$
R_X(t_1, t_2) = E[X(t_1)X(t_2)], \qquad \text{for } t_1, t_2 \in J.
$$

For a random process $\{X(t), t \in J\}$, the **autocovariance function** or, simply, the **covariance function**, $C_X(t_1, t_2)$, is defined by

$$
\begin{aligned}
C_X(t_1, t_2) &= \text{Cov}\big(X(t_1), X(t_2)\big) \\
&= R_X(t_1, t_2) - \mu_X(t_1)\mu_X(t_2), \qquad \text{for } t_1, t_2 \in J.
\end{aligned}
$$

Note that if we let $t_1 = t_2 = t$, we obtain

$$
\begin{aligned}
R_X(t, t) &= E[X(t)X(t)] \\
&= E[X(t)^2], \qquad \text{for } t \in J;
\end{aligned}
$$

$$C_X(t,t) = \text{Cov}\big(X(t), X(t)\big)$$
$$= \text{Var}\big(X(t)\big), \qquad \text{for } t \in J.$$

If $t_1 \neq t_2$, then the covariance function $C_X(t_1, t_2)$ gives us some information about how $X(t_1)$ and $X(t_2)$ are statistically related. In particular, note that

$$C_X(t_1, t_2) = E\left[\Big(X(t_1) - E[X(t_1)]\Big)\Big(X(t_2) - E[X(t_2)]\Big)\right].$$

Intuitively, $C_X(t_1, t_2)$ shows how $X(t_1)$ and $X(t_2)$ move relative to each other. If large values of $X(t_1)$ tend to imply large values of $X(t_2)$, then $\big(X(t_1) - E[X(t_1)]\big)\big(X(t_2) - E[X(t_2)]\big)$ is positive on average. In this case, $C_X(t_1, t_2)$ is positive, and we say $X(t_1)$ and $X(t_2)$ are positively correlated. On the other hand, if large values of $X(t_1)$ imply small values of $X(t_2)$, then $\big(X(t_1) - E[X(t_1)]\big)\big(X(t_2) - E[X(t_2)]\big)$ is negative on average, and we say $X(t_1)$ and $X(t_2)$ are negatively correlated. If $C_X(t_1, t_2) = 0$, then $X(t_1)$ and $X(t_2)$ are uncorrelated.

Example 10.5. Find the correlation functions and covariance functions for the random processes given in Examples 10.1 and 10.2.

Solution: For $\big\{X_n, n = 0, 1, 2, \cdots\big\}$ given in Example 10.1, we have

$$R_X(m,n) = E[X_m X_n]$$
$$= 10^6 E[Y^m Y^n] \qquad \big(\text{where } Y = 1 + R \quad \sim \quad Uniform(1.04, 1.05)\big)$$
$$= 10^6 \int_{1.04}^{1.05} 100 y^{(m+n)} \; dy \qquad \text{(by LOTUS)}$$
$$= \frac{10^8}{m+n+1}\left[y^{m+n+1}\right]_{1.04}^{1.05}$$
$$= \frac{10^8}{m+n+1}\left[(1.05)^{m+n+1} - (1.04)^{m+n+1}\right], \qquad \text{for all } m, n \in \{0, 1, 2, \cdots\}.$$

To find the covariance function, we write

$$C_X(m,n) = R_X(m,n) - E[X_m]E[X_n]$$
$$= \frac{10^8}{m+n+1}\left[(1.05)^{m+n+1} - (1.04)^{m+n+1}\right]$$
$$\quad - \frac{10^{10}}{(m+1)(n+1)}\left[(1.05)^{m+1} - (1.04)^{m+1}\right]\left[(1.05)^{n+1} - (1.04)^{n+1}\right].$$

For $\big\{X(t), t \in [0, \infty)\big\}$ given in Example 10.2, we have

$$R_X(t_1, t_2) = E[X(t_1)X(t_2)]$$
$$= E[(A + Bt_1)(A + Bt_2)]$$
$$= E[A^2] + E[AB](t_1 + t_2) + E[B^2]t_1 t_2$$
$$= 2 + E[A]E[B](t_1 + t_2) + 2t_1 t_2 \qquad \text{(since } A \text{ and } B \text{ are independent)}$$
$$= 2 + t_1 + t_2 + 2t_1 t_2, \qquad \text{for all } t_1, t_2 \in [0, \infty).$$

Finally, to find the covariance function for $X(t)$, we can write

$$C_X(t_1, t_2) = R_X(t_1, t_2) - E[X(t_1)]E[X(t_2)]$$
$$= 2 + t_1 + t_2 + 2t_1t_2 - (1 + t_1)(1 + t_2)$$
$$= 1 + t_1t_2, \qquad \text{for all } t_1, t_2 \in [0, \infty).$$

10.1.3 Multiple Random Processes

We often need to study more than one random process. For example, when investing in the stock market you consider several different stocks and you are interested in how they are related. In particular, you might be interested in finding out whether two stocks are positively or negatively correlated. A useful idea in these situations is to look at **cross-correlation** and **cross-covariance** functions.

For two random processes $\{X(t), t \in J\}$ and $\{Y(t), t \in J\}$:

- the **cross-correlation** function $R_{XY}(t_1, t_2)$, is defined by

$$R_{XY}(t_1, t_2) = E[X(t_1)Y(t_2)], \qquad \text{for } t_1, t_2 \in J;$$

- the **cross-covariance** function $C_{XY}(t_1, t_2)$, is defined by

$$C_{XY}(t_1, t_2) = \text{Cov}\big(X(t_1), Y(t_2)\big)$$
$$= R_{XY}(t_1, t_2) - \mu_X(t_1)\mu_Y(t_2), \qquad \text{for } t_1, t_2 \in J.$$

To get an idea about these concepts suppose that $X(t)$ is the price of oil (per gallon) and $Y(t)$ is the price of gasoline (per gallon) at time t. Since gasoline is produced from oil, as oil prices increase, the gasoline prices tend to increase, too. Thus, we conclude that $X(t)$ and $Y(t)$ should be positively correlated (at least for the same t, i.e., $C_{XY}(t, t) > 0$).

Example 10.6. Let A, B, and C be independent normal $N(1, 1)$ random variables. Let $\{X(t), t \in [0, \infty)\}$ be defined as

$$X(t) = A + Bt, \qquad \text{for all } t \in [0, \infty).$$

Also, let $\{Y(t), t \in [0, \infty)\}$ be defined as

$$Y(t) = A + Ct, \qquad \text{for all } t \in [0, \infty).$$

Find $R_{XY}(t_1, t_2)$ and $C_{XY}(t_1, t_2)$, for $t_1, t_2 \in [0, \infty)$.

Solution: First, note that

$$\mu_X(t) = E[X(t)]$$
$$= EA + EB \cdot t$$
$$= 1 + t, \qquad \text{for all } t \in [0, \infty).$$

Similarly,

$$\begin{aligned}
\mu_Y(t) &= E[Y(t)] \\
&= EA + EC \cdot t \\
&= 1 + t, \qquad \text{for all } t \in [0, \infty).
\end{aligned}$$

To find $R_{XY}(t_1, t_2)$ for $t_1, t_2 \in [0, \infty)$, we write

$$\begin{aligned}
R_{XY}(t_1, t_2) &= E[X(t_1)Y(t_2)] \\
&= E\big[(A + Bt_1)(A + Ct_2)\big] \\
&= E\big[A^2 + ACt_2 + BAt_1 + BCt_1t_2\big] \\
&= E[A^2] + E[AC]t_2 + E[BA]t_1 + E[BC]t_1t_2 \\
&= E[A^2] + E[A]E[C]t_2 + E[B]E[A]t_1 + E[B]E[C]t_1t_2, \qquad \text{(by independence)} \\
&= 2 + t_1 + t_2 + t_1t_2.
\end{aligned}$$

To find $C_{XY}(t_1, t_2)$ for $t_1, t_2 \in [0, \infty)$, we write

$$\begin{aligned}
C_{XY}(t_1, t_2) &= R_{XY}(t_1, t_2) - \mu_X(t_1)\mu_Y(t_2) \\
&= \big(2 + t_1 + t_2 + t_1t_2\big) - \big(1 + t_1\big)\big(1 + t_2\big) \\
&= 1.
\end{aligned}$$

Independent Random Processes:

We have seen independence for random variables. In particular, remember that random variables X_1, X_2,...,X_n are independent if, for all $(x_1, x_2, ..., x_n) \in \mathbb{R}^n$, we have

$$F_{X_1, X_2, ..., X_n}(x_1, x_2, ..., x_n) = F_{X_1}(x_1)F_{X_2}(x_2)...F_{X_n}(x_n).$$

Now, note that a random process is a collection of random variables. Thus, we can define the concept of independence for random processes, too. In particular, if for two random processes $X(t)$ and $Y(t)$, the random variables $X(t_i)$ are independent from the random variables $Y(t_j)$, we say that the two random processes are independent. More precisely, we have the following definition:

Two random processes $\{X(t), t \in J\}$ and $\{Y(t), t \in J'\}$ are said to be **independent** if, for all

$$t_1, t_2, \ldots, t_m \in J$$
$$\text{and}$$
$$t_1', t_2', \ldots, t_n' \in J',$$

the set of random variables

$$X(t_1), X(t_2), \cdots, X(t_m)$$

are independent of the set of random variables

$$Y(t_1'), Y(t_2'), \cdots, Y(t_n').$$

The above definition implies that for all real numbers $x_1, x_2, \cdots, x_m$ and $y_1, y_2, \cdots, y_n$, we have

$$F_{X(t_1), X(t_2), \cdots, X(t_m), Y(t_1'), Y(t_2'), \cdots, Y(t_n')}(x_1, x_2, \cdots, x_m, y_1, y_2, \cdots, y_n)$$
$$= F_{X(t_1), X(t_2), \cdots, X(t_m)}(x_1, x_2, \cdots, x_m) \cdot F_{Y(t_1'), Y(t_2'), \cdots, Y(t_n')}(y_1, y_2, \cdots, y_n).$$

The above equation might seem complicated; however, in many real-life applications we can often argue that two random processes are independent by looking at the problem structure. For example, in engineering we can reasonably assume that the thermal noise processes in two separate systems are independent. Note that if two random processes $X(t)$ and $Y(t)$ are independent, then their covariance function, $C_{XY}(t_1, t_2)$, for all t_1 and t_2 is given by

$$C_{XY}(t_1, t_2) = \text{Cov}\big(X(t_1), Y(t_2)\big)$$
$$= 0 \qquad \text{(since } X(t_1) \text{ and } Y(t_2) \text{ are independent).}$$

10.1.4   Stationary Processes

We can classify random processes based on many different criteria. One of the important questions that we can ask about a random process is whether it is a **stationary** process. Intuitively, a random process $\{X(t), t \in J\}$ is stationary if its statistical properties do not change by time. For example, for a stationary process, $X(t)$ and $X(t + \Delta)$ have the same probability distributions. In particular, we have

$$F_{X(t)}(x) = F_{X(t+\Delta)}(x), \qquad \text{for all } t, t + \Delta \in J.$$

More generally, for a stationary process, the joint distribution of $X(t_1)$ and $X(t_2)$ is the same as the joint distribution of $X(t_1 + \Delta)$ and $X(t_2 + \Delta)$. For example, if you have a stationary process $X(t)$, then

$$P\bigg(\Big(X(t_1), X(t_2)\Big) \in A\bigg) = P\bigg(\Big(X(t_1 + \Delta), X(t_2 + \Delta)\Big) \in A\bigg),$$

for any set $A \in \mathbb{R}^2$. In sum, a random process is stationary if *a time shift does not change its statistical properties*. Here is a formal definition of stationarity of continuous-time processes.

A continuous-time random process $\{X(t), t \in \mathbb{R}\}$ is **strict-sense stationary** or simply **stationary** if, for all $t_1, t_2, \cdots, t_r \in \mathbb{R}$ and all $\Delta \in \mathbb{R}$, the joint CDF of

$$X(t_1), X(t_2), \cdots, X(t_r)$$

is the same as the joint CDF of

$$X(t_1 + \Delta), X(t_2 + \Delta), \cdots, X(t_r + \Delta).$$

That is, for all real numbers $x_1, x_2, \cdots, x_r$, we have

$$F_{X(t_1)X(t_2)\cdots X(t_r)}(x_1, x_2, \cdots, x_r) = F_{X(t_1+\Delta)X(t_2+\Delta)\cdots X(t_r+\Delta)}(x_1, x_2, \cdots, x_r).$$

We can provide similar definition for discrete-time processes.

A discrete-time random process $\{X(n), n \in \mathbb{Z}\}$ is **strict-sense stationary** or simply **stationary**, if for all $n_1, n_2, \cdots, n_r \in \mathbb{Z}$ and all $D \in \mathbb{Z}$, the joint CDF of

$$X(n_1), X(n_2), \cdots, X(n_r)$$

is the same as the joint CDF of

$$X(n_1 + D), X(n_2 + D), \cdots, X(n_r + D).$$

That is, for all real numbers $x_1, x_2, \cdots, x_r$, we have

$$F_{X(n_1)X(n_2)\cdots X(n_r)}(x_1, x_2, \cdots, x_n) = F_{X(n_1+D)X(n_2+D)\cdots X(n_r+D)}(x_1, x_2, \cdots, x_r).$$

Example 10.7. Consider the discrete-time random process $\{X(n), n \in \mathbb{Z} \cdots\}$, in which the $X(n)$'s are i.i.d. with CDF $F_{X(n)}(x) = F(x)$. Show that this is a (strict-sense) stationary process.

Solution: Intuitively, since $X(n)$'s are i.i.d., we expect that as time evolves the probabilistic behavior of the process does not change. Therefore, this must be a stationary process. To show this rigorously, we can argue as follows. For all real numbers $x_1, x_2, \cdots, x_r$ and all distinct integers $n_1, n_2, \cdots, n_r$, we have

$$
\begin{aligned}
F_{X(n_1)X(n_2)\cdots X(n_r)}&(x_1, x_2, \cdots, x_r) \\
&= F_{X(n_1)}(x_1)F_{X(n_2)}(x_2)\cdots F_{X(n_r)}(x_r) \quad \text{(since the } X(n_i)\text{'s are independent)} \\
&= F(x_1)F(x_2)\cdots F(x_r) \quad \text{(since } F_{X(n_i)}(x) = F(x)).
\end{aligned}
$$

We also have

$$
\begin{aligned}
F_{X(n_1+D)X(n_2+D)\cdots X(n_r+D)}&(x_1, x_2, \cdots, x_r) \\
&= F_{X(n_1+D)}(x_1)F_{X(n_2+D)}(x_2)\cdots F_{X(n_r+D)}(x_r) \quad \text{(since the } X(n_i + D)\text{'s are independent)} \\
&= F(x_1)F(x_2)\cdots F(x_n) \quad \text{(since } F_{X(n_i+D)}(x) = F(x)).
\end{aligned}
$$

In practice, it is desirable if a random process $X(t)$ is stationary. In particular, if a process is stationary, then its analysis is usually simpler as the probabilistic properties do not change by time. For example, suppose that you need to do forecasting about the future of a process $X(t)$. If you know the process is stationary, you can observe the past, which will normally give you a lot of information about how the process will behave in the future.

However, it turns out that many real-life processes are not strict-sense stationary. Even if a process is strict-sense stationary, it might be difficult to prove it. Fortunately, it is often enough to show a "weaker" form of stationarity than the one defined above.

Weak-Sense Stationary Processes:

Here, we define one of the most common forms of stationarity that is widely used in practice. A random process is called **weak-sense stationary** or **wide-sense stationary** (**WSS**) if its mean function and its correlation function do not change by shifts in time. More precisely, $X(t)$ is WSS if, for all $t_1, t_2 \in \mathbb{R}$ and all $\Delta \in \mathbb{R}$,

1. $E[X(t_1)] = E[X(t_2)]$,

2. $E[X(t_1)X(t_2)] = E[X(t_1 + \Delta)X(t_2 + \Delta)]$.

Note that the first condition states that the mean function $\mu_X(t)$ is not a function of time, t, thus we can write $\mu_X(t) = \mu_X$. The second condition states that the correlation function $R_X(t_1, t_2)$ is only a function of $\tau = t_1 - t_2$, and not t_1 and t_2 individually. Thus, we can write $R_X(t_1, t_2) = R_X(t_1 - t_2) = R_X(\tau)$. Therefore, we can provide the following definition.

A continuous-time random process $\{X(t), t \in \mathbb{R}\}$ is **weak-sense stationary** or **wide-sense stationary** (**WSS**) if

1. $\mu_X(t) = \mu_X$, for all $t \in \mathbb{R}$,

2. $R_X(t_1, t_2) = R_X(t_1 - t_2)$, for all $t_1, t_2 \in \mathbb{R}$.

We can provide a similar definition for discrete-time WSS processes.

A discrete-time random process $\{X(n), n \in \mathbb{Z}\}$ is **weak-sense stationary** or **wide-sense stationary (WSS)** if

1. $\mu_X(n) = \mu_X$, for all $n \in \mathbb{Z}$,

2. $R_X(n_1, n_2) = R_X(n_1 - n_2)$, for all $n_1, n_2 \in \mathbb{Z}$.

Example 10.8. Consider the random process $\{X(t), t \in \mathbb{R}\}$ defined as

$$X(t) = \cos(t + U),$$

where $U \sim Uniform(0, 2\pi)$. Show that $X(t)$ is a WSS process.

Solution: We need to check two conditions:

1. $\mu_X(t) = \mu_X$, for all $t \in \mathbb{R}$, and

2. $R_X(t_1, t_2) = R_X(t_1 - t_2)$, for all $t_1, t_2 \in \mathbb{R}$.

We have

$$\begin{aligned}
\mu_X(t) &= E[X(t)] \\
&= E[\cos(t + U)] \\
&= \int_0^{2\pi} \cos(t + u) \frac{1}{2\pi} \, du \\
&= 0, \qquad \text{for all } t \in \mathbb{R}.
\end{aligned}$$

We can also find $R_X(t_1, t_2)$ as follows

$$\begin{aligned}
R_X(t_1, t_2) &= E[X(t_1)X(t_2)] \\
&= E[\cos(t_1 + U)\cos(t_2 + U)] \\
&= E\left[\frac{1}{2}\cos(t_1 + t_2 + 2U) + \frac{1}{2}\cos(t_1 - t_2)\right] \\
&= E\left[\frac{1}{2}\cos(t_1 + t_2 + 2U)\right] + E\left[\frac{1}{2}\cos(t_1 - t_2)\right] \\
&= \int_0^{2\pi} \cos(t_1 + t_2 + u)\frac{1}{2\pi} \, du + \frac{1}{2}\cos(t_1 - t_2) \\
&= 0 + \frac{1}{2}\cos(t_1 - t_2) \\
&= \frac{1}{2}\cos(t_1 - t_2), \qquad \text{for all } t_1, t_2 \in \mathbb{R}.
\end{aligned}$$

As we see, both conditions are satisfied, thus $X(t)$ is a WSS process.

Since for WSS random processes, $R_X(t_1, t_2) = R_X(t_1 - t_2)$, we usually denote the correlation function by $R_X(\tau)$, where $\tau = t_1 - t_2$. Thus, for a WSS process, we can write

$$R_X(\tau) = E[X(t)X(t - \tau)] = E[X(t + \tau)X(t)] \tag{10.1}$$

As we will see in Section 10.2, $R_X(\tau)$ is a very useful tool when we do frequency domain analysis. Here, we would like to study some properties of $R_X(\tau)$ for WSS signals. Let $\{X(t), t \in \mathbb{R}\}$ be a WSS process with correlation function $R_X(\tau)$. Then, we can write

$$R_X(0) = E[X(t)^2].$$

The quantity $E[X(t)^2]$ is called the **expected (average) power** in $X(t)$ at time t. For a WSS process, the expected power is not a function of time. Since $X(t)^2 \geq 0$, we conclude that $R_X(0) \geq 0$.

$$R_X(0) = E[X(t)^2] \geq 0$$

Next, let's consider $R_X(-\tau)$. We have

$$
\begin{aligned}
R_X(-\tau) &= E[X(t)X(t + \tau)] && \text{(by definition (Equation 10.1))} \\
&= E[X(t + \tau)X(t)] \\
&= R_X(\tau) && \text{(Equation 10.1)}
\end{aligned}
$$

Thus, we conclude that $R_X(\tau)$ is an **even function**.

$$R_X(\tau) = R_X(-\tau), \qquad \text{for all } \tau \in \mathbb{R}.$$

Finally, we would like to show that $R_X(\tau)$ takes its maximum value at $\tau = 0$. That is, $X(t)$ and $X(t + \tau)$ have the highest correlation when $\tau = 0$.

$$|R_X(\tau)| \leq R_X(0), \qquad \text{for all } \tau \in \mathbb{R}.$$

The proof can be done using the Cauchy-Schwarz inequality: For any two random variables X and Y, we have

$$|EXY| \leq \sqrt{E[X^2]E[Y^2]},$$

where equality holds if and only if $X = \alpha Y$ for some constant $\alpha \in \mathbb{R}$. Now, if we choose $X = X(t)$ and $Y = X(t - \tau)$, we obtain

$$\begin{aligned}
|E[X(t)X(t-\tau)]| &\leq \sqrt{E[X(t)^2]E[X(t-\tau)^2]} \\
&= \sqrt{R_X(0)R_X(0)} \\
&= R_X(0).
\end{aligned}$$

Therefore, we conclude that $|R_X(\tau)| \leq R_X(0)$. Considering these properties, Figure 10.4 shows some possible shapes for $R_X(\tau)$.

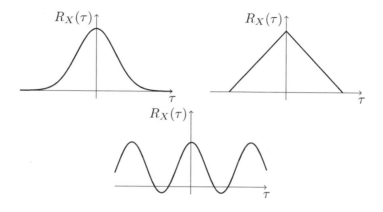

Figure 10.4: Some possible shapes for $R_X(\tau)$.

Jointly Wide-Sense Stationary Processes:

We often work with multiple random processes, so we extend the concept of wide-sense stationarity to more than one process. More specifically, we can talk about *jointly wide-sense stationary* processes.

Two random processes $\{X(t), t \in \mathbb{R}\}$ and $\{Y(t), t \in \mathbb{R}\}$ are said to be **jointly wide-sense stationary** if

1. $X(t)$ and $Y(t)$ are each wide-sense stationary.

2. $R_{XY}(t_1, t_2) = R_{XY}(t_1 - t_2)$.

Example 10.9. Let $X(t)$ and $Y(t)$ be two jointly WSS random processes. Consider the random process $Z(t)$ defined as

$$Z(t) = X(t) + Y(t).$$

Show that $Z(t)$ is WSS.

Solution: Since $X(t)$ and $Y(t)$ are jointly WSS, we conclude

1. $\mu_X(t) = \mu_X$, $\mu_Y(t) = \mu_Y$,

2. $R_X(t_1, t_2) = R_X(t_1 - t_2)$, $R_Y(t_1, t_2) = R_Y(t_1 - t_2)$,

3. $R_{XY}(t_1, t_2) = R_{XY}(t_1 - t_2)$.

Therefore, we have

$$\begin{aligned} \mu_Z(t) &= E[X(t) + Y(t)] \\ &= E[X(t)] + E[Y(t)] \\ &= \mu_X + \mu_Y. \end{aligned}$$

$$\begin{aligned} R_Z(t_1, t_2) &= E\left[\left(X(t_1) + Y(t_1)\right)\left(X(t_2) + Y(t_2)\right)\right] \\ &= E[X(t_1)X(t_2)] + E[X(t_1)Y(t_2)] + E[Y(t_1)X(t_2)]E[Y(t_1)Y(t_2)] \\ &= R_X(t_1 - t_2) + R_{XY}(t_1 - t_2) + R_{YX}(t_1 - t_2) + R_Y(t_1 - t_2). \end{aligned}$$

Cyclostationary Processes:

Some practical random processes have a periodic structure. That is, the statistical properties are repeated every T units of time (e.g., every T seconds). In other words, the random variables

$$X(t_1), X(t_2), \cdots, X(t_r)$$

have the same joint CDF as the random variables

$$X(t_1 + T), X(t_2 + T), \cdots, X(t_r + T).$$

Such random variables are called **cyclostationary**. For example, consider the random process $\left\{X(t), t \in \mathbb{R}\right\}$ defined as

$$X(t) = A\cos(\omega t),$$

where A is a random variable. Here, we have

$$\begin{aligned} X\left(t + \frac{2\pi}{\omega}\right) &= A\cos(\omega t + 2\pi) \\ &= A\cos(\omega t) = X(t). \end{aligned}$$

We conclude $X(t)$ is in fact a periodic signal with period $T = \frac{2\pi}{\omega}$. Therefore, the statistical properties of $X(t)$ do not change by shifting the time by T units, so $X(t)$ is a cyclostationary random process with period $T = \frac{2\pi}{\omega}$. Similarly, we can define wide-sense cyclostationary random processes.

A continuous-time random process $\{X(t), t \in \mathbb{R}\}$ is **cyclostationary** if there exists a positive real number T such that, for all $t_1, t_2, \cdots, t_r \in \mathbb{R}$, the joint CDF of

$$X(t_1), X(t_2), \cdots, X(t_r)$$

is the same as the joint CDF of

$$X(t_1 + T), X(t_2 + T), \cdots, X(t_r + T).$$

A continuous-time random process $\{X(t), t \in \mathbb{R}\}$ is **weak-sense cyclostationary** or **wide-sense cyclostationary** if there exists a positive real number T such that

1. $\mu_X(t + T) = \mu_X(t)$, for all $t \in \mathbb{R}$;

2. $R_X(t_1 + T, t_2 + T) = R_X(t_1, t_2)$, for all $t_1, t_2 \in \mathbb{R}$.

Similarly, you can define cyclostationary discrete-time processes. For example, a discrete-time random process $\{X(n), n \in \mathbb{Z}\}$ is wide-sense cyclostationary if there exists $M \in \mathbb{N}$ such that

1. $\mu_X(n + M) = \mu_X$, for all $n \in \mathbb{Z}$;

2. $R_X(n_1 + M, n_2 + M) = R_X(n_1, n_2)$, for all $n_1, n_2 \in \mathbb{Z}$.

Derivatives and Integrals of Random Processes:

Many real-life systems are described by differential equations. To analyze such systems when randomness is involved, we often need to differentiate or integrate the random processes that are present in the system. You have seen concepts such as continuity, differentiability, and integrability in calculus for deterministic signals (deterministic functions). Here, we need to extend those concepts to random processes. Without going much into mathematical technicalities, here we would like to provide some guidelines on how to deal with derivatives and integrals of random processes.

Let $X(t)$ be a continuous-time random process. We say that $X(t)$ is **mean-square continuous** at time t if

$$\lim_{\delta \to 0} E\left[\left| X(t + \delta) - X(t) \right|^2 \right] = 0.$$

Note that mean-square continuity does not mean that every possible realization of $X(t)$ is a continuous function. It roughly means that the difference $X(t + \delta) - X(t)$ is small on average.

Example 10.10. The Poisson process is discussed in detail in Chapter 11. If $X(t)$ is a Poisson process with intensity λ, then for all $t > s \geq 0$, we have

$$X(t) - X(s) \sim Poisson\big(\lambda(t - s)\big).$$

Show that $X(t)$ is mean-square continuous at any time $t \geq 0$.

Solution: We have

$$X(t + \delta) - X(t) \sim Poisson(\lambda\delta).$$

Thus,

$$\lim_{\delta \to 0} E[|X(t + \delta) - X(t)|^2] = \lim_{\delta \to 0} \lambda\delta + (\lambda\delta)^2$$
$$= 0.$$

It is worth noting that there are jumps in a Poisson process; however, those jumps are not very "dense" in time, so the random process is still continuous in the mean-square sense. Figure 10.5 shows a possible realization of a Poisson process.

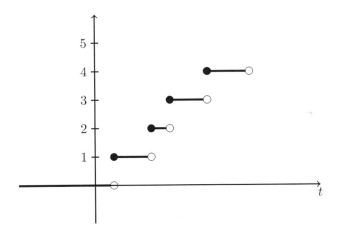

Figure 10.5: A possible sample function of a Poisson process.

We can similarly talk about mean-square differentiability and mean-square integrability. If $X(t)$ is a random process, the derivative of $X(t)$,

$$Y(t) = \frac{d}{dt}X(t),$$

is also a random process. For nice and smooth processes, the derivative can be obtained in a natural way. For example, if you have a random process defined as

$$X(t) = A + Bt + Ct^2, \qquad \text{for all } t \in [0, \infty),$$

where A, B, and C are random variables, then the derivative of $X(t)$ can be written as

$$X'(t) = B + 2Ct, \qquad \text{for all } t \in [0, \infty).$$

Without trying to go much into mathematical technicalities, here we would like to provide some guidelines on how to deal with derivatives and integrals of random processes (assuming some mild regularity conditions are satisfied). A key point to note is that differentiation and integration are linear operations. This, for example, means that you can often interchange integration and expectation. More specifically, you can write

$$E\left[\int_0^t X(u)du\right] = \int_0^t E[X(u)]du.$$

Similarly, if the derivative of $X(t)$ is well-defined, we can write

$$E\left[\frac{d}{dt}X(t)\right] = \frac{d}{dt}E[X(t)].$$

Example 10.11. Consider a random process $X(t)$ and its derivative, $X'(t) = \frac{d}{dt}X(t)$. Assuming that the derivatives are well-defined, show that

$$R_{XX'}(t_1, t_2) = \frac{\partial}{\partial t_2}R_X(t_1, t_2).$$

Solution: We have

$$
\begin{aligned}
R_{XX'}(t_1, t_2) &= E[X(t_1)X'(t_2)] \\
&= E\left[X(t_1)\frac{d}{dt_2}X(t_2)\right] \\
&= E\left[\frac{\partial}{\partial t_2}\Big(X(t_1)X(t_2)\Big)\right] \\
&= \frac{\partial}{\partial t_2}E\big[X(t_1)X(t_2)\big] \\
&= \frac{\partial}{\partial t_2}R_X(t_1, t_2).
\end{aligned}
$$

10.1.5   Gaussian Random Processes

Here, we will briefly introduce normal (Gaussian) random processes. We will discuss some examples of Gaussian processes in more detail later on. Many important practical random processes are subclasses of normal random processes.

First, let us remember a few facts about Gaussian random vectors. As we saw before, random variables X_1, X_2,..., X_n are said to be **jointly normal** if, for all $a_1, a_2,..., a_n \in \mathbb{R}$, the random variable

$$a_1 X_1 + a_2 X_2 + ... + a_n X_n$$

is a normal random variable. Also, a random vector

$$\mathbf{X} = \begin{bmatrix} X_1 \\ X_2 \\ \vdots \\ X_n \end{bmatrix}$$

is said to be **normal** or **Gaussian** if the random variables X_1, X_2,..., X_n are jointly normal. An important property of jointly normal random variables is that their joint PDF is completely determined by their mean and covariance matrices. More specifically, for a normal random vector $\mathbf{X}$ with mean $\mathbf{m}$ and covariance matrix $\mathbf{C}$, the PDF is given by

$$f_{\mathbf{X}}(\mathbf{x}) = \frac{1}{(2\pi)^{\frac{n}{2}}\sqrt{\det \mathbf{C}}} \exp\left\{-\frac{1}{2}(\mathbf{x}-\mathbf{m})^T \mathbf{C}^{-1}(\mathbf{x}-\mathbf{m})\right\}.$$

Now, let us define Gaussian random processes.

A random process $\{X(t), t \in J\}$ is said to be a **Gaussian (normal) random process** if, for all

$$t_1, t_2, \ldots, t_n \in J,$$

the random variables $X(t_1)$, $X(t_2)$,..., $X(t_n)$ are jointly normal.

Example 10.12. Let $X(t)$ be a zero-mean WSS Gaussian process with $R_X(\tau) = e^{-\tau^2}$, for all $\tau \in \mathbb{R}$.

1. Find $P\big(X(1) < 1\big)$.

2. Find $P\big(X(1) + X(2) < 1\big)$.

Solution:

1. $X(1)$ is a normal random variable with mean $E[X(1)] = 0$ and variance

$$\begin{aligned}\text{Var}\big(X(1)\big) &= E[X(1)^2] \\ &= R_X(0) = 1.\end{aligned}$$

Thus,

$$\begin{aligned}P\big(X(1) < 1\big) &= \Phi\left(\frac{1-0}{1}\right) \\ &= \Phi(1) \approx 0.84\end{aligned}$$

2. Let $Y = X(1) + X(2)$. Then, Y is a normal random variable. We have

$$\begin{aligned}EY &= E[X(1)] + E[X(2)] \\ &= 0;\end{aligned}$$

$$\text{Var}(Y) = \text{Var}\big(X(1)\big) + \text{Var}\big(X(2)\big) + 2\text{Cov}\big(X(1), X(2)\big).$$

Note that

$$\begin{aligned}\text{Var}\big(X(1)\big) &= E[X(1)^2] - E[X(1)]^2 \\ &= R_X(0) - \mu_X^2 \\ &= 1 - 0 = 1 = \text{Var}\big(X(2)\big);\end{aligned}$$

$$\text{Cov}\big(X(1), X(2)\big) = E[X(1)X(2)] - E[X(1)]E[X(2)]$$
$$= R_X(-1) - \mu_X^2$$
$$= e^{-1} - 0 = \frac{1}{e}.$$

Therefore,

$$\text{Var}(Y) = 2 + \frac{2}{e}.$$

We conclude $Y \sim N(0, 2 + \frac{2}{e})$. Thus,

$$P\big(Y < 1\big) = \Phi\left(\frac{1 - 0}{\sqrt{2 + \frac{2}{e}}}\right)$$
$$= \Phi(0.6046) \approx 0.73$$

An important property of normal random processes is that wide-sense stationarity and strict-sense stationarity are equivalent for these processes. More specifically, we can state the following theorem.

Theorem 10.1. Consider the Gaussian random processes $\{X(t), t \in \mathbb{R}\}$. If $X(t)$ is WSS, then $X(t)$ is a stationary process.

Proof. We need to show that, for all $t_1, t_2, \cdots, t_r \in \mathbb{R}$ and all $\Delta \in \mathbb{R}$, the joint CDF of

$$X(t_1), X(t_2), \cdots, X(t_r)$$

is the same as the joint CDF of

$$X(t_1 + \Delta), X(t_2 + \Delta), \cdots, X(t_r + \Delta).$$

Since these random variables are jointly Gaussian, it suffices to show that the mean vectors and the covariance matrices are the same. To see this, note that $X(t)$ is a WSS process, so

$$\mu_X(t_i) = \mu_X(t_j) = \mu_X, \qquad \text{for all } i, j,$$

and

$$C_X(t_i + \Delta, t_j + \Delta) = C_X(t_i, t_j) = C_X(t_i - t_j), \qquad \text{for all } i, j.$$

From the above, we conclude that the mean vector and the covariance matrix of

$$X(t_1), X(t_2), \cdots, X(t_r)$$

is the same as the mean vector and the covariance matrix of

$$X(t_1 + \Delta), X(t_2 + \Delta), \cdots, X(t_r + \Delta).$$

■

Similarly, we can define jointly Gaussian random processes.

Two random processes $\{X(t), t \in J\}$ and $\{Y(t), t \in J'\}$ are said to be **jointly Gaussian (normal)**, if for all

$$t_1, t_2, \ldots, t_m \in J$$
$$\text{and}$$
$$t_1', t_2', \ldots, t_n' \in J',$$

the random variables

$$X(t_1), X(t_2), \cdots, X(t_m), Y(t_1'), Y(t_2'), \cdots, Y(t_n')$$

are jointly normal.

Note that from the properties of jointly normal random variables, we can conclude that if two jointly Gaussian random processes $X(t)$ and $Y(t)$ are uncorrelated, i.e.,

$$C_{XY}(t_1, t_2) = 0, \qquad \text{for all } t_1, t_2,$$

then $X(t)$ and $Y(t)$ are two independent random processes.

10.1.6 Solved Problems

1. Let Y_1, Y_2, Y_3, $\cdots$ be a sequence of i.i.d. random variables with mean $EY_i = 0$ and $\text{Var}(Y_i) = 4$. Define the discrete-time random process $\{X(n), n \in \mathbb{N}\}$ as

$$X(n) = Y_1 + Y_2 + \cdots + Y_n, \qquad \text{for all } n \in \mathbb{N}.$$

Find $\mu_X(n)$ and $R_X(m, n)$, for all $n, m \in \mathbb{N}$.

Solution: We have

$$\begin{aligned}
\mu_X(n) &= E[X(n)] \\
&= E[Y_1 + Y_2 + \cdots + Y_n] \\
&= E[Y_1] + E[Y_2] + \cdots + E[Y_n] \\
&= 0.
\end{aligned}$$

Let $m \leq n$, then

$$
\begin{aligned}
R_X(m, n) &= E[X(m)X(n)] \\
&= E\left[X(m)\big(X(m) + Y_{m+1} + Y_{m+2} + \cdots + Y_n\big)\right] \\
&= E[X(m)^2] + E[X(m)]E[Y_{m+1} + Y_{m+2} + \cdots + Y_n] \\
&= E[X(m)^2] + 0 \\
&= \mathrm{Var}\big(X(m)\big) \\
&= \mathrm{Var}\big(Y_1\big) + \mathrm{Var}\big(Y_2\big) + \cdots + \mathrm{Var}\big(Y_m\big) \\
&= 4m.
\end{aligned}
$$

Similarly, for $m \geq n$, we have

$$
\begin{aligned}
R_X(m, n) &= E[X(m)X(n)] \\
&= 4n.
\end{aligned}
$$

We conclude

$$
R_X(m, n) = 4 \min(m, n).
$$

2. For any $k \in \mathbb{Z}$, define the function $g_k(t)$ as

$$
g_k(t) = \begin{cases} 1 & k < t \leq k+1 \\ 0 & \text{otherwise} \end{cases}
$$

Now, consider the continuous-time random process $\{X(t), t \in \mathbb{R}\}$ defined as

$$
X(t) = \sum_{k=-\infty}^{+\infty} A_k g_k(t),
$$

where A_1, A_2, $\cdots$ are i.i.d. random variables with $EA_k = 1$ and $\mathrm{Var}(A_k) = 1$. Find $\mu_X(t)$, $R_X(s, t)$, and $C_X(s, t)$ for all $s, t \in \mathbb{R}$.

Solution: Note that, for any $k \in \mathbb{Z}$, $g(t) = 0$ outside of the interval $(k, k+1]$. Thus, if $k < t \leq k+1$, we can write

$$
X(t) = A_k.
$$

Thus,

$$
\begin{aligned}
\mu_X(t) &= E[X(t)] \\
&= E[A_k] = 1.
\end{aligned}
$$

So, $\mu_X(t) = 1$ for all $t \in \mathbb{R}$.

Now consider two real numbers s and t. If for some $k \in \mathbb{Z}$, we have

$$k < s, t \leq k+1,$$

then

$$
\begin{aligned}
R_X(s,t) &= E[X(s)X(t)] \\
&= E[A_k^2] = 1 + 1 = 2.
\end{aligned}
$$

On the other hand, if s and t are in two different subintervals of $\mathbb{R}$, that is if

$$k < s \leq k+1, \qquad \text{and} \qquad l < t \leq l+1,$$

where k and l are two different integers, then

$$
\begin{aligned}
R_X(s,t) &= E[X(s)X(t)] \\
&= E[A_k A_l] = E[A_k]E[A_l] = 1.
\end{aligned}
$$

To find $C_X(s,t)$, note that if

$$k < s, t \leq k+1,$$

then

$$
\begin{aligned}
C_X(s,t) &= R_X(s,t) - E[X(s)]E[X(t)] \\
&= 2 - 1 \cdot 1 = 1.
\end{aligned}
$$

On the other hand, if

$$k < s \leq k+1, \qquad \text{and} \qquad l < t \leq l+1,$$

where k and l are two different integers, then

$$
\begin{aligned}
C_X(s,t) &= R_X(s,t) - E[X(s)]E[X(t)] \\
&= 1 - 1 \cdot 1 = 0.
\end{aligned}
$$

3. Let $X(t)$ be a continuous-time WSS process with mean $\mu_X = 1$ and

$$
R_X(\tau) = \begin{cases} 3 - |\tau| & -2 \leq \tau \leq 2 \\ 1 & \text{otherwise} \end{cases}
$$

(a) Find the expected power in $X(t)$.

(b) Find $E\left[\left(X(1) + X(2) + X(3)\right)^2\right]$.

Solution:

(a) The expected power in $X(t)$ at time t is $E[X(t)^2]$, which is given by

$$R_X(0) = 3.$$

(b) We have

$$E\left[\left(X(1) + X(2) + X(3)\right)^2\right] = E\Big[X(1)^2 + X(2)^2 + X(3)^2$$

$$+\, 2X(1)X(2) + 2X(1)X(3) + 2X(2)X(3)\Big]$$

$$= 3R_X(0) + 2R_X(-1) + 2R_X(-2) + 2R_X(-1)$$
$$= 3 \cdot 3 + 2 \cdot 2 + 2 \cdot 1 + 2 \cdot 2$$
$$= 19.$$

4. Let $X(t)$ be a continuous-time WSS process with mean $\mu_X = 0$ and

$$R_X(\tau) = \delta(\tau),$$

where $\delta(\tau)$ is the Dirac delta function. We define the random process $Y(t)$ as

$$Y(t) = \int_{t-2}^{t} X(u)du.$$

(a) Find $\mu_Y(t) = E[Y(t)]$.
(b) Find $R_{XY}(t_1, t_2)$.

Solution:

(a) We have

$$\mu_Y(t) = E\left[\int_{t-2}^{t} X(u)du\right]$$

$$= \int_{t-2}^{t} E[X(u)] \ du$$

$$= \int_{t-2}^{t} 0 \ du$$

$$= 0.$$

(b)

$$R_{XY}(t_1, t_2) = E\left[X(t_1) \int_{t_2-2}^{t_2} X(u)du\right]$$

$$= E\left[\int_{t_2-2}^{t_2} X(t_1)X(u)du\right]$$

$$= \int_{t_2-2}^{t_2} R_X(t_1 - u) \ du$$

$$= \int_{t_2-2}^{t_2} \delta(t_1 - u) \ du$$

$$= \begin{cases} 1 & t_2 - 2 < t_1 < t_2 \\ \\ 0 & \text{otherwise} \end{cases}$$

5. Let $X(t)$ be a Gaussian process with $\mu_X(t) = t$, and $R_X(t_1, t_2) = 1 + 2t_1t_2$, for all $t, t_1, t_2 \in \mathbb{R}$. Find $P(2X(1) + X(2) < 3)$.

Solution: Let $Y = 2X(1) + X(2)$. Then, Y is a normal random variable. We have

$$EY = 2E[X(1)] + E[X(2)]$$
$$= 2 \cdot 1 + 2 = 4.$$

$$\text{Var}(Y) = 4\text{Var}(X(1)) + \text{Var}(X(2)) + 4\text{Cov}(X(1), X(2)).$$

Note that

$$\text{Var}(X(1)) = E[X(1)^2] - E[X(1)]^2$$
$$= R_X(1, 1) - \mu_X(1)^2$$
$$= 1 + 2 \cdot 1 \cdot 1 - 1 = 2.$$

$$\text{Var}(X(2)) = E[X(2)^2] - E[X(2)]^2$$
$$= R_X(2, 2) - \mu_X(2)^2$$
$$= 1 + 2 \cdot 2 \cdot 2 - 4 = 5.$$

$$\text{Cov}(X(1), X(2)) = E[X(1)X(2)] - E[X(1)]E[X(2)]$$
$$= R_X(1, 2) - \mu_X(1)\mu_X(2)$$
$$= 1 + 2 \cdot 1 \cdot 2 - 1 \cdot 2 = 3.$$

Therefore,

$$\text{Var}(Y) = 4 \cdot 2 + 5 + 4 \cdot 3 = 25.$$

We conclude $Y \sim N(4, 25)$. Thus,

$$P(Y < 3) = \Phi\left(\frac{3-4}{5}\right)$$
$$= \Phi(-0.2) \approx 0.42$$

10.2   Processing of Random Signals

In this section, we will study what happens when a WSS random signal passes through a *linear time-invariant (LTI)* system. Such scenarios are encountered in many real-life systems, specifically in communications and signal processing. A main result of this section is that if the input to an LTI system is a WSS process, then the output is also a WSS process. Moreover, the input and output are jointly WSS. To better understand these systems, we will discuss the study of random signals in the frequency domain.

10.2.1   Power Spectral Density

So far, we have studied random processes in the time domain. It is often very useful to study random processes in the frequency domain as well. To do this, we need to use the **Fourier** transform. Here, we will assume that you are familiar with the Fourier transform. A brief review of the Fourier transform and its properties is given in the appendix.

Consider a WSS random process $X(t)$ with autocorrelation function $R_X(\tau)$. We define the *Power Spectral Density (PSD)* of $X(t)$ as the Fourier transform of $R_X(\tau)$. We show the PSD of $X(t)$, by $S_X(f)$. More specifically, we can write

$$S_X(f) = \mathcal{F}\{R_X(\tau)\} = \int_{-\infty}^{\infty} R_X(\tau) e^{-2j\pi f\tau} \, d\tau,$$

where $j = \sqrt{-1}$.

Power Spectral Density (PSD)

$$S_X(f) = \mathcal{F}\{R_X(\tau)\} = \int_{-\infty}^{\infty} R_X(\tau) e^{-2j\pi f\tau} \, d\tau, \qquad \text{where } j = \sqrt{-1}.$$

From this definition, we can conclude that $R_X(\tau)$ can be obtained by the inverse Fourier transform of $S_X(f)$. That is

$$R_X(\tau) = \mathcal{F}^{-1}\{S_X(f)\} = \int_{-\infty}^{\infty} S_X(f)e^{2j\pi f\tau} \, df.$$

As we have seen before, if $X(t)$ is a real-valued random process, then $R_X(\tau)$ is an even, real-valued function of τ. From the properties of the Fourier transform, we conclude that $S_X(f)$ is also real-valued and an even function of f. Also, from what we will discuss later on, we can conclude that $S_X(f)$ is non-negative for all f.

$$S_X(-f) = S_X(f), \text{ for all } f;$$
$$S_X(f) \geq 0, \text{ for all } f.$$

Before going any further, let's try to understand the idea behind the PSD. To do so, let's choose $\tau = 0$. We know that expected power in $X(t)$ is given by

$$E[X(t)^2] = R_X(0) = \int_{-\infty}^{\infty} S_X(f)e^{2j\pi f\cdot 0} \, df$$
$$= \int_{-\infty}^{\infty} S_X(f) \, df.$$

We conclude that the expected power in $X(t)$ can be obtained by integrating the PSD of $X(t)$. This fact helps us to understand why $S_X(f)$ is called the power spectral density. In fact, as we will see shortly, we can find the expected power of $X(t)$ in a specific frequency range by integrating the PSD over that specific range.

The expected power in $X(t)$ can be obtained as

$$E[X(t)^2] = R_X(0) = \int_{-\infty}^{\infty} S_X(f) \, df.$$

Example 10.13. Consider a WSS random process $X(t)$ with

$$R_X(\tau) = e^{-a|\tau|},$$

where a is a positive real number. Find the PSD of $X(t)$.

Solution: We need to find the Fourier transform of $R_X(\tau)$. We can do this by looking at a

Fourier transform table or by finding the Fourier transform directly as follows.

$$S_X(f) = \mathcal{F}\{R_X(\tau)\}$$

$$= \int_{-\infty}^{\infty} e^{-a|\tau|} e^{-2j\pi f\tau} \ d\tau$$

$$= \int_{-\infty}^{0} e^{a\tau} e^{-2j\pi f\tau} \ d\tau + \int_{0}^{\infty} e^{-a\tau} e^{-2j\pi f\tau} \ d\tau$$

$$= \frac{1}{a - j2\pi f} + \frac{1}{a + j2\pi f}$$

$$= \frac{2a}{a^2 + 4\pi^2 f^2}.$$

Cross Spectral Density:

For two jointly WSS random processes $X(t)$ and $Y(t)$, we define the *cross spectral density* $S_{XY}(f)$ as the Fourier transform of the cross-correlation function $R_{XY}(\tau)$,

$$S_{XY}(f) = \mathcal{F}\{R_{XY}(\tau)\} = \int_{-\infty}^{\infty} R_{XY}(\tau)e^{-2j\pi f\tau} \ d\tau.$$

10.2.2   Linear Time-Invariant (LTI) Systems with Random Inputs

Linear Time-Invariant (LTI) Systems:

A **linear time-invariant (LTI)** system can be represented by its **impulse response** (Figure 10.6). More specifically, if $X(t)$ is the input signal to the system, the output, $Y(t)$, can be written as

$$Y(t) = \int_{-\infty}^{\infty} h(\alpha)X(t - \alpha) \ d\alpha = \int_{-\infty}^{\infty} X(\alpha)h(t - \alpha) \ d\alpha.$$

The above integral is called the *convolution* of h and X, and we write

$$Y(t) = h(t) * X(t) = X(t) * h(t).$$

Note that as the name suggests, the impulse response can be obtained if the input to the system is chosen to be the unit impulse function (delta function) $x(t) = \delta(t)$. For discrete-time systems, the output can be written as (Figure 10.6)

$$Y(n) = h(n) * X(n) = X(n) * h(n)$$

$$= \sum_{k=-\infty}^{\infty} h(k)X(n - k) = \sum_{k=-\infty}^{\infty} X(k)h(n - k).$$

The discrete-time unit impulse function is defined as

$$\delta(n) = \begin{cases} 1 & n = 0 \\ 0 & \text{otherwise} \end{cases}$$

For the rest of this chapter, we mainly focus on continuous-time signals.

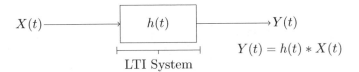

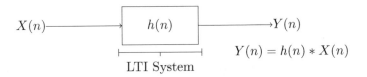

Figure 10.6: LTI systems.

LTI Systems with Random Inputs:

Consider an LTI system with impulse response $h(t)$. Let $X(t)$ be a WSS random process. If $X(t)$ is the input of the system, then the output, $Y(t)$, is also a random process. More specifically, we can write

$$Y(t) = h(t) * X(t)$$
$$= \int_{-\infty}^{\infty} h(\alpha)X(t - \alpha) \ d\alpha.$$

Here, our goal is to show that $X(t)$ and $Y(t)$ are jointly WSS processes. Let's first start by calculating the mean function of $Y(t)$, $\mu_Y(t)$. We have

$$\mu_Y(t) = E[Y(t)] = E\left[\int_{-\infty}^{\infty} h(\alpha)X(t - \alpha) \ d\alpha\right]$$
$$= \int_{-\infty}^{\infty} h(\alpha)E[X(t - \alpha)] \ d\alpha$$
$$= \int_{-\infty}^{\infty} h(\alpha)\mu_X \ d\alpha$$
$$= \mu_X \int_{-\infty}^{\infty} h(\alpha) \ d\alpha.$$

We note that $\mu_Y(t)$ is not a function of t, so we can write

$$\mu_Y(t) = \mu_Y = \mu_X \int_{-\infty}^{\infty} h(\alpha) \ d\alpha.$$

Let's next find the cross-correlation function, $R_{XY}(t_1, t_2)$. We have

$$R_{XY}(t_1, t_2) = E[X(t_1)Y(t_2)] = E\left[X(t_1)\int_{-\infty}^{\infty} h(\alpha)X(t_2 - \alpha)\ d\alpha\right]$$

$$= E\left[\int_{-\infty}^{\infty} h(\alpha)X(t_1)X(t_2 - \alpha)\ d\alpha\right]$$

$$= \int_{-\infty}^{\infty} h(\alpha)E[X(t_1)X(t_2 - \alpha)]\ d\alpha$$

$$= \int_{-\infty}^{\infty} h(\alpha)R_X(t_1, t_2 - \alpha)\ d\alpha$$

$$= \int_{-\infty}^{\infty} h(\alpha)R_X(t_1 - t_2 + \alpha)\ d\alpha \qquad \text{(since } X(t) \text{ is WSS).}$$

We note that $R_{XY}(t_1, t_2)$ is only a function of $\tau = t_1 - t_2$, so we may write

$$R_{XY}(\tau) = \int_{-\infty}^{\infty} h(\alpha)R_X(\tau + \alpha)\ d\alpha$$

$$= h(\tau) * R_X(-\tau) = h(-\tau) * R_X(\tau).$$

Similarly, you can show that

$$R_Y(\tau) = h(\tau) * h(-\tau) * R_X(\tau).$$

This has been shown in the Solved Problems section. From the above results we conclude that $X(t)$ and $Y(t)$ are jointly WSS. The following theorem summarizes the results.

Theorem 10.2. Let $X(t)$ be a WSS random process and $Y(t)$ be given by

$$Y(t) = h(t) * X(t),$$

where $h(t)$ is the impulse response of the system. Then $X(t)$ and $Y(t)$ are jointly WSS. Moreover,

1. $\mu_Y(t) = \mu_Y = \mu_X \int_{-\infty}^{\infty} h(\alpha)\ d\alpha$;

2. $R_{XY}(\tau) = h(-\tau) * R_X(\tau) = \int_{-\infty}^{\infty} h(-\alpha)R_X(t - \alpha)\ d\alpha$;

3. $R_Y(\tau) = h(\tau) * h(-\tau) * R_X(\tau)$.

Frequency Domain Analysis:

Let's now rewrite the statement of Theorem 10.2 in the frequency domain. Let $H(f)$ be the Fourier transform of $h(t)$,

$$H(f) = \mathcal{F}\{h(t)\} = \int_{-\infty}^{\infty} h(t)e^{-2j\pi ft}\ dt.$$

$H(f)$ is called the **transfer function** of the system. We can rewrite

$$\mu_Y = \mu_X \int_{-\infty}^{\infty} h(\alpha) \ d\alpha$$

as

$$\mu_Y = \mu_X H(0)$$

Since $h(t)$ is assumed to be a real signal, we have

$$\mathcal{F}\{h(-t)\} = H(-f) = H^*(f),$$

where $*$ shows the complex conjugate. By taking the Fourier transform from both sides of $R_{XY}(\tau) = R_X(\tau) * h(-\tau)$, we conclude

$$S_{XY}(f) = S_X(f)H(-f) = S_X(f)H^*(f).$$

Finally, by taking the Fourier transform from both sides of $R_Y(\tau) = h(\tau) * h(-\tau) * R_X(\tau)$, we conclude

$$S_Y(f) = S_X(f)H^*(f)H(f)$$
$$= S_X(f)|H(f)|^2.$$

$$S_Y(f) = S_X(f)|H(f)|^2$$

Example 10.14. Let $X(t)$ be a zero-mean WSS process with $R_X(\tau) = e^{-|\tau|}$. $X(t)$ is input to an LTI system with

$$|H(f)| = \begin{cases} \sqrt{1 + 4\pi^2 f^2} & |f| < 2 \\ 0 & \text{otherwise} \end{cases}$$

Let $Y(t)$ be the output.

(a) Find $\mu_Y(t) = E[Y(t)]$.

(b) Find $R_Y(\tau)$.

(c) Find $E[Y(t)^2]$.

Solution: Note that since $X(t)$ is WSS, $X(t)$ and $Y(t)$ are jointly WSS, and therefore $Y(t)$ is WSS.

(a) To find $\mu_Y(t)$, we can write

$$\mu_Y = \mu_X H(0)$$
$$= 0 \cdot 1 = 0.$$

(b) To find $R_Y(\tau)$, we first find $S_Y(f)$.

$$S_Y(f) = S_X(f)|H(f)|^2.$$

From Fourier transform tables, we can see that

$$S_X(f) = \mathcal{F}\{e^{-|\tau|}\}$$
$$= \frac{2}{1 + (2\pi f)^2}.$$

Then, we can find $S_Y(f)$ as

$$S_Y(f) = S_X(f)|H(f)|^2$$
$$= \begin{cases} 2 & |f| < 2 \\ 0 & \text{otherwise} \end{cases}$$

We can now find $R_Y(\tau)$ by taking the inverse Fourier transform of $S_Y(f)$.

$$R_Y(\tau) = 8 \operatorname{sinc}(4\tau),$$

where

$$\operatorname{sinc}(f) = \frac{\sin(\pi f)}{\pi f}.$$

(c) We have

$$E[Y(t)^2] = R_Y(0) = 8.$$

10.2.3   Power in a Frequency Band

Here, we would like show that if you integrate $S_X(f)$ over a frequency range, you will obtain the expected power in $X(t)$ in that frequency range. Let's first define what we mean by the expected power "in a frequency range."

Consider a WSS random process $X(t)$ that goes through an LTI system with the following transfer function (Figure 10.7):

$$H(f) = \begin{cases} 1 & f_1 < |f| < f_2 \\ 0 & \text{otherwise} \end{cases}$$

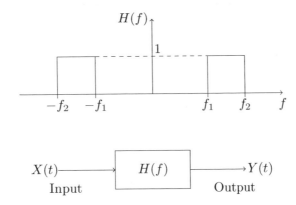

Figure 10.7: A bandpass filter

This is in fact a bandpass filter. This filter eliminates every frequency outside of the frequency band $f_1 < |f| < f_2$. Thus, the resulting random process $Y(t)$ is a filtered version of $X(t)$ in which frequency components in the frequency band $f_1 < |f| < f_2$ are preserved. The expected power in $Y(t)$ is said to be the expected power in $X(t)$ in the frequency range $f_1 < |f| < f_2$.

Now, let's find the expected power in $Y(t)$. We have

$$S_Y(f) = S_X(f)|H(f)|^2 = \begin{cases} S_X(f) & f_1 < |f| < f_2 \\ 0 & \text{otherwise} \end{cases}$$

Thus, the power in $Y(t)$ is

$$E[Y(t)^2] = \int_{-\infty}^{\infty} S_Y(f) \, df$$
$$= \int_{-f_2}^{-f_1} S_X(f) \, df + \int_{f_1}^{f_2} S_X(f) \, df$$
$$= 2 \int_{f_1}^{f_2} S_X(f) \, df \qquad (\text{since } S_X(-f) = S_X(f))$$

Therefore, we conclude that, if we integrate $S_X(f)$ over the frequency range $f_1 < |f| < f_2$, we will obtain the expected power in $X(t)$ in that frequency range. That is why $S_X(f)$ is called the *power spectral density* of $X(t)$.

Gaussian Processes through LTI Systems:

Let $X(t)$ be a stationary Gaussian random process that goes through an LTI system with impulse response $h(t)$. Then, the output process is given by

$$Y(t) = h(t) * X(t)$$
$$= \int_{-\infty}^{\infty} h(\alpha)X(t - \alpha) \, d\alpha.$$

For each t, you can think of the above integral as a limit of a sum. Now, since the different sums of jointly normal random variables are also jointly normal, you can argue that $Y(t)$ is also a Gaussian random process. Indeed, we can conclude that $X(t)$ and $Y(t)$ are jointly normal. Note that, for Gaussian processes, stationarity and wide-sense stationarity are equal.

> Let $X(t)$ be a stationary Gaussian process. If $X(t)$ is the input to an LTI system, then the output random process, $Y(t)$, is also a stationary Gaussian process. Moreover, $X(t)$ and $Y(t)$ are jointly Gaussian.

Example 10.15. Let $X(t)$ be a zero-mean Gaussian random process with $R_X(\tau) = 8 \ \text{sinc}(4\tau)$. Suppose that $X(t)$ is input to an LTI system with transfer function

$$H(f) = \begin{cases} \frac{1}{2} & |f| < 1 \\ 0 & \text{otherwise} \end{cases}$$

If $Y(t)$ is the output, find $P(Y(2) < 1|Y(1) = 1)$.

Solution: Since $X(t)$ is a WSS Gaussian process, $Y(t)$ is also a WSS Gaussian process. Thus, it suffices to find μ_Y and $R_Y(\tau)$. Since $\mu_X = 0$, we have

$$\mu_Y = \mu_X H(0) = 0.$$

Also, note that

$$S_X(f) = \mathcal{F}\{R_X(\tau)\}$$
$$= \begin{cases} 2 & |f| < 2 \\ 0 & \text{otherwise} \end{cases}$$

We can then find $S_Y(f)$ as

$$S_Y(f) = S_X(f)|H(f)|^2$$
$$= \begin{cases} \frac{1}{2} & |f| < 1 \\ 0 & \text{otherwise} \end{cases}$$

Thus, $R_Y(\tau)$ is given by

$$R_Y(\tau) = \mathcal{F}^{-1}\{S_X(f)\}$$
$$= \text{sinc}(2\tau).$$

Therefore,

$$E[Y(t)^2] = R_Y(0) = 1.$$

We conclude that $Y(t) \sim N(0,1)$, for all t. Since $Y(1)$ and $Y(2)$ are jointly Gaussian, to determine their joint PDF, it only remains to find their covariance. We have

$$
\begin{aligned}
E[Y(1)Y(2)] &= R_Y(-1) \\
&= \text{sinc}(-2) \\
&= \frac{\sin(-2\pi)}{-2\pi} \\
&= 0.
\end{aligned}
$$

Since $E[Y(1)] = E[Y(2)] = 0$, we conclude that $Y(1)$ and $Y(2)$ are uncorrelated. Since $Y(1)$ and $Y(2)$ are jointly normal, we conclude that they are independent, so

$$
\begin{aligned}
P(Y(2) < 1 | Y(1) = 1) &= P(Y(2) < 1) \\
&= \Phi(1) \approx 0.84
\end{aligned}
$$

10.2.4   White Noise

A very commonly-used random process is *white noise*. White noise is often used to model the thermal noise in electronic systems. By definition, the random process $X(t)$ is called white noise if $S_X(f)$ is constant for all frequencies. By convention, the constant is usually denoted by $\frac{N_0}{2}$.

The random process $X(t)$ is called a **white noise** process if

$$
S_X(f) = \frac{N_0}{2}, \qquad \text{for all } f.
$$

Before going any further, let's calculate the expected power in $X(t)$. We have

$$
\begin{aligned}
E\left[X(t)^2\right] &= \int_{-\infty}^{\infty} S_X(f) \; df \\
&= \int_{-\infty}^{\infty} \frac{N_0}{2} \; df = \infty.
\end{aligned}
$$

Thus, white noise, as defined above, has infinite power! In reality, white noise is in fact an approximation to the noise that is observed in real systems. To better understand the idea, consider the PSDs shown in Figure 10.8.

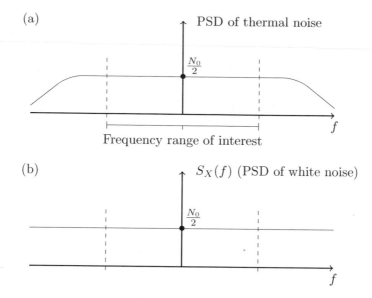

Figure 10.8: Part (a): PSD of thermal noise; Part (b): PSD of white noise.

Part (a) in the figure shows what the real PSD of a thermal noise might look like. As we see, the PSD is not constant for all frequencies; however, it is approximately constant over the frequency range that we are interested in. In other words, real systems are bandlimited and work on a limited range of frequencies. For the frequency range that we are interested in, the two PSDs (the PSD in Part (a) and the PSD of the white noise, shown in Part (b)) are approximately the same.

The thermal noise in electronic systems is usually modeled as a white Gaussian noise process. It is usually assumed that it has zero mean $\mu_X = 0$ and is Gaussian.

The random process $X(t)$ is called a **white Gaussian noise** process if $X(t)$ is a stationary Gaussian random process with zero mean, $\mu_X = 0$, and flat power spectral density,

$$S_X(f) = \frac{N_0}{2}, \qquad \text{for all } f.$$

Since the PSD of a white noise process is given by $S_X(f) = \frac{N_0}{2}$, its autocorrelation function is given by

$$R_X(\tau) = \mathcal{F}^{-1}\left\{\frac{N_0}{2}\right\}$$

$$= \frac{N_0}{2}\delta(\tau),$$

where $\delta(\tau)$ is the dirac delta function

$$\delta(x) = \begin{cases} \infty & x = 0 \\ 0 & \text{otherwise} \end{cases}$$

This again confirms that white noise has infinite power, $E[X(t)^2] = R_X(0)$. We also note that $R_X(\tau) = 0$ for any $\tau \neq 0$. This means that $X(t_1)$ and $X(t_2)$ are uncorrelated for any $t_1 \neq t_2$. Therefore, for a white Gaussian noise, $X(t_1)$ and $X(t_2)$ are independent for any $t_1 \neq t_2$. White Gaussian noise can be described as the "derivative" of *Brownian motion*. Brownian motion is an important random process that will be discussed in the next chapter.

Example 10.16. Let $X(t)$ be a white Gaussian noise process that is input to an LTI system with transfer function

$$|H(f)| = \begin{cases} 2 & 1 < |f| < 2 \\ 0 & \text{otherwise} \end{cases}$$

If $Y(t)$ is the output, find $P(Y(1) < \sqrt{N_0})$.

Solution: Since $X(t)$ is a zero-mean Gaussian process, $Y(t)$ is also a zero-mean Gaussian process. $S_Y(f)$ is given by

$$\frac{N_0}{2}|H(f)|^2 = \begin{cases} 2N_0 & 1 < |f| < 2 \\ 0 & \text{otherwise} \end{cases}$$

Therefore,

$$E[Y(t)^2] = \int_{-\infty}^{\infty} S_Y(f) \; df$$

$$= 4N_0.$$

Thus,

$$Y(t) \sim N(0, 4N_0).$$

To find $P(Y(1) < \sqrt{N_0})$, we can write

$$P(Y(1) < \sqrt{N_0}) = \Phi\left(\frac{\sqrt{N_0}}{\sqrt{4N_0}}\right)$$

$$= \Phi\left(\frac{1}{2}\right) \approx 0.69$$

10.2.5　Solved Problems

1. Consider a WSS random process $X(t)$ with

$$R_X(\tau) = \begin{cases} 1 - |\tau| & -1 \leq \tau \leq 1 \\ \\ 0 & \text{otherwise} \end{cases}$$

Find the PSD of $X(t)$, and $E[X(t)^2]$.

Solution: First, we have

$$E[X(t)^2] = R_X(0) = 1.$$

We can write triangular function, $R_X(\tau) = \Lambda(\tau)$, as

$$R_X(\tau) = \Pi(\tau) * \Pi(\tau),$$

where

$$\Pi(\tau) = \begin{cases} 1 & -\frac{1}{2} \leq \tau \leq \frac{1}{2} \\ \\ 0 & \text{otherwise} \end{cases}$$

Thus, we conclude

$$\begin{aligned} S_X(f) &= \mathcal{F}\{R_X(\tau)\} \\ &= \mathcal{F}\{\Pi(\tau) * \Pi(\tau)\} \\ &= \mathcal{F}\{\Pi(\tau)\} \cdot \mathcal{F}\{\Pi(\tau)\} \\ &= \left[\text{sinc}(f)\right]^2. \end{aligned}$$

2. Let $X(t)$ be a random process with mean function $\mu_X(t)$ and autocorrelation function $R_X(s, t)$ ($X(t)$ is not necessarily a WSS process). Let $Y(t)$ be given by

$$Y(t) = h(t) * X(t),$$

where $h(t)$ is the impulse response of the system. Show that

(a) $\mu_Y(t) = \mu_X(t) * h(t)$.

(b) $R_{XY}(t_1, t_2) = h(t_2) * R_X(t_1, t_2) = \int_{-\infty}^{\infty} h(\alpha) R_X(t_1, t_2 - \alpha)\ d\alpha$.

Solution:

(a) We have

$$\mu_Y(t) = E[Y(t)] = E\left[\int_{-\infty}^{\infty} h(\alpha)X(t-\alpha)\ d\alpha\right]$$

$$= \int_{-\infty}^{\infty} h(\alpha)E[X(t-\alpha)]\ d\alpha$$

$$= \int_{-\infty}^{\infty} h(\alpha)\mu_X(t-\alpha)\ d\alpha$$

$$= \mu_X(t) * h(t).$$

(b) We have

$$R_{XY}(t_1, t_2) = E[X(t_1)Y(t_2)] = E\left[X(t_1)\int_{-\infty}^{\infty} h(\alpha)X(t_2-\alpha)\ d\alpha\right]$$

$$= E\left[\int_{-\infty}^{\infty} h(\alpha)X(t_1)X(t_2-\alpha)\ d\alpha\right]$$

$$= \int_{-\infty}^{\infty} h(\alpha)E[X(t_1)X(t_2-\alpha)]\ d\alpha$$

$$= \int_{-\infty}^{\infty} h(\alpha)R_X(t_1, t_2-\alpha)\ d\alpha.$$

3. Prove the third part of Theorem 10.2: Let $X(t)$ be a WSS random process and $Y(t)$ be given by

$$Y(t) = h(t) * X(t),$$

where $h(t)$ is the impulse response of the system. Show that

$$R_Y(s,t) = R_Y(s-t) = \int_{-\infty}^{\infty}\int_{-\infty}^{\infty} h(\alpha)h(\beta)R_X(s-t-\alpha+\beta)\ d\alpha d\beta.$$

Also, show that we can rewrite the above integral as $R_Y(\tau) = h(\tau) * h(-\tau) * R_X(\tau)$.

Solution:

$$R_Y(s,t) = E[X(s)Y(t)]$$

$$= E\left[\int_{-\infty}^{\infty} h(\alpha)X(s-\alpha)\ d\alpha \int_{-\infty}^{\infty} h(\beta)X(s-\beta)\ d\beta\right]$$

$$= \int_{-\infty}^{\infty}\int_{-\infty}^{\infty} h(\alpha)h(\beta)E[X(s-\alpha)X(t-\beta)]\ d\alpha\ d\beta$$

$$= \int_{-\infty}^{\infty}\int_{-\infty}^{\infty} h(\alpha)h(\beta)R_X(s-t-\alpha+\beta)\ d\alpha\ d\beta.$$

We now compute $h(\tau) * h(-\tau) * R_X(\tau)$. First, let $g(\tau) = h(\tau) * h(-\tau)$. Note that

$$g(\tau) = h(\tau) * h(-\tau)$$
$$= \int_{-\infty}^{\infty} h(\alpha) h(\alpha - \tau) \ d\alpha.$$

Thus, we have

$$g(\tau) * R_X(\tau) = \int_{-\infty}^{\infty} g(\theta) R_X(\theta - \tau) \ d\theta$$
$$= \int_{-\infty}^{\infty} \left[\int_{-\infty}^{\infty} h(\alpha) h(\alpha - \theta) \ d\alpha \right] R_X(\theta - \tau) \ d\theta$$
$$= \int_{-\infty}^{\infty} \int_{-\infty}^{\infty} h(\alpha) h(\alpha - \theta) R_X(\theta - \tau) \ d\alpha \ d\theta$$
$$= \int_{-\infty}^{\infty} \int_{-\infty}^{\infty} h(\alpha) h(\beta) R_X(\alpha - \beta - \tau) \ d\alpha \ d\beta$$
$$= \int_{-\infty}^{\infty} \int_{-\infty}^{\infty} h(\alpha) h(\beta) R_X(\tau - \alpha + \beta) \ d\alpha \ d\beta \qquad \left(\text{since } R_X(-\tau) = R_X(\tau) \right).$$

4. Let $X(t)$ be a WSS random process. Assuming that $S_X(f)$ is continuous at f_1, show that $S_X(f_1) \geq 0$.

Solution: Let $f_1 \in \mathbb{R}$. Suppose that $X(t)$ goes through an LTI system with the following transfer function

$$H(f) = \begin{cases} 1 & f_1 < |f| < f_1 + \Delta \\ 0 & \text{otherwise} \end{cases}$$

where Δ is chosen to be very small. The PSD of $Y(t)$ is given by

$$S_Y(f) = S_X(f)|H(f)|^2 = \begin{cases} S_X(f) & f_1 < |f| < f_1 + \Delta \\ 0 & \text{otherwise} \end{cases}$$

Thus, the power in $Y(t)$ is

$$E[Y(t)^2] = \int_{-\infty}^{\infty} S_Y(f) \ df$$
$$= 2 \int_{f_1}^{f_1 + \Delta} S_X(f) \ df$$
$$\approx 2\Delta S_X(f_1).$$

Since $E[Y(t)^2] \geq 0$, we conclude that $S_X(f_1) \geq 0$.

5. Let $X(t)$ be a white Gaussian noise with $S_X(f) = \frac{N_0}{2}$. Assume that $X(t)$ is input to an LTI system with

$$h(t) = e^{-t}u(t).$$

Let $Y(t)$ be the output.

(a) Find $S_Y(f)$.

(b) Find $R_Y(\tau)$.

(c) Find $E[Y(t)^2]$.

Solution: First, note that

$$H(f) = \mathcal{F}\{h(t)\}$$
$$= \frac{1}{1 + j2\pi f}.$$

(a) To find $S_Y(f)$, we can write

$$S_Y(f) = S_X(f)|H(f)|^2$$
$$= \frac{N_0/2}{1 + (2\pi f)^2}.$$

(b) To find $R_Y(\tau)$, we can write

$$R_Y(\tau) = \mathcal{F}^{-1}\{S_Y(f)\}$$
$$= \frac{N_0}{4}e^{-|\tau|}.$$

(c) We have

$$E[Y(t)^2] = R_Y(0)$$
$$= \frac{N_0}{4}.$$

10.3   End of Chapter Problems

1. Let $\{X_n, n \in \mathbb{Z}\}$ be a discrete-time random process, defined as

$$X_n = 2\cos\left(\frac{\pi n}{8} + \Phi\right),$$

   where $\Phi \sim Uniform(0, 2\pi)$.

   (a) Find the mean function, $\mu_X(n)$.
   (b) Find the correlation function $R_X(m, n)$.
   (c) Is X_n a WSS process?

2. Let $\{X(t), t \in \mathbb{R}\}$ be a continuous-time random process, defined as

$$X(t) = A\cos(2t + \Phi),$$

   where $A \sim U(0, 1)$ and $\Phi \sim U(0, 2\pi)$ are two independent random variables.

   (a) Find the mean function $\mu_X(t)$.
   (b) Find the correlation function $R_X(t_1, t_2)$.
   (c) Is $X(t)$ a WSS process?

3. Let $\{X(n), n \in \mathbb{Z}\}$ be a WSS discrete-time random process with $\mu_X(n) = 1$ and $R_X(m, n) = e^{-(m-n)^2}$. Define the random process $Z(n)$ as

$$Z(n) = X(n) + X(n - 1), \qquad \text{for all } n \in \mathbb{Z}.$$

   (a) Find the mean function of $Z(n)$, $\mu_Z(n)$.
   (b) Find the autocorrelation function of $Z(n)$, $R_Z(m, n)$.
   (c) Is $Z(n)$ a WSS random process?

4. Let $g : \mathbb{R} \mapsto \mathbb{R}$ be a periodic function with period T, i.e.,

$$g(t + T) = g(t), \qquad \text{for all } t \in \mathbb{R}.$$

   Define the random process $\{X(t), t \in \mathbb{R}\}$ as

$$X(t) = g(t + U), \qquad \text{for all } t \in \mathbb{R},$$

   where $U \sim Uniform(0, T)$. Show that $X(t)$ is a WSS random process.

5. Let $\{X(t), t \in \mathbb{R}\}$ and $\{Y(t), t \in \mathbb{R}\}$ be two independent random processes. Let $Z(t)$ be defined as

$$Z(t) = X(t)Y(t), \qquad \text{for all } t \in \mathbb{R}.$$

   Prove the following statements:

   (a) $\mu_Z(t) = \mu_X(t)\mu_Y(t)$, for all $t \in \mathbb{R}$.
   (b) $R_Z(t_1, t_2) = R_X(t_1, t_2)R_Y(t_1, t_2)$, for all $t \in \mathbb{R}$.

(c) If $X(t)$ and $Y(t)$ are WSS, then they are jointly WSS.

(d) If $X(t)$ and $Y(t)$ are WSS, then $Z(t)$ is also WSS.

(e) If $X(t)$ and $Y(t)$ are WSS, then $X(t)$ and $Z(t)$ are jointly WSS.

6. Let $X(t)$ be a Gaussian process such that for all $t > s \geq 0$ we have

$$X(t) - X(s) \sim N(0, t - s).$$

Show that $X(t)$ is mean-square continuous at any time $t \geq 0$.

7. Let $X(t)$ be a WSS Gaussian random process with $\mu_X(t) = 1$ and $R_X(\tau) = 1 + 4\text{sinc}(\tau)$.

(a) Find $P(1 < X(1) < 2)$.

(b) Find $P(1 < X(1) < 2, X(2) < 3)$.

8. Let $X(t)$ be a Gaussian random process with $\mu_X(t) = 0$ and $R_X(t_1, t_2) = \min(t_1, t_2)$. Find $P(X(4) < 3 | X(1) = 1)$.

9. Let $\{X(t), t \in \mathbb{R}\}$ be a continuous-time random process, defined as

$$X(t) = \sum_{k=0}^{n} A_k t^k,$$

where $A_0, A_1, \cdots, A_n$ are i.i.d. $N(0, 1)$ random variables and n is a fixed positive integer.

(a) Find the mean function $\mu_X(t)$.

(b) Find the correlation function $R_X(t_1, t_2)$.

(c) Is $X(t)$ a WSS process?

(d) Find $P(X(1) < 1)$. Assume $n = 10$.

(e) Is $X(t)$ a Gaussian process?

10. (Complex Random Processes) In some applications, we need to work with complex-valued random processes. More specifically, a complex random process $X(t)$ can be written as

$$X(t) = X_r(t) + jX_i(t),$$

where $X_r(t)$ and $X_i(t)$ are two real-valued random processes and $j = \sqrt{-1}$. We define the mean function and the autocorrelation function as

$$\begin{aligned}
\mu_X(t) &= E[X(t)] \\
&= E[X_r(t)] + jE[X_i(t)] \\
&\doteq \mu_{X_r}(t) + j\mu_{X_i}(t);
\end{aligned}$$

$$\begin{aligned}
R_X(t_1, t_2) &= E[X(t_1)X^*(t_2)] \\
&= E\left[\left(X_r(t_1) + jX_i(t_1)\right)\left(X_r(t_2) - jX_i(t_2)\right)\right].
\end{aligned}$$

Let $X(t)$ be a complex-valued random process defined as

$$X(t) = Ae^{j(\omega t + \Phi)},$$

where $\Phi \sim Uniform(0, 2\pi)$, and A is a random variable independent of Φ with $EA = \mu$ and $\text{Var}(A) = \sigma^2$.

(a) Find the mean function of $X(t)$, $\mu_X(t)$.

(b) Find the autocorrelation function of $X(t)$, $R_X(t_1, t_2)$.

11. (Time Averages) Let $\{X(t), t \in \mathbb{R}\}$ be a continuous-time random process. The time average mean of $X(t)$ is defined as[2]

$$\langle X(t) \rangle = \lim_{T \to \infty} \left[\frac{1}{2T} \int_{-T}^{T} X(t) dt \right].$$

Consider the random process $\{X(t), t \in \mathbb{R}\}$ defined as

$$X(t) = \cos(t + U),$$

where $U \sim Uniform(0, 2\pi)$. Find $\langle X(t) \rangle$.

12. (Ergodicity) Let $X(t)$ be a WSS process. We say that $X(t)$ is *mean ergodic* if $\langle X(t) \rangle$ (defined above) is equal to μ_X.

Let A_0, A_1, A_{-1}, A_2, A_{-2}, $\cdots$ be a sequence of i.i.d. random variables with mean $EA_i = \mu < \infty$. Define the random process $\{X(t), t \in \mathbb{R}\}$ as

$$X(t) = \sum_{k=-\infty}^{\infty} A_k g(t - k),$$

where, $g(t)$ is given by

$$g(t) = \begin{cases} 1 & 0 \le t < 1 \\ 0 & \text{otherwise} \end{cases}$$

Show that $X(t)$ is mean ergodic.

13. Let $\{X(t), t \in \mathbb{R}\}$ be a WSS random process. Show that for any $\alpha > 0$, we have

$$P\left(|X(t + \tau) - X(t)| > \alpha\right) \le \frac{2R_X(0) - 2R_X(\tau)}{\alpha^2}.$$

14. Let $\{X(t), t \in \mathbb{R}\}$ be a WSS random process. Suppose that $R_X(\tau) = R_X(0)$ for some $\tau > 0$. Show that, for any t, we have

$$X(t + \tau) = X(t), \qquad \text{with probability one.}$$

15. Let $X(t)$ be a real-valued WSS random process with autocorrelation function $R_X(\tau)$. Show that the Power Spectral Density (PSD) of $X(t)$ is given by

$$S_X(f) = \int_{-\infty}^{\infty} R_X(\tau) \cos(2\pi f \tau) \, d\tau.$$

---

[2] Assuming that the limit exists in mean-square sense.

16. Let $X(t)$ and $Y(t)$ be real-valued jointly WSS random processes. Show that

$$S_{YX}(f) = S_{XY}^*(f),$$

where, $*$ shows the complex conjugate.

17. Let $X(t)$ be a WSS process with autocorrelation function

$$R_X(\tau) = \frac{1}{1 + \pi^2 \tau^2}.$$

Assume that $X(t)$ is input to a low-pass filter with frequency response

$$H(f) = \begin{cases} 3 & |f| < 2 \\ 0 & \text{otherwise} \end{cases}$$

Let $Y(t)$ be the output.

(a) Find $S_X(f)$.

(b) Find $S_{XY}(f)$.

(c) Find $S_Y(f)$.

(d) Find $E[Y(t)^2]$.

18. Let $X(t)$ be a WSS process with autocorrelation function

$$R_X(\tau) = 1 + \delta(\tau).$$

Assume that $X(t)$ is input to an LTI system with impulse response

$$h(t) = e^{-t} u(t).$$

Let $Y(t)$ be the output.

(a) Find $S_X(f)$.

(b) Find $S_{XY}(f)$.

(c) Find $R_{XY}(\tau)$.

(d) Find $S_Y(f)$.

(e) Find $R_Y(\tau)$.

(f) Find $E[Y(t)^2]$.

19. Let $X(t)$ be a zero-mean WSS Gaussian random process with $R_X(\tau) = e^{-\pi \tau^2}$. Suppose that $X(t)$ is input to an LTI system with transfer function

$$|H(f)| = e^{-\frac{3}{2} \pi f^2}.$$

Let $Y(t)$ be the output.

(a) Find μ_Y.

(b) Find $R_Y(\tau)$ and $\text{Var}(Y(t))$.

(c) Find $E[Y(3)|Y(1) = -1]$.

(d) Find $\text{Var}(Y(3)|Y(1) = -1)$.

(e) Find $P(Y(3) < 0|Y(1) = -1)$.

20. Let $X(t)$ be a white Gaussian noise with $S_X(f) = \frac{N_0}{2}$. Assume that $X(t)$ is input to a bandpass filter with frequency response

$$H(f) = \begin{cases} 2 & 1 < |f| < 3 \\ 0 & \text{otherwise} \end{cases}$$

Let $Y(t)$ be the output.

(a) Find $S_Y(f)$.

(b) Find $R_Y(\tau)$.

(c) Find $E[Y(t)^2]$.

Chapter 11

Some Important Random Processes

In the previous chapter, we discussed a general theory of random processes. In this chapter, we will focus on some specific random processes that are used frequently in applications. More specifically, we will discuss *the Poisson process, Markov chains*, and *Brownian Motion (the Wiener process)*. Most of the discussion in this chapter is self-contained in the sense that it depends very lightly on the material of the previous chapter.

11.1 Poisson Processes

11.1.1 Counting Processes

In some problems, we count the occurrences of some types of events. In such scenarios, we are dealing with a *counting process*. For example, you might have a random process $N(t)$ that shows the number of customers who arrive at a supermarket by time t starting from time 0. For such a process, we usually assume $N(0) = 0$, so as time passes and customers arrive, $N(t)$ takes positive integer values.

Definition 11.1. A random process $\{N(t), t \in [0, \infty)\}$ is said to be a **counting process** if $N(t)$ is the number of events occurred from time 0 up to and including time t. For a counting process, we assume

1. $N(0) = 0$;

2. $N(t) \in \{0, 1, 2, \cdots\}$, for all $t \in [0, \infty)$;

3. for $0 \leq s < t$, $N(t) - N(s)$ shows the number of events that occur in the interval $(s, t]$.

Since counting processes have been used to model arrivals (such as the supermarket example above), we usually refer to the occurrence of each event as an "arrival." For example, if $N(t)$ is the number of accidents in a city up to time t, we still refer to each accident as an arrival. Figure 11.1 shows a possible realization and the corresponding sample function of a counting process.

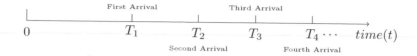

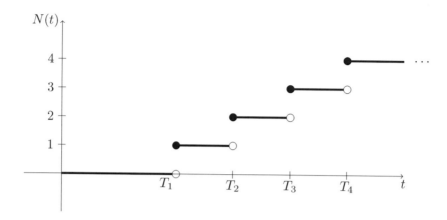

Figure 11.1: A possible realization and the corresponding sample path of a counting process.

By the above definition, the only sources of randomness are the arrival times T_i. Before introducing the Poisson process, we would like to provide two definitions.

Definition 11.2. Let $\{X(t), t \in [0, \infty)\}$ be a continuous-time random process. We say that $X(t)$ has **independent increments** if, for all $0 \leq t_1 < t_2 < t_3 \cdots < t_n$, the random variables

$$X(t_2) - X(t_1), \ \ X(t_3) - X(t_2), \ \ \cdots, \ \ X(t_n) - X(t_{n-1})$$

are independent.

Note that for a counting process, $N(t_i) - N(t_{i-1})$ is the number of arrivals in the interval $(t_{i-1}, t_i]$. Thus, a counting process has independent increments if the numbers of arrivals in non-overlapping (disjoint) intervals

$$(t_1, t_2], (t_2, t_3], \ \ \cdots, \ \ (t_{n-1}, t_n]$$

are independent. Having independent increments simplifies analysis of a counting process. For example, suppose that we would like to find the probability of having 2 arrivals in the interval $(1, 2]$, and 3 arrivals in the interval $(3, 5]$. Since the two intervals $(1, 2]$ and $(3, 5]$ are disjoint, we can write

$$P\bigg(2 \text{ arrivals in } (1, 2] \quad \text{and} \quad 3 \text{ arrivals in } (3, 5]\bigg) =$$

$$P\bigg(2 \text{ arrivals in } (1, 2]\bigg) \cdot P\bigg(3 \text{ arrivals in } (3, 5]\bigg).$$

Here is another useful definition.

Definition 11.3. Let $\{X(t), t \in [0, \infty)\}$ be a continuous-time random process. We say that $X(t)$ has **stationary increments** if, for all $t_2 > t_1 \geq 0$, and all $r > 0$, the two random variables $X(t_2) - X(t_1)$ and $X(t_2 + r) - X(t_1 + r)$ have the same distributions. In other words, the distribution of the difference depends only on the length of the interval $(t_1, t_2]$, and not on the exact location of the interval on the real line.

Note that for a counting process $N(t)$, $N(t_2) - N(t_1)$ is the number of arrivals in the interval $(t_1, t_2]$. We also assume $N(0) = 0$. Therefore, a counting process has stationary increments if for all $t_2 > t_1 \geq 0$, $N(t_2) - N(t_1)$ has the same distribution as $N(t_2 - t_1)$. This means that the distribution of the number of arrivals in any interval depends only on the length of the interval, and not on the exact location of the interval on the real line.

A counting process has **independent increments** if the numbers of arrivals in non-overlapping (disjoint) intervals are independent.

A counting process has **stationary increments** if, for all $t_2 > t_1 \geq 0$, $N(t_2) - N(t_1)$ has the same distribution as $N(t_2 - t_1)$.

11.1.2 Basic Concepts of the Poisson Process

The Poisson process is one of the most widely-used counting processes. It is usually used in scenarios where we are counting the occurrences of certain events that appear to happen at a certain rate, but completely at random (without a certain structure). For example, suppose that from historical data, we know that earthquakes occur in a certain area with a rate of 2 per month. Other than this information, the timings of earthquakes seem to be completely random. Thus, we conclude that the Poisson process might be a good model for earthquakes. In practice, the Poisson process or its extensions have been used to model [24]

- the number of car accidents at a site or in an area;

- the location of users in a wireless network;

- the requests for individual documents on a web server;

- the outbreak of wars;

- photons landing on a photodiode.

Poisson random variable: Here, we briefly review some properties of the Poisson random variable that we have discussed in the previous chapters. Remember that a discrete random variable X is said to be a *Poisson* random variable with parameter μ, shown as $X \sim Poisson(\mu)$, if its range is $R_X = \{0, 1, 2, 3, ...\}$, and its PMF is given by

$$P_X(k) = \begin{cases} \frac{e^{-\mu}\mu^k}{k!} & \text{for } k \in R_X \\ 0 & \text{otherwise} \end{cases}$$

Here are some useful facts that we have seen before:

1. If $X \sim Poisson(\mu)$, then $EX = \mu$, and $\mathrm{Var}(X) = \mu$.

2. If $X_i \sim Poisson(\mu_i)$, for $i = 1, 2, \cdots, n$, and the X_i's are independent, then

$$X_1 + X_2 + \cdots + X_n \sim Poisson(\mu_1 + \mu_2 + \cdots + \mu_n).$$

3. The Poisson distribution can be viewed as the limit of binomial distribution.

   Theorem 11.1. Let $Y_n \sim Binomial\big(n, p = p(n)\big)$. Let $\mu > 0$ be a fixed real number, and $\lim_{n \to \infty} np = \mu$. Then, the PMF of Y_n converges to a $Poisson(\mu)$ PMF, as $n \to \infty$. That is, for any $k \in \{0, 1, 2, ...\}$, we have

$$\lim_{n \to \infty} P_{Y_n}(k) = \frac{e^{-\mu} \mu^k}{k!}.$$

Poisson Process as the Limit of a Bernoulli Process:

Suppose that we would like to model the arrival of events that happen completely at random at a rate λ per unit time. Here is one way to do this. At time $t = 0$, we have no arrivals yet, so $N(0) = 0$. We now divide the half-line $[0, \infty)$ to tiny subintervals of length δ as shown in Figure 11.2.

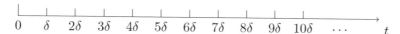

Figure 11.2: Dividing the half-line $[0, \infty)$ to tiny subintervals of length δ.

Each subinterval corresponds to a time slot of length δ. Thus, the intervals are $(0, \delta]$, $(\delta, 2\delta]$, $(2\delta, 3\delta]$, $\cdots$. More generally, the kth interval is $\big((k-1)\delta, k\delta\big]$. We assume that in each time slot, we toss a coin for which $P(H) = p = \lambda\delta$. If the coin lands heads up, we say that we have an arrival in that subinterval. Otherwise, we say that we have no arrival in that interval. Figure 11.3 shows this process. Here, we have an arrival at time $t = k\delta$, if the kth coin flip results in a heads.

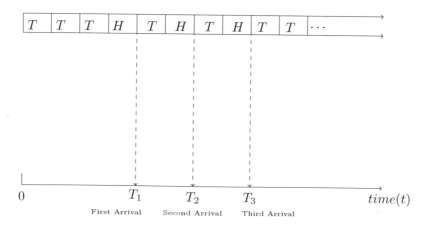

Figure 11.3: Poisson process as a limit of a Bernoulli process.

Now, let $N(t)$ be defined as the number of arrivals (number of heads) from time 0 to time t. There are $n \approx \frac{t}{\delta}$ time slots in the interval $(0, t]$. Thus, $N(t)$ is the number of heads in n coin flips. We conclude that $N(t) \sim Binomial(n, p)$. Note that here $p = \lambda\delta$, so

$$np = n\lambda\delta$$
$$= \frac{t}{\delta} \cdot \lambda\delta$$
$$= \lambda t.$$

Thus, by Theorem 11.1, as $\delta \to 0$, the PMF of $N(t)$ converges to a Poisson distribution with rate λt. More generally, we can argue that the number of arrivals in any interval of length τ follows a $Poisson(\lambda\tau)$ distribution as $\delta \to 0$.

Consider several non-overlapping intervals. The number of arrivals in each interval is determined by the results of the coin flips for that interval. Since different coin flips are independent, we conclude that the above counting process has independent increments.

Definition of the Poisson Process:

The above construction can be made mathematically rigorous. The resulting random process is called a Poisson process with rate (or intensity) λ. Here is a formal definition of the Poisson process.

The Poisson Process

Let $\lambda > 0$ be fixed. The counting process $\{N(t), t \in [0, \infty)\}$ is called a **Poisson process** with **rate** λ if all the following conditions hold:

1. $N(0) = 0$;

2. $N(t)$ has independent increments;

3. the number of arrivals in any interval of length $\tau > 0$ has $Poisson(\lambda\tau)$ distribution.

Note that from the above definition, we conclude that in a Poisson process, the distribution of the number of arrivals in any interval depends only on the length of the interval, and not on the exact location of the interval on the real line. Therefore the *Poisson process has stationary increments*.

Example 11.1. The number of customers arriving at a grocery store can be modeled by a Poisson process with intensity $\lambda = 10$ customers per hour.

1. Find the probability that there are 2 customers between 10:00 and 10:20.

2. Find the probability that there are 3 customers between 10:00 and 10:20 and 7 customers between 10:20 and 11.

Solution:

1. Here, $\lambda = 10$ and the interval between 10:00 and 10:20 has length $\tau = \frac{1}{3}$ hours. Thus, if X is the number of arrivals in that interval, we can write $X \sim Poisson(10/3)$. Therefore,

$$P(X = 2) = \frac{e^{-\frac{10}{3}} \left(\frac{10}{3}\right)^2}{2!}$$
$$\approx 0.2$$

2. Here, we have two non-overlapping intervals $I_1 = (10:00a.m., 10:20a.m.]$ and $I_2 = (10:20a.m., 11a.m.]$. Thus, we can write

$$P\left(3 \text{ arrivals in } I_1 \quad \text{and} \quad 7 \text{ arrivals in } I_2\right) =$$
$$P\left(3 \text{ arrivals in } I_1\right) \cdot P\left(7 \text{ arrivals in } I_2\right).$$

Since the lengths of the intervals are $\tau_1 = 1/3$ and $\tau_2 = 2/3$ respectively, we obtain $\lambda\tau_1 = 10/3$ and $\lambda\tau_2 = 20/3$. Thus, we have

$$P\left(3 \text{ arrivals in } I_1 \quad \text{and} \quad 7 \text{ arrivals in } I_2\right) = \frac{e^{-\frac{10}{3}} \left(\frac{10}{3}\right)^3}{3!} \cdot \frac{e^{-\frac{20}{3}} \left(\frac{20}{3}\right)^7}{7!}$$
$$\approx 0.0325$$

Second Definition of the Poisson Process:

Let $N(t)$ be a Poisson process with rate λ. Consider a very short interval of length Δ. Then, the number of arrivals in this interval has the same distribution as $N(\Delta)$. In particular, we can write

$$P(N(\Delta) = 0) = e^{-\lambda\Delta}$$
$$= 1 - \lambda\Delta + \frac{\lambda^2}{2}\Delta^2 - \cdots \quad \text{(Taylor Series)}.$$

Note that if Δ is small, the terms that include second or higher powers of Δ are negligible compared to Δ. We write this as

$$P(N(\Delta) = 0) = 1 - \lambda\Delta + o(\Delta) \tag{11.1}$$

Here $o(\Delta)$ shows a function that is negligible compared to Δ, as $\Delta \to 0$. More precisely, $g(\Delta) = o(\Delta)$ means that

$$\lim_{\Delta \to 0} \frac{g(\Delta)}{\Delta} = 0.$$

Now, let us look at the probability of having one arrival in an interval of length Δ.

$$P(N(\Delta) = 1) = e^{-\lambda\Delta}\lambda\Delta$$

$$= \lambda\Delta\left(1 - \lambda\Delta + \frac{\lambda^2}{2}\Delta^2 - \cdots\right) \qquad \text{(Taylor Series)}$$

$$= \lambda\Delta + \left(-\lambda^2\Delta^2 + \frac{\lambda^3}{2}\Delta^3 \cdots\right)$$

$$= \lambda\Delta + o(\Delta).$$

We conclude that

$$P(N(\Delta) = 1) = \lambda\Delta + o(\Delta) \qquad (11.2)$$

Similarly, we can show that

$$P(N(\Delta) \geq 2) = o(\Delta) \qquad (11.3)$$

In fact, equations 11.1, 11.2, and 11.3 give us another way to define a Poisson process.

The Second Definition of the Poisson Process

Let $\lambda > 0$ be fixed. The counting process $\{N(t), t \in [0, \infty)\}$ is called a **Poisson process** with **rate** λ if all the following conditions hold:

1. $N(0) = 0$;

2. $N(t)$ has <u>independent</u> and <u>stationary</u> increments;

3. we have

$$P(N(\Delta) = 0) = 1 - \lambda\Delta + o(\Delta),$$
$$P(N(\Delta) = 1) = \lambda\Delta + o(\Delta),$$
$$P(N(\Delta) \geq 2) = o(\Delta).$$

We have already shown that any Poisson process satisfies the above definition. To show that the above definition is equivalent to our original definition, we also need to show that any process that satisfies the above definition also satisfies the original definition. A method to show this is outlined in the End of Chapter Problems.

Arrival and Interarrival Times:

Let $N(t)$ be a Poisson process with rate λ. Let X_1 be the time of the first arrival. Then,

$$P(X_1 > t) = P\big(\text{no arrival in } (0, t]\big)$$
$$= e^{-\lambda t}.$$

We conclude that

$$F_{X_1}(t) = \begin{cases} 1 - e^{-\lambda t} & t > 0 \\ \\ 0 & \text{otherwise} \end{cases}$$

Therefore, $X_1 \sim Exponential(\lambda)$. Let X_2 be the time elapsed between the first and the second arrival (Figure 11.4).

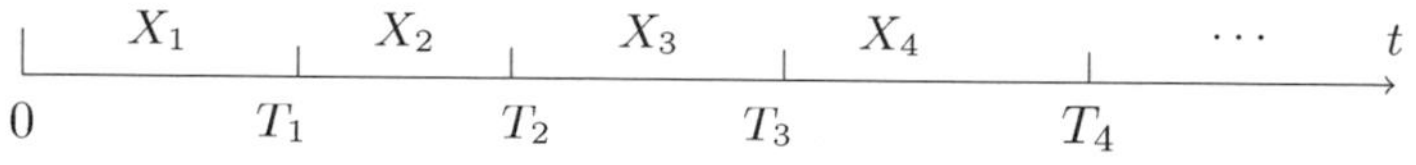

Figure 11.4: The random variables X_1, X_2, $\cdots$ are called the interarrival times of the counting process $N(t)$.

Let $s > 0$ and $t > 0$. Note that the two intervals $(0, s]$ and $(s, s + t]$ are disjoint. We can write

$$\begin{aligned} P(X_2 > t | X_1 = s) &= P\big(\text{no arrival in } (s, s + t] | X_1 = s\big) \\ &= P\big(\text{no arrivals in } (s, s + t]\big) \qquad \text{(independent increments)} \\ &= e^{-\lambda t}. \end{aligned}$$

We conclude that $X_2 \sim Exponential(\lambda)$, and that X_1 and X_2 are independent. The random variables X_1, X_2, $\cdots$ are called the **interarrival times** of the counting process $N(t)$. Similarly, we can argue that all X_i's are independent and $X_i \sim Exponential(\lambda)$ for $i = 1, 2, 3, \cdots$.

Interarrival Times for Poisson Processes

If $N(t)$ is a Poisson process with rate λ, then the interarrival times X_1, X_2, $\cdots$ are independent and

$$X_i \sim Exponential(\lambda), \qquad \text{for } i = 1, 2, 3, \cdots.$$

Remember that if X is exponential with parameter $\lambda > 0$, then X is a *memoryless* random variable, that is

$$P(X > x + a | X > a) = P(X > x), \qquad \text{for } a, x \geq 0.$$

Thinking of the Poisson process, the memoryless property of the interarrival times is consistent with the independent increment property of the Poisson distribution. In some sense, both are implying that the number of arrivals in non-overlapping intervals are independent. To better understand this issue, let's look at an example.

Example 11.2. Let $N(t)$ be a Poisson process with intensity $\lambda = 2$, and let X_1, X_2, $\cdots$ be the corresponding interarrival times.

(a) Find the probability that the first arrival occurs after $t = 0.5$, i.e., $P(X_1 > 0.5)$.

(b) Given that we have had no arrivals before $t = 1$, find $P(X_1 > 3)$.

(c) Given that the third arrival occurred at time $t = 2$, find the probability that the fourth arrival occurs after $t = 4$.

(d) I start watching the process at time $t = 10$. Let T be the time of the first arrival that I see. In other words, T is the first arrival after $t = 10$. Find ET and $\text{Var}(T)$.

(e) I start watching the process at time $t = 10$. Let T be the time of the first arrival that I see. Find the conditional expectation and the conditional variance of T given that I am informed that the last arrival occurred at time $t = 9$.

Solution:

(a) Since $X_1 \sim Exponential(2)$, we can write

$$P(X_1 > 0.5) = e^{-(2 \times 0.5)}$$
$$\approx 0.37$$

Another way to solve this is to note that

$$P(X_1 > 0.5) = P(\text{no arrivals in } (0, 0.5]) = e^{-(2 \times 0.5)} \approx 0.37$$

(b) We can write

$$P(X_1 > 3 | X_1 > 1) = P(X_1 > 2) \qquad \text{(memoryless property)}$$
$$= e^{-2 \times 2}$$
$$\approx 0.0183$$

Another way to solve this is to note that the number of arrivals in $(1, 3]$ is independent of the arrivals before $t = 1$. Thus,

$$P(X_1 > 3 | X_1 > 1) = P\big(\text{no arrivals in } (1, 3] \mid \text{no arrivals in } (0, 1]\big)$$
$$= P\big(\text{no arrivals in } (1, 3]\big) \qquad \text{(independent increments)}$$
$$= e^{-2 \times 2}$$
$$\approx 0.0183$$

(c) The time between the third and the fourth arrival is $X_4 \sim Exponential(2)$. Thus, the desired conditional probability is equal to

$$P(X_4 > 2 | X_1 + X_2 + X_3 = 2) = P(X_4 > 2) \qquad \text{(independence of the } X_i\text{'s)}$$
$$= e^{-2 \times 2}$$
$$\approx 0.0183$$

(d) When I start watching the process at time $t = 10$, I will see a Poisson process. Thus, the time of the first arrival from $t = 10$ is $Exponential(2)$. In other words, we can write

$$T = 10 + X,$$

where $X \sim Exponential(2)$. Thus,

$$ET = 10 + EX$$
$$= 10 + \frac{1}{2} = \frac{21}{2},$$

$$\text{Var}(T) = \text{Var}(X)$$
$$= \frac{1}{4}.$$

(e) Arrivals before $t = 10$ are independent of arrivals after $t = 10$. Thus, knowing that the last arrival occurred at time $t = 9$ does not impact the distribution of the first arrival after $t = 10$. Thus, if A is the event that the last arrival occurred at $t = 9$, we can write

$$E[T|A] = E[T]$$
$$= \frac{21}{2},$$

$$\text{Var}(T|A) = \text{Var}(T)$$
$$= \frac{1}{4}.$$

Now that we know the distribution of the interarrival times, we can find the distribution of arrival times

$$T_1 = X_1,$$
$$T_2 = X_1 + X_2,$$
$$T_3 = X_1 + X_2 + X_3,$$

$$\vdots$$

More specifically, T_n is the sum of n independent $Exponential(\lambda)$ random variables. In previous chapters we have seen that if $T_n = X_1 + X_2 + \cdots + X_n$, where the X_i's are independent $Exponential(\lambda)$ random variables, then $T_n \sim Gamma(n, \lambda)$. This has been shown using MGFs. Note that here $n \in \mathbb{N}$. The $Gamma(n, \lambda)$ is also called **Erlang** distribution, i.e, we can write

$$T_n \sim Erlang(n, \lambda) = Gamma(n, \lambda), \qquad \text{for } n = 1, 2, 3, \cdots.$$

The PDF of T_n, for $n = 1, 2, 3, \cdots$, is given by

$$f_{T_n}(t) = \frac{\lambda^n t^{n-1} e^{-\lambda t}}{(n-1)!}, \qquad \text{for } t > 0.$$

Remember that if $X \sim Exponential(\lambda)$, then

$$E[X] = \frac{1}{\lambda},$$

$$\text{Var}(X) = \frac{1}{\lambda^2}.$$

Since $T_n = X_1 + X_2 + \cdots + X_n$, we conclude that

$$E[T_n] = nEX_1 = \frac{n}{\lambda},$$

$$\text{Var}(T_n) = n\text{Var}(X_n) = \frac{n}{\lambda^2}.$$

Note that the arrival times are not independent. In particular, we must have $T_1 \leq T_2 \leq T_3 \leq \cdots$.

Arrival Times for Poisson Processes

If $N(t)$ is a Poisson process with rate λ, then the arrival times T_1, T_2, $\cdots$ have $Gamma(n, \lambda)$ distribution. In particular, for $n = 1, 2, 3, \cdots$, we have

$$E[T_n] = \frac{n}{\lambda}, \quad \text{and} \quad \text{Var}(T_n) = \frac{n}{\lambda^2}.$$

The above discussion suggests a way to simulate (generate) a Poisson process with rate λ. We first generate i.i.d. random variables X_1, X_2, X_3, $\cdots$, where $X_i \sim Exponential(\lambda)$. Then the arrival times are given by

$$T_1 = X_1,$$
$$T_2 = X_1 + X_2,$$
$$T_3 = X_1 + X_2 + X_3,$$

$$\vdots$$

11.1.3 Merging and Splitting Poisson Processes

Merging Independent Poisson Processes:

Let $N_1(t)$ and $N_2(t)$ be two independent Poisson processes with rates λ_1 and λ_2 respectively. Let us define $N(t) = N_1(t) + N_2(t)$. That is, the random process $N(t)$ is obtained by combining the arrivals in $N_1(t)$ and $N_2(t)$ (Figure 11.5). We claim that $N(t)$ is a Poisson process with rate $\lambda = \lambda_1 + \lambda_2$. To see this, first note that

$$N(0) = N_1(0) + N_2(0)$$
$$= 0 + 0 = 0.$$

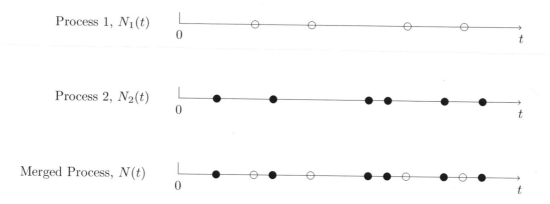

Figure 11.5: Merging two Poisson processes $N_1(t)$ and $N_2(t)$.

Next, since $N_1(t)$ and $N_2(t)$ are independent and both have independent increments, we conclude that $N(t)$ also has independent increments. Finally, consider an interval of length τ, i.e, $I = (t, t + \tau]$. Then the numbers of arrivals in I associated with $N_1(t)$ and $N_2(t)$ are $Poisson(\lambda_1 \tau)$ and $Poisson(\lambda_2 \tau)$ and they are independent. Therefore, the number of arrivals in I associated with $N(t)$ is $Poisson\big((\lambda_1 + \lambda_2)\tau\big)$ (sum of two independent Poisson random variables).

Merging Independent Poisson Processes

Let $N_1(t)$, $N_2(t)$, $\cdots$, $N_m(t)$ be m independent Poisson processes with rates λ_1, λ_2, $\cdots$, λ_m. Let also

$$N(t) = N_1(t) + N_2(t) + \cdots + N_m(t), \quad \text{for all } t \in [0, \infty).$$

Then, $N(t)$ is a Poisson process with rate $\lambda_1 + \lambda_2 + \cdots + \lambda_m$.

Splitting (Thinning) of Poisson Processes:

Here, we will talk about splitting a Poisson process into two independent Poisson processes. The idea will be better understood if we look at a concrete example.

Example 11.3. Suppose that the number of customers visiting a fast food restaurant in a given time interval I is $N \sim Poisson(\mu)$. Assume that each customer purchases a drink with probability p, independently from other customers, and independently from the value of N. Let X be the number of customers who purchase drinks in that time interval. Also, let Y be the number of customers that do not purchase drinks; so $X + Y = N$.

(a) Find the marginal PMFs of X and Y.

(b) Find the joint PMF of X and Y.

(c) Are X and Y independent?

Solution:

(a) First, note that $R_X = R_Y = \{0, 1, 2, ...\}$. Also, given $N = n$, X is a sum of n independent *Bernoulli*(p) random variables. Thus, given $N = n$, X has a binomial distribution with parameters n and p, so

$$X|N = n \quad \sim \quad Binomial(n, p),$$
$$Y|N = n \quad \sim \quad Binomial(n, q = 1 - p).$$

We have

$$P_X(k) = \sum_{n=0}^{\infty} P(X = k|N = n)P_N(n) \qquad \text{(law of total probability)}$$

$$= \sum_{n=k}^{\infty} \binom{n}{k} p^k q^{n-k} e^{-\mu} \frac{\mu^n}{n!}$$

$$= \sum_{n=k}^{\infty} \frac{p^k q^{n-k} e^{-\mu} \mu^n}{k!(n-k)!}$$

$$= \frac{e^{-\mu}(\mu p)^k}{k!} \sum_{n=k}^{\infty} \frac{(\mu q)^{n-k}}{(n-k)!}$$

$$= \frac{e^{-\mu}(\mu p)^k}{k!} e^{\mu q} \qquad \text{(Taylor series for } e^x)$$

$$= \frac{e^{-\mu p}(\mu p)^k}{k!}, \qquad \text{for } k = 0, 1, 2, ...$$

Thus, we conclude that

$$X \quad \sim \quad Poisson(\mu p).$$

Similarly, we obtain

$$Y \quad \sim \quad Poisson(\mu q).$$

(b) To find the joint PMF of X and Y, we can also use the law of total probability:

$$P_{XY}(i, j) = \sum_{n=0}^{\infty} P(X = i, Y = j|N = n)P_N(n) \qquad \text{(law of total probability)}.$$

However, note that $P(X = i, Y = j | N = n) = 0$ if $N \neq i + j$, thus

$$
\begin{aligned}
P_{XY}(i,j) &= P(X = i, Y = j | N = i + j)P_N(i + j) \\
&= P(X = i | N = i + j)P_N(i + j) \\
&= \binom{i + j}{i} p^i q^j e^{-\mu} \frac{\mu^{i+j}}{(i + j)!} \\
&= \frac{e^{-\mu}(\mu p)^i (\mu q)^j}{i! j!} \\
&= \frac{e^{-\mu p}(\mu p)^i}{i!} \cdot \frac{e^{-\mu q}(\mu q)^j}{j!} \\
&= P_X(i)P_Y(j).
\end{aligned}
$$

(c) X and Y are independent since, as we saw above,

$$
P_{XY}(i,j) = P_X(i)P_Y(j).
$$

The above example was given for a specific interval I, in which a Poisson random variable N was split to two independent Poisson random variables X and Y. However, the argument can be used to show the same result for splitting a Poisson process to two independent Poisson processes. More specifically, we have the following result.

Splitting a Poisson Process

Let $N(t)$ be a Poisson process with rate λ. Here, we divide $N(t)$ to two processes $N_1(t)$ and $N_2(t)$ in the following way (Figure 11.6). For each arrival, a coin with $P(H) = p$ is tossed. If the coin lands heads up, the arrival is sent to the first process ($N_1(t)$), otherwise it is sent to the second process. The coin tosses are independent of each other and are independent of $N(t)$. Then,

1. $N_1(t)$ is a Poisson process with rate λp;

2. $N_2(t)$ is a Poisson process with rate $\lambda(1 - p)$;

3. $N_1(t)$ and $N_2(t)$ are independent.

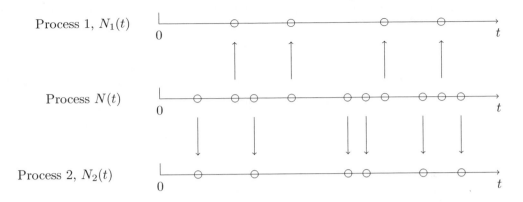

Figure 11.6: Splitting a Poisson process to two independent Poisson processes.

11.1.4 Nonhomogeneous Poisson Processes

Let $N(t)$ be the number of customers arriving at a fast food restaurant by time t. We think that the customers arrive somewhat randomly, so we might want to model $N(t)$ as a Poisson process. However, we notice that this process does not have stationary increments. For example, we note that the arrival rate of customers is larger during lunch time compared to, say, 4 p.m. In such scenarios, we might model $N(t)$ as a *nonhomogeneous Poisson process*. Such a process has all the properties of a Poisson process, except for the fact that its rate is a function of time, i.e., $\lambda = \lambda(t)$.

Nonhomogeneous Poisson Process

Let $\lambda(t) : [0, \infty) \mapsto [0, \infty)$ be an integrable function. The counting process $\{N(t), t \in [0, \infty)\}$ is called a **nonhomogeneous Poisson process** with **rate** $\lambda(t)$ if all the following conditions hold.

1. $N(0) = 0$;

2. $N(t)$ has independent increments;

3. for any $t \in [0, \infty)$, we have

$$P(N(t + \Delta) - N(t) = 0) = 1 - \lambda(t)\Delta + o(\Delta),$$
$$P(N(t + \Delta) - N(t) = 1) = \lambda(t)\Delta + o(\Delta),$$
$$P(N(t + \Delta) - N(t) \geq 2) = o(\Delta).$$

For a nonhomogeneous Poisson process with rate $\lambda(t)$, the number of arrivals in any interval is a Poisson random variable; however, its parameter can depend on the location of the interval.

More specifically, we can write

$$N(t+s) - N(t) \sim Poisson\left(\int_t^{t+s} \lambda(\alpha)d\alpha\right).$$

11.1.5   Solved Problems

1. Let $\{N(t), t \in [0, \infty)\}$ be a Poisson process with rate $\lambda = 0.5$.

   (a) Find the probability of no arrivals in $(3, 5]$.

   (b) Find the probability that there is exactly one arrival in each of the following intervals: $(0, 1]$, $(1, 2]$, $(2, 3]$, and $(3, 4]$.

   Solution:

   (a) If Y is the number arrivals in $(3, 5]$, then $Y \sim Poisson(\mu = 0.5 \times 2)$. Therefore,

   $$P(Y = 0) = e^{-1}$$
   $$= 0.37$$

   (b) Let Y_1, Y_2, Y_3 and Y_4 be the numbers of arrivals in the intervals $(0, 1]$, $(1, 2]$, $(2, 3]$, and $(3, 4]$. Then $Y_i \sim Poisson(0.5)$ and Y_i's are independent, so

   $$P(Y_1 = 1, Y_2 = 1, Y_3 = 1, Y_4 = 1) = P(Y_1 = 1) \cdot P(Y_2 = 1) \cdot P(Y_3 = 1) \cdot P(Y_4 = 1)$$
   $$= \left[0.5e^{-0.5}\right]^4$$
   $$\approx 8.5 \times 10^{-3}.$$

2. Let $\{N(t), t \in [0, \infty)\}$ be a Poisson process with rate λ. Find the probability that there are two arrivals in $(0, 2]$ and three arrivals in $(1, 4]$.

   Solution: Note that the two intervals $(0, 2]$ and $(1, 4]$ are not disjoint. Thus, we cannot multiply the probabilities for each interval to obtain the desired probability. In particular,

   $$(0, 2] \cap (1, 4] = (1, 2].$$

Let X, Y, and Z be the numbers of arrivals in $(0, 1]$, $(1, 2]$, and $(2, 4]$ respectively. Then X, Y, and Z are independent, and

$$X \sim Poisson(\lambda \cdot 1),$$
$$Y \sim Poisson(\lambda \cdot 1),$$
$$Z \sim Poisson(\lambda \cdot 2).$$

Let A be the event that there are two arrivals in $(0, 2]$ and three arrivals in $(1, 4]$. We can use the law of total probability to obtain $P(A)$. In particular,

$$P(A) = P(X + Y = 2 \text{ and } Y + Z = 3)$$
$$= \sum_{k=0}^{\infty} P(X + Y = 2 \text{ and } Y + Z = 3 | Y = k) P(Y = k)$$
$$= P(X = 2, Z = 3 | Y = 0) P(Y = 0) + P(X = 1, Z = 2 | Y = 1) P(Y = 1) +$$
$$+ P(X = 0, Z = 1 | Y = 2) P(Y = 2)$$
$$= P(X = 2, Z = 3) P(Y = 0) + P(X = 1, Z = 2) P(Y = 1) +$$
$$P(X = 0, Z = 1) P(Y = 2)$$
$$= P(X = 2) P(Z = 3) P(Y = 0) + P(X = 1) P(Z = 2) P(Y = 1) +$$
$$P(X = 0) P(Z = 1) P(Y = 2)$$
$$= \left(\frac{e^{-\lambda} \lambda^2}{2} \right) \cdot \left(\frac{e^{-2\lambda} (2\lambda)^3}{6} \right) \cdot \left(e^{-\lambda} \right) + \left(\lambda e^{-\lambda} \right) \cdot \left(\frac{e^{-2\lambda} (2\lambda)^2}{2} \right) \cdot \left(\lambda e^{-\lambda} \right) +$$
$$\left(e^{-\lambda} \right) \cdot \left(e^{-2\lambda} (2\lambda) \right) \cdot \left(\frac{e^{-\lambda} \lambda^2}{2} \right).$$

3. Let $\{N(t), t \in [0, \infty)\}$ be a Poisson Process with rate λ. Find its covariance function

$$C_N(t_1, t_2) = \text{Cov}\big(N(t_1), N(t_2)\big), \qquad \text{for } t_1, t_2 \in [0, \infty)$$

Solution: Let's assume $t_1 \geq t_2 \geq 0$. Then, by the independent increment property of the Poisson process, the two random variables $N(t_1) - N(t_2)$ and $N(t_2)$ are independent. We can write

$$C_N(t_1, t_2) = \text{Cov}\big(N(t_1), N(t_2)\big)$$
$$= \text{Cov}\big(N(t_1) - N(t_2) + N(t_2), N(t_2)\big)$$
$$= \text{Cov}\big(N(t_1) - N(t_2), N(t_2)\big) + \text{Cov}\big(N(t_2), N(t_2)\big)$$
$$= \text{Cov}\big(N(t_2), N(t_2)\big)$$
$$= \text{Var}\big(N(t_2)\big)$$
$$= \lambda t_2, \qquad \text{since } N(t_2) \sim Poisson(\lambda t_2).$$

Similarly, if $t_2 \geq t_1 \geq 0$, we conclude

$$C_N(t_1, t_2) = \lambda t_1.$$

Therefore, we can write

$$C_N(t_1, t_2) = \lambda \min(t_1, t_2), \qquad \text{for } t_1, t_2 \in [0, \infty).$$

4. Let $\{N(t), t \in [0, \infty)\}$ be a Poisson process with rate λ, and X_1 be its first arrival time. Show that given $N(t) = 1$, then X_1 is uniformly distributed in $(0, t]$. That is, show that

$$P(X_1 \leq x | N(t) = 1) = \frac{x}{t}, \qquad \text{for } 0 \leq x \leq t.$$

Solution: For $0 \leq x \leq t$, we can write

$$P(X_1 \leq x | N(t) = 1) = \frac{P(X_1 \leq x, N(t) = 1)}{P(N(t) = 1)}.$$

We know that

$$P(N(t) = 1) = \lambda t e^{-\lambda t},$$

and

$$P(X_1 \leq x, N(t) = 1) = P\left(\text{one arrival in } (0, x] \quad \text{and} \quad \text{no arrivals in } (x, t] \right)$$

$$= \left[\lambda x e^{-\lambda x} \right] \cdot \left[e^{-\lambda(t-x)} \right]$$

$$= \lambda x e^{-\lambda t}.$$

Thus,

$$P(X_1 \leq x | N(t) = 1) = \frac{x}{t}, \qquad \text{for } 0 \leq x \leq t.$$

Note: The above result can be generalized for n arrivals. That is, given that $N(t) = n$, the n arrival times have the same joint CDF as the order statistics of n independent $Uniform(0, t)$ random variables. This fact is discussed more in detail in the End of Chapter Problems.

5. Let $N_1(t)$ and $N_2(t)$ be two independent Poisson processes with rates $\lambda_1 = 1$ and $\lambda_2 = 2$, respectively. Let $N(t)$ be the merged process $N(t) = N_1(t) + N_2(t)$.

(a) Find the probability that $N(1) = 2$ and $N(2) = 5$.

(b) Given that $N(1) = 2$, find the probability that $N_1(1) = 1$.

Solution: $N(t)$ is a Poisson process with rate $\lambda = 1 + 2 = 3$.

(a) We have

$$P(N(1) = 2, N(2) = 5) = P\left(\underline{\text{two}} \text{ arrivals in } (0, 1] \text{ and } \underline{\text{three}} \text{ arrivals in } (1, 2] \right)$$
$$= \left[\frac{e^{-3} 3^2}{2!} \right] \cdot \left[\frac{e^{-3} 3^3}{3!} \right]$$
$$\approx .05$$

(b)

$$P(N_1(1) = 1 | N(1) = 2) = \frac{P\big(N_1(1) = 1, N(1) = 2\big)}{P(N(1) = 2)}$$
$$= \frac{P\big(N_1(1) = 1, N_2(1) = 1\big)}{P(N(1) = 2)}$$
$$= \frac{P\big(N_1(1) = 1\big) \cdot P\big(N_2(1) = 1\big)}{P(N(1) = 2)}$$
$$= \left[e^{-1} \cdot 2e^{-2} \right] / \left[\frac{e^{-3} 3^2}{2!} \right]$$
$$= \frac{4}{9}.$$

6. Let $N_1(t)$ and $N_2(t)$ be two independent Poisson processes with rates $\lambda_1 = 1$ and $\lambda_2 = 2$, respectively. Find the probability that the second arrival in $N_1(t)$ occurs before the third arrival in $N_2(t)$. *Hint:* One way to solve this problem is to think of $N_1(t)$ and $N_2(t)$ as two processes obtained from splitting a Poisson process.

Solution: Let $N(t)$ be a Poisson process with rate $\lambda = 1 + 2 = 3$. We split $N(t)$ into two processes $N_1(t)$ and $N_2(t)$ in the following way. For each arrival, a coin with $P(H) = \frac{1}{3}$ is tossed. If the coin lands heads up, the arrival is sent to the first process ($N_1(t)$), otherwise it is sent to the second process. The coin tosses are independent of each other and are independent of $N(t)$. Then

(a) $N_1(t)$ is a Poisson process with rate $\lambda p = 1$;

(b) $N_2(t)$ is a Poisson process with rate $\lambda(1 - p) = 2$;

(c) $N_1(t)$ and $N_2(t)$ are independent.

Thus, $N_1(t)$ and $N_2(t)$ have the same probabilistic properties as the ones stated in the problem. We can now restate the probability that the second arrival in $N_1(t)$ occurs before the third arrival in $N_2(t)$ as the probability of observing at least two heads in four coin tosses, which is

$$\sum_{k=2}^{4} \binom{4}{k} \left(\frac{1}{3}\right)^k \left(\frac{2}{3}\right)^{4-k}.$$

11.2 Discrete-Time Markov Chains

11.2.1 Introduction

Consider a discrete-time random process $\{X_m, m = 0, 1, 2, \dots\}$. In the very simple case where the X_m's are independent, the analysis of this process is relatively straightforward. In this case, there is no "memory" in the system, so each X_m can be looked at independently from previous ones.

However, for many real life processes, the independence assumption is not valid. For example, if X_m is the stock price of a company at time $m \in \{0, 1, 2, ..\}$, then it is reasonable to assume that the X_m's are dependent. Therefore, we need to develop models where the value of X_m depends on the previous values. In a Markov chain, X_{m+1} depends on X_m, but given X_m, it does not depend on the other previous values X_0, X_1, $\cdots$, X_{m-1}. That is, conditioned on X_m, the random variable X_{m+1} is independent of the random variables X_0, X_1, $\cdots$, X_{m-1}.

Markov chains are usually used to model the evolution of "states" in probabilistic systems. More specifically, consider a system with the set of possible states $S = \{s_1, s_2, ...\}$. Without loss of generality, the states are usually chosen to be $0, 1, 2, \cdots$, or $1, 2, 3, \cdots$, depending on which one is more convenient for a particular problem. If $X_n = i$, we say that the system is in state i at time n. The idea behind Markov chains is usually summarized as follows: "conditioned on the current state, the past and the future states are independent."

For example, suppose that we are modeling a queue at a bank. The number of people in the queue is a non-negative integer. Here, the state of the system can be defined as the number of people in the queue. More specifically, if X_n shows the number of people in the queue at time n, then $X_n \in S = \{0, 1, 2, ...\}$. The set S is called the *state space* of the Markov chain. Let us now provide a formal definition for discrete-time Markov chains.

Discrete-Time Markov Chains

Consider the random process $\{X_n, n = 0, 1, 2, \cdots\}$, where $R_{X_i} = S \subset \{0, 1, 2, \cdots\}$. We say that this process is a **Markov chain** if

$$P(X_{m+1} = j | X_m = i, X_{m-1} = i_{m-1}, \cdots, X_0 = i_0) = P(X_{m+1} = j | X_m = i),$$

for all $m, j, i, i_0, i_1, \cdots i_{m-1}$.

If the number of states is finite, e.g., $S = \{0, 1, 2, \cdots, r\}$, we call it a **finite** Markov chain.

If $X_n = j$, we say that the process is in state j. The numbers $P(X_{m+1} = j | X_m = i)$ are called the **transition probabilities**. We assume that the transition probabilities do not depend on time. That is, $P(X_{m+1} = j | X_m = i)$ does not depend on m. Thus, we can define

---

$$p_{ij} = P(X_{m+1} = j | X_m = i).$$

In particular, we have

$$p_{ij} = P(X_1 = j | X_0 = i) = P(X_2 = j | X_1 = i) = P(X_3 = j | X_2 = i) = \cdots.$$

In other words, if the process is in state i, it will next make a transition to state j with probability p_{ij}.

11.2.2   State Transition Matrix and Diagram

We often list the transition probabilities in a matrix. The matrix is called the **state transition matrix** or **transition probability matrix** and is usually shown by P. Assuming the states are $1, 2, \cdots, r$, then the state transition matrix is given by

$$P = \begin{bmatrix} p_{11} & p_{12} & \cdots & p_{1r} \\ p_{21} & p_{22} & \cdots & p_{2r} \\ \vdots & \vdots & \vdots & \vdots \\ p_{r1} & p_{r2} & \cdots & p_{rr} \end{bmatrix}.$$

Note that $p_{ij} \geq 0$, and for all i, we have

$$\sum_{k=1}^{r} p_{ik} = \sum_{k=1}^{r} P(X_{m+1} = k | X_m = i)$$

$$= 1.$$

This is because, given that we are in state i, the next state must be one of the possible states. Thus, when we sum over all the possible values of k, we should get one. That is, the rows of any state transition matrix must sum to one.

State Transition Diagram:

A Markov chain is usually shown by a **state transition diagram**. Consider a Markov chain with three possible states 1, 2, and 3 and the following transition probabilities

$$P = \begin{bmatrix} \frac{1}{4} & \frac{1}{2} & \frac{1}{4} \\ \frac{1}{3} & 0 & \frac{2}{3} \\ \frac{1}{2} & 0 & \frac{1}{2} \end{bmatrix}.$$

Figure 11.7 shows the state transition diagram for the above Markov chain. In this diagram, there are three possible states 1, 2, and 3, and the arrows from each state to other states show the transition probabilities p_{ij}. When there is no arrow from state i to state j, it means that $p_{ij} = 0$.

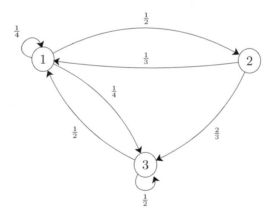

Figure 11.7: A state transition diagram.

Example 11.4. Consider the Markov chain shown in Figure 11.7.

(a) Find $P(X_4 = 3|X_3 = 2)$.

(b) Find $P(X_3 = 1|X_2 = 1)$.

(c) If we know $P(X_0 = 1) = \frac{1}{3}$, find $P(X_0 = 1, X_1 = 2)$.

(d) If we know $P(X_0 = 1) = \frac{1}{3}$, find $P(X_0 = 1, X_1 = 2, X_2 = 3)$.

Solution:

(a) By definition

$$P(X_4 = 3|X_3 = 2) = p_{23} = \frac{2}{3}.$$

(b) By definition

$$P(X_3 = 1|X_2 = 1) = p_{11} = \frac{1}{4}.$$

(c) We can write

$$P(X_0 = 1, X_1 = 2) = P(X_0 = 1)P(X_1 = 2|X_0 = 1)$$
$$= \frac{1}{3} \cdot p_{12}$$
$$= \frac{1}{3} \cdot \frac{1}{2} = \frac{1}{6}.$$

(d) We can write

$$P(X_0 = 1, X_1 = 2, X_2 = 3)$$
$$= P(X_0 = 1)P(X_1 = 2|X_0 = 1)P(X_2 = 3|X_1 = 2, X_0 = 1)$$
$$= P(X_0 = 1)P(X_1 = 2|X_0 = 1)P(X_2 = 3|X_1 = 2) \quad \text{(by Markov property)}$$
$$= \frac{1}{3} \cdot p_{12} \cdot p_{23}$$
$$= \frac{1}{3} \cdot \frac{1}{2} \cdot \frac{2}{3}$$
$$= \frac{1}{9}.$$

11.2.3   Probability Distributions

State Probability Distributions:

Consider a Markov chain $\{X_n, n = 0, 1, 2, ...\}$, where $X_n \in S = \{1, 2, \cdots, r\}$. Suppose that we know the probability distribution of X_0. More specifically, define the row vector $\pi^{(0)}$ as

$$\pi^{(0)} = \begin{bmatrix} P(X_0 = 1) & P(X_0 = 2) & \cdots & P(X_0 = r) \end{bmatrix}.$$

How can we obtain the probability distribution of X_1, X_2, $\cdots$? We can use the law of total probability. More specifically, for any $j \in S$, we can write

$$P(X_1 = j) = \sum_{k=1}^{r} P(X_1 = j|X_0 = k)P(X_0 = k)$$
$$= \sum_{k=1}^{r} p_{kj} P(X_0 = k).$$

If we generally define

$$\pi^{(n)} = \begin{bmatrix} P(X_n = 1) & P(X_n = 2) & \cdots & P(X_n = r) \end{bmatrix},$$

we can rewrite the above result in the form of matrix multiplication

$$\pi^{(1)} = \pi^{(0)} P,$$

where P is the state transition matrix. Similarly, we can write

$$\pi^{(2)} = \pi^{(1)} P = \pi^{(0)} P^2.$$

More generally, we can write

$$\pi^{(n+1)} = \pi^{(n)} P, \qquad \text{for } n = 0, 1, 2, \cdots ;$$
$$\pi^{(n)} = \pi^{(0)} P^n, \qquad \text{for } n = 0, 1, 2, \cdots .$$

Example 11.5. Consider a system that can be in one of two possible states, $S = \{0, 1\}$. In particular, suppose that the transition matrix is given by

$$P = \begin{bmatrix} \frac{1}{2} & \frac{1}{2} \\ \frac{1}{3} & \frac{2}{3} \end{bmatrix}.$$

Suppose that the system is in state 0 at time $n = 0$, i.e., $X_0 = 0$.

(a) Draw the state transition diagram.

(b) Find the probability that the system is in state 1 at time $n = 3$.

   Solution:

(a) The state transition diagram is shown in Figure 11.8.

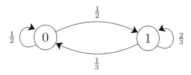

Figure 11.8: A state transition diagram.

(b) Here, we know

$$\pi^{(0)} = \begin{bmatrix} P(X_0 = 0) & P(X_0 = 1) \end{bmatrix}$$
$$= \begin{bmatrix} 1 & 0 \end{bmatrix}.$$

Thus,

$$\pi^{(3)} = \pi^{(0)} P^3$$

$$= \begin{bmatrix} 1 & 0 \end{bmatrix} \begin{bmatrix} \frac{1}{2} & \frac{1}{2} \\ \frac{1}{3} & \frac{2}{3} \end{bmatrix}^3$$

$$= \begin{bmatrix} \frac{29}{72} & \frac{43}{72} \end{bmatrix}.$$

Thus, the probability that the system is in state 1 at time $n = 3$ is $\frac{43}{72}$.

n-Step Transition Probabilities:

Consider a Markov chain $\{X_n, n = 0, 1, 2, ...\}$, where $X_n \in S$. If $X_0 = i$, then $X_1 = j$ with probability p_{ij}. That is, p_{ij} gives us the probability of going from state i to state j in one step. Now suppose that we are interested in finding the probability of going from state i to state j in two steps, i.e.,

$$p_{ij}^{(2)} = P(X_2 = j | X_0 = i).$$

We can find this probability by applying the law of total probability. In particular, we argue that X_1 can take one of the possible values in S. Thus, we can write

$$p_{ij}^{(2)} = P(X_2 = j | X_0 = i) = \sum_{k \in S} P(X_2 = j | X_1 = k, X_0 = i) P(X_1 = k | X_0 = i)$$

$$= \sum_{k \in S} P(X_2 = j | X_1 = k) P(X_1 = k | X_0 = i) \quad \text{(by Markov property)}$$

$$= \sum_{k \in S} p_{kj} p_{ik}.$$

We conclude

$$p_{ij}^{(2)} = P(X_2 = j | X_0 = i) = \sum_{k \in S} p_{ik} p_{kj} \qquad (11.4)$$

We can explain the above formula as follows. In order to get to state j, we need to pass through some intermediate state k. The probability of this event is $p_{ik}p_{kj}$. To obtain $p_{ij}^{(2)}$, we sum over all possible intermediate states. Accordingly, we can define the two-step transition matrix as follows:

$$P^{(2)} = \begin{bmatrix} p_{11}^{(2)} & p_{12}^{(2)} & \cdots & p_{1r}^{(2)} \\ p_{21}^{(2)} & p_{22}^{(2)} & \cdots & p_{2r}^{(2)} \\ \vdots & \vdots & \vdots & \vdots \\ p_{r1}^{(2)} & p_{r2}^{(2)} & \cdots & p_{rr}^{(2)} \end{bmatrix}.$$

Looking at Equation 11.4, we notice that $p_{ij}^{(2)}$ is in fact the element in the ith row and jth column of the matix

$$P^2 = \begin{bmatrix} p_{11} & p_{12} & \cdots & p_{1r} \\ p_{21} & p_{22} & \cdots & p_{2r} \\ \vdots & \vdots & \vdots & \vdots \\ p_{r1} & p_{r2} & \cdots & p_{rr} \end{bmatrix} \cdot \begin{bmatrix} p_{11} & p_{12} & \cdots & p_{1r} \\ p_{21} & p_{22} & \cdots & p_{2r} \\ \vdots & \vdots & \vdots & \vdots \\ p_{r1} & p_{r2} & \cdots & p_{rr} \end{bmatrix}.$$

Thus, we conclude that the two-step transition matrix can be obtained by squaring the state transition matrix, i.e.,

$$P^{(2)} = P^2.$$

More generally, we can define the n-step transition probabilities $p_{ij}^{(n)}$ as

$$p_{ij}^{(n)} = P(X_n = j | X_0 = i), \qquad \text{for } n = 0, 1, 2, \cdots, \tag{11.5}$$

and the n-step transition matrix, $P^{(n)}$, as

$$P^{(n)} = \begin{bmatrix} p_{11}^{(n)} & p_{12}^{(n)} & \cdots & p_{1r}^{(n)} \\ p_{21}^{(n)} & p_{22}^{(n)} & \cdots & p_{2r}^{(n)} \\ \vdots & \vdots & \vdots & \vdots \\ p_{r1}^{(n)} & p_{r2}^{(n)} & \cdots & p_{rr}^{(n)} \end{bmatrix}.$$

We can now generalize Equation 11.4. Let m and n be two positive integers and assume $X_0 = i$. In order to get to state j in $(m + n)$ steps, the chain will be at some intermediate state k after m steps. To obtain $p_{ij}^{(m+n)}$, we sum over all possible intermediate states:

$$\begin{aligned} p_{ij}^{(m+n)} &= P(X_{m+n} = j | X_0 = i) \\ &= \sum_{k \in S} p_{ik}^{(m)} p_{kj}^{(n)}. \end{aligned}$$

The above equation is called the **Chapman-Kolmogorov equation**. Similar to the case of two-step transition probabilities, we can show that $P^{(n)} = P^n$, for $n = 1, 2, 3, \cdots$.

The Chapman-Kolmogorov equation can be written as

$$\begin{aligned} p_{ij}^{(m+n)} &= P(X_{m+n} = j | X_0 = i) \\ &= \sum_{k \in S} p_{ik}^{(m)} p_{kj}^{(n)}. \end{aligned}$$

The n-step transition matrix is given by

$$P^{(n)} = P^n, \qquad \text{for } n = 1, 2, 3, \cdots.$$

11.2.4 Classification of States

To better understand Markov chains, we need to introduce some definitions. The first definition concerns the accessibility of states from each other: if it is possible to go from state i to state j, we say that state j is *accessible* from state i. In particular, we can provide the following definitions.

We say that state j is **accessible** from state i, written as $i \to j$, if $p_{ij}^{(n)} > 0$ for some n. We assume every state is accessible from itself since $p_{ii}^{(0)} = 1$.

Two states i and j are said to **communicate**, written as $i \leftrightarrow j$, if they are **accessible** from each other. In other words,

$$i \leftrightarrow j \quad \text{means} \quad i \to j \text{ and } j \to i.$$

Communication is an *equivalence* relation. That means that

- every state communicates with itself, $i \leftrightarrow i$;

- if $i \leftrightarrow j$, then $j \leftrightarrow i$;

- if $i \leftrightarrow j$ and $j \leftrightarrow k$, then $i \leftrightarrow k$.

Therefore, the states of a Markov chain can be partitioned into communicating *classes* such that only members of the same class communicate with each other. That is, two states i and j belong to the same class if and only if $i \leftrightarrow j$.

Example 11.6. Consider the Markov chain shown in Figure 11.9. It is assumed that when there is an arrow from state i to state j, then $p_{ij} > 0$. Find the equivalence classes for this Markov chain.

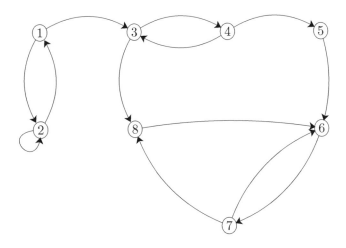

Figure 11.9: A state transition diagram.

Solution: There are four communicating classes in this Markov chain. Looking at Figure 11.10, we notice that states 1 and 2 communicate with each other, but they do not communicate with any other nodes in the graph. Similarly, nodes 3 and 4 communicate with each other, but they do not communicate with any other nodes in the graph. State 5 does not communicate with any other states, so it by itself is a class. Finally, states 6, 7, and 8 construct another class. Thus, here are the classes:

$$\text{Class } 1 = \{\text{state } 1, \text{state } 2\},$$
$$\text{Class } 2 = \{\text{state } 3, \text{state } 4\},$$
$$\text{Class } 3 = \{\text{state } 5\},$$
$$\text{Class } 4 = \{\text{state } 6, \text{state } 7, \text{state } 8\}.$$

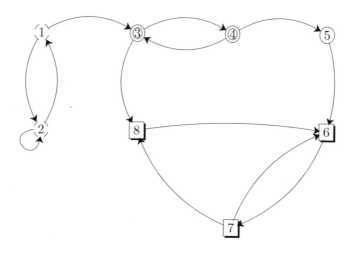

Figure 11.10: Equivalence classes.

A Markov chain is said to be *irreducible* if it has only one communicating class. As we will see shortly, irreducibility is a desirable property in the sense that it can simplify analysis of the limiting behavior.

A Markov chain is said to be **irreducible** if all states communicate with each other.

Looking at Figure 11.10, we notice that there are two kinds of classes. In particular, if at any time the Markov chain enters Class 4, it will always stay in that class. On the other hand, for other classes this is not true. For example, if $X_0 = 1$, then the Markov chain might stay in Class 1 for a while, but at some point, it will leave that class and it will never return to that class again. The states in Class 4 are called *recurrent* states, while the other states in this chain are called *transient*.

In general, a state is said to be recurrent if, any time that we leave that state, we will return to that state in the future with probability one. On the other hand, if the probability of returning is less than one, the state is called transient. Here, we provide a formal definition:

For any state i, we define

$$f_{ii} = P(X_n = i, \text{ for some } n \geq 1 | X_0 = i).$$

State i is **recurrent** if $f_{ii} = 1$, and it is **transient** if $f_{ii} < 1$.

It is relatively easy to show that if two states are in the same class, either both of them are recurrent, or both of them are transient. Thus, we can extend the above definitions to classes. A class is said to be recurrent if the states in that class are recurrent. If, on the other hand, the states are transient, the class is called transient. In general, a Markov chain might consist of several transient classes as well as several recurrent classes.

Consider a Markov chain and assume $X_0 = i$. If i is a recurrent state, then the chain will return to state i any time it leaves that state. Therefore, the chain will visit state i an infinite number of times. On the other hand, if i is a transient state, the chain will return to state i with probability $f_{ii} < 1$. Thus, in that case, the total number of visits to state i will be a Geometric random variable with parameter $1 - f_{ii}$.

Consider a discrete-time Markov chain. Let V be the total number of visits to state i.

(a) If i is a recurrent state, then

$$P(V = \infty | X_0 = i) = 1.$$

(b) If i is a transient state, then

$$V | X_0 = i \; \sim \; Geometric(1 - f_{ii}).$$

Example 11.7. Show that in a finite Markov chain, there is at least one recurrent class.

Solution: Consider a finite Markov chain with r states, $S = \{1, 2, \cdots, r\}$. Suppose that all states are transient. Then, starting from time 0, the chain might visit state 1 several times, but at some point the chain will leave state 1 and will never return to it. That is, there exists an integer $M_1 > 0$ such that $X_n \neq 1$, for all $n \geq M_1$. Similarly, there exists an integer $M_2 > 0$ such that $X_n \neq 2$, for all $n \geq M_2$, and so on. Now, if you choose

$$n \geq \max\{M_1, M_2, \cdots, M_r\},$$

then X_n cannot be equal to any of the states $1, 2, \cdots, r$. This is a contradiction, so we conclude that there must be at least one recurrent state, which means that there must be at least one recurrent class.

Periodicity:

Consider the Markov chain shown in Figure 11.11. There is a periodic pattern in this chain. Starting from state 0, we only return to 0 at times $n = 3, 6, \cdots$. In other words, $p_{00}^{(n)} = 0$, if n is not divisible by 3. Such a state is called a *periodic* state with period $d(0) = 3$.

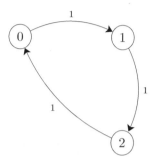

Figure 11.11: A state transition diagram.

The **period** of a state i is the largest integer d satisfying the following property: $p_{ii}^{(n)} = 0$, whenever n is not divisible by d. The period of i is shown by $d(i)$. If $p_{ii}^{(n)} = 0$, for all $n > 0$, then we let $d(i) = \infty$.

- If $d(i) > 1$, we say that state i is **periodic**.

- If $d(i) = 1$, we say that state i is **aperiodic**.

You can show that all states in the same communicating class have the same period. A class is said to be periodic if its states are periodic. Similarly, a class is said to be aperiodic if its states are aperiodic. Finally, a Markov chain is said to be aperiodic if all of its states are aperiodic.

If $i \leftrightarrow j$, then $d(i) = d(j)$.

Why is periodicity important? As we will see shortly, it plays a role when we discuss limiting distributions. It turns out that in a typical problem, we are given an irreducible Markov chain, and we need to check if it is aperiodic.

How do we check that a Markov chain is aperiodic? Here is a useful method. Remember that two numbers m and l are said to be *co-prime* if their greatest common divisor (gcd) is 1, i.e., $\gcd(l, m) = 1$. Now, suppose that we can find two co-prime numbers l and m such that

$p_{ii}^{(l)} > 0$ and $p_{ii}^{(m)} > 0$. That is, we can go from state i to itself in l steps, and also in m steps. Then, we can conclude state i is aperiodic. If we have an irreducible Markov chain, this means that the chain is aperiodic. Since the number 1 is co-prime to every integer, any state with a self-transition is aperiodic.

Consider a finite irreducible Markov chain X_n:

(a) If there is a self-transition in the chain ($p_{ii} > 0$ for some i), then the chain is aperiodic.

(b) Suppose that you can go from state i to state i in l steps, i.e., $p_{ii}^{(l)} > 0$. Also suppose that $p_{ii}^{(m)} > 0$. If $\gcd(l, m) = 1$, then state i is aperiodic.

(c) The chain is aperiodic if and only if there exists a positive integer n such that all elements of the matrix P^n are strictly positive, i.e.,

$$p_{ij}^{(n)} > 0, \quad \text{for all } i, j \in S.$$

Example 11.8. Consider the Markov chain in Example 11.6.

(a) Is Class 1 = {state 1, state 2} aperiodic?

(b) Is Class 2 = {state 3, state 4} aperiodic?

(c) Is Class 4 = {state 6, state 7, state 8} aperiodic?

Solution:

(a) Class 1 = {state 1, state 2} is aperiodic since it has a self-transition, $p_{22} > 0$.

(b) Class 2 = {state 3, state 4} is periodic with period 2.

(c) Class 4 = {state 6, state 7, state 8} is aperiodic. For example, note that we can go from state 6 to state 6 in two steps $(6-7-6)$ and in three steps $(6-7-8-6)$. Since $\gcd(2, 3) = 1$, we conclude state 6 and its class are aperiodic.

11.2.5 Using the Law of Total Probability with Recursion

A very useful technique in the analysis of Markov chains is using the law of total probability. In fact, we have already used this when finding n-step transition probabilities. In this section, we will use this technique to find **absorption probabilities, mean hitting times,** and **mean return times.** We will introduce this technique by looking at an example. We will then provide the general formulas. You should try to understand the main idea here. This way, you do not need to memorize any formulas. Let's consider the Markov chain shown in Figure 11.12.

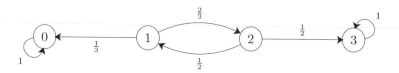

Figure 11.12: A state transition diagram.

The state transition matrix of this Markov chain is given by the following matrix.

$$P = \begin{bmatrix} 1 & 0 & 0 & 0 \\ \frac{1}{3} & 0 & \frac{2}{3} & 0 \\ 0 & \frac{1}{2} & 0 & \frac{1}{2} \\ 0 & 0 & 0 & 1 \end{bmatrix}.$$

Before going any further, let's identify the classes in this Markov chain.

Example 11.9. For the Markov chain given in Figure 11.12, answer the following questions: How many classes are there? For each class, mention if it is recurrent or transient.

Solution: There are three classes: Class 1 consists of one state, state 0, which is a recurrent state. Class two consists of two states, states 1 and 2, both of which are transient. Finally, class three consists of one state, state 3, which is a recurrent state.

Note that states 0 and 3 have the following property: once you enter those states, you never leave them. For this reason, we call them **absorbing** states. For our example here, there are two absorbing states. The process will eventually get absorbed in one of them. The first question that we would like to address deals with finding absorption probabilities.

Absorption Probabilities:

Consider the Markov chain in Figure 11.12. Let's define a_i as the absorption probability in state 0 if we start from state i. More specifically,

$$a_0 = P(\text{absorption in } 0|X_0 = 0),$$
$$a_1 = P(\text{absorption in } 0|X_0 = 1),$$
$$a_2 = P(\text{absorption in } 0|X_0 = 2),$$
$$a_3 = P(\text{absorption in } 0|X_0 = 3).$$

By the above definition, we have $a_0 = 1$ and $a_3 = 0$. To find the values of a_1 and a_2, we apply the law of total probability with recursion. The main idea is the following: if $X_n = i$, then the next state will be $X_{n+1} = k$ with probability p_{ik}. Thus, we can write

$$a_i = \sum_k a_k p_{ik}, \qquad \text{for } i = 0, 1, 2, 3 \tag{11.6}$$

Solving the above equations will give us the values of a_1 and a_2. More specifically, using Equation 11.6, we obtain

$$a_0 = a_0,$$

$$a_1 = \frac{1}{3}a_0 + \frac{2}{3}a_2,$$

$$a_2 = \frac{1}{2}a_1 + \frac{1}{2}a_3,$$

$$a_3 = a_3.$$

We also know $a_0 = 1$ and $a_3 = 0$. Solving for a_1 and a_2, we obtain

$$a_1 = \frac{1}{2},$$

$$a_2 = \frac{1}{4}.$$

Let's now define b_i as the absorption probability in state 3 if we start from state i. Since $a_i + b_i = 1$, we conclude

$$b_0 = 0, \quad b_1 = \frac{1}{2}, \quad b_2 = \frac{3}{4}, \quad b_3 = 1.$$

Nevertheless, for practice, let's find the b_i's directly.

Example 11.10. Consider the Markov chain in Figure 11.12. Let's define b_i as the absorption probability in state 3 if we start from state i. Use the above procedure to obtain b_i for $i = 0, 1, 2, 3$.

Solution: From the definition of b_i and the Markov chain graph, we have $b_0 = 0$ and $b_3 = 1$. Writing Equation 11.6 for $i = 1, 2$, we obtain

$$b_1 = \frac{1}{3}b_0 + \frac{2}{3}b_2$$

$$= \frac{2}{3}b_2,$$

$$b_2 = \frac{1}{2}b_1 + \frac{1}{2}b_3$$

$$= \frac{1}{2}b_1 + \frac{1}{2}.$$

Solving the above equations, we obtain

$$b_1 = \frac{1}{2},$$

$$b_2 = \frac{3}{4}.$$

Absorption Probabilities

Consider a finite Markov chain $\{X_n, n = 0, 1, 2, \cdots\}$ with state space $S = \{0, 1, 2, \cdots, r\}$. Suppose that all states are either absorbing or transient. Let $l \in S$ be an absorbing state. Define

$$a_i = P(\text{absorption in } l | X_0 = i), \qquad \text{for all } i \in S.$$

By the above definition, we have $a_l = 1$, and $a_j = 0$ if j is any other absorbing state. To find the unknown values of a_i's, we can use the following equations

$$a_i = \sum_k a_k p_{ik}, \qquad \text{for } i \in S.$$

In general, a finite Markov chain might have several transient as well as several recurrent classes. As n increases, the chain will get absorbed in one of the recurrent classes and it will stay there forever. We can use the above procedure to find the probability that the chain will get absorbed in each of the recurrent classes. In particular, we can replace each recurrent class with one absorbing state. Then, the resulting chain consists of only transient and absorbing states. We can then follow the above procedure to find absorption probabilities. An example of this procedure is provided in the Solved Problems Section (See Problem 2 in Section 11.2.7).

Mean Hitting Times:

We now would like to study the expected time until the process hits a certain set of states for the first time. Again, consider the Markov chain in Figure 11.12. Let's define t_i as the number of steps needed until the chain hits state 0 or state 3, given that $X_0 = i$. In other words, t_i is the expected time (number of steps) until the chain is absorbed in 0 or 3, given that $X_0 = i$. By this definition, we have $t_0 = t_3 = 0$.

To find t_1 and t_2, we use the law of total probability with recursion as before. For example, if $X_0 = 1$, then after one step, we have $X_1 = 0$ or $X_1 = 2$. Thus, we can write

$$t_1 = 1 + \frac{1}{3}t_0 + \frac{2}{3}t_2$$
$$= 1 + \frac{2}{3}t_2.$$

Similarly, we can write

$$t_2 = 1 + \frac{1}{2}t_1 + \frac{1}{2}t_3$$
$$= 1 + \frac{1}{2}t_1.$$

Solving the above equations, we obtain

$$t_1 = \frac{5}{2}, \qquad t_2 = \frac{9}{4}.$$

Generally, let $A \subset S$ be a set of states. The above procedure can be used to find the expected time until the chain first hits one of the states in the set A.

Mean Hitting Times

Consider a finite Markov chain $\{X_n, n = 0, 1, 2, \cdots\}$ with state space $S = \{0, 1, 2, \cdots, r\}$. Let $A \subset S$ be a set of states. Let T be the first time the chain visits a state in A. For all $i \in S$, define

$$t_i = E[T|X_0 = i].$$

By the above definition, we have $t_j = 0$, for all $j \in A$. To find the unknown values of t_i's, we can use the following equations

$$t_i = 1 + \sum_k t_k p_{ik}, \qquad \text{for } i \in S - A.$$

Mean Return Times:

Another interesting random variable is the *first return time*. In particular, assuming the chain is in state l, we consider the expected time (number of steps) needed until the chain returns to state l. For example, consider a Markov chain for which $X_0 = 2$. If the chain gets the values

$$X_0 = 2, \ X_1 = 1, \ X_2 = 4, \ X_3 = 3, \ X_4 = 2, \ X_5 = 3, \ X_6 = 2, \ X_7 = 3, \ \cdots,$$

then the first return to state 2 occurs at time $n = 4$. Thus, the first return time to state 2 is equal to 4 for this example. Here, we are interested in the expected value of the first return time. In particular, assuming $X_0 = l$, let's define r_l as the expected number of steps needed until the chain returns to state l. To make the definition more precise, let's define

$$R_l = \min\{n \geq 1 : X_n = l\}.$$

Then,

$$r_l = E[R_l|X_0 = l].$$

Note that by definition, $R_l \geq 1$, so we conclude $r_l \geq 1$. In fact, $r_l = 1$ if and only if l is an absorbing state (i.e., $p_{ll} = 1$).

As before, we can apply the law of total probability to obtain r_l. Again, let's define t_k as the expected time until the chain hits state l for the first time, given that $X_0 = k$. We have already seen how to find t_k's (mean hitting times). Using the law of total probability, we can write

$$r_l = 1 + \sum_k p_{lk} t_k.$$

Let's look at an example to see how we can find the mean return time.

Example 11.11. Consider the Markov chain shown in Figure 11.13. Let t_k be the expected number of steps until the chain hits state 1 for the first time, given that $X_0 = k$. Clearly, $t_1 = 0$. Also, let r_1 be the mean return time to state 1.

1. Find t_2 and t_3.

2. Find r_1.

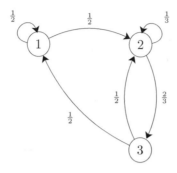

Figure 11.13: A state transition diagram.

Solution:

1. To find t_2 and t_3, we use the law of total probability with recursion as before. For example, if $X_0 = 2$, then after one step, we have $X_1 = 2$ or $X_1 = 3$. Thus, we can write

$$t_2 = 1 + \frac{1}{3}t_2 + \frac{2}{3}t_3.$$

Similarly, we can write

$$t_3 = 1 + \frac{1}{2}t_1 + \frac{1}{2}t_2$$
$$= 1 + \frac{1}{2}t_2.$$

Solving the above equations, we obtain

$$t_2 = 5, \qquad t_3 = \frac{7}{2}.$$

2. To find r_1, we note that if $X_0 = 1$, then $X_1 = 1$ or $X_1 = 2$. We can write

$$r_1 = 1 + \frac{1}{2} \cdot t_1 + \frac{1}{2}t_2$$
$$= 1 + \frac{1}{2} \cdot 0 + \frac{1}{2} \cdot 5$$
$$= \frac{7}{2}.$$

Here, we summarize the formulas for finding the mean return times. As we mentioned before, there is no need to memorize these formulas once you understand how they are derived.

<u>Mean Return Times</u>

Consider a finite irreducible Markov chain $\{X_n, n = 0, 1, 2, \cdots\}$ with state space $S = \{0, 1, 2, \cdots, r\}$. Let $l \in S$ be a state. Let r_l be the **mean return time** to state l. Then

$$r_l = 1 + \sum_k t_k p_{lk},$$

where t_k is the expected time until the chain hits state l given $X_0 = k$. Specifically,

$$t_l = 0,$$
$$t_k = 1 + \sum_j t_j p_{kj}, \qquad \text{for } k \neq l.$$

11.2.6 Stationary and Limiting Distributions

Here, we would like to discuss long-term behavior of Markov chains. In particular, we would like to know the fraction of times that the Markov chain spends in each state as n becomes large. More specifically, we would like to study the distributions

$$\pi^{(n)} = \begin{bmatrix} P(X_n = 0) & P(X_n = 1) & \cdots \end{bmatrix}$$

as $n \to \infty$. To better understand the subject, we will first look at an example and then provide a general analysis.

Example 11.12. Consider a Markov chain with two possible states, $S = \{0, 1\}$. In particular, suppose that the transition matrix is given by

$$P = \begin{bmatrix} 1 - a & a \\ b & 1 - b \end{bmatrix},$$

where a and b are two real numbers in the interval $[0, 1]$ such that $0 < a + b < 2$. Suppose that the system is in state 0 at time $n = 0$ with probability α, i.e.,

$$\pi^{(0)} = \begin{bmatrix} P(X_0 = 0) & P(X_0 = 1) \end{bmatrix} = \begin{bmatrix} \alpha & 1 - \alpha \end{bmatrix},$$

where $\alpha \in [0, 1]$.

(a) Using induction (or any other method), show that

$$P^n = \frac{1}{a + b} \begin{bmatrix} b & a \\ b & a \end{bmatrix} + \frac{(1 - a - b)^n}{a + b} \begin{bmatrix} a & -a \\ -b & b \end{bmatrix}.$$

(b) Show that

$$\lim_{n\to\infty} P^n = \frac{1}{a+b} \begin{bmatrix} b & a \\ b & a \end{bmatrix}.$$

(c) Show that

$$\lim_{n\to\infty} \pi^{(n)} = \begin{bmatrix} \frac{b}{a+b} & \frac{a}{a+b} \end{bmatrix}.$$

Solution:

(a) For $n = 1$, we have

$$P^1 = \begin{bmatrix} 1-a & a \\ b & 1-b \end{bmatrix}$$

$$= \frac{1}{a+b} \begin{bmatrix} b & a \\ b & a \end{bmatrix} + \frac{1-a-b}{a+b} \begin{bmatrix} a & -a \\ -b & b \end{bmatrix}.$$

Assuming that the statement of the problem is true for n, we can write P^{n+1} as

$$P^{n+1} = P^n P = \frac{1}{a+b} \left(\begin{bmatrix} b & a \\ b & a \end{bmatrix} + (1-a-b)^n \begin{bmatrix} a & -a \\ -b & b \end{bmatrix} \right) \cdot \begin{bmatrix} 1-a & a \\ b & 1-b \end{bmatrix}$$

$$= \frac{1}{a+b} \begin{bmatrix} b & a \\ b & a \end{bmatrix} + \frac{(1-a-b)^{n+1}}{a+b} \begin{bmatrix} a & -a \\ -b & b \end{bmatrix},$$

which completes the proof.

(b) By assumption $0 < a + b < 2$, which implies $-1 < 1 - a - b < 1$. Thus,

$$\lim_{n\to\infty} (1 - a - b)^n = 0.$$

Therefore,

$$\lim_{n\to\infty} P^n = \frac{1}{a+b} \begin{bmatrix} b & a \\ b & a \end{bmatrix}.$$

(c) We have

$$\lim_{n\to\infty} \pi^{(n)} = \lim_{n\to\infty} \left[\pi^{(0)} P^n \right]$$

$$= \pi^{(0)} \lim_{n\to\infty} P^n$$

$$= \begin{bmatrix} \alpha & 1-\alpha \end{bmatrix} \cdot \frac{1}{a+b} \begin{bmatrix} b & a \\ b & a \end{bmatrix}$$

$$= \begin{bmatrix} \frac{b}{a+b} & \frac{a}{a+b} \end{bmatrix}.$$

In the above example, the vector

$$\lim_{n\to\infty} \pi^{(n)} = \begin{bmatrix} \frac{b}{a+b} & \frac{a}{a+b} \end{bmatrix}$$

is called the *limiting distribution* of the Markov chain. Note that the limiting distribution does not depend on the initial probabilities α and $1 - \alpha$. In other words, the initial state (X_0) does not matter as n becomes large. Thus, for $i = 1, 2$, we can write

$$\lim_{n\to\infty} P(X_n = 0 | X_0 = i) = \frac{b}{a+b},$$
$$\lim_{n\to\infty} P(X_n = 1 | X_0 = i) = \frac{a}{a+b}.$$

Remember that we show $P(X_n = j | X_0 = i)$ by $P_{ij}^{(n)}$, which is the entry in the ith row and jth column of P^n.

Limiting Distributions

The probability distribution $\pi = [\pi_0, \pi_1, \pi_2, \cdots]$ is called the **limiting distribution** of the Markov chain X_n if

$$\pi_j = \lim_{n\to\infty} P(X_n = j | X_0 = i)$$

for all $i, j \in S$, and we have

$$\sum_{j \in S} \pi_j = 1.$$

By the above definition, when a limiting distribution exists, it does not depend on the initial state ($X_0 = i$), so we can write

$$\pi_j = \lim_{n\to\infty} P(X_n = j), \text{ for all } j \in S.$$

So far we have shown that the Markov chain in Example 11.12 has the following limiting distribution:

$$\pi = \begin{bmatrix} \pi_0 & \pi_1 \end{bmatrix} = \begin{bmatrix} \frac{b}{a+b} & \frac{a}{a+b} \end{bmatrix}.$$

Let's now look at mean return times for this Markov chain.

Example 11.13. Consider a Markov chain in Example 11.12: a Markov chain with two possible states, $S = \{0, 1\}$, and the transition matrix

$$P = \begin{bmatrix} 1 - a & a \\ b & 1 - b \end{bmatrix},$$

where a and b are two real numbers in the interval $[0,1]$ such that $0 < a+b < 2$. Find the mean return times, r_0 and r_1, for this Markov chain.

Solution: We can use the method of the law of total probability that we explained before to find the mean return times (Example 11.11). We can also find r_0 and r_1 directly as follows:

Let R be the first return time to state 0, i.e., $r_0 = E[R|X_0 = 0]$. If $X_0 = 0$, then $X_1 = 0$ with probability $1 - a$, and $X_1 = 1$ with probability a. Thus, using the law of total probability, and assuming $X_0 = 0$, we can write

$$r_0 = E[R|X_1 = 0]P(X_1 = 0) + E[R|X_1 = 1]P(X_1 = 1)$$
$$= E[R|X_1 = 0] \cdot (1 - a) + E[R|X_1 = 1] \cdot a.$$

If $X_1 = 0$, then $R = 1$, so

$$E[R|X_1 = 0] = 1.$$

If $X_1 = 1$, then $R \sim 1 + Geometric(b)$, so

$$E[R|X_1 = 1] = 1 + E[Geometric(b)]$$
$$= 1 + \frac{1}{b}.$$

We conclude

$$r_0 = E[R|X_1 = 0]P(X_1 = 0) + E[R|X_1 = 1]P(X_1 = 1)$$
$$= 1 \cdot (1 - a) + \left(1 + \frac{1}{b}\right) \cdot a$$
$$= \frac{a + b}{b}.$$

Similarly, we can obtain the mean return time to state 1:

$$r_1 = \frac{a + b}{a}.$$

We notice that for this example, the mean return times are given by the inverse of the limiting probabilities. In particular, we have

$$r_0 = \frac{1}{\pi_0}, \qquad r_1 = \frac{1}{\pi_1}.$$

As we will see shortly, this is not a coincidence. In fact, we can explain this intuitively. The larger the π_i is, the smaller the r_i will be. For example, if $\pi_i = \frac{1}{4}$, we conclude that the chain is in state i one-fourth of the time. In this case, $r_i = 4$, which means that on average it takes the chain four time units to go back to state i.

The two-state Markov chain discussed above is a "nice" one in the sense that it has a well-defined limiting behavior that does not depend on the initial probability distribution (PMF of X_0). However, not all Markov chains are like that. For example, consider the same Markov chain; however, choose $a = b = 1$. In this case, the chain has a periodic behavior, i.e.,

$$X_{n+2} = X_n, \qquad \text{for all } n.$$

In particular,

$$X_n = \begin{cases} X_0 & \text{if } n \text{ is even} \\ X_1 & \text{if } n \text{ is odd} \end{cases}$$

In this case, the distribution of X_n does not converge to a single PMF. Also, the distribution of X_n depends on the initial distribution.

As another example, if we choose $a = b = 0$, the chain will consist of two disconnected nodes. In this case,

$$X_n = X_0, \qquad \text{for all } n.$$

Here again, the PMF of X_n depends on the initial distribution.

Now, the question that arises here is: when does a Markov chain have a limiting distribution (that does not depend on the initial PMF)? We will next discuss this question. We will first consider finite Markov chains and then discuss infinite Markov chains.

Finite Markov Chains:

Here, we consider Markov chains with a finite number of states. In general, a finite Markov chain can consist of several transient as well as recurrent states. As n becomes large, the chain will enter a recurrent class and it will stay there forever. Therefore, when studying long-run behaviors we focus only on the recurrent classes.

If a finite Markov chain has more than one recurrent class, then the chain will get absorbed in one of the recurrent classes. Thus, the first question is: in which recurrent class does the chain get absorbed? We have already seen how to address this when we discussed absorption probabilities (see Section 11.2.5, and Problem 2 of in Section 11.2.7).

Thus, we can limit our attention to the case where our Markov chain consists of one recurrent class. In other words, we have an irreducible Markov chain. Note that as we showed in Example 11.7, in any finite Markov chain, there is at least one recurrent class. Therefore, in finite irreducible chains, all states are recurrent.

It turns out that in this case the Markov chain has a well-defined limiting behavior if it is aperiodic (states have period 1). How do we find the limiting distribution? The trick is to find a *stationary distribution*. Here is the idea: If $\pi = [\pi_1, \pi_2, \cdots]$ is a limiting distribution for a Markov chain, then we have

$$\pi = \lim_{n \to \infty} \pi^{(n)}$$
$$= \lim_{n \to \infty} \left[\pi^{(0)} P^n \right].$$

Similarly, we can write

$$\begin{aligned}
\pi &= \lim_{n\to\infty} \pi^{(n+1)} \\
&= \lim_{n\to\infty} \left[\pi^{(0)} P^{n+1} \right] \\
&= \lim_{n\to\infty} \left[\pi^{(0)} P^n P \right] \\
&= \left[\lim_{n\to\infty} \pi^{(0)} P^n \right] P \\
&= \pi P.
\end{aligned}$$

We can explain the equation $\pi = \pi P$ intuitively: Suppose that X_n has distribution π. As we saw before, πP gives the probability distribution of X_{n+1}. If we have $\pi = \pi P$, we conclude that X_n and X_{n+1} have the same distribution. In other words, the chain has reached its *steady-state* (limiting) distribution. We can equivalently write $\pi = \pi P$ as

$$\pi_j = \sum_{k\in S} \pi_k P_{kj}, \quad \text{for all } j \in S.$$

The righthand side gives the probability of going to state j in the next step. When we equate both sides, we are implying that the probability of being in state j in the next step is the same as the probability of being in state j now.

Example 11.14. Consider a Markov chain in Example 11.12: a Markov chain with two possible states, $S = \{0, 1\}$, and the transition matrix

$$P = \begin{bmatrix} 1-a & a \\ b & 1-b \end{bmatrix},$$

where a and b are two real numbers in the interval $[0, 1]$ such that $0 < a + b < 2$. Using

$$\pi = \pi P,$$

find the limiting distribution of this Markov chain.

Solution: Let $\pi = [\pi_0, \pi_1]$. Then, we can write

$$\begin{aligned}
[\pi_0, \pi_1] = \pi P = [\pi_0, \pi_1] &\begin{bmatrix} 1-a & a \\ b & 1-b \end{bmatrix} \\
&= \left[\pi_0(1-a) + \pi_1 b \quad \pi_0 a + \pi_1(1-b) \right].
\end{aligned}$$

We obtain two equations; however, they both simplify to

$$\pi_0 a = \pi_1 b.$$

We remember that π must be a valid probability distribution, i.e., $\pi_0 + \pi_1 = 1$. Thus, we can obtain a unique solution, i.e.,

$$\pi = \begin{bmatrix} \pi_0 & \pi_1 \end{bmatrix} = \begin{bmatrix} \frac{b}{a+b} & \frac{a}{a+b} \end{bmatrix}$$

which is the same answer that we obtained previously.

We now summarize the above discussion in the following theorem.

Theorem 11.2. Consider a finite Markov chain $\{X_n, n = 0, 1, 2, ...\}$ where $X_n \in S = \{0, 1, 2, \cdots, r\}$. Assume that the chain is <u>irreducible</u> and <u>aperiodic</u>. Then,

1. The set of equations

$$\pi = \pi P,$$

$$\sum_{j \in S} \pi_j = 1$$

   has a unique solution.

2. The unique solution to the above equations is the limiting distribution of the Markov chain, i.e.,

$$\pi_j = \lim_{n \to \infty} P(X_n = j | X_0 = i),$$

   for all $i, j \in S$.

3. We have

$$r_j = \frac{1}{\pi_j}, \qquad \text{for all } j \in S,$$

   where r_j is the mean return time to state j.

In practice, if we are given a finite irreducible Markov chain with states $0, 1, 2, \cdots, r$, we first find a stationary distribution. That is, we find a probability distribution π that satisfies

$$\pi_j = \sum_{k \in S} \pi_k P_{kj}, \qquad \text{for all } j \in S,$$

$$\sum_{j \in S} \pi_j = 1.$$

In this case, if the chain is also aperiodic, we conclude that the stationary distribution is a limiting distribution.

Example 11.15. Consider the Markov chain shown in Figure 11.14.

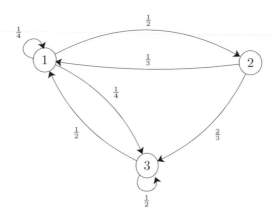

Figure 11.14: A state transition diagram.

(a) Is this chain irreducible?

(b) Is this chain aperiodic?

(c) Find the stationary distribution for this chain.

(d) Is the stationary distribution a limiting distribution for the chain?

Solution:

(a) The chain is irreducible since we can go from any state to any other states in a finite number of steps.

(b) Since there is a self-transition, i.e., $p_{11} > 0$, we conclude that the chain is aperiodic.

(c) To find the stationary distribution, we need to solve

$$\pi_1 = \frac{1}{4}\pi_1 + \frac{1}{3}\pi_2 + \frac{1}{2}\pi_3,$$
$$\pi_2 = \frac{1}{2}\pi_1,$$
$$\pi_3 = \frac{1}{4}\pi_1 + \frac{2}{3}\pi_2 + \frac{1}{2}\pi_3,$$
$$\pi_1 + \pi_2 + \pi_3 = 1.$$

We find

$$\pi_1 = \frac{3}{8}, \quad \pi_2 = \frac{3}{16}, \quad \pi_3 = \frac{7}{16}.$$

(d) Since the chain is irreducible and aperiodic, we conclude that the above stationary distribution is a limiting distribution.

Countably Infinite Markov Chains:

When a Markov chain has an infinite (but countable) number of states, we need to distinguish between two types of recurrent states: *positive* recurrent and *null* recurrent states.

Remember that if state i is recurrent, then that state will be visited an infinite number of times (any time that we visit that state, we will return to it with probability one in the future). We previously defined r_i as the expected number of transitions between visits to state i. Consider a recurrent state i. If $r_i < \infty$, then state i is a *positive* recurrent state. Otherwise, it is called *null* recurrent.

Let i be a recurrent state. Assuming $X_0 = i$, let R_i be the number of transitions needed to return to state i, i.e.,

$$R_i = \min\{n \geq 1 : X_n = i\}.$$

If $r_i = E[R_i|X_0 = i] < \infty$, then i is said to be **positive recurrent**. If $E[R_i|X_0 = i] = \infty$, then i is said to be **null recurrent**.

Theorem 11.3. Consider an infinite Markov chain $\{X_n, n = 0, 1, 2, ...\}$ where $X_n \in S = \{0, 1, 2, \cdots\}$. Assume that the chain is <u>irreducible</u> and <u>aperiodic</u>. Then, one of the following cases can occur:

1. All states are <u>transient</u>, and

$$\lim_{n \to \infty} P(X_n = j|X_0 = i) = 0, \text{ for all } i, j.$$

2. All states are <u>null recurrent</u>, and

$$\lim_{n \to \infty} P(X_n = j|X_0 = i) = 0, \text{ for all } i, j.$$

3. All states are <u>positive recurrent</u>. In this case, there exists a limiting distribution, $\pi = [\pi_0, \pi_1, \cdots]$, where

$$\pi_j = \lim_{n \to \infty} P(X_n = j|X_0 = i) > 0,$$

for all $i, j \in S$. The limiting distribution is the unique solution to the equations

$$\pi_j = \sum_{k=0}^{\infty} \pi_k P_{kj}, \qquad \text{for } j = 0, 1, 2, \cdots,$$

$$\sum_{j=0}^{\infty} \pi_j = 1.$$

We also have

$$r_j = \frac{1}{\pi_j}, \qquad \text{for all } j = 0, 1, 2, \cdots,$$

where r_j is the mean return time to state j.

How do we use the above theorem? Consider an infinite Markov chain $\{X_n, n = 0, 1, 2, ...\}$, where $X_n \in S = \{0, 1, 2, \cdots\}$. Assume that the chain is irreducible and aperiodic. We first try to find a stationary distribution π by solving the equations

$$\pi_j = \sum_{k=0}^{\infty} \pi_k P_{kj}, \qquad \text{for } j = 0, 1, 2, \cdots,$$

$$\sum_{j=0}^{\infty} \pi_j = 1.$$

If the above equations have a unique solution, we conclude that the chain is positive recurrent and the stationary distribution is the limiting distribution of this chain. On the other hand, if no stationary solution exists, we conclude that the chain is either transient or null recurrent, so

$$\lim_{n \to \infty} P(X_n = j | X_0 = i) = 0, \text{ for all } i, j.$$

Example 11.16. Consider the Markov chain shown in Figure 11.15. Assume that $0 < p < \frac{1}{2}$. Does this chain have a limiting distribution?

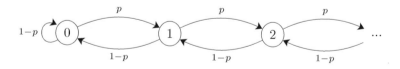

Figure 11.15: A state transition diagram.

Solution: This chain is irreducible since all states communicate with each other. It is also aperiodic since it includes a self-transition, $P_{00} > 0$. Let's write the equations for a stationary distribution. For state 0, we can write

$$\pi_0 = (1 - p)\pi_0 + (1 - p)\pi_1,$$

which results in

$$\pi_1 = \frac{p}{1 - p}\pi_0.$$

For state 1, we can write

$$\pi_1 = p\pi_0 + (1 - p)\pi_2$$
$$= (1 - p)\pi_1 + (1 - p)\pi_2,$$

which results in

$$\pi_2 = \frac{p}{1 - p}\pi_1.$$

Similarly, for any $j \in \{1, 2, \cdots \}$, we obtain

$$\pi_j = \alpha\pi_{j-1},$$

where $\alpha = \frac{p}{1-p}$. Note that since $0 < p < \frac{1}{2}$, we conclude that $0 < \alpha < 1$. We obtain

$$\pi_j = \alpha^j \pi_0, \qquad \text{for } j = 1, 2, \cdots .$$

Finally, we must have

$$1 = \sum_{j=0}^{\infty} \pi_j$$

$$= \sum_{j=0}^{\infty} \alpha^j \pi_0, \qquad \text{(where } 0 < \alpha < 1\text{)}$$

$$= \frac{1}{1 - \alpha}\pi_0 \qquad \text{(geometric series).}$$

Thus, $\pi_0 = 1 - \alpha$. Therefore, the stationary distribution is given by

$$\pi_j = (1 - \alpha)\alpha^j, \qquad \text{for } j = 0, 1, 2, \cdots .$$

Since this chain is irreducible and aperiodic and we have found a stationary distribution, we conclude that all states are positive recurrent and $\pi = [\pi_0, \pi_1, \cdots]$ is the limiting distribution.

11.2.7   Solved Problems

1. Consider the Markov chain with three states, $S = \{1, 2, 3\}$, that has the following transition matrix

$$P = \begin{bmatrix} \frac{1}{2} & \frac{1}{4} & \frac{1}{4} \\ \frac{1}{3} & 0 & \frac{2}{3} \\ \frac{1}{2} & \frac{1}{2} & 0 \end{bmatrix}.$$

(a) Draw the state transition diagram for this chain.

(b) If we know $P(X_1 = 1) = P(X_1 = 2) = \frac{1}{4}$, find $P(X_1 = 3, X_2 = 2, X_3 = 1)$.

Solution:

(a) The state transition diagram is shown in Figure 11.16

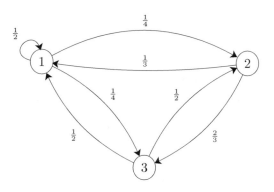

Figure 11.16: A state transition diagram.

(b) First, we obtain

$$P(X_1 = 3) = 1 - P(X_1 = 1) - P(X_1 = 2)$$
$$= 1 - \frac{1}{4} - \frac{1}{4}$$
$$= \frac{1}{2}.$$

We can now write

$$P(X_1 = 3, X_2 = 2, X_3 = 1) = P(X_1 = 3) \cdot p_{32} \cdot p_{21}$$
$$= \frac{1}{2} \cdot \frac{1}{2} \cdot \frac{1}{3}$$
$$= \frac{1}{12}.$$

2. Consider the Markov chain in Figure 11.17. There are two recurrent classes, $R_1 = \{1, 2\}$, and $R_2 = \{5, 6, 7\}$. Assuming $X_0 = 3$, find the probability that the chain gets absorbed in R_1.

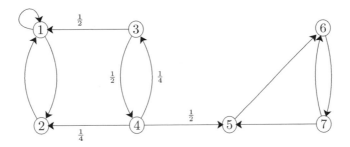

Figure 11.17: A state transition diagram.

Solution: Here, we can replace each recurrent class with one absorbing state. The resulting state diagram is shown in Figure 11.18

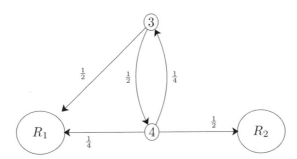

Figure 11.18: The state transition diagram in which we have replaced each recurrent class with one absorbing state

Now we can apply our standard methodology to find the probability of absorption in state R_1. In particular, define

$$a_i = P(\text{absorption in } R_1 | X_0 = i), \qquad \text{for all } i \in S.$$

By the above definition, we have $a_{R_1} = 1$, and $a_{R_2} = 0$. To find the unknown values of a_i's, we can use the following equations

$$a_i = \sum_k a_k p_{ik}, \qquad \text{for } i \in S.$$

We obtain

$$a_3 = \frac{1}{2}a_{R_1} + \frac{1}{2}a_4$$
$$= \frac{1}{2} + \frac{1}{2}a_4,$$

$$a_4 = \frac{1}{4}a_{R_1} + \frac{1}{4}a_3 + \frac{1}{2}a_{R_2}$$
$$= \frac{1}{4} + \frac{1}{4}a_3.$$

Solving the above equations, we obtain

$$a_3 = \frac{5}{7}, \qquad a_4 = \frac{3}{7}.$$

Therefore, if $X_0 = 3$, the chain will end up in class R_1 with probability $a_3 = \frac{5}{7}$.

3. Consider the Markov chain of Example 2. Again assume $X_0 = 3$. We would like to find the expected time (number of steps) until the chain gets absorbed in R_1 or R_2. More specifically, let T be the absorption time, i.e., the first time the chain visits a state in R_1 or R_2. We would like to find $E[T|X_0 = 3]$.

Solution: Here we follow our standard procedure for finding mean hitting times. Consider Figure 11.18. Let T be the first time the chain visits R_1 or R_2. For all $i \in S$, define

$$t_i = E[T|X_0 = i].$$

By the above definition, we have $t_{R_1} = t_{R_2} = 0$. To find t_3 and t_4, we can use the following equations

$$t_i = 1 + \sum_k t_k p_{ik}, \qquad \text{for } i = 3, 4.$$

Specifically, we obtain

$$t_3 = 1 + \frac{1}{2}t_{R_1} + \frac{1}{2}t_4$$
$$= 1 + \frac{1}{2}t_4,$$

$$t_4 = 1 + \frac{1}{4}t_{R_1} + \frac{1}{4}t_3 + \frac{1}{2}t_{R_2}$$
$$= 1 + \frac{1}{4}t_3.$$

Solving the above equations, we obtain

$$t_3 = \frac{12}{7}, \qquad t_4 = \frac{10}{7}.$$

Therefore, if $X_0 = 3$, it will take on average $\frac{12}{7}$ steps until the chain gets absorbed in R_1 or R_2.

4. Consider the Markov chain shown in Figure 11.19. Assume $X_0 = 1$, and let R be the first time that the chain returns to state 1, i.e.,

$$R = \min\{n \geq 1 : X_n = 1\}.$$

Find $E[R|X_0 = 1]$.

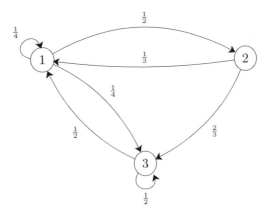

Figure 11.19: A state transition diagram.

Solution: In this question, we are asked to find the mean return time to state 1. Let r_1 be the mean return time to state 1, i.e., $r_1 = E[R|X_0 = 1]$. Then

$$r_1 = 1 + \sum_k t_k p_{1k},$$

where t_k is the expected time until the chain hits state 1 given $X_0 = k$. Specifically,

$$t_1 = 0,$$
$$t_k = 1 + \sum_j t_j p_{kj}, \qquad \text{for } k \neq 1.$$

So, let's first find t_k's. We obtain

$$t_2 = 1 + \frac{1}{3}t_1 + \frac{2}{3}t_3$$
$$= 1 + \frac{2}{3}t_3,$$

$$t_3 = 1 + \frac{1}{2}t_3 + \frac{1}{2}t_1$$
$$= 1 + \frac{1}{2}t_3.$$

Solving the above equations, we obtain

$$t_3 = 2, \qquad t_2 = \frac{7}{3}.$$

Now, we can write

$$r_1 = 1 + \frac{1}{4}t_1 + \frac{1}{2}t_2 + \frac{1}{4}t_3$$
$$= 1 + \frac{1}{4} \cdot 0 + \frac{1}{2} \cdot \frac{7}{3} + \frac{1}{4} \cdot 2$$
$$= \frac{8}{3}.$$

5. Consider the Markov chain shown in Figure 11.20.

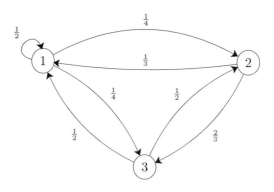

Figure 11.20: A state transition diagram.

(a) Is this chain irreducible?

(b) Is this chain aperiodic?

(c) Find the stationary distribution for this chain.

(d) Is the stationary distribution a limiting distribution for the chain?

Solution:

(a) The chain is irreducible since we can go from any state to any other states in a finite number of steps.

(b) The chain is aperiodic since there is a self-transition, i.e., $p_{11} > 0$.

(c) To find the stationary distribution, we need to solve

$$\pi_1 = \frac{1}{2}\pi_1 + \frac{1}{3}\pi_2 + \frac{1}{2}\pi_3,$$
$$\pi_2 = \frac{1}{4}\pi_1 + \frac{1}{2}\pi_3,$$
$$\pi_3 = \frac{1}{4}\pi_1 + \frac{2}{3}\pi_2,$$
$$\pi_1 + \pi_2 + \pi_3 = 1.$$

We find

$$\pi_1 \approx 0.457, \quad \pi_2 \approx 0.257, \quad \pi_3 \approx 0.286$$

(d) The above stationary distribution is a limiting distribution for the chain because the chain is irreducible and aperiodic.

6. Consider the Markov chain shown in Figure 11.21. Assume that $\frac{1}{2} < p < 1$. Does this chain have a limiting distribution? For all $i, j \in \{0, 1, 2, \cdots\}$, find

$$\lim_{n \to \infty} P(X_n = j | X_0 = i).$$

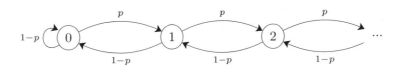

Figure 11.21: A state transition diagram.

Solution: This chain is irreducible since all states communicate with each other. It is also aperiodic since it includes a self-transition, $P_{00} > 0$. Let's write the equations for a stationary distribution. For state 0, we can write

$$\pi_0 = (1-p)\pi_0 + (1-p)\pi_1,$$

which results in

$$\pi_1 = \frac{p}{1-p}\pi_0.$$

For state 1, we can write

$$\pi_1 = p\pi_0 + (1-p)\pi_2$$
$$= (1-p)\pi_1 + (1-p)\pi_2,$$

which results in

$$\pi_2 = \frac{p}{1-p}\pi_1.$$

Similarly, for any $j \in \{1, 2, \cdots\}$, we obtain

$$\pi_j = \alpha\pi_{j-1},$$

where $\alpha = \frac{p}{1-p}$. Note that since $\frac{1}{2} < p < 1$, we conclude that $\alpha > 1$. We obtain

$$\pi_j = \alpha^j \pi_0, \qquad \text{for } j = 1, 2, \cdots.$$

Finally, we must have

$$1 = \sum_{j=0}^{\infty} \pi_j$$
$$= \sum_{j=0}^{\infty} \alpha^j \pi_0, \qquad\qquad (\text{where } \alpha > 1)$$
$$= \infty\pi_0.$$

Therefore, the above equation cannot be satisfied if $\pi_0 > 0$. If $\pi_0 = 0$, then all π_j's must be zero, so they cannot sum to 1. We conclude that there is no stationary distribution. This means that either all states are transient, or all states are null recurrent. In either case, we have

$$\lim_{n\to\infty} P(X_n = j | X_0 = i) = 0, \text{ for all } i, j.$$

We will see how to figure out if the states are transient or null recurrent in the End of Chapter Problems (see Problem 15 in Section 11.5).

11.3 Continuous-Time Markov Chains

11.3.1 Introduction

So far, we have discussed discrete-time Markov chains in which the chain jumps from the current state to the next state after one unit time. That is, the time that the chain spends in each state is a positive integer. It is equal to 1 if the state does not have a self-transition ($p_{ii} = 0$), or it is a $Geometric(1 - p_{ii})$ random variable if $p_{ii} > 0$. Here, we would like to discuss continuous-time Markov chains where the time spent in each state is a continuous random variable.

More specifically, we will consider a random process $\{X(t), t \in [0, \infty)\}$. Again, we assume that we have a countable state space $S \subset \{0, 1, 2, \cdots\}$. If $X(0) = i$, then $X(t)$ stays in state i for a random amount of time, say T_1, where T_1 is a continuous random variable. At time T_1, the process jumps to a new state j and will spend a random amount of time T_2 in that state, and so on. As it will be clear shortly, the random variables T_1, T_2, $\cdots$ have exponential distribution. The probability of going from state i to state j is shown by p_{ij}.

Similar to discrete-time Markov chains, we would like to have the Markov property, i.e., conditioned on the current value of $X(t)$, the past and the future values of the process must be independent. We can express the Markov property as follows: for all $0 \leq t_1 < t_2 < \cdots < t_n < t_{n+1}$, we must have

$$P\big(X(t_{n+1}) = j | X(t_n) = i, X(t_{n-1}) = i_{n-1}, \cdots, X(t_1) = i_1\big) = P\big(X(t_{n+1}) = j | X(t_n) = i\big).$$

In particular, suppose that at time t, we know that $X(t) = i$. To make any prediction about the future, it should not matter how long the process has been in state i. Thus, the time that the process spends in each state must have a "memoryless" property. As it has been discussed in previous chapters (e.g, Chapter 4), the exponential distribution is the only continuous distribution with that property. Thus, the time that a continuous-time Markov chain spends in state i (called the **holding time**) will have $Exponential(\lambda_i)$ distribution, where λ_i is a nonnegative real number. We further assume that the λ_i's are bounded, i.e., there exists a real number $M < \infty$ such that $\lambda_i < M$, for all $i \in S$.

Thus, a continuous Markov chain has two components. First, we have a discrete-time Markov chain, called the **jump chain** or the **the embedded Markov chain**, that gives us the transition probabilities p_{ij}. Second, for each state we have a **holding time parameter λ_i** that controls the amount of time spent in each state.

Note that if i is not an absorbing state, we can assume that i does not have a self-transition, i.e., $p_{ii} = 0$. The reason is that if we go from state i to state i, it's as if we never left that state. On the other hand, if i is an absorbing state, we have $p_{ii} = 1$, and $p_{ij} = 0$, for all $i \neq j$. In this case, we have $\lambda_i = 0$, which means that the chain will spend an infinite amount of time in the absorbing state i.

Continuous-Time Markov Chains

A continuous-time Markov chain $X(t)$ is defined by two components: a *jump chain*, and a set of *holding time parameters* λ_i. The jump chain consists of a countable set of states $S \subset \{0, 1, 2, \cdots\}$ along with transition probabilities p_{ij}. We assume $p_{ii} = 0$, for all non-absorbing states $i \in S$. We assume

1. if $X(t) = i$, the time until the state changes has *Exponential*(λ_i) distribution;

2. if $X(t) = i$, the next state will be j with probability p_{ij}.

The process satisfies the Markov property. That is, for all $0 \le t_1 < t_2 < \cdots < t_n < t_{n+1}$, we have

$$P\left(X(t_{n+1}) = j \,\middle|\, X(t_n) = i, X(t_{n-1}) = i_{n-1}, \cdots, X(t_1) = i_1\right)$$
$$= P\left(X(t_{n+1}) = j \,\middle|\, X(t_n) = i\right).$$

Let's define the transition probability $P_{ij}(t)$ as

$$P_{ij}(t) = P(X(t + s) = j | X(s) = i)$$
$$= P(X(t) = j | X(0) = i), \qquad \text{for all } s, t \in [0, \infty).$$

We can then define the *transition matrix*, $P(t)$. Assuming the states are $1, 2, \cdots, r$, then the state transition matrix for any $t \ge 0$ is given by

$$P(t) = \begin{bmatrix} p_{11}(t) & p_{12}(t) & \cdots & p_{1r}(t) \\ p_{21}(t) & p_{22}(t) & \cdots & p_{2r}(t) \\ \vdots & \vdots & \vdots & \vdots \\ p_{r1}(t) & p_{r2}(t) & \cdots & p_{rr}(t) \end{bmatrix}.$$

Let's look at an example.

Example 11.17. Consider a continuous Markov chain with two states $S = \{0, 1\}$. Assume the holding time parameters are given by $\lambda_0 = \lambda_1 = \lambda > 0$. That is, the time that the chain spends in each state before going to the other state has an *Exponential*(λ) distribution.

(a) Draw the state diagram of the embedded (jump) chain.

(b) Find the transition matrix $P(t)$. *Hint:* You might want to use the following identities

$$\sinh(x) = \frac{e^x - e^{-x}}{2} = \sum_{n=0}^{\infty} \frac{x^{2n+1}}{(2n+1)!},$$

$$\cosh(x) = \frac{e^x + e^{-x}}{2} = \sum_{n=0}^{\infty} \frac{x^{2n}}{(2n)!}.$$

Solution:

(a) There are two states in the chain and none of them are absorbing (since $\lambda_i > 0$). Since we do not allow self-transitions, the jump chain must have the following transition matrix:

$$P = \begin{bmatrix} 0 & 1 \\ 1 & 0 \end{bmatrix}.$$

The state transition diagram of the jump chain is shown in Figure 11.22.

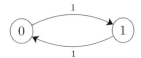

Figure 11.22: The jump chain of the continuous-time Markov chain in Example 11.17.

(b) This Markov chain has a simple structure. Let's find $P_{00}(t)$. By definition

$$P_{00}(t) = P(X(t) = 0 | X(0) = 0), \qquad \text{for all } t \in [0, \infty).$$

Assuming that $X(0) = 0$, $X(t)$ will be 0 if and only if we have an even number of transitions in the time interval $[0, t]$. The time between each transition is an *Exponential*(λ) random variable. Thus, the transitions occur according to a Poisson process with parameter λ. We have

$$\begin{aligned}
P_{00}(t) &= P(X(t) = 0 | X(0) = 0) \\
&= P\big(\text{an even number of arrivals in } [0, t]\big) \\
&= \sum_{n=0}^{\infty} e^{-\lambda t} \frac{(\lambda t)^{2n}}{(2n)!} \\
&= e^{-\lambda t} \sum_{n=0}^{\infty} \frac{(\lambda t)^{2n}}{(2n)!} \\
&= e^{-\lambda t} \left[\frac{e^{\lambda t} + e^{-\lambda t}}{2} \right] \qquad \text{(by the hint)} \\
&= \frac{1}{2} + \frac{1}{2} e^{-2\lambda t}.
\end{aligned}$$

Next, we obtain

$$\begin{aligned}
P_{01}(t) &= 1 - P_{00}(t) \\
&= \frac{1}{2} - \frac{1}{2} e^{-2\lambda t}.
\end{aligned}$$

Finally, because of the symmetry in the problem, we have

$$P_{11}(t) = P_{00}(t)$$
$$= \frac{1}{2} + \frac{1}{2}e^{-2\lambda t},$$

$$P_{10}(t) = P_{01}(t)$$
$$= \frac{1}{2} - \frac{1}{2}e^{-2\lambda t}.$$

Thus, the transition matrix for any $t \geq 0$ is given by

$$P(t) = \begin{bmatrix} \frac{1}{2} + \frac{1}{2}e^{-2\lambda t} & \frac{1}{2} - \frac{1}{2}e^{-2\lambda t} \\ \frac{1}{2} - \frac{1}{2}e^{-2\lambda t} & \frac{1}{2} + \frac{1}{2}e^{-2\lambda t} \end{bmatrix}.$$

If we let t go to infinity in the above example, the transition matrix becomes

$$\lim_{t \to \infty} P(t) = \begin{bmatrix} \frac{1}{2} & \frac{1}{2} \\ \frac{1}{2} & \frac{1}{2} \end{bmatrix},$$

which means that for $i = 0, 1$, we have

$$\lim_{t \to \infty} P(X(t) = 0 | X(0) = i) = \frac{1}{2},$$
$$\lim_{t \to \infty} P(X(t) = 1 | X(0) = i) = \frac{1}{2}.$$

In fact, $\pi = [\frac{1}{2}, \frac{1}{2}]$ is a limiting distribution of this chain. We will discuss limiting distributions shortly. Before doing that let's talk about some properties of the transition matrices. First, similar to the discrete-time analysis, the rows of the transition matrix must sum to 1:

$$\sum_{j \in S} p_{ij}(t) = 1, \qquad \text{for all } t \geq 0.$$

Next, we note that by definition

$$P_{ii}(0) = P(X(0) = i | X(0) = i)$$
$$= 1, \qquad \text{for all } i.$$

$$P_{ij}(0) = P(X(0) = j | X(0) = i)$$
$$= 0, \qquad \text{for all } i \neq j.$$

We conclude that $P(0)$ is equal to the identity matrix, $P(0) = I$. Finally, we can obtain the Chapman-Kolmogorov equation by applying the law of total probability and by using the

Markov property:

$$\begin{aligned}
P_{ij}(s+t) &= P(X(s+t) = j | X(0) = i) \\
&= \sum_{k \in S} P(X(s) = k | X(0) = i) P(X(s+t) = j | X(s) = k) \\
&= \sum_{k \in S} P_{ik}(s) P_{kj}(t), \qquad \text{for all } s, t \geq 0.
\end{aligned}$$

The above equation can be written in the matrix form as follows

$$P(s+t) = P(s)P(t), \qquad \text{for all } s, t \geq 0.$$

Transition Matrix

For a continuous-time Markov chain, we define the **transition matrix** $P(t)$. The (i, j)th entry of the transition matrix is given by

$$P_{ij}(t) = P(X(t) = j | X(0) = i).$$

The transition matrix satisfies the following properties:

1. $P(0)$ is equal to the identity matrix, $P(0) = I$;

2. the rows of the transition matrix must sum to 1,

$$\sum_{j \in S} p_{ij}(t) = 1, \qquad \text{for all } t \geq 0;$$

3. for all $s, t \geq 0$, we have

$$P(s+t) = P(s)P(t).$$

11.3.2   Stationary and Limiting Distributions

Here we introduce *stationary* distributions for continuous Markov chains. As in the case of discrete-time Markov chains, for "nice" chains, a unique stationary distribution exists and it is equal to the *limiting distribution*. Remember that for discrete-time Markov chains, stationary distributions are obtained by solving $\pi = \pi P$. We have a similar definition for continuous-time Markov chains.

Let $X(t)$ be a continuous-time Markov chain with transition matrix $P(t)$ and state space $S = \{0, 1, 2, \cdots\}$. A probability distribution π on S, i.e, a vector $\pi = [\pi_0, \pi_1, \pi_2, \cdots]$, where $\pi_i \in [0, 1]$ and

$$\sum_{i \in S} \pi_i = 1,$$

is said to be a **stationary distribution** for $X(t)$ if

$$\pi = \pi P(t), \qquad \text{for all } t \geq 0.$$

The intuition here is exactly the same as in the case of discrete-time chains. If the probability distribution of $X(0)$ is π, then the distribution of $X(t)$ is also given by π, for any $t \geq 0$.

Example 11.18. Consider the continuous Markov chain of Example 11.17: A chain with two states $S = \{0, 1\}$ and $\lambda_0 = \lambda_1 = \lambda > 0$. In that example, we found that the transition matrix for any $t \geq 0$ is given by

$$P(t) = \begin{bmatrix} \frac{1}{2} + \frac{1}{2}e^{-2\lambda t} & \frac{1}{2} - \frac{1}{2}e^{-2\lambda t} \\ \frac{1}{2} - \frac{1}{2}e^{-2\lambda t} & \frac{1}{2} + \frac{1}{2}e^{-2\lambda t} \end{bmatrix}.$$

Find the stationary distribution π for this chain.

Solution: For $\pi = [\pi_0, \pi_1]$, we obtain

$$\pi P(t) = [\pi_0, \pi_1] \begin{bmatrix} \frac{1}{2} + \frac{1}{2}e^{-2\lambda t} & \frac{1}{2} - \frac{1}{2}e^{-2\lambda t} \\ \frac{1}{2} - \frac{1}{2}e^{-2\lambda t} & \frac{1}{2} + \frac{1}{2}e^{-2\lambda t} \end{bmatrix} = [\pi_0, \pi_1].$$

We also need

$$\pi_0 + \pi_1 = 1.$$

Solving the above equations, we obtain

$$\pi_0 = \pi_1 = \frac{1}{2}.$$

Similar to the case of discrete-time Markov chains, we are interested in *limiting distributions* for continuous-time Markov chains.

Limiting Distributions

The probability distribution $\pi = [\pi_0, \pi_1, \pi_2, \cdots]$ is called the **limiting distribution** of the continuous-time Markov chain $X(t)$ if

$$\pi_j = \lim_{t \to \infty} P(X(t) = j | X(0) = i)$$

for all $i, j \in S$, and we have

$$\sum_{j \in S} \pi_j = 1.$$

As we will see shortly, for "nice" chains, there exists a unique stationary distribution which will be equal to the limiting distribution. In theory, we can find the stationary (and limiting) distribution by solving $\pi P(t) = \pi$, or by finding $\lim_{t \to \infty} P(t)$. However, in practice, finding $P(t)$ itself is usually very difficult. It is easier if we think in terms of the jump (embedded) chain. The following intuitive argument gives us the idea of how to obtain the limiting distribution of a continuous Markov chain from the limiting distribution of the corresponding jump chain.

Suppose that $\tilde{\pi} = [\tilde{\pi}_0, \tilde{\pi}_1, \tilde{\pi}_2, \cdots]$ is the limiting distribution of the jump chain. That is, the discrete-time Markov chain associated with the jump chain will spend a fraction $\tilde{\pi}_j$ of time in state j in the long run. Note that, for the corresponding continuous-time Markov chain, any time that the chain visits state j, it spends on average $\frac{1}{\lambda_j}$ time units in that state. Thus, we can obtain the limiting distribution of the continuous-time Markov chain by multiplying each $\tilde{\pi}_j$ by $\frac{1}{\lambda_j}$. We also need to normalize (divide by $\sum \frac{\tilde{\pi}_k}{\lambda_k}$) to get a valid probability distribution. The following theorem states this result more accurately.[1]

---

[1]It is worth noting that in the discrete-time case, we worried about periodicity. However, for continuous-time Markov chains, this is not an issue. This is because the times could any take positive real values and will not be multiples of a specific period.

Theorem 11.4. Let $\{X(t), t \geq 0\}$ be a continuous-time Markov chain with an irreducible positive recurrent jump chain. Suppose that the unique stationary distribution of the jump chain is given by

$$\tilde{\pi} = \left[\tilde{\pi}_0, \tilde{\pi}_1, \tilde{\pi}_2, \cdots\right].$$

Further assume that

$$0 < \sum_{k \in S} \frac{\tilde{\pi}_k}{\lambda_k} < \infty.$$

Then,

$$\pi_j = \lim_{t \to \infty} P(X(t) = j | X(0) = i) = \frac{\frac{\tilde{\pi}_j}{\lambda_j}}{\sum_{k \in S} \frac{\tilde{\pi}_k}{\lambda_k}}.$$

for all $i, j \in S$. That is, $\pi = [\pi_0, \pi_1, \pi_2, \cdots]$ is the limiting distribution of $X(t)$.

Example 11.19. Consider a continuous-time Markov chain $X(t)$ that has the jump chain shown in Figure 11.23. Assume the holding time parameters are given by $\lambda_1 = 2$, $\lambda_2 = 1$, and $\lambda_3 = 3$. Find the limiting distribution for $X(t)$.

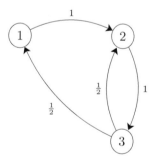

Figure 11.23: The jump chain for the Markov chain of Example 11.19.

Solution: We first note that the jump chain is irreducible. In particular, the transition matrix of the jump chain is given by

$$P = \begin{bmatrix} 0 & 1 & 0 \\ 0 & 0 & 1 \\ \frac{1}{2} & \frac{1}{2} & 0 \end{bmatrix}.$$

The next step is to find the stationary distribution for the jump chain by solving $\tilde{\pi} P = \tilde{\pi}$. We obtain

$$\tilde{\pi} = \frac{1}{5}[1, \ 2, \ 2].$$

Finally, we can obtain the limiting distribution of $X(t)$ using

$$\pi_j = \frac{\frac{\tilde{\pi}_j}{\lambda_j}}{\sum_{k \in S} \frac{\tilde{\pi}_k}{\lambda_k}}.$$

We obtain

$$\pi_1 = \frac{\frac{\tilde{\pi}_1}{\lambda_1}}{\frac{\tilde{\pi}_1}{\lambda_1} + \frac{\tilde{\pi}_2}{\lambda_2} + \frac{\tilde{\pi}_3}{\lambda_3}}$$

$$= \frac{\frac{1}{2}}{\frac{1}{2} + \frac{2}{1} + \frac{2}{3}}$$

$$= \frac{3}{19}.$$

$$\pi_2 = \frac{\frac{\tilde{\pi}_2}{\lambda_2}}{\frac{\tilde{\pi}_1}{\lambda_1} + \frac{\tilde{\pi}_2}{\lambda_2} + \frac{\tilde{\pi}_3}{\lambda_3}}$$

$$= \frac{\frac{2}{1}}{\frac{1}{2} + \frac{2}{1} + \frac{2}{3}}$$

$$= \frac{12}{19}.$$

$$\pi_3 = \frac{\frac{\tilde{\pi}_3}{\lambda_3}}{\frac{\tilde{\pi}_1}{\lambda_1} + \frac{\tilde{\pi}_2}{\lambda_2} + \frac{\tilde{\pi}_3}{\lambda_3}}$$

$$= \frac{\frac{2}{3}}{\frac{1}{2} + \frac{2}{1} + \frac{2}{3}}$$

$$= \frac{4}{19}.$$

Thus, we conclude that $\pi = \frac{1}{19}[3, 12, 4]$ is the limiting distribution of $X(t)$.

11.3.3 The Generator Matrix

Here, we introduce the *generator matrix*. The generator matrix, usually shown by G, gives us an alternative way of analyzing continuous-time Markov chains. Consider a continuous-time Markov chain $X(t)$. Assume $X(0) = i$. The chain will jump to the next state at time T_1, where $T_1 \sim Exponential(\lambda_i)$. In particular, for a very small $\delta > 0$, we can write

$$P(T_1 < \delta) = 1 - e^{-\lambda_i \delta}$$

$$\approx 1 - (1 - \lambda_i \delta)$$

$$= \lambda_i \delta.$$

Thus, in a short interval of length δ, the probability of leaving state i is approximately $\lambda_i \delta$. For this reason, λ_i is often called **the transition rate out of state** i. Formally, we can write

$$\lambda_i = \lim_{\delta \to 0^+} \left[\frac{P(X(\delta) \neq i | X(0) = i)}{\delta} \right] \tag{11.7}$$

Since we go from state i to state j with probability p_{ij}, we call the quantity $g_{ij} = \lambda_i p_{ij}$, **the transition rate from state i to state j.**

Here, we introduce the generator matrix, G, whose (i,j)th element is g_{ij}, when $i \neq j$. We choose the diagonal elements (g_{ii}) of G such that the rows of G sum to 0. That is, we let

$$g_{ii} = -\sum_{j \neq i} g_{ij}$$

$$= -\sum_{j \neq i} \lambda_i p_{ij}$$

$$= -\lambda_i \sum_{j \neq i} p_{ij}$$

$$= -\lambda_i.$$

The last equality resulted as follows: If $\lambda_i = 0$, then clearly

$$\lambda_i \sum_{j \neq i} p_{ij} = \lambda_i = 0.$$

If $\lambda_i \neq 0$, then $p_{ii} = 0$ (no self-transitions), so

$$\sum_{j \neq i} p_{ij} = 1.$$

It turns out the generator matrix is useful in analyzing continuous-time Markov chains.

Example 11.20. Explain why the following approximations hold:

(a) $p_{jj}(\delta) \approx 1 + g_{jj}\delta$, for all $j \in S$.

(b) $p_{kj}(\delta) \approx \delta g_{kj}$, for $k \neq j$.

   Solution: Let δ be small. Equation 11.7 can be written as

$$p_{jj}(\delta) \approx 1 - \lambda_j \delta$$
$$= 1 + g_{jj}\delta.$$

Also, we can approximate $p_{kj}(\delta) = P(X(\delta) = j | X(0) = k)$ as follows. This probability is approximately equal to the probability that we have a single transition from state k to state j in the interval $[0, \delta]$. Note that the probability of more than one transition is negligible if δ is small (refer to the Poisson process section). Thus, we can write

$$p_{kj}(\delta) = P(X(\delta) = j | X(0) = k)$$
$$\approx P(X(\delta) \neq k | X(0) = k) p_{kj}$$
$$\approx \lambda_k \delta p_{kj}$$
$$= \delta g_{kj}, \text{ for } k \neq j.$$

We can state the above approximations more precisely as

(a) $g_{jj} = -\lim_{\delta \to 0^+} \left[\frac{1 - p_{jj}(\delta)}{\delta} \right]$, for all $j \in S$;

(b) $g_{kj} = \lim_{\delta \to 0^+} \left[\frac{p_{kj}(\delta)}{\delta} \right]$, for $k \neq j$.

The Generator Matrix

For a continuous-time Markov chain, we define the **generator matrix** G. The (i,j)th entry of the transition matrix is given by

$$g_{ij} = \begin{cases} \lambda_i p_{ij} & \text{if } i \neq j \\ -\lambda_i & \text{if } i = j \end{cases}.$$

Example 11.21. Consider the continuous Markov chain of Example 11.17: A chain with two states $S = \{0, 1\}$ and $\lambda_0 = \lambda_1 = \lambda > 0$. In that example, we found that the transition matrix for any $t \geq 0$ is given by

$$P(t) = \begin{bmatrix} \frac{1}{2} + \frac{1}{2}e^{-2\lambda t} & \frac{1}{2} - \frac{1}{2}e^{-2\lambda t} \\ \frac{1}{2} - \frac{1}{2}e^{-2\lambda t} & \frac{1}{2} + \frac{1}{2}e^{-2\lambda t} \end{bmatrix}.$$

(a) Find the generator matrix G.

(b) Show that for any $t \geq 0$, we have

$$P'(t) = P(t)G = GP(t),$$

where $P'(t)$ is the derivative of $P(t)$.

Solution:

(a) First, we have

$$g_{00} = -\lambda_0$$
$$= -\lambda,$$
$$g_{11} = -\lambda_1$$
$$= -\lambda.$$

The transition matrix for the corresponding jump chain is given by

$$P = \begin{bmatrix} p_{00} & p_{01} \\ p_{10} & p_{11} \end{bmatrix} = \begin{bmatrix} 0 & 1 \\ 1 & 0 \end{bmatrix}.$$

Therefore, we have

$$g_{01} = \lambda_0 p_{01}$$
$$= \lambda,$$
$$g_{10} = \lambda_1 p_{10}$$
$$= \lambda.$$

Thus, the generator matrix is given by

$$G = \begin{bmatrix} -\lambda & \lambda \\ \lambda & -\lambda \end{bmatrix}.$$

(b) We have

$$P'(t) = \begin{bmatrix} -\lambda e^{-2\lambda t} & \lambda e^{-2\lambda t} \\ \lambda e^{-2\lambda t} & -\lambda e^{-2\lambda t} \end{bmatrix},$$

where $P'(t)$ is the derivative of $P(t)$. We also have

$$P(t)G = \begin{bmatrix} \frac{1}{2} + \frac{1}{2}e^{-2\lambda t} & \frac{1}{2} - \frac{1}{2}e^{-2\lambda t} \\ \frac{1}{2} - \frac{1}{2}e^{-2\lambda t} & \frac{1}{2} + \frac{1}{2}e^{-2\lambda t} \end{bmatrix} \begin{bmatrix} -\lambda & \lambda \\ \lambda & -\lambda \end{bmatrix} = \begin{bmatrix} -\lambda e^{-2\lambda t} & \lambda e^{-2\lambda t} \\ \lambda e^{-2\lambda t} & -\lambda e^{-2\lambda t} \end{bmatrix},$$

$$GP(t) = \begin{bmatrix} -\lambda & \lambda \\ \lambda & -\lambda \end{bmatrix} \begin{bmatrix} \frac{1}{2} + \frac{1}{2}e^{-2\lambda t} & \frac{1}{2} - \frac{1}{2}e^{-2\lambda t} \\ \frac{1}{2} - \frac{1}{2}e^{-2\lambda t} & \frac{1}{2} + \frac{1}{2}e^{-2\lambda t} \end{bmatrix} = \begin{bmatrix} -\lambda e^{-2\lambda t} & \lambda e^{-2\lambda t} \\ \lambda e^{-2\lambda t} & -\lambda e^{-2\lambda t} \end{bmatrix}.$$

We conclude

$$P'(t) = P(t)G = GP(t).$$

The equation $P'(t) = P(t)G = GP(t)$ (in the above example) is in fact true in general. To see the proof idea, we can argue as follows. Let δ be small. By Example 11.20, we have

$$p_{jj}(\delta) \approx 1 + g_{jj}\delta,$$
$$p_{kj}(\delta) \approx \delta g_{kj}, \quad \text{for } k \neq j.$$

Using the Chapman-Kolmogorov equation, we can write

$$P_{ij}(t + \delta) = \sum_{k \in S} P_{ik}(t)p_{kj}(\delta)$$

$$= p_{ij}(t)p_{jj}(\delta) + \sum_{k \neq j} P_{ik}(t)p_{kj}(\delta)$$

$$\approx p_{ij}(t)(1 + g_{jj}\delta) + \sum_{k \neq j} P_{ik}(t)\delta g_{kj}$$

$$= p_{ij}(t) + \delta p_{ij}(t)g_{jj} + \delta \sum_{k \neq j} P_{ik}(t)g_{kj}$$

$$= p_{ij}(t) + \delta \sum_{k \in S} P_{ik}(t)g_{kj}.$$

Thus,

$$\frac{P_{ij}(t + \delta) - p_{ij}(t)}{\delta} \approx \sum_{k \in S} P_{ik}(t)g_{kj},$$

which is the (i, j)th element of $P(t)G$. The above argument can be made rigorous.

Forward and Backward Equations

The **forward equations** state that

$$P'(t) = P(t)G,$$

which is equivalent to

$$p'_{ij}(t) = \sum_{k \in S} p_{ik}(t)g_{kj}, \text{ for all } i, j \in S.$$

The **backward equations** state that

$$P'(t) = GP(t),$$

which is equivalent to

$$p'_{ij}(t) = \sum_{k \in S} g_{ik}p_{kj}(t), \text{ for all } i, j \in S.$$

One of the main uses of the generator matrix is finding the stationary distribution. So far, we have seen how to find the stationary distribution using the jump chain. The following result tells us how to find the stationary matrix using the generator matrix.

> Consider a continuous Markov chain $X(t)$ with the state space S and the generator Matrix G. The probability distribution π on S is a stationary distribution for $X(t)$ if and only if it satisfies
>
> $$\pi G = 0.$$

Proof. For simplicity, let's assume that S is finite, i.e., $\pi = [\pi_0, \pi_1, \cdots, \pi_r]$, for some $r \in \mathbb{N}$. If π is a stationary distribution, then $\pi = \pi P(t)$. Differentiating both sides, we obtain

$$
\begin{aligned}
0 &= \frac{d}{dt}[\pi P(t)] \\
&= \pi P'(t) \\
&= \pi G P(t) \qquad \text{(backward equations)}
\end{aligned}
$$

Now, let $t = 0$ and remember that $P(0) = I$, the identity matrix. We obtain

$$0 = \pi G P(0) = \pi G.$$

Next, let π be a probability distribution on S that satisfies $\pi G = 0$. Then, by backward equations,

$$P'(t) = G P(t).$$

Multiplying both sides by π, we obtain

$$\pi P'(t) = \pi G P(t) = 0.$$

Note that $\pi P'(t)$ is the derivative of $\pi P(t)$. Thus, we conclude $\pi P(t)$ does not depend on t. In particular, for any $t \geq 0$, we have

$$\pi P(t) = \pi P(0) = \pi.$$

Therefore, π is a stationary distribution.

■

Example 11.22. The generator matrix for the continuous Markov chain of Example 11.17 is given by

$$
G = \begin{bmatrix} -\lambda & \lambda \\ \lambda & -\lambda \end{bmatrix}.
$$

Find the stationary distribution for this chain by solving $\pi G = 0$.

Solution: We obtain

$$\pi G = [\pi_0, \pi_1] \begin{bmatrix} -\lambda & \lambda \\ \lambda & -\lambda \end{bmatrix} = 0.$$

which results in

$$\pi_0 = \pi_1.$$

We also need

$$\pi_0 + \pi_1 = 1.$$

Solving the above equations, we obtain

$$\pi_0 = \pi_1 = \frac{1}{2}.$$

Transition Rate Diagram:

A continuous-time Markov chain can be shown by its **transition rate diagram**. In this diagram, the values g_{ij} are shown on the edges. The values of g_{ii}'s are not usually shown because they are implied by the other values, i.e.,

$$g_{ii} = -\sum_{j \neq i} g_{ij}.$$

For example, Figure 11.24 shows the transition rate diagram for the following generator matrix

$$G = \begin{bmatrix} -5 & 5 & 0 \\ 1 & -2 & 1 \\ 3 & 1 & -4 \end{bmatrix}, \tag{11.8}$$

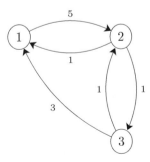

Figure 11.24: The transition rate diagram for the continuous-time Markov chain defined by Equation 11.8.

11.3.4   Solved Problems

1. Consider a continuous-time Markov chain $X(t)$ with the jump chain shown in Figure11.25. [2] Assume $\lambda_1 = 2$, $\lambda_2 = 3$, and $\lambda_3 = 4$.

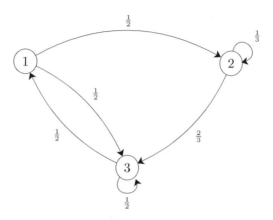

Figure 11.25: The jump chain for the Markov chain of Problem 1.

(a) Find the stationary distribution of the jump chain $\tilde{\pi} = \left[\tilde{\pi}_1, \tilde{\pi}_2, \tilde{\pi}_3\right]$.

(b) Using $\tilde{\pi}$, find the stationary distribution for $X(t)$.

Solution:

(a) To find the stationary distribution of the jump chain, $\tilde{\pi} = \left[\tilde{\pi}_1, \tilde{\pi}_2, \tilde{\pi}_3\right]$, we need to solve

$$\tilde{\pi}_1 = \frac{1}{2}\tilde{\pi}_3,$$

$$\tilde{\pi}_2 = \frac{1}{2}\tilde{\pi}_1 + \frac{1}{3}\tilde{\pi}_2$$

$$\tilde{\pi}_3 = \frac{1}{2}\tilde{\pi}_1 + \frac{2}{3}\tilde{\pi}_2 + \frac{1}{2}\tilde{\pi}_3,$$

$$\tilde{\pi}_1 + \tilde{\pi}_2 + \tilde{\pi}_3 = 1.$$

We find

$$\tilde{\pi}_1 = \frac{4}{15}, \quad \tilde{\pi}_2 = \frac{3}{15}, \quad \tilde{\pi}_3 = \frac{8}{15}.$$

---

[2]Although this jump chain has self-transitions, you can still use the discussed methods. In fact, $X(t)$ can be equivalently shown by a jump chain with no self-transitions along with appropriate holding times.

(b) We have obtained

$$\tilde{\pi} = \frac{1}{15}[4, \ 3, \ 8].$$

We can find the limiting distribution of $X(t)$ using

$$\pi_j = \frac{\frac{\tilde{\pi}_j}{\lambda_j}}{\sum_{k \in S} \frac{\tilde{\pi}_k}{\lambda_k}}.$$

We obtain

$$\pi_1 = \frac{\frac{\tilde{\pi}_1}{\lambda_1}}{\frac{\tilde{\pi}_1}{\lambda_1} + \frac{\tilde{\pi}_2}{\lambda_2} + \frac{\tilde{\pi}_3}{\lambda_3}}$$

$$= \frac{\frac{4}{2}}{\frac{4}{2} + \frac{3}{3} + \frac{8}{4}}$$

$$= \frac{2}{5}.$$

$$\pi_2 = \frac{\frac{\tilde{\pi}_2}{\lambda_2}}{\frac{\tilde{\pi}_1}{\lambda_1} + \frac{\tilde{\pi}_2}{\lambda_2} + \frac{\tilde{\pi}_3}{\lambda_3}}$$

$$= \frac{\frac{3}{3}}{\frac{4}{2} + \frac{3}{3} + \frac{8}{4}}$$

$$= \frac{1}{5}.$$

$$\pi_3 = \frac{\frac{\tilde{\pi}_3}{\lambda_3}}{\frac{\tilde{\pi}_1}{\lambda_1} + \frac{\tilde{\pi}_2}{\lambda_2} + \frac{\tilde{\pi}_3}{\lambda_3}}$$

$$= \frac{\frac{8}{4}}{\frac{4}{2} + \frac{3}{3} + \frac{8}{4}}$$

$$= \frac{2}{5}.$$

Thus, we conclude that $\pi = \frac{1}{5}[2, 1, 2]$ is the limiting distribution of $X(t)$.

2. Consider a continuous-time Markov chain $X(t)$ that has the jump chain shown in Figure 11.26 (this is the same Markov chain given in Example 11.19). Assume $\lambda_1 = 2$, $\lambda_2 = 1$, and $\lambda_3 = 3$.

   (a) Find the generator matrix for this chain.

   (b) Find the limiting distribution for $X(t)$ by solving $\pi G = 0$.

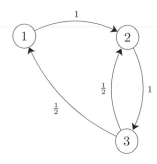

Figure 11.26: The jump chain for the Markov chain of Problem 2.

Solution: The jump chain is irreducible and the transition matrix of the jump chain is given by

$$P = \begin{bmatrix} 0 & 1 & 0 \\ 0 & 0 & 1 \\ \frac{1}{2} & \frac{1}{2} & 0 \end{bmatrix}.$$

The generator matrix can be obtained using

$$g_{ij} = \begin{cases} \lambda_i p_{ij} & \text{if } i \neq j \\[2mm] -\lambda_i & \text{if } i = j \end{cases}$$

We obtain

$$G = \begin{bmatrix} -2 & 2 & 0 \\ 0 & -1 & 1 \\ \frac{3}{2} & \frac{3}{2} & -3 \end{bmatrix}.$$

Solving

$$\pi G = 0, \qquad \text{and} \qquad \pi_1 + \pi_2 + \pi_3 = 1$$

we obtain $\pi = \frac{1}{19}[3, 12, 4]$, which is the same answer that we obtained in Example 11.19.

3. (**A queuing system**) Suppose that customers arrive according to a Poisson process with rate λ at a service center that has a single server. Customers are served one at a time in order of arrival. Service times are assumed to be i.i.d. *Exponential*(μ) random variables and independent of the arrival process. Customers leave the system after being served. Our goal in this problem is to model the above system as a continuous-time Markov

chain. Let $X(t)$ be the number of customers in the system at time t, so the state space is $S = \{0, 1, 2, \cdots\}$. Assume $i > 0$. If the system is in state i at time t, then the next state would either be $i + 1$ (if a new customers arrive) or state $i - 1$ (if a customer leaves).

(a) Suppose that the system is in state 0, so there are no customers in the system and the next transition will be to state 1. Let T_0 be the time until the next transition. Show that $T_0 \sim Exponential(\lambda)$.

(b) Suppose that the system is currently in state i, where $i > 0$. Let T_i be the time until the next transition. Show that $T_i \sim Exponential(\lambda + \mu)$.

(c) Suppose that the system is at state i. Find the probability that the next transition will be to state $i + 1$.

(d) Draw the jump chain, and provide the holding time parameters λ_i.

(e) Find the Generator matrix.

(f) Draw the transition rate diagram.

Solution: Note that to solve this problem, we use several results from the Poisson process Section. In particular, you might want to review merging and splitting of Poisson processes before reading the solution to this problem.

(a) If there are no customers in the system, the next transition occurs when a new customer arrives. Since the customers arrive according to a Poisson process, and the interarrival times in the Poisson process have $Exponential(\lambda)$ distribution, we conclude $T_0 \sim Exponential(\lambda)$.

(b) Suppose that the system is in state i, where $i > 0$. Thus, there is a customer being served. We assume the service times have $Exponential(\mu)$ distribution. The next transition occurs either when a new customer arrives, or when the service time of the current customer is ended. Thus, we can express T_i as

$$T_i = \min(X, Y),$$

where $X \sim Exponential(\lambda)$ and $Y \sim Exponential(\mu)$, and X and Y are independent. We claim that $T_i \sim Exponential(\lambda + \mu)$. One way to see this is as follows. Here, you can imagine two independent Poisson processes. The first one is the customer arrival process. The second one is the process that has interarrival times equal to the service times. Now, the merged process has rate $\lambda + \mu$. Since T_i can be thought of the first arrival in the merged process, we conclude that $T_i \sim Exponential(\lambda + \mu)$.

(c) Suppose that the system is at state i. We would like to find the probability that the next transition will be to state $i + 1$, shown by $p_{i,i+1}$. Again consider the two Poisson processes defined above. We can model this system as follows. Each arrival in the merged process is of type 1 (customer arrival) with probability $\frac{\lambda}{\lambda+\mu}$, or of type 2 (customer departure) with probability $\frac{\mu}{\lambda+\mu}$.

The probability $p_{i,i+1}$ is the probability that the first arrival in the merged process is of type 1. This happens with probability $\frac{\lambda}{\lambda+\mu}$, so we conclude

$$P_{i,i+1} = \frac{\lambda}{\lambda+\mu},$$

$$p_{i,i-1} = 1 - p_{i,i+1} = \frac{\mu}{\lambda+\mu}.$$

(d) From the above, we can draw the jump chain as in Figure 11.27

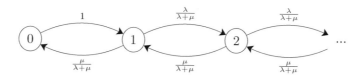

Figure 11.27: The jump chain for the above queuing system.

The holding time parameters, λ_i's, are given by

$$\lambda_0 = \lambda,$$
$$\lambda_i = \lambda + \mu, \qquad \text{for } i = 1, 2, \cdots.$$

(e) The generator matrix can be obtained using

$$g_{ij} = \begin{cases} \lambda_i p_{ij} & \text{if } i \neq j \\ -\lambda_i & \text{if } i = j \end{cases}$$

We obtain

$$G = \begin{bmatrix} -\lambda & \lambda & 0 & 0 & \cdots \\ \mu & -(\mu+\lambda) & \lambda & 0 & \cdots \\ 0 & \mu & -(\mu+\lambda) & \lambda & \cdots \\ \vdots & \vdots & \vdots & \vdots & \end{bmatrix}.$$

(f) Remember that in the transition rate diagram, the values g_{ij} are shown on the edges (the values of g_{ii}'s are not usually shown). The transition rate diagram for this chain is shown in Figure 11.28.

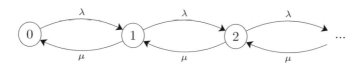

Figure 11.28: The transition rate diagram for the above queuing system.

11.4   Brownian Motion (Wiener Process)

Brownian motion is another widely-used random process. It has been used in engineering, finance, and physical sciences. It is a Gaussian random process and it has been used to model motion of particles suspended in a fluid, percentage changes in the stock prices, integrated white noise, etc. Figure 11.29 shows a sample path of Brownian motion.

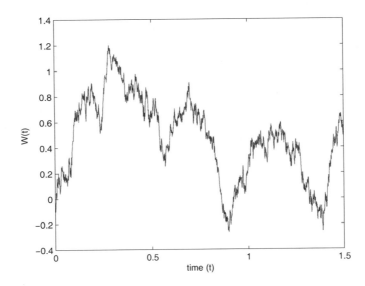

Figure 11.29: A possible realization of Brownian motion.

In this section, we provide a very brief introduction to Brownian motion. It is worth noting that in order to have a deep understanding of Brownian motion, one needs to understand *Itō calculus*, a topic that is beyond the scope of this book. A good place to start learning Itō calculus is [25].

11.4.1   Brownian Motion as the Limit of a Symmetric Random Walk

Here, we introduce a construction of Brownian motion from a symmetric random walk. Divide the half-line $[0, \infty)$ to tiny subintervals of length δ as shown in Figure 11.30.

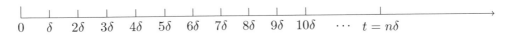

Figure 11.30: Dividing the half-line $[0, \infty)$ to tiny subintervals of length δ.

Each subinterval corresponds to a time slot of length δ. Thus, the intervals are $(0, \delta]$, $(\delta, 2\delta]$, $(2\delta, 3\delta]$, $\cdots$. More generally, the kth interval is $\big((k-1)\delta, k\delta\big]$. We assume that in each time slot, we toss a fair coin. We define the random variables X_i as follows. $X_i = \sqrt{\delta}$ if the kth coin toss results in heads, and $X_i = -\sqrt{\delta}$ if the kth coin toss results in tails. Thus,

$$
X_i = \begin{cases}
\sqrt{\delta} & \text{with probability } \frac{1}{2} \\[2mm]
-\sqrt{\delta} & \text{with probability } \frac{1}{2}
\end{cases}
$$

Moreover, the X_i's are independent. Note that

$$
E[X_i] = 0,
$$
$$
\mathrm{Var}(X_i) = \delta.
$$

Now, we would like to define the process $W(t)$ as follows. We let $W(0) = 0$. At time $t = n\delta$, the value of $W(t)$ is given by

$$
W(t) = W(n\delta) = \sum_{i=1}^{n} X_i.
$$

Since $W(t)$ is the sum of n i.i.d. random variables, we know how to find $E[W(t)]$ and $\mathrm{Var}(W(t))$. In particular,

$$
E[W(t)] = \sum_{i=1}^{n} E[X_i]
$$
$$
= 0,
$$
$$
\mathrm{Var}(W(t)) = \sum_{i=1}^{n} \mathrm{Var}(X_i)
$$
$$
= n\mathrm{Var}(X_1)
$$
$$
= n\delta
$$
$$
= t.
$$

For any $t \in (0, \infty)$, as n goes to ∞, δ goes to 0. By the central limit theorem, $W(t)$ will become a normal random variable,

$$
W(t) \sim N(0, t).
$$

Since the coin tosses are independent, we conclude that $W(t)$ has *independent increments*. That is, for all $0 \le t_1 < t_2 < t_3 \cdots < t_n$, the random variables

$$
W(t_2) - W(t_1), \quad W(t_3) - W(t_2), \quad \cdots, \quad W(t_n) - W(t_{n-1})
$$

are independent.

Remember that we say that a random process $X(t)$ has *stationary increments* if, for all $t_2 > t_1 \ge 0$, and all $r > 0$, the two random variables $X(t_2) - X(t_1)$ and $X(t_2 + r) - X(t_1 + r)$ have the same distributions. In other words, the distribution of the difference depends only on the length of the interval $(t_1, t_2]$, and not on the exact location of the interval on the real line. We now claim that the random process $W(t)$, defined above, has stationary increments. To see this, we argue as follows.

For $0 \le t_1 < t_2$, if we have $t_1 = n_1 \delta$ and $t_2 = n_2 \delta$, we obtain

$$W(t_1) = W(n_1 \delta) = \sum_{i=1}^{n_1} X_i,$$

$$W(t_2) = W(n_2 \delta) = \sum_{i=1}^{n_2} X_i.$$

Then, we can write

$$W(t_2) - W(t_1) = \sum_{i=n_1+1}^{n_2} X_i.$$

Therefore, we conclude

$$E[W(t_2) - W(t_1)] = \sum_{i=n_1+1}^{n_2} E[X_i]$$
$$= 0,$$
$$\mathrm{Var}(W(t_2) - W(t_1)) = \sum_{i=n_1+1}^{n_2} \mathrm{Var}(X_i)$$
$$= (n_2 - n_1)\mathrm{Var}(X_1)$$
$$= (n_2 - n_1)\delta$$
$$= t_2 - t_1.$$

Therefore, for any $0 \le t_1 < t_2$, the distribution of $W(t_2) - W(t_1)$ only depends on the lengths of the interval $[t_1, t_2]$, i.e., how many coin tosses are in that interval. In particular, for any $0 \le t_1 < t_2$, the distribution of $W(t_2) - W(t_1)$ converges to $N(0, t_2 - t_1)$. Therefore, we conclude that $W(t)$ has *stationary increments*.

The above construction can be made more rigorous. The random process $W(t)$ is called the standard Brownian motion or the standard Wiener process. Brownian motion has continuous sample paths, i.e., $W(t)$ is a continuous function of t (See Figure 11.29). However, it can be shown that it is nowhere differentiable.

11.4.2 Definition and Some Properties

Here, we provide a more formal definition for Brownian Motion.

Standard Brownian Motion

A Gaussian random process $\{W(t), t \in [0, \infty)\}$ is called a (standard) **Brownian motion** or a (standard) **Wiener process** if

1. W(0)=0;

2. for all $0 \leq t_1 < t_2$, $W(t_2) - W(t_1) \sim N(0, t_2 - t_1)$;

3. W(t) has independent increments. That is, for all $0 \leq t_1 < t_2 < t_3 \cdots < t_n$, the random variables

$$W(t_2) - W(t_1), \quad W(t_3) - W(t_2), \quad \cdots, \quad W(t_n) - W(t_{n-1})$$

are independent;

4. W(t) has continuous sample paths.

A more general process is obtained if we define $X(t) = \mu + \sigma W(t)$. In this case, $X(t)$ is a Brownian motion with

$$E[X(t)] = \mu, \qquad \text{Var}(X(t)) = \sigma^2 t.$$

Nevertheless, since $X(t)$ is obtained by simply shifting and scaling $W(t)$, it suffices to study properties of the standard Brownian motion, $W(t)$.

Example 11.23. Let $W(t)$ be a standard Brownian motion. For all $s, t \in [0, \infty)$, find

$$C_W(s, t) = \text{Cov}(W(s), W(t)).$$

Solution: Let's assume $s \leq t$. Then, we have

$$\begin{aligned}
\text{Cov}\big(W(s), W(t)\big) &= \text{Cov}\big(W(s), W(s) + W(t) - W(s)\big) \\
&= \text{Cov}\big(W(s), W(s)\big) + \text{Cov}\big(W(s), W(t) - W(s)\big) \\
&= \text{Var}\big(W(s)\big) + \text{Cov}\big(W(s), W(t) - W(s)\big) \\
&= s + \text{Cov}\big(W(s), W(t) - W(s)\big).
\end{aligned}$$

Brownian motion has independent increments, so the two random variables $W(s) = W(s) - W(0)$ and $W(t) - W(s)$ are independent. Therefore, $\text{Cov}\big(W(s), W(t) - W(s)\big) = 0$. We conclude

$$\text{Cov}\big(W(s), W(t)\big) = s.$$

Similarly, if $t \leq s$, we obtain

$$\text{Cov}\big(W(s), W(t)\big) = t.$$

We conclude

$$\text{Cov}\big(W(s), W(t)\big) = \min(s, t), \qquad \text{for all } s, t.$$

If $W(t)$ is a standard Brownian motion, we have

$$\text{Cov}(W(s), W(t)) = \min(s, t), \qquad \text{for all } s, t.$$

Example 11.24. Let $W(t)$ be a standard Brownian motion.

(a) Find $P(1 < W(1) < 2)$.

(b) Find $P(W(2) < 3 | W(1) = 1)$.

Solution:

(a) We have $W(1) \sim N(0, 1)$. Thus,

$$P(1 < W(1) < 2) = \Phi(2) - \Phi(1)$$
$$\approx 0.136$$

(b) Note that $W(2) = W(1) + W(2) - W(1)$. Also, note that $W(1)$ and $W(2) - W(1)$ are independent, and

$$W(2) - W(1) \sim N(0, 1).$$

We conclude that

$$W(2) | W(1) = 1 \ \sim \ N(1, 1).$$

Thus,

$$P(W(2) < 3 | W(1) = 1) = \Phi\left(\frac{3 - 1}{1}\right)$$
$$= \Phi(2) \approx 0.98$$

11.4.3 Solved Problems

1. Let $W(t)$ be a standard Brownian motion. Find $P(W(1) + W(2) > 2)$.

Solution: Let $X = W(1) + W(2)$. Since $W(t)$ is a Gaussian process, X is a normal random variable.

$$EX = E[W(1)] + E[W(2)] = 0,$$

$$\text{Var}(X) = \text{Var}(W(1)) + \text{Var}(W(2)) + 2\text{Cov}(W(1), W(2))$$
$$= 1 + 2 + 2 \cdot 1$$
$$= 5.$$

We conclude

$$X \sim N(0, 5).$$

Thus,

$$P(X > 2) = 1 - \Phi\left(\frac{2-0}{\sqrt{5}}\right)$$

$$\approx 0.186$$

2. Let $W(t)$ be a standard Brownian motion, and $0 \leq s < t$. Find the conditional PDF of $W(s)$ given $W(t) = a$.

Solution: It is useful to remember the following result from the previous chapters: Suppose X and Y are jointly normal random variables with parameters μ_X, σ_X^2, μ_Y, σ_Y^2, and ρ. Then, given $X = x$, Y is normally distributed with

$$E[Y|X = x] = \mu_Y + \rho\sigma_Y \frac{x - \mu_X}{\sigma_X},$$

$$\text{Var}(Y|X = x) = (1 - \rho^2)\sigma_Y^2.$$

Now, if we let $X = W(t)$ and $Y = W(s)$, we have $X \sim N(0, t)$ and $Y \sim N(0, s)$ and

$$\rho = \frac{\text{Cov}(X, Y)}{\sigma_x \sigma_Y}$$

$$= \frac{\min(s, t)}{\sqrt{t}\sqrt{s}}$$

$$= \frac{s}{\sqrt{t}\sqrt{s}}$$

$$= \sqrt{\frac{s}{t}}.$$

We conclude that

$$E[Y|X = a] = \frac{s}{t}a,$$

$$\text{Var}(Y|X = a) = s\left(1 - \frac{s}{t}\right).$$

Therefore,

$$W(s)|W(t) = a \quad \sim \quad N\left(\frac{s}{t}a, s\left(1 - \frac{s}{t}\right)\right).$$

3. (**Geometric Brownian Motion**) Let $W(t)$ be a standard Brownian motion. Define

$$X(t) = \exp\{W(t)\}, \qquad \text{for all t } \in [0, \infty).$$

(a) Find $E[X(t)]$, for all $t \in [0, \infty)$.

(b) Find $\text{Var}(X(t))$, for all $t \in [0, \infty)$.

(c) Let $0 \le s \le t$. Find $\text{Cov}(X(s), X(t))$.

Solution: It is useful to remember the MGF of the normal distribution. In particular, if $X \sim N(\mu, \sigma)$, then

$$M_X(s) = E[e^{sX}] = \exp\left\{s\mu + \frac{\sigma^2 s^2}{2}\right\}, \qquad \text{for all} \quad s \in \mathbb{R}.$$

(a) We have

$$E[X(t)] = E[e^{W(t)}], \qquad (\text{where } W(t) \sim N(0, t))$$
$$= \exp\left\{\frac{t}{2}\right\}.$$

(b) We have

$$E[X^2(t)] = E[e^{2W(t)}], \qquad (\text{where } W(t) \sim N(0, t))$$
$$= \exp\{2t\}.$$

Thus,

$$\text{Var}(X(t)) = E[X^2(t)] - E[X(t)]^2$$
$$= \exp\{2t\} - \exp\{t\}.$$

(c) Let $0 \le s \le t$. Then, we have

$$\text{Cov}(X(s), X(t)) = E[X(s)X(t)] - E[X(s)]E[X(t)]$$
$$= E[X(s)X(t)] - \exp\left\{\frac{s+t}{2}\right\}.$$

To find $E[X(s)X(t)]$, we can write

$$E[X(s)X(t)] = E\left[\exp\{W(s)\}\exp\{W(t)\}\right]$$
$$= E\left[\exp\{W(s)\}\exp\{W(s) + W(t) - W(s)\}\right]$$
$$= E\left[\exp\{2W(s)\}\exp\{W(t) - W(s)\}\right]$$
$$= E\left[\exp\{2W(s)\}\right]E\left[\exp\{W(t) - W(s)\}\right]$$
$$= \exp\{2s\}\exp\left\{\frac{t-s}{2}\right\}$$
$$= \exp\left\{\frac{3s+t}{2}\right\}.$$

We conclude, for $0 \leq s \leq t$,

$$\mathrm{Cov}(X(s), X(t)) = \exp\left\{\frac{3s+t}{2}\right\} - \exp\left\{\frac{s+t}{2}\right\}.$$

11.5   End of Chapter Problems

1. The number of orders arriving at a service facility can be modeled by a Poisson process with intensity $\lambda = 10$ orders per hour.

   (a) Find the probability that there are no orders between 10:30 and 11.

   (b) Find the probability that there are 3 orders between 10:30 and 11 and 7 orders between 11:30 and 12.

2. Let $\{N(t), t \in [0, \infty)\}$ be a Poisson process with rate λ. Find the probability that there are two arrivals in $(0, 2]$ or three arrivals in $(4, 7]$.

3. Let $X \sim Poisson(\mu_1)$ and $Y \sim Poisson(\mu_2)$ be two independent random variables. Define $Z = X + Y$. Show that

$$X | Z = n \sim Binomial\left(n, \frac{\mu_1}{\mu_1 + \mu_2}\right).$$

4. Let $N(t)$ be a Poisson process with rate λ. Let $0 < s < t$. Show that given $N(t) = n$, $N(s)$ is a binomial random variable with parameters n and $p = \frac{s}{t}$.

5. Let $N_1(t)$ and $N_2(t)$ be two independent Poisson processes with rate λ_1 and λ_2 respectively. Let $N(t) = N_1(t) + N_2(t)$ be the merged process. Show that given $N(t) = n$, $N_1(t) \sim Binomial\left(n, \frac{\lambda_1}{\lambda_1 + \lambda_2}\right)$.

   Note: We can interpret this result as follows: Any arrival in the merged process belongs to $N_1(t)$ with probability $\frac{\lambda_1}{\lambda_1 + \lambda_2}$ and belongs to $N_2(t)$ with probability $\frac{\lambda_2}{\lambda_1 + \lambda_2}$ independent of other arrivals.

6. In this problem, our goal is to complete the proof of the equivalence of the first and the second definitions of the Poisson process. More specifically, suppose that the counting process $\{N(t), t \in [0, \infty)\}$ satisfies all the following conditions:

   (a) $N(0) = 0$.

   (b) $N(t)$ has <u>independent</u> and <u>stationary</u> increments.

   (c) We have

$$P(N(\Delta) = 0) = 1 - \lambda\Delta + o(\Delta),$$
$$P(N(\Delta) = 1) = \lambda\Delta + o(\Delta),$$
$$P(N(\Delta) \geq 2) = o(\Delta).$$

   We would like to show that $N(t) \sim Poisson(\lambda t)$. To this, for any $k \in \{0, 1, 2, \cdots\}$, define the function

$$g_k(t) = P(N(t) = k).$$

   (a) Show that for any $\Delta > 0$, we have

$$g_0(t + \Delta) = g_0(t)[1 - \lambda\Delta + o(\Delta)].$$

(b) Using Part (a), show that

$$\frac{g_0'(t)}{g_0(t)} = -\lambda.$$

(c) By solving the above differential equation and using the fact that $g_0(0) = 1$, conclude that

$$g_0(t) = e^{-\lambda t}.$$

(d) For $k \geq 1$, show that

$$g_k(t + \Delta) = g_k(t)(1 - \lambda\Delta) + g_{k-1}(t)\lambda\Delta + o(\Delta).$$

(e) Using the previous part show that

$$g_k'(t) = -\lambda g_k(t) + \lambda g_{k-1}(t),$$

which is equivalent to

$$\frac{d}{dt}\left[e^{\lambda t} g_k(t)\right] = \lambda e^{\lambda t} g_{k-1}(t).$$

(f) Check that the function

$$g_k(t) = \frac{e^{-\lambda t}(\lambda t)^k}{k!}$$

satisfies the above differential equation for any $k \geq 1$. In fact, this is the only solution that satisfies $g_0(t) = e^{-\lambda t}$, and $g_k(0) = 0$ for $k \geq 1$.

7. Let $\{N(t), t \in [0, \infty)\}$ be a Poisson process with rate λ. Let T_1, T_2, $\cdots$ be the arrival times for this process. Show that

$$f_{T_1, T_2, \ldots, T_n}(t_1, t_2, \cdots, t_n) = \lambda^n e^{-\lambda t_n}, \qquad \text{for } 0 < t_1 < t_2 < \cdots < t_n.$$

Hint: One way to show the above result is to show that for sufficiently small Δ_i, we have

$$P\left(t_1 \leq T_1 < t_1 + \Delta_1, t_2 \leq T_2 < t_2 + \Delta_2, \ldots, t_n \leq T_n < t_n + \Delta_n\right) \approx$$
$$\lambda^n e^{-\lambda t_n} \Delta_1 \Delta_2 \cdots \Delta_n, \qquad \text{for } 0 < t_1 < t_2 < \cdots < t_n.$$

8. Let $\{N(t), t \in [0, \infty)\}$ be a Poisson process with rate λ. Show the following: given that $N(t) = n$, the n arrival times have the same joint CDF as the order statistics of n independent $Uniform(0, t)$ random variables. To show this you can show that

$$f_{T_1, T_2, \ldots, T_n \mid N(t)=n}(t_1, t_2, \cdots, t_n) = \frac{n!}{t^n}, \qquad \text{for } 0 < t_1 < t_2 < \cdots < t_n < t.$$

9. Let $\{N(t), t \in [0, \infty)\}$ be a Poisson process with rate λ. Let T_1, T_2, $\cdots$ be the arrival times for this process. Find

$$E[T_1 + T_2 + \cdots + T_{10}|N(4) = 10].$$

Hint: Use the result of Problem 8.

10. Two teams A and B play a soccer match. The number of goals scored by Team A is modeled by a Poisson process $N_1(t)$ with rate $\lambda_1 = 0.02$ goals per minute, and the number of goals scored by Team B is modeled by a Poisson process $N_2(t)$ with rate $\lambda_2 = 0.03$ goals per minute. The two processes are assumed to be independent. Let $N(t)$ be the total number of goals in the game up to and including time t. The game lasts for 90 minutes.

(a) Find the probability that no goals are scored, i.e., the game ends with a 0-0 draw.

(b) Find the probability that at least two goals are scored in the game.

(c) Find the probability of the final score being

Team $A : 1$, Team $B : 2$

(d) Find the probability that they draw.

11. In Problem 10, find the probability that Team B scores the first goal. That is, find the probability that at least one goal is scored in the game and the first goal is scored by Team B.

12. Let $\{N(t), t \in [0, \infty)\}$ be a Poisson process with rate λ. Let $p : [0, \infty) \mapsto [0, 1]$ be a function. Here we divide $N(t)$ to two processes $N_1(t)$ and $N_2(t)$ in the following way. For each arrival, a coin with $P(H) = p(t)$ is tossed. If the coin lands heads up, the arrival is sent to the first process ($N_1(t)$), otherwise it is sent to the second process. The coin tosses are independent of each other and are independent of $N(t)$. Show that $N_1(t)$ is a nonhomogeneous Poisson process with rate $\lambda(t) = \lambda p(t)$.

13. Consider the Markov chain with three states $S = \{1, 2, 3\}$, that has the state transition diagram is shown in Figure 11.31.

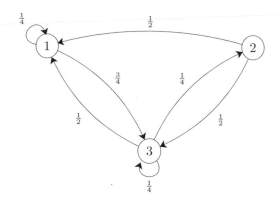

Figure 11.31: A state transition diagram.

Suppose $P(X_1 = 1) = \frac{1}{2}$ and $P(X_1 = 2) = \frac{1}{4}$.

(a) Find the state transition matrix for this chain.

(b) Find $P(X_1 = 3, X_2 = 2, X_3 = 1)$.

(c) Find $P(X_1 = 3, X_3 = 1)$.

14. Let $\alpha_0, \alpha_1, \cdots$ be a sequence of nonnegative numbers such that

$$\sum_{j=0}^{\infty} \alpha_j = 1.$$

Consider a Markov chain $X_0, X_1, X_2, \cdots$ with the state space $S = \{0, 1, 2, \cdots\}$ such that

$$p_{ij} = \alpha_j, \qquad \text{for all } j \in S.$$

Show that $X_1, X_2, \cdots$ is a sequence of i.i.d random variables.

15. Let X_n be a discrete-time Markov chain. Remember that, by definition, $p_{ii}^{(n)} = P(X_n = i | X_0 = i)$. Show that state i is recurrent if and only if

$$\sum_{n=1}^{\infty} p_{ii}^{(n)} = \infty.$$

16. Consider the Markov chain in Figure 11.32. There are two recurrent classes, $R_1 = \{1, 2\}$, and $R_2 = \{5, 6, 7\}$. Assuming $X_0 = 4$, find the probability that the chain gets absorbed to R_1.

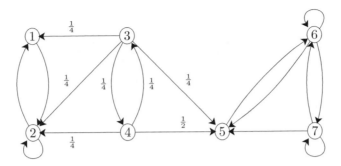

Figure 11.32: A state transition diagram.

17. Consider the Markov chain of Problem 16. Again assume $X_0 = 4$. We would like to find the expected time (number of steps) until the chain gets absorbed in R_1 or R_2. More specifically, let T be the absorption time, i.e., the first time the chain visits a state in R_1 or R_2. We would like to find $E[T|X_0 = 4]$.

18. Consider the Markov chain shown in Figure 11.33. Assume $X_0 = 2$, and let N be the first time that the chain returns to state 2, i.e.,

$$N = \min\{n \geq 1 : X_n = 2\}.$$

Find $E[N|X_0 = 2]$.

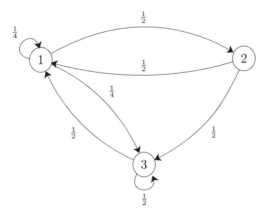

Figure 11.33: A state transition diagram.

19. Consider the Markov chain shown in Figure 11.34.

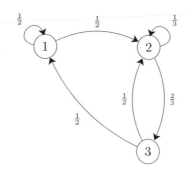

Figure 11.34: A state transition diagram.

(a) Is this chain irreducible?

(b) Is this chain aperiodic?

(c) Find the stationary distribution for this chain.

(d) Is the stationary distribution a limiting distribution for the chain?

20. (**Random Walk**) Consider the Markov chain shown in Figure 11.35.

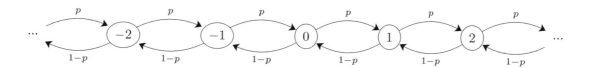

Figure 11.35: Simple random walk.

This is known as the *simple random walk*. Show that

$$p_{00}^{(2n)} = \binom{2n}{n} p^n (1-p)^n,$$
$$p_{00}^{(2n+1)} = 0.$$

Note: Using Stirling's formula, it can be shown that

$$\sum_{k=1}^{\infty} p_{00}^{(k)} = \sum_{n=1}^{\infty} \binom{2n}{n} p^n (1-p)^n$$

is finite if and only if $p \neq \frac{1}{2}$. Thus, we conclude that the simple random walk is recurrent if $p = \frac{1}{2}$ and is transient if $p \neq \frac{1}{2}$ (see Problem 15).

21. Consider the Markov chain shown in Figure 11.36. Assume that $0 < p < q$. Does this chain have a limiting distribution? For all $i, j \in \{0, 1, 2, \cdots\}$, find

$$\lim_{n \to \infty} P(X_n = j | X_0 = i).$$

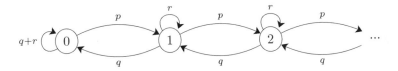

Figure 11.36: A state transition diagram.

22. Consider the Markov chain shown in Figure 11.37. Assume that $p > q > 0$. Does this chain have a limiting distribution? For all $i, j \in \{0, 1, 2, \cdots\}$, find

$$\lim_{n \to \infty} P(X_n = j | X_0 = i).$$

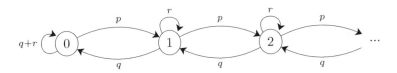

Figure 11.37: A state transition diagram.

23. (**Gambler's Ruin Problem**) Two gamblers, call them Gambler A and Gambler B, play repeatedly. In each round, A wins 1 dollar with probability p or loses 1 dollar with probability $q = 1 - p$ (thus, equivalently, in each round B wins 1 dollar with probability $q = 1 - p$ and loses 1 dollar with probability p). We assume different rounds are independent. Suppose that initially A has i dollars and B has $N - i$ dollars. The game ends when one of the gamblers runs out of money (in which case the other gambler will have N dollars). Our goal is to find p_i, the probability that A wins the game given that he has initially i dollars.

   (a) Define a Markov chain as follows: The chain is in state i if the Gambler A has i dollars. Here, the state space is $S = \{0, 1, \cdots, N\}$. Draw the state transition diagram of this chain.

   (b) Let a_i be the probability of absorption to state N (the probability that A wins) given that $X_0 = i$. Show that

$$a_0 = 0,$$
$$a_N = 1,$$
$$a_{i+1} - a_i = \frac{q}{p}(a_i - a_{i-1}), \qquad \text{for } i = 1, 2, \cdots, N-1.$$

(c) Show that

$$a_i = \left[1 + \frac{q}{p} + \left(\frac{q}{p}\right)^2 + \cdots + \left(\frac{q}{p}\right)^{i-1}\right] a_1, \text{ for } i = 1, 2, \cdots, N.$$

(d) Find a_i for any $i \in \{0, 1, 2, \cdots, N\}$. Consider two cases: $p = \frac{1}{2}$ and $p \neq \frac{1}{2}$.

24. Let $N = 4$ and $i = 2$ in the gambler's ruin problem (Problem 23). Find the expected number of rounds the gamblers play until one of them wins the game.

25. The Poisson process is a continuous-time Markov chain. Specifically, let $N(t)$ be a Poisson process with rate λ.

   (a) Draw the state transition diagram of the corresponding jump chain.
   (b) What are the rates λ_i for this chain?

26. Consider a continuous-time Markov chain $X(t)$ that has the jump chain shown in Figure 11.38. Assume $\lambda_1 = \lambda_2 = \lambda_3$, and $\lambda_4 = 2\lambda_1$.

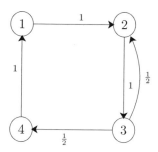

Figure 11.38: The jump chain for the Markov chain of Problem 26.

   (a) Find the stationary distribution of the jump chain $\tilde{\pi} = \left[\tilde{\pi}_1, \tilde{\pi}_2, \tilde{\pi}_3, \tilde{\pi}_4\right]$.
   (b) Using $\tilde{\pi}$, find the stationary distribution for $X(t)$.

27. Consider a continuous-time Markov chain $X(t)$ that has the jump chain shown in Figure 11.39. Assume $\lambda_1 = 1$, $\lambda_2 = 2$, and $\lambda_3 = 4$.

   (a) Find the generator matrix for this chain.
   (b) Find the limiting distribution for $X(t)$ by solving $\pi G = 0$.

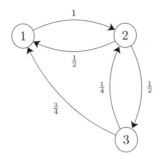

Figure 11.39: The jump chain for the Markov chain of Problem 2.

28. Consider the queuing system of Problem 3 in the Solved Problems Section (Section 11.3.4). Specifically, in that problem we found the following generator matrix and transition rate diagram:

$$G = \begin{bmatrix} -\lambda & \lambda & 0 & 0 & \cdots \\ \mu & -(\mu + \lambda) & \lambda & 0 & \cdots \\ 0 & \mu & -(\mu + \lambda) & \lambda & \cdots \\ \vdots & \vdots & \vdots & \vdots & \end{bmatrix}.$$

The transition rate diagram is shown in Figure 11.40

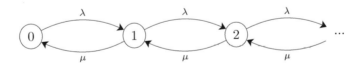

Figure 11.40: The transition rate diagram for the above queuing system.

Assume that $0 < \lambda < \mu$. Find the stationary distribution for this queueing system.

29. Let $W(t)$ be the standard Brownian motion.

(a) Find $P(-1 < W(1) < 1)$.

(b) Find $P(1 < W(2) + W(3) < 2)$.

(c) Find $P(W(1) > 2 | W(2) = 1)$.

30. Let $W(t)$ be a standard Brownian motion. Find

$$P\big(0 < W(1) + W(2) < 2, 3W(1) - 2W(2) > 0\big).$$

31. (Brownian Bridge) Let $W(t)$ be a standard Brownian motion. Define

$$X(t) = W(t) - tW(1), \qquad \text{for all } t \in [0, \infty).$$

Note that $X(0) = X(1) = 0$. Find $\text{Cov}(X(s), X(t))$, for $0 \le s \le t \le 1$.

32. (Correlated Brownian Motions) Let $W(t)$ and $U(t)$ be two independent standard Brownian motions. Let $-1 \le \rho \le 1$. Define the random process $X(t)$ as

$$X(t) = \rho W(t) + \sqrt{1 - \rho^2} U(t), \qquad \text{for all } t \in [0, \infty).$$

(a) Show that $X(t)$ is a standard Brownian motion.

(b) Find the covariance and correlation coefficient of $X(t)$ and $W(t)$. That is, find $\text{Cov}(X(t), W(t))$ and $\rho(X(t), W(t))$.

33. (Hitting Times for Brownian Motion) Let $W(t)$ be a standard Brownian motion. Let $a > 0$. Define T_a as the first time that $W(t) = a$. That is

$$T_a = \min\{t : W(t) = a\}.$$

(a) Show that for any $t \ge 0$, we have

$$P(W(t) \ge a) = P(W(t) \ge a | T_a \le t) P(T_a \le t).$$

(b) Using Part (a), show that

$$P(T_a \le t) = 2 \left[1 - \Phi \left(\frac{a}{\sqrt{t}} \right) \right].$$

(c) Using Part (b), show that the PDF of T_a is given by

$$f_{T_a}(t) = \frac{a}{t\sqrt{2\pi t}} \exp\left\{ -\frac{a^2}{2t} \right\}.$$

Note: By symmetry of Brownian motion, we conclude that for any $a \ne 0$, we have

$$f_{T_a}(t) = \frac{|a|}{t\sqrt{2\pi t}} \exp\left\{ -\frac{a^2}{2t} \right\}.$$

Chapter 12

Introduction to Simulation Using MATLAB (Available Online)

Chapter 13

Introduction to Simulation Using R (Available Online)

Chapter 14

Recursive Methods (Available Online)

Appendix

Some Important Distributions

Discrete Distributions

$$X \sim Bernoulli(p)$$

PMF:

$$P_X(k) = \begin{cases} p & \text{for } k = 1 \\ 1 - p & \text{for } k = 0 \end{cases}$$

CDF:

$$F_X(x) = \begin{cases} 0 & \text{for } x < 0 \\ 1 - p & \text{for } 0 \leq x < 1 \\ 1 & \text{for } 1 \leq x \end{cases}$$

Moment Generating Function (MGF):

$$M_X(s) = 1 - p + pe^s$$

Characteristic Function:

$$\phi_X(\omega) = 1 - p + pe^{i\omega}$$

Expected Value:

$$EX = p$$

Variance:

$$\text{Var}(X) = p(1 - p)$$

$$X \sim Binomial(n, p)$$

PMF:

$$P_X(k) = \binom{n}{k} p^k (1 - p)^{n-k} \quad \text{for } k = 0, 1, 2, \cdots, n$$

Moment Generating Function (MGF):

$$M_X(s) = (1 - p + pe^s)^n$$

Characteristic Function:

$$\phi_X(\omega) = (1 - p + pe^{i\omega})^n$$

Expected Value:

$$EX = np$$

Variance:

$$\text{Var}(X) = np(1 - p)$$

MATLAB:

$$\text{R} = \text{binornd}(n, p)$$

$$X \sim Geometric(p)$$

PMF:

$$P_X(k) = p(1 - p)^{k-1} \quad \text{for } k = 1, 2, 3, \ldots$$

CDF:

$$F_X(x) = 1 - (1 - p)^{\lfloor x \rfloor} \quad \text{for } x \geq 0$$

Moment Generating Function (MGF):

$$M_X(s) = \frac{pe^s}{1 - (1 - p)e^s} \quad \text{for } s < -\ln(1 - p)$$

Characteristic Function:

$$\phi_X(\omega) = \frac{pe^{i\omega}}{1 - (1 - p)e^{i\omega}}$$

Expected Value:

$$EX = \frac{1}{p}$$

Variance:

$$\text{Var}(X) = \frac{1 - p}{p^2}$$

MATLAB:

$$\text{R} = \text{geornd}(p) + 1$$

$$X \sim Pascal(m, p) \text{ (Negative Binomial)}$$

PMF:

$$P_X(k) = \binom{k-1}{m-1} p^m (1-p)^{k-m} \quad \text{for } k = m, m+1, m+2, m+3, ...$$

Moment Generating Function (MGF):

$$M_X(s) = \left(\frac{pe^s}{1 - (1-p)e^s} \right)^m \quad \text{for} \quad s < -\log(1-p)$$

Characteristic Function:

$$\phi_X(\omega) = \left(\frac{pe^{i\omega}}{1 - (1-p)e^{i\omega}} \right)^m$$

Expected Value:

$$EX = \frac{m}{p}$$

Variance:

$$\text{Var}(X) = \frac{m(1-p)}{p^2}$$

MATLAB:

$$\text{R = nbinrnd}(m, p) + 1$$

$$X \sim Hypergeometric(b, r, k)$$

PMF:

$$P_X(x) = \frac{\binom{b}{x}\binom{r}{k-x}}{\binom{b+r}{k}} \quad \text{for } x = \max(0, k-r), \max(0, k-r) + 1, ..., \min(k, b)$$

Expected Value:

$$EX = \frac{kb}{b+r}$$

Variance:

$$\text{Var}(X) = \frac{kbr}{(b+r)^2} \frac{b+r-k}{b+r-1}$$

MATLAB:

$$\text{R = hygernd}(b+r, b, k)$$

$$X \sim Poisson(\lambda)$$

PMF:

$$P_X(k) = \frac{e^{-\lambda}\lambda^k}{k!} \quad \text{for } k = 0, 1, 2, \cdots$$

Moment Generating Function (MGF):

$$M_X(s) = e^{\lambda(e^s - 1)}$$

Characteristic Function:

$$\phi_X(\omega) = e^{\lambda\left(e^{i\omega} - 1\right)}$$

Expected Value:

$$EX = \lambda$$

Variance:

$$\text{Var}(X) = \lambda$$

MATLAB:

$$\text{R = poissrnd}(\lambda)$$

Continuous Distributions

$$X \sim Exponential(\lambda)$$

PDF:

$$f_X(x) = \lambda e^{-\lambda x}, \quad x > 0$$

CDF:

$$F_X(x) = 1 - e^{-\lambda x}, \quad x > 0$$

Moment Generating Function (MGF):

$$M_X(s) = \left(1 - \frac{s}{\lambda}\right)^{-1} \quad \text{for} \quad s < \lambda$$

Characteristic Function:

$$\phi_X(\omega) = \left(1 - \frac{i\omega}{\lambda}\right)^{-1}$$

Expected Value:

$$EX = \frac{1}{\lambda}$$

Variance:

$$\text{Var}(X) = \frac{1}{\lambda^2}$$

MATLAB:

$$R = \text{exprnd}(\mu), \text{ where } \mu = \tfrac{1}{\lambda}.$$

$$X \sim Laplace(\mu, b)$$

PDF:

$$f_X(x) = \frac{1}{2b} \exp\left(-\frac{|x-\mu|}{b}\right) = \begin{cases} \frac{1}{2b} \exp\left(\frac{x-\mu}{b}\right) & \text{if } x < \mu \\ \frac{1}{2b} \exp\left(-\frac{x-\mu}{b}\right) & \text{if } x \geq \mu \end{cases}$$

CDF:

$$F_X(x) = \begin{cases} \frac{1}{2} \exp\left(\frac{x-\mu}{b}\right) & \text{if } x < \mu \\ 1 - \frac{1}{2} \exp\left(-\frac{x-\mu}{b}\right) & \text{if } x \geq \mu \end{cases}$$

Moment Generating Function (MGF):

$$M_X(s) = \frac{e^{\mu s}}{1 - b^2 s^2} \quad \text{for} \quad |s| < \frac{1}{b}$$

Characteristic Function:

$$\phi_X(\omega) = \frac{e^{\mu i \omega}}{1 + b^2 \omega^2}$$

Expected Value:

$$EX = \mu$$

Variance:

$$\text{Var}(X) = 2b^2$$

Gaussian Distribution, $X \sim N(\mu, \sigma^2)$

PDF:

$$f_X(x) = \frac{1}{\sigma\sqrt{2\pi}} e^{-\frac{(x-\mu)^2}{2\sigma^2}}$$

CDF:

$$F_X(x) = \Phi\left(\frac{x - \mu}{\sigma}\right)$$

Moment Generating Function (MGF):

$$M_X(s) = e^{\mu s + \frac{1}{2}\sigma^2 s^2}$$

Characteristic Function:

$$\phi_X(\omega) = e^{i\mu\omega - \frac{1}{2}\sigma^2 \omega^2}$$

Expected Value:

$$EX = \mu$$

Variance:

$$\text{Var}(X) = \sigma^2$$

MATLAB:

$$Z = \text{randn}, \ R = \text{normrnd}(\mu, \sigma)$$

$X \sim Beta(a, b)$

PDF:

$$f_X(x) = \frac{\Gamma(a+b)}{\Gamma(a)\Gamma(b)} x^{(a-1)} (1-x)^{(b-1)}, \quad \text{for } 0 \le x \le 1$$

Moment Generating Function (MGF):

$$M_X(s) = 1 + \sum_{k=1}^{\infty} \left(\prod_{r=0}^{k-1} \frac{a+r}{a+b+r} \right) \frac{s^k}{k!}$$

Expected Value:

$$EX = \frac{a}{a+b}$$

Variance:

$$\text{Var}(X) = \frac{ab}{(a+b)^2(a+b+1)}$$

MATLAB:

$$R = \text{betarnd}(a,b)$$

Chi-squared, $X \sim \chi^2(n)$

Note:

$$\chi^2(n) = Gamma\left(\frac{n}{2}, \frac{1}{2}\right)$$

PDF:

$$f_X(x) = \frac{1}{2^{\frac{n}{2}}\Gamma\left(\frac{n}{2}\right)} x^{\frac{n}{2}-1} e^{-\frac{x}{2}}, \qquad \text{for } x > 0.$$

Moment Generating Function (MGF):

$$M_X(s) = (1-2s)^{-\frac{n}{2}} \quad \text{for} \quad s < \frac{1}{2}$$

Characteristic Function:

$$\phi_X(\omega) = (1-2i\omega)^{-\frac{n}{2}}$$

Expected Value:

$$EX = n$$

Variance:

$$\text{Var}(X) = 2n$$

MATLAB:

$$R = \text{chi2rnd}(n)$$

The t-Distribution, $X \sim T(n)$

PDF:

$$f_X(x) = \frac{\Gamma(\frac{n+1}{2})}{\sqrt{n\pi}\Gamma\left(\frac{n}{2}\right)} \left(1 + \frac{x^2}{n}\right)^{-\frac{n+1}{2}}$$

Moment Generating Function (MGF):

undefined

Expected Value:

$$EX = 0$$

Variance:

$$\text{Var}(X) = \frac{n}{n-2} \quad \text{for} \quad n > 2, \quad \infty \quad \text{for } 1 < n \le 2, \quad \text{undefined} \quad \text{otherwise}$$

MATLAB:

$$R = \text{trnd}(n)$$

$$X \sim Gamma(\alpha, \lambda)$$

PDF:

$$f_X(x) = \frac{\lambda^\alpha x^{\alpha-1} e^{-\lambda x}}{\Gamma(\alpha)}, \quad x > 0$$

Moment Generating Function (MGF):

$$M_X(s) = \left(1 - \frac{s}{\lambda}\right)^{-\alpha} \quad \text{for} \quad s < \lambda$$

Expected Value:

$$EX = \frac{\alpha}{\lambda}$$

Variance:

$$\text{Var}(X) = \frac{\alpha}{\lambda^2}$$

MATLAB:

$$R = \text{gamrnd}(\alpha, \lambda)$$

$$X \sim Erlang(k, \lambda) \left[= Gamma(k, \lambda)\right], \, k > 0 \text{ is an integer}$$

PDF:

$$f_X(x) = \frac{\lambda^k x^{k-1} e^{-\lambda x}}{(k-1)!}, \quad x > 0$$

Moment Generating Function (MGF):

$$M_X(s) = \left(1 - \frac{s}{\lambda}\right)^{-k} \quad \text{for} \quad s < \lambda$$

Expected Value:

$$EX = \frac{k}{\lambda}$$

Variance:

$$\text{Var}(X) = \frac{k}{\lambda^2}$$

$$X \sim Uniform(a,b)$$

PDF:

$$f_X(x) = \frac{1}{b-a}, \quad x \in [a,b]$$

CDF:

$$F_X(x) = \begin{cases} 0 & x < a \\ \frac{x-a}{b-a} & x \in [a,b) \\ 1 & \text{for } x \geq b \end{cases}$$

Moment Generating Function (MGF):

$$M_X(s) = \begin{cases} \frac{e^{sb} - e^{sa}}{s(b-a)} & s \neq 0 \\ 1 & s = 0 \end{cases}$$

Characteristic Function:

$$\phi_X(\omega) = \frac{e^{i\omega b} - e^{i\omega a}}{i\omega(b-a)}$$

Expected Value:

$$EX = \frac{1}{2}(a+b)$$

Variance:

$$\text{Var}(X) = \frac{1}{12}(b-a)^2$$

MATLAB:

$$U = \text{rand or } R = \text{unifrnd}(a,b)$$

Review of the Fourier Transform

Here, we briefly review some properties of the Fourier transform. For a deterministic function $x(t)$ the Fourier transform (if exists) is defined as

$$\mathcal{F}\{x(t)\} = \int_{-\infty}^{\infty} x(t)e^{-2\pi i f t} \, dt,$$

where $i = \sqrt{-1}$. The Fourier transform of $x(t)$ is a function of f, so we can show it by $X(f) = \mathcal{F}\{x(t)\}$. We can obtain $x(t)$ from its fourier transform $X(f)$ using

$$x(t) = \mathcal{F}^{-1}\{X(f)\} = \int_{-\infty}^{\infty} X(f)e^{2\pi i f t} \, df.$$

In general $X(f)$ is a complex-valued function, i.e., we can write $X(f) : \mathbb{R} \mapsto \mathbb{C}$.

Fourier Transform

Fourier transform

$$X(f) = \mathcal{F}\{x(t)\} = \int_{-\infty}^{\infty} x(t)e^{-i2\pi f t} dt$$

Inversion formula

$$x(t) = \mathcal{F}^{-1}\{X(f)\} = \int_{-\infty}^{\infty} X(f)e^{i2\pi f t} df$$

When working with Fourier transform, it is often useful to use tables. There are two tables given in the next pages. One gives the Fourier transform for some important functions and the other provides general properties of the Fourier transform. Using these tables, we can find the Fourier transform for many other functions.

(1) $\operatorname{sinc}(x) = \frac{\sin(\pi x)}{\pi x}$

(2) $\Pi(x) = I_{[-\frac{1}{2}, \frac{1}{2}]}(x)$

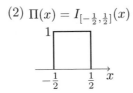

(3) $u(x)$

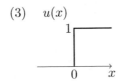

(4) $\Lambda(x)$

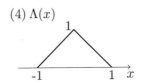

Figure 14.1: Some common functions.

Table of Fourier Transform Pairs

x(t)	Fourier Transform $X(f)$
$\delta(t)$	1
1	$\delta(f)$
$\delta(t-a)$	$e^{-i2\pi fa}$
$e^{i2\pi at}$	$\delta(f-a)$
$\cos(2\pi at)$	$\frac{1}{2}\delta(f-a) + \frac{1}{2}\delta(f+a)$
$\sin(2\pi at)$	$-\frac{1}{2i}\delta(f+a) + \frac{1}{2i}\delta(f-a)$
$\Pi(t)$	$\text{sinc}(f)$
$\text{sinc}(t)$	$\Pi(f)$
$\Lambda(t)$	$\text{sinc}^2(f)$
$\text{sinc}^2(t)$	$\Lambda(f)$
$e^{-at}u(t), a > 0$	$\frac{1}{a+i2\pi f}$
$te^{-at}u(t), a > 0$	$\frac{1}{(a+i2\pi f)^2}$
$e^{-a|t|}$	$\frac{2a}{a^2+(2\pi f)^2}$
$\frac{2a}{a^2+t^2}$	$2\pi e^{-2\pi a|f|}$
$e^{-\pi t^2}$	$e^{-\pi f^2}$
$u(t)$	$\frac{1}{2}\delta(f) + \frac{1}{i2\pi f}$
$\text{sgn}(t)$	$\frac{1}{i\pi f}$

Table of Fourier Transform Properties

Function	Fourier Transform
$ax_1(t) + bx_2(t)$	$aX_1(f) + bX_2(f)$
$x(at)$	$\frac{1}{|a|}X(\frac{f}{a})$
$x(t-a)$	$e^{-i2\pi fa}X(f)$
$e^{i2\pi at}x(t)$	$X(f-a)$
$x(t) * y(t)$	$X(f)Y(f)$
$x(t)y(t)$	$X(f) * Y(f)$
$\frac{d}{dt}x(t)$	$i2\pi f X(f)$
$tx(t)$	$\left(\frac{i}{2\pi}\right)\frac{d}{df}X(f)$
$\int_{-\infty}^{t} x(u)du$	$\frac{X(f)}{i2\pi f} + \frac{1}{2}X(0)\delta(f)$
$X(t) = \mathcal{F}\{x(t)\}\big|_{f=t}$	$x(-f) = \mathcal{F}^{-1}\{X(f)\}\big|_{t=-f}$

Index

Bibliography

[1] http://en.wikipedia.org/wiki/Boy_or_Girl_paradox

[2] http://en.wikipedia.org/wiki/Law_of_total_expectation#cite_note-1

[3] http://en.wikipedia.org/wiki/Law_of_total_variance

[4] http://en.wikipedia.org/wiki/Young's_inequality

[5] http://en.wikipedia.org/wiki/False_positive_paradox

[6] Y. Suhov and M. Kelbert, *Probability and Statistics by Example*. Cambridge University Press, 2005.

[7] L. Mlodinow, *The Drunkard's Walk*. Pantheon, 2008.

[8] http://en.wikipedia.org/wiki/Coupon_collector's_problem

[9] http://en.wikipedia.org/wiki/St._Petersburg_paradox

[10] http://en.wikipedia.org/wiki/Gamma_function

[11] P. Erdös and A. Rényi, *On the Evolution of Random Graphs*. Publications of the Mathematical Institute of the Hungarian Academy of Sciences, 5, 17-61, 1960.

[12] http://en.wikipedia.org/wiki/Erd%C5%91s%E2%80%93R%C3%A9nyi_model

[13] http://en.wikipedia.org/wiki/Boole's_inequality

[14] http://en.wikipedia.org/wiki/Laplace_distribution

[15] http://en.wikipedia.org/wiki/Pareto_distribution

[16] http://en.wikipedia.org/wiki/Chebyshev's_inequality

[17] http://en.wikipedia.org/wiki/Minkowski_inequality

[18] http://en.wikipedia.org/wiki/Sampling_(statistics)

[19] N. Etemadi, *An Elementary Proof of the Strong Law of large numbers*. Z. Wahrsch. Verw. Gebiete, 55(1981):119–122, 1981.

[20] Sheldon Ross, *A First Course in Probability*. Printice Hall, Upper Saddle River, New Jersey 07458, Eighth Edition, 2010.

[21] http://en.wikipedia.org/wiki/Overfitting

[22] http://en.wikipedia.org/wiki/Heteroscedasticity

[23] http://en.wikipedia.org/wiki/Multicollinearity

[24] http://en.wikipedia.org/wiki/Poisson_process

[25] Ubbo F Wiersema, *Brownian Motion Calculus*. John Wiley & Sons, Ltd, 2008.

Made in the USA
Monee, IL
05 January 2023

24424462R00409